AF769534

Microbial BioEnergy: Hydrogen Production

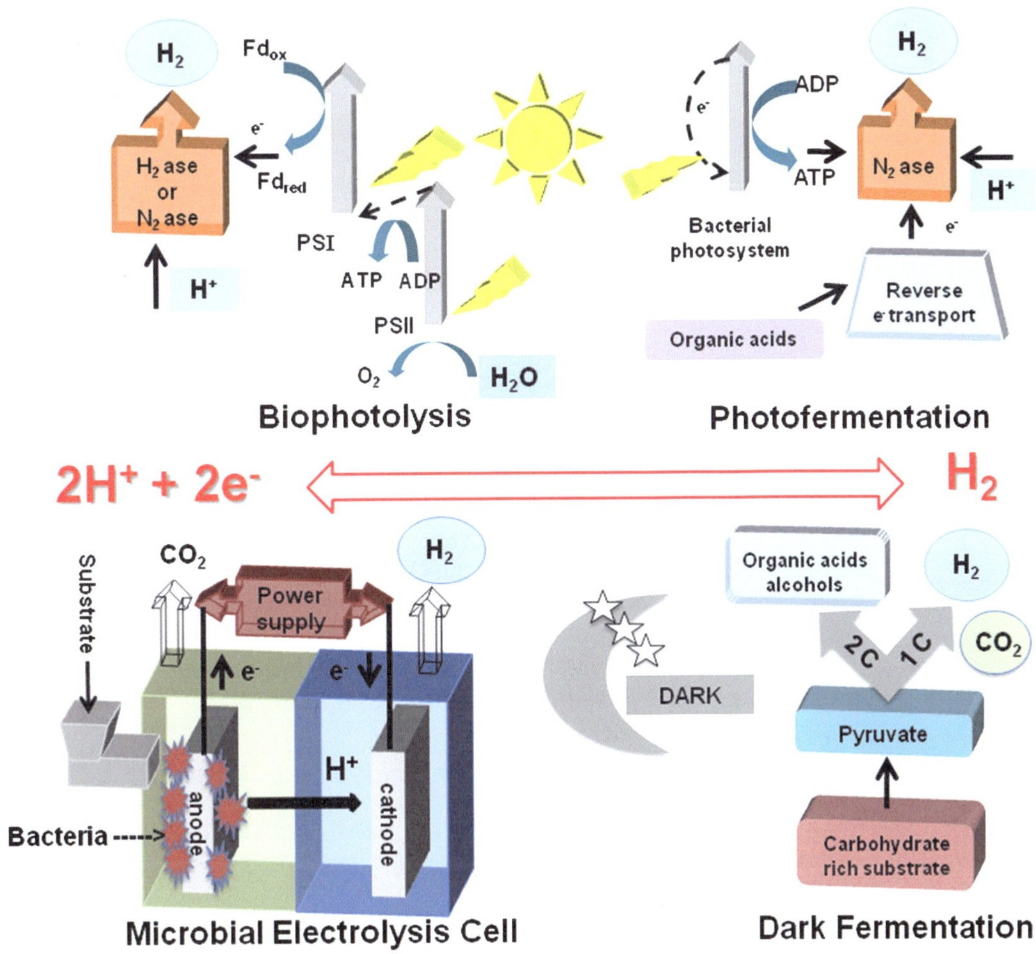

Different Ways for BioHydrogen Production The four possible ways for producing H_2, by exploiting microbial activities, are shown here. *Biophotolysis*: H_2 production by microalgae (through H_2-ase) or Cyanobacteria (through H_2-ase or N_2-ase) by using low potential reductants derived from either water or stored sugars *via* the photosynthetic machinery. *Photofermentation*: H_2 production by anoxygenic photosynthetic bacteria (through N_2-ase) by using reductants obtained from the oxidation of organic compounds as well as solar energy used through photosynthesis. *Dark fermentation*: H_2 production by mesophilic or thermophilic chemoheterotrophic bacteria (through H_2-ase) by using reductants and energy obtained from the oxidation of organic compounds. *Microbial Electrolysis Cell (MEC):* H_2 production by means of cathodic proton reduction with applied potential exploiting the low redox potential produced by exoelectrogenic bacteria at the anode. This figure is adapted from Fig. 1.3 in Chap. 1 of this book.

Advances in Photosynthesis and Respiration
Including Bioenergy and Related Processes

VOLUME 38

The book series ADVANCES IN PHOTOSYNTHESIS AND RESPIRATION *Including Bioenergy and Related Processes* provides a comprehensive and state-of-the-art account of research in photosynthesis, respiration and related processes. Virtually all life on our planet Earth ultimately depends on photosynthetic energy capture and conversion to energy-rich organic molecules. These are used for food, fuel, and fiber. Photosynthesis is the source of almost all bioenergy on Earth. The fuel and energy uses of photosynthesized products and processes have become an important area of study, and competition between food and fuel has led to resurgence in photosynthesis research. This series of books spans topics from physics to agronomy and medicine; from femtosecond processes through season-long production to evolutionary changes over the course of the history of the Earth; from the photophysics of light absorption, excitation energy transfer in the antenna to the reaction centers, where the highly-efficient primary conversion of light energy to charge separation occurs, through the electrochemistry of intermediate electron transfer, to the physiology of whole organisms and ecosystems; and from X-ray crystallography of proteins to the morphology of organelles and intact organisms. In addition to photosynthesis in natural systems, genetic engineering of photosynthesis and artificial photosynthesis is included in this series. The goal of the series is to offer beginning researchers, advanced undergraduate students, graduate students, and even research specialists, a comprehensive, up-to-date picture of the remarkable advances across the full scope of research on photosynthesis and related energy processes. The purpose of this series is to improve understanding of photosynthesis and respiration at many levels both to improve basic understanding of these important processes and to enhance our ability to use photosynthesis for the improvement of the human condition.

For further volumes:
www.springer.com/series/5599

Microbial BioEnergy: Hydrogen Production

Edited by

Davide Zannoni
University of Bologna
Bologna
Italy

and

Roberto De Philippis
University of Florence
Florence
Italy

Springer

Editors
Davide Zannoni
Department of Pharmacy
and BioTechnology (FaBiT)
University of Bologna
Bologna, Italy
davide.zannoni@unibo.it

Roberto De Philippis
Department of AgriFood Production
and Environmental Sciences
University of Florence
Florence, Italy

Institute of Chemistry of Organometallic
Compounds (ICCOM), CNR, Sesto
Fiorentino, Florence, Italy
roberto.dephilippis@unifi.it

ISSN 1572-0233 ISSN 2215-0102 (electronic)
ISBN 978-94-024-0612-2 ISBN 978-94-017-8554-9 (eBook)
DOI 10.1007/978-94-017-8554-9
Springer Dordrecht Heidelberg New York London

Printed on acid-free paper

Springer is part of Springer Science+Business Media (www.springer.com)

This book is dedicated to the memory of

Hans Gaffron (1902–1979) and **Howard Gest (1921–2012)**

pioneers of the microbial based hydrogen gas production

From the Series Editors

Advances in Photosynthesis and Respiration Including Bioenergy and Related Processes

Volume 38: Microbial BioEnergy: Hydrogen Production

We are delighted to announce the publication of volume 38 in this series. The title of our series *Advances in Photosynthesis and Respiration* was already updated in volume 35 to include the subtitle: *Including Bioenergy and Related Processes*. Earlier, the front cover of each volume had a distinctive white background and color palette; from volume 35, it has been changed to a web-friendly green background; and each volume begins with a unique figure, representing the book. Further, the publisher, Springer, makes the front matter of all of the volumes freely available online. Links to each volume are given under "Our Books: Published Volumes." Readers may also notice that this volume and the past few volumes have had color figures integrated into the chapters, instead of being collected in one section of the book. This improvement was possible because of changes in how the books are produced. Another change is that references to chapters in books are now being tracked by bibliographic services. This will help authors provide evidence of the importance of their work. We hope that these updates will maintain the importance of these edited volumes in the dissemination of the science of photosynthesis and bioenergy.

We are delighted to announce that volume 38 is the first one to deal with the new direction "*Including Bioenergy and Related Processes.*" We are indeed fortunate to have two distinguished authorities with us as editors of this new volume 38: Davide Zannoni, and Roberto De Philippis. Zannoni is a Professor of General Microbiology, at the University of Bologna, Italy; he is an authority on the structure and the function of membrane redox-complexes in microbes, and a pioneer of bioenergetics and genomics of microbial remediation of metals in many systems. More importantly, his present focus is on the production of hydrogen by thermophilic bacteria and electricity from microbial systems. De Philippis is an Associate Professor of Microbial Biotechnology at the University of Florence, Italy. He is an authority on exopolysaccharide-producing cyanobacteria and their biotechnological exploitation and, more importantly, in the last 15 years he has been deeply involved in studies on the photofermentative production of hydrogen indoors and outdoors and on the efficient conversion of light energy into hydrogen energy.

This Book: Volume 38

Microbial BioEnergy: Hydrogen Production is a comprehensive book covering most of the processes important for the microbial hydrogen production. It provides a broad coverage of this emerging research field and, in our opinion, it should be accessible to advanced undergraduates, graduate students, and researchers needing to broaden their knowledge on the photosynthetic and fermentation processes applied to hydrogen gas generation. For biologists, biochemists, biophysicists and microbiologists, this volume provides a solid and quick starting base to get into biotechnological problems of "microbial bioenergy." We believe that this volume will also be of interest to teachers of advanced undergraduate and graduate students in chemical engineering and biotechnology needing a single reference book on the latest understanding of the critical aspects of microbial bioenergy production.

The *Preface* of this book appropriately states "Solar energy is the source of most of the living organisms on Earth so that the overall efficiency of oxygenic and/or non-oxygenic photosynthesis, when used to generate biomass, bioenergy and biofuels, is a critical point to be considered." This volume in our series, however, not only provides a comprehensive view of the current understanding of the photosynthetic mechanisms linked to bio-hydrogen production but also extends this view to the anaerobic-dark processes involved in transforming the solar-generated biomass into bio-hydrogen along with an in-depth coverage of both structural and functional aspects of the main enzymes involved, such as *nitrogenases* and *hydrogenases*.

In our opinion, this book has been appropriately dedicated to Hans Gaffron (1902–1979) and Howard Gest (1921–2012), founders of the microbial based hydrogen gas production technologies. One of us (Govindjee) is fortunate to have known personally both these giants of photosynthesis research. Among the various other discoveries, Gaffron was the first to observe, in 1942, hydrogen production by green algae under sulfur starvation, while Gest was the first to describe, in 1949, hydrogen-production by purple non-sulfur phototrophic bacteria (See below for information on "Discoveries in Photosynthesis" volume 20, in our series, where these discoveries are described).

Authors

The current book contains 15 chapters written by 42 international authors from 10 different countries (Australia, Canada, France, Germany, Italy, Portugal, Russia, Spain, Turkey, USA). We give special thanks to each and every author for their valuable contribution to the successful production of this unique book:

Francisco Gabriel **A**cién-Fernández (Spain; Chap. 13); Alessandra Adessi (Italy; Chap. 12); Giacomo Antonioni (Italy; Chap. 15); Sara E. **B**lumer-Schuette (USA; Chap. 8); Hermann Bothe (Germany; Chap. 6); Martina **C**appelletti (Italy; Chap. 9); Jonathan M. Conway (USA; Chap. 8); Roberto **D**e Philippis (Italy; Chap. 12); Alexandra Dubini (USA; Chap. 5); Carrie **E**ckert (USA; Chap. 5); Ela Eroglu (Australia; Chap. 11); Inci Eroglu (Turkey; Chap. 11); José M. **F**ernández-Sevilla (Spain; Chap. 13); Juan C. Fontecilla-Camps (France; Chap. 2); Dario Frascari (Italy; Chap. 15); Maria L. **G**hirardi (USA; Chap. 5); Ufuk Gündüz (Turkey; Chap. 11); Patrick C. **H**allenbeck (Canada; Chap. 1); Robert M. **K**elly (USA; Chap. 8); Paul W. King (USA; Chap. 5); Sergey Kosourov (Russia; Chap. 14); Pierre-Pol **L**iebgott (France; Chap. 3); Pin-Ching **M**aness (USA; Chap. 5); James B. McKinlay (USA; Chap. 7); Emilio Molina-Grima (Spain; Chap. 13); David W. Mulder (USA; Chap. 5); William E. **N**ewton (USA; Chap. 6); Paulo **O**liveira (Portugal; Chap. 4); Bernard Ollivier (France; Chap. 9); Ebru Özgür (Turkey; Chap. 11); Catarina C. **P**acheco (Portugal; Chap. 4); Anne Postec (France; Chap. 9); John M. **R**egan (USA; Chap. 10); Marc Rousset (France; Chap. 3); Paula **T**amagnini (Portugal; Chap. 4); Anatoly Tsygankov (Russia; Chap. 14); Anne **V**olbeda (France; Chap. 2); Hengjing **Y**an (USA; Chap. 10); Jianping Yu (USA; Chap. 5); Meral Yücel (Turkey; Chap. 11); Davide **Z**annoni (Italy; Chap. 9); Jeffrey V. Zurawski (USA; Chap. 8).

Our Books: Published Volumes

We list below information on all the 37 volumes that have been published thus far (see http://www.springer.com/series/5599 for the web site of the series). Electronic access to individual chapters depends on subscription (ask your librarian) but Springer provides free downloadable front matter as well as indexes at the above site. As of July, 2011, Tables of Contents have been available for all the volumes. The available web sites of the books in the Series are listed below.

- **Volume 37 (2014)** Photosynthesis in Bryophytes and Early Land Plants, edited by David T. Hanson and Steven K. Rice, from USA. Eighteen chapters, approx. 500 pp, Hardcover, ISBN: 978-94-007-6987-8 (HB) ISBN 978-94-007-6988-5 (e-book) [http://www.springer.com/life+sciences/plant+sciences/book/978-94-007-6987-8]
- **Volume 36 (2013) Plastid Development in Leaves During Growth and Senescence**, edited by Basanti Biswal, Karin Krupinska and Udaya Biswal, from India and Germany. Twenty-eight chapters, 837 pp, Hardcover, ISBN:978-94-007-5723-3 (HB) ISBN 978-94-XXXXX (e-book) [http://www.springer.com/life+sciences/plant+sciences/book/978-94-007-5723-3]
- **Volume 35 (2012) Genomics of Chloroplasts and Mitochondria**, edited by Ralph Bock and Volker Knoop, from Germany. Nineteen chapters, 475 pp, Hardcover, ISBN: 978-94-007-2919-3 (HB) ISBN 978-94-007-2920-9 (e-book) [http://www.springer.com/life+sciences/plant+sciences/book/978-94-007-2919-3]
- **Volume 34 (2012) Photosynthesis – Plastid Biology, Energy Conversion and Carbon Assimilation**, edited by Julian Eaton-Rye, Baishnab C. Tripathy, and Thomas D. Sharkey, from New Zealand, India, and USA. Thirty-three chapters, 854 pp, Hardcover, ISBN: 978-94-007-1578-3 (HB) ISBN 978-94-007-1579-0 (e-book) [http://www.springer.com/life+sciences/plant+sciences/book/978-94-007-1578-3]
- **Volume 33 (2012): Functional Genomics and Evolution of Photosynthetic Systems**, edited by Robert L. Burnap and Willem F. J. Vermaas, from USA. Fifteen chapters, 428 pp, ISBN: 978-94-007- 1532–5 [http://www.springer.com/life+sciences/book/978-94-007-1532-5]
- **Volume 32 (2011): C_4 Photosynthesis and Related CO_2 Concentrating Mechanisms**, edited by Agepati S. Raghavendra and Rowan Sage, from India and Canada. Nineteen chapters, 425 pp, Hardcover, ISBN: 978-90-481-9406-3 [http://www.springer.com/life+sciences/plant+sciences/book/978-90-481-9406-3]
- **Volume 31 (2010): The Chloroplast: Basics and Applications**, edited by Constantin Rebeiz (USA), Christoph Benning (USA), Hans J. Bohnert (USA), Henry Daniell (USA), J. Kenneth Hoober (USA), Hartmut K. Lichtenthaler (Germany), Archie R. Portis (USA), and Baishnab C. Tripathy (India). Twenty-five chapters, 451 pp, Hardcover, ISBN: 978-90-481-8530-6 [http://www.springer.com/life+sciences/plant+sciences/book/978-90-481-8530-6]
- **Volume 30 (2009): Lipids in Photosynthesis: Essential and Regulatory Functions**, edited by Hajime Wada and Norio Murata, both from Japan. Twenty chapters, 506 pp, Hardcover, ISBN: 978-90-481-2862-4; e-book, ISBN: 978-90-481-2863-1 [http://www.springer.com/life+sciences/plant+sciences/book/978-90-481-2862-4]
- **Volume 29 (2009): Photosynthesis in Silico: Understanding Complexity from Molecules**, edited by Agu Laisk, Ladislav Nedbal, and Govindjee, from Estonia, The Czech Republic, and USA. Twenty chapters, 525 pp, Hardcover, ISBN: 978-1-4020-9236-7 [http://www.springer.com/life+sciences/plant+sciences/book/978-1-4020-9236-7]
- **Volume 28 (2009): The Purple Phototrophic Bacteria**, edited by C. Neil Hunter, Fevzi Daldal, Marion C. Thurnauer and J. Thomas Beatty, from UK, USA and Canada. Forty-eight chapters, 1053 pp, Hardcover, ISBN: 978-1-4020-8814-8 [http://www.springer.com/life+sciences/plant+sciences/book/978-1-4020-8814-8]
- **Volume 27 (2008): Sulfur Metabolism in Phototrophic Organisms**, edited by Christiane Dahl, Rüdiger Hell, David Knaff and Thomas Leustek, from Germany and USA.

Twenty-four chapters, 551 pp, Hardcover, ISBN: 978-4020-6862-1 [http://www.springer.com/life+sciences/plant+sciences/book/978-1-4020-6862-1]

- **Volume 26 (2008): Biophysical Techniques Photosynthesis**, Volume II, edited by Thijs Aartsma and Jörg Matysik, both from The Netherlands. Twenty-four chapters, 548 pp, Hardcover, ISBN: 978-1-4020-8249-8 [http://www.springer.com/life+sciences/plant+sciences/book/978-1-4020-8249-8]
- **Volume 25 (2006): Chlorophylls and Bacteriochlorophylls: Biochemistry, Biophysics, Functions and Applications**, edited by Bernhard Grimm, Robert J. Porra, Wolfhart Rüdiger, and Hugo Scheer, from Germany and Australia. Thirty-seven chapters, 603 pp, Hardcover, ISBN: 978-1-40204515-8 [http://www.springer.com/life+sciences/plant+sciences/book/978-1-4020-4515-8]
- **Volume 24 (2006): Photosystem I: The Light-Driven Plastocyanin: Ferredoxin Oxidoreductase**, edited by John H. Golbeck, from USA. Forty chapters, 716 pp, Hardcover, ISBN: 978-1-40204255-3 [http://www.springer.com/life+sciences/plant+sciences/book/978-1-4020-4255-3]
- **Volume 23 (2006): The Structure and Function of Plastids**, edited by Robert R. Wise and J. Kenneth Hoober, from USA. Twenty-seven chapters, 575 pp, Softcover, ISBN: 978-1-4020-6570–6; Hardcover, ISBN: 978-1-4020-4060-3 [http://www.springer.com/life+sciences/plant+sciences/book/978-1-4020-4060-3]
- **Volume 22 (2005): Photosystem II: The Light-Driven Water: Plastoquinone Oxidoreductase**, edited by Thomas J. Wydrzynski and Kimiyuki Satoh, from Australia and Japan. Thirty-four chapters, 786 pp, Hardcover, ISBN: 978-1-4020-4249-2 [http://www.springer.com/life+sciences/plant+sciences/book/978-1-4020-4249-2]
- **Volume 21 (2005): Photoprotection, Photoinhibition, Gene Regulation, and Environment**, edited by Barbara Demmig-Adams, William W. Adams III and Autar K. Mattoo, from USA. Twenty-one chapters, 380 pp, Hardcover, ISBN: 978-14020-3564-7 [http://www.springer.com/life+sciences/plant+sciences/book/978-1-4020-3564-7]
- **Volume 20 (2006): Discoveries in Photosynthesis**, edited by Govindjee, J. Thomas Beatty, Howard Gest and John F. Allen, from USA, Canada and UK. One hundred and eleven chapters, 1304 pp, Hardcover, ISBN: 978-1-4020-3323-0 [http://www.springer.com/life+sciences/plant+sciences/book/978-1-4020-3323-0]
- **Volume 19 (2004): Chlorophyll *a* Fluorescence: A Signature of Photosynthesis**, edited by George C. Papageorgiou and Govindjee, from Greece and USA. Thirty-one chapters, 820 pp, Hardcover, ISBN: 978-1-4020-3217-2 [http://www.springer.com/life+sciences/biochemistry+%26+biophysics/book/978-1-4020-3217-2]
- **Volume 18 (2005): Plant Respiration: From Cell to Ecosystem**, edited by Hans Lambers and Miquel Ribas-Carbo, from Australia and Spain. Thirteen chapters, 250 pp, Hardcover, ISBN: 978-14020-3588-3 [http://www.springer.com/life+sciences/plant+sciences/book/978-1-4020-3588-3]
- **Volume 17 (2004): Plant Mitochondria: From Genome to Function**, edited by David Day, A. Harvey Millar and James Whelan, from Australia. Fourteen chapters, 325 pp, Hardcover, ISBN: 978-1-4020-2399-6 [http://www.springerlink.com/content/978-1-7923-2399-6]
- **Volume 16 (2004): Respiration in Archaea and Bacteria: Diversity of Prokaryotic Respiratory Systems**, edited by Davide Zannoni, from Italy. Thirteen chapters, 310 pp, Hardcover, ISBN: 978-14020-2002-5 [http://www.springer.com/life+sciences/plant+sciences/book/978-1-4020-2002-5]
- **Volume 15 (2004): Respiration in Archaea and Bacteria: Diversity of Prokaryotic Electron Transport Carriers**, edited by

Davide Zannoni, from Italy. Thirteen chapters, 350 pp, Hardcover, ISBN: 978-1-4020-2001-8 [http://www.springer.com/life+sciences/biochemistry+%26+biophysics/book/978-1-4020-2001-8]

- **Volume 14 (2004): Photosynthesis in Algae**, edited by Anthony W. Larkum, Susan Douglas and John A. Raven, from Australia, Canada and UK. Nineteen chapters, 500 pp, Hardcover, ISBN: 978-0-7923-6333-0 [http://link.springer.com/book/10.1007/978-94-007-1038-2/page/1]
- **Volume 13 (2003): Light-Harvesting Antennas in Photosynthesis**, edited by Beverley R. Green and William W. Parson, from Canada and USA. Seventeen chapters, 544 pp, Hardcover, ISBN: 978-07923-6335-4 [http://www.springer.com/life+sciences/plant+sciences/book/978-0-7923-6335-4?otherVersion=978-90-481-5468-5]
- **Volume 12 (2003): Photosynthetic Nitrogen Assimilation and Associated Carbon and Respiratory Metabolism**, edited by Christine H. Foyer and Graham Noctor, from UK and France. Sixteen chapters, 304 pp, Hardcover, ISBN: 978-07923-6336-1 [http://www.springer.com/life+sciences/plant+sciences/book/978-0-7923-6336-1]
- **Volume 11 (2001): Regulation of Photosynthesis**, edited by Eva-Mari Aro and Bertil Andersson, from Finland and Sweden. Thirty-two chapters, 640 pp, Hardcover, ISBN: 978-0-7923-6332-3 [http://www.springer.com/life+sciences/plant+sciences/book/978-0-7923-6332-3]
- **Volume 10 (2001): Photosynthesis: Photobiochemistry and Photobiophysics**, edited by Bacon Ke, from USA. Thirty-six chapters, 792 pp, Softcover, ISBN: 978-0-7923-6791-8; Hardcover: ISBN: 978-0-7923-6334-7 [http://www.springer.com/life+sciences/plant+sciences/book/978-0-7923-6334-7]
- **Volume 9 (2000): Photosynthesis: Physiology and Metabolism**, edited by Richard C. Leegood, Thomas D. Sharkey and Susanne von Caemmerer, from UK, USA and Australia. Twenty-four chapters, 644 pp, Hardcover, ISBN: 978-07923-6143-5 [http://www.springer.com/life+sciences/plant+sciences/book/978-0-7923-6143-5]
- **Volume 8 (1999): The Photochemistry of Carotenoids**, edited by Harry A. Frank, Andrew J. Young, George Britton and Richard J. Cogdell, from USA and UK. Twenty chapters, 420 pp, Hardcover, ISBN: 978-0-7923-5942-5 [http://www.springer.com/life+sciences/plant+sciences/book/978-0-7923-5942-5]
- **Volume 7 (1998): The Molecular Biology of Chloroplasts and Mitochondria in *Chlamydomonas***, edited by Jean David Rochaix, Michel Goldschmidt-Clermont and Sabeeha Merchant, from Switzerland and USA. Thirty-six chapters, 760 pp, Hardcover, ISBN: 978-0-7923-5174-0 [http://www.springer.com/life+sciences/plant+sciences/book/978-0-7923-5174-0]
- **Volume 6 (1998): Lipids in Photosynthesis: Structure, Function and Genetics**, edited by Paul-André Siegenthaler and Norio Murata, from Switzerland and Japan. Fifteen chapters, 332 pp, Hardcover, ISBN: 978-0-7923-5173-3 [http://www.springer.com/life+sciences/plant+sciences/book/978-0-7923-5173-3]
- **Volume 5 (1997): Photosynthesis and the Environment**, edited by Neil R. Baker, from UK. Twenty chapters, 508 pp, Hardcover, ISBN: 978-07923-4316-5 [http://www.springer.com/life+sciences/plant+sciences/book/978-0-7923-4316-5]
- **Volume 4 (1996): Oxygenic Photosynthesis: The Light Reactions**, edited by Donald R. Ort and Charles F. Yocum, from USA. Thirty-four chapters, 696 pp, Softcover: ISBN: 978-0-7923-3684–6; Hardcover, ISBN: 978-0-7923-3683-9 [http://www.springer.com/life+sciences/plant+sciences/book/978-0-7923-3683-9]
- **Volume 3 (1996): Biophysical Techniques in Photosynthesis**, edited by Jan Amesz and Arnold J. Hoff, from The Netherlands.

Twenty-four chapters, 426 pp, Hardcover, ISBN: 978-0-7923-3642-6 [http://www.springer.com/life+sciences/plant+sciences/book/978-0-7923-3642-6]

- **Volume 2 (1995): Anoxygenic Photosynthetic Bacteria**, edited by Robert E. Blankenship, Michael T. Madigan and Carl E. Bauer, from USA. Sixty-two chapters, 1331 pp, Hardcover, ISBN: 978-0-7923-3682-8 [http://www.springer.com/life+sciences/plant+sciences/book/978-0-7923-3681-5]
- **Volume 1 (1994): The Molecular Biology of Cyanobacteria**, edited by Donald R. Bryant, from USA. Twenty-eight chapters, 916 pp, Hardcover, ISBN: 978-1-4020-2400-9 [http://www.springerlink.com/content/978-1-4020-2400-9]

Further information on these books and ordering instructions can be found at <http://www.springer.com/series/5599>. References to all chapters of volumes 1–31 can be found at <http://www.life.illinois.edu/govindjee/g/References.html>. (For volumes 32–35, pdf files of the entire Front Matter are also available here.)

Special 25 % discounts are available to members of the International Society of Photosynthesis Research, ISPR <http://www.photosynthesisresearch.org/>. See <http://www.springer.com/ispr>.

Future Advances in Photosynthesis and Respiration and Other Related Books

The readers of the current series are encouraged to watch for the publication of the forthcoming books (not necessarily arranged in the order of future appearance):

- The Structural Basis of Biological Energy Generation (Editor: Martin Hohmann-Marriott) [This book is almost ready to go to the typesetters]
- Canopy Photosynthesis: From Basics to Applications (Editors: Kouki Hikosaka, Ülo Niinemets and Niels P. R. Anten)
- Non-Photochemical Quenching (NPQ) and Energy Dissipation in Plants, Algae and Cyanobacteria (Editors: Barbara Demmig-Adams, Győző Garab and Govindjee)
- Cytochrome Complexes: Evolution, Structures, Energy Transduction, and Signaling (Editors: William Cramer and Toivo Kallas)
- Photosynthesis for Bioenergy (Editors: Elizabeth A. Ainsworth and Stephen P. Long)

In addition to the above contracted books, the following topics are under consideration:

- Algae, Cyanobacteria: Biofuel and Bioenergy
- Artificial Photosynthesis
- ATP Synthase
- Bacterial Respiration II
- Carotenoids II
- Cyanobacteria II
- Ecophysiology
- Evolution of Photosynthesis
- Global Aspects of Photosynthesis
- Green Bacteria and Heliobacteria
- Interactions between Photosynthesis and other Metabolic Processes
- Limits of Photosynthesis: Where do we go from here
- Photosynthesis, Biomass and Bioenergy
- Photosynthesis under Abiotic and Biotic Stress
- Plant Respiration II

If you have any interest in editing/co-editing any of the above listed books, or being an author, please send an e-mail to Tom Sharkey (tsharkey@msu.edu) and/or to Govindjee at gov@illinois.edu. Suggestions for additional topics are also welcome.

In view of the interdisciplinary character of research in photosynthesis and respiration, it is our earnest hope that this series of books will be used in educating students and researchers not only in Plant Sciences, Molecular and Cell Biology, Integrative Biology, Biotechnology, Agricultural Sciences, Microbiology, Biochemistry, Chemical Biology, Biological Physics, and Biophysics, but also in Bioengineering, Chemistry, and Physics.

Acknowledgments

We take this opportunity to thank and congratulate Davide Zannoni and Roberto De Philippis for their outstanding editorial work; they havc, indeed, done a fantastic job, not only in editing, but also in organizing this book for all of us, and for their highly professional dealing with the reviewing process. We thank all the 42 authors of this book (see the list above); without their authoritative chapters, there would be no such volume. We give special thanks to A. Lakshmi Praba, SPi Global, India for directing the typesetting of this book; her expertise has been crucial in bringing this book to completion. We owe Jacco Flipsen, Andre Tournois, and Ineke Ravesloot (of Springer) thanks for their friendly working relation with us that led to the production of this book.

October 24, 2013

Govindjee

Department of Plant Biology,
Department of Biochemistry and Center
of Biophysics & Quantitative Biology
University of Illinois at Urbana-Champaign,
Urbana, IL 61801, USA
gov@illinois.edu

Thomas D. Sharkey
Department of Biochemistry
and Molecular Biology
Michigan State University,
East Lansing, MI 48824, USA
tsharkey@msu.edu

Series Editors

The above photograph shows the 1965 Z-Scheme that Govindjee had made when he, with Eugene Rabinowitch, had coined the term P680 for the reaction center of Photosystem II. From left to right: Rebecca Slattery (a graduate student of Donald R. Ort), Robert Koester (a graduate student of Lisa Ainsworth), and Govindjee. Photo by Joan Huber taken in May, 2013.

Govindjee, who uses one name only, was born on October 24, 1932, in Allahabad, India. Since 1999, he has been Professor Emeritus of Biochemistry, Biophysics, and Plant Biology at the University of Illinois at Urbana-Champaign (UIUC), Urbana, IL, USA. He obtained his B.Sc. (Chemistry and Biology) and M.Sc. (Botany: Plant Physiology) in 1952 and 1954 from the University of Allahabad. Govindjee studied "Photosynthesis" at the UIUC under two pioneers of photosynthesis, Robert Emerson and Eugene Rabinowitch, obtaining his Ph.D. in 1960, in Biophysics. He is best known for his research on excitation energy transfer, light emission, primary photochemistry, and electron transfer in "Photosystem II" (PS II, water-plastoquinone oxidoreductase). His early research included the discovery of a short-wavelength form of chlorophyll (Chl) *a* functioning in the Chl *b*-containing system, now called PS II; of the two-light effect in Chl *a* fluorescence; and, with his wife Rajni, of the two-light effect (Emerson enhancement) in NADP reduction in chloroplasts. His major achievements, with many collaborators, include an understanding of the basic relationships between Chl *a* fluorescence and photosynthetic reactions; a unique role of bicarbonate/carbonate on the electron acceptor side of PS II, particularly in the protonation events involving the Q_B binding region in PSII; molecular understanding of thermoluminescence in plants, algae and cyanobacteria; the first picosecond measurements on the primary photochemistry of PS II; and the first use of Chl *a* fluorescence lifetime measurements in understanding photoprotection, by plants and algae, against excess light. His current focus is on the "History of Photosynthesis Research," on "Photosynthesis Education," and on "What to Learn from Natural Photosynthesis to do Artificial Photosynthesis." He has served on the faculty of the UIUC for approximately 40 years. Govindjee's honors include: Fellow of the American Association of Advancement of Science (AAAS); Distinguished Lecturer of the School of Life Sciences, UIUC; Fellow and Lifetime Member of the National Academy of Sciences (India); President of the American Society for Photobiology (1980–1981); Fulbright Scholar (1956), Fulbright Senior Lecturer (1998) and Fulbright Specialist (2012); Honorary President of the 2004 International Photosynthesis Congress (Montréal, Canada); the first recipient of the Lifetime Achievement Award of the Rebeiz Foundation for Basic Biology, 2006; 2007 recipient of the Communication Award of the International Society of Photosynthesis Research (ISPR); 2008 recipient of the Liberal Arts & Sciences (LAS) Lifetime Achievement Award of the UIUC. Further, Govindjee was honored **(1)** in 2007, through two special volumes of *Photosynthesis Research*, celebrating his 75th birthday and for his 50-year dedicated research in "Photosynthesis" (Guest Editor: Julian Eaton-Rye); **(2)** in 2008, through a special International Symposium on "Photosynthesis in a Global Perspective," at the University of Indore, India; **(3)** in 2012, through dedication to him of volume 34 of this series, celebrating his 80th year; **(4)** in 2012, through another book *Photosynthesis: Overviews on Recent Progress and Future Perspectives*, honoring him for *his outstanding research and teaching of photosynthesis and for being a global leader for stimulating photosynthesis research throughout the world*; and **(5)** in 2013, through two special volumes of *Photosynthesis Research*, on "Photosynthesis Education", which also celebrates his 80th birthday (Guest Editors: Suleyman Allakhverdiev; Jian-Ren Shen and Gerald Edwards). Govindjee is coauthor of *Photosynthesis* (John Wiley, 1969) and editor of many books, published by several publishers including Academic Press and Springer. Since 2007, each year Govindjee and Rajni Govindjee Award for Excellence in Biological Sciences is being given to graduate students by the Department of Plant Biology at the UIUC. Starting in 2014, these awards will alternate between Department of Biochemistry (even years) and Department of Plant Biology (odd years). For further information on Govindjee, see his website http://www.life.illinois.edu/govindjee.

Thomas D. (Tom) Sharkey obtained his Bachelor's degree in Biology in 1974 from Lyman Briggs College, a residential science college at Michigan State University, East Lansing, Michigan, USA. After 2 years as a research technician in the federally funded Plant Research Laboratory at Michigan State University under the mentorship of Prof. Klaus Raschke, Tom entered the Ph.D. program in the same lab, and graduated in 1980. Postdoctoral research was carried out with Prof. Graham Farquhar at the Australian National University, in Canberra, where he co-authored a landmark review on photosynthesis and stomatal conductance that continues to receive much attention 30 years after its publication. For 5 years, Tom worked at the Desert Research Institute together with Prof. Barry Osmond, followed by 20 years as a Professor of Botany at the University of Wisconsin in Madison. In 2008, Tom became Professor and Chair of the Department of Biochemistry and Molecular Biology at Michigan State University. Tom's research interests center on the biochemistry and biophysics of gas exchange between plants and the atmosphere. Photosynthetic gas exchange, especially carbon dioxide uptake and use, and isoprene emission from plants, are the two major research topics in his laboratory. Among his contributions are measurements of the carbon dioxide concentration inside leaves, studies of the resistance to diffusion of carbon dioxide within the mesophyll of leaves of C_3 plants, and an exhaustive study of short-term feedback effects on carbon metabolism. As part of the study of short-term feedback effects, Tom's research group demonstrated that maltose is the major form of carbon export from chloroplasts at night, and made significant contributions to the elucidation of the pathway by which leaf starch is converted to sucrose at night. In the isoprene research field, Tom is recognized as the leading advocate for thermotolerance of photosynthesis as the explanation for why plants emit isoprene. In addition, his laboratory has cloned many of the genes that underlie isoprene synthesis, and he has published many papers on the biochemical regulation of isoprene synthesis. Tom has coedited three books: T.D. Sharkey, E.A. Holland and H.A. Mooney (Eds.) *Trace Gas Emissions from Plants*, Academic, San Diego, CA, 1991; R.C. Leegood, T.D. Sharkey, and S. von Caemmerer (Eds.) *Physiology and Metabolism*, Advances in Photosynthesis (and Respiration), Volume 9 of this Series, Kluwer (now Springer), Dordrecht, 2000; and Volume 34 of this series *Photosynthesis: Plastid Biology, Energy Conversion and Carbon Assimilation*, Advances in Photosynthesis and Respiration *Including Bioenergy and Related Processes,* Julian J. Eaton-Rye, Baishnab C. Tripathy and Thomas D. Sharkey (Eds.) Springer. Tom joined the series founder Govindjee as Series Co-editor from Volume 31 of this series. Tom is currently the Chairperson of the Department of Biochemistry and Molecular Biology, Michigan State University, East Lansing, Michigan. For further information see his web page at: <http://www.bmb.msu.edu/faculty/sharkey/Sharkey/index.html>

Contents

Part II: Applied Aspects in Biohydrogen Production

Preface

There is a general consensus in considering the use of fossil fuels (petroleum, natural gas and coal) as the cause of serious environmental problems. Because the amount of energy derived from these fossil reserves is close to 80 % of the entire World's energy consumption, there is a pressing need of new, non-polluting and renewable energy sources (Report of the International Energy Agency, 2010). Although hydrogen (H_2) is not a primary energy source, it has been considered a promising alternative to fossil fuels. By definition, an energy source is such only if useful energy can be directly extracted or recovered from it; in this respect, H_2 is an "energy carrier" as it is derived from an energy reservoir and it can be used, like electricity, for the "transport" of energy from the production site to the sites of its use. The main consequence of this feature is that hydrogen can be produced only by consuming primary energy sources, which at the moment are mainly fossil fuels. However, there are at least two properties making the use of hydrogen quite attractive, namely: its large presence in nature, even if usually linked with other atoms, e.g. with oxygen in water or with oxygen, carbon and other elements in organic compounds, and the possibility to use it without releasing pollutants or greenhouse gases (GHGs) in the atmosphere. Interestingly, hydrogen can also be produced by exploiting the metabolic features of several microorganisms in a carbon neutral process that has been called "biological hydrogen production". Bio-hydrogen production is also characterized by important advantages over the thermochemical and electrochemical techniques currently utilized or under study. Indeed, microbiological processes can produce hydrogen using renewable resources, in carbon neutral processes operating at room temperature and pressure, and with low environmental impact. A negative aspect of the microbial hydrogen production in natural habitats is the fact that although a large number of bacteria, belonging to different taxonomic groups, possess the capability to produce hydrogen, free hydrogen of biological origin can hardly be captured in nature because hydrogen-producing and hydrogen-consuming microorganisms live in the same natural environments.

Bio-hydrogen production has been known for almost a century and research directed at applying this process to a practical means of hydrogen fuel production has been carried out for over a quarter of a century. A milestone in bio-hydrogen research was the observation by Hans Gaffron, while working at the University of Chicago in 1939, that algae can generate hydrogen by both fermentation and photochemistry (H. Gaffron (1939) Reduction of CO_2 with H_2 in green plants. *Nature* 143:204–205). Ten years later, Gest and Kamen (H. Gest and M.D. Kamen (1949) Photoproduction of molecular hydrogen by *Rhodospirillum rubrum*. *Science* 109: 558–559) discovered the light-dependent production of hydrogen in parallel to nitrogen fixation by the facultative photosynthetic bacterium *Rhodospirillum rubrum*. Notably, in "*Memoir of a 1949 railway journey with photosynthetic bacteria*" (Photosynthesis Res 61: 91–96), H. Gest (1999) commented on this discovery as "*A serendipic observation at the Hopkins Marine Station of Stanford University in 1948 led to the discovery that anoxygenic photosynthetic bacteria can fix molecular nitrogen … and generate hydrogen*". One of us (Zannoni D), while working as an associate researcher at the St. Louis University Medical School in 1978, was fortunate enough to have known personally H. Gest, Professor of Microbiology at the University of Bloomington (Indiana). He remembers that they had a long discussion on the way to define what it is now recognized as the "accessory oxidant-dependent fermentation in photosynthetic bacteria" (See: *The Phototrophic Bacteria: Anaerobic Life in the Light*, J.G. Ormerod Ed., 1983, vol 4, University of California Press, Blackwell Sc. Pub.). Enormous advances have been made since then on genetics, biochemistry, and biotechnological applications of photosynthetic bacteria and the present book, entitled *Microbial BioEnergy: Hydrogen Production* is a compendium overviewing most of the processes important for the microbial hydrogen production including bacterial hydrogen photo-generation.

The book begins with a chapter on bioenergy from microorganisms describing some of the challenges in meeting future energy needs in order to address climate changes through the development of bioenergy (Chap. 1). Critical factors in mature technologies and future directions in nascent technologies are also reviewed. The volume includes a section (Chaps. 2, 3, 4, and 5) covering structural, molecular, and functional aspects of hydrogenases as efficient biological catalysts for the production of molecular hydrogen and, consequently, its oxidation a way to get rid of the excess reducing power in cyanobacteria and green algae. As cyanobacteria are unique organisms that accommodate both oxygenic photosynthesis and nitrogen fixation, they are extensively covered in Chap. 6 with respect to their production of ammonium concomitantly with hydrogen formation. Solar energy is also used by photosynthetic purple non-sulfur bacteria to generate hydrogen gas from organic sources via the enzyme nitrogenase. Chapter 7 focuses on the advances that have been made in hydrogen generation through the use of systems biology approaches such as genomics, transcriptomics and ^{13}C-fluxomics in *Rhodopseudomonas palustris* CGA009. Chapters 8 and 9 cover two emerging research fields in hydrogen production: the use of thermophilic and hyperthermophilic microorganisms of the genera *Caldicellulosiruptor* and *Thermotoga*. As these genera utilize an extraordinary array of substrates that are converted by dark-fermentation to hydrogen at efficiencies approaching the "Thauer limit" of 4 mol H_2/mol glucose, the availability of several genome sequences and their metabolic features open new perspectives for biohydrogen generation. Bioelectrochemical systems coupled to indirect hydrogen production are reviewed in Chap. 8. These systems are not subjected to the hydrogen yield constraints and have been proven to work with any biodegradable organic substrate. Chapters 11, 12, 13, and 14 are mostly dedicated to photobioreactors using purple non-sulfur bacteria and microalgae. This section of the book examines in detail how hydrogen production depends on various kinds of organic wastes, on the photosynthetic efficiency and light distribution. The basic principles for designing photobioreactors in mass culture for biofuel are also examined along with the advantages and limitations of immobilized cell-systems for hydrogen photoproduction. The volume ends with a chapter (Chap. 15) dealing with the unconventional concept that if hydrogen is used as an energy carrier, there are consistent benefits to be expected, depending on how hydrogen is generated. The existing technical problems lying ahead for the creation of an apparent "Hydrogen Based Society" are examined and it is concluded that they will be solved within a reasonable period of time.

Following the suggestion of the Series editors, Govindjee and Tom Sharkey, we are deeply honored to dedicate this book to Hans Gaffron (1902–1979) and Howard Gest (1921–2012).

Davide Zannoni
Department of Pharmacy
and BioTechnology (FaBiT)
University of Bologna
Irnerio 42, 40126 Bologna, Italy
davide.zannoni@unibo.it

Roberto De Philippis
Department of AgriFood Production
and Environmental Sciences
P.zzale delle Cascine 24,
50144 Firenze, Italy

Institute of Chemistry of Organometallic
Compounds (ICCOM), CNR, Sesto
Fiorentino, Florence, Italy
roberto.dephilippis@unifi.it

The Editors

Davide Zannoni, Professor of General Microbiology, received the doctoral degree in Biological Sciences, in 1973, from the University of Bologna, Italy; his thesis was on the Bioenergetics of the facultative photosynthetic bacterium *Rhodobacter (Rb.) capsulatus*. During 1977–1978, he was a research fellow of the North Atlantic Treaty Organization (NATO) at the St. Louis Medical School, Department of Biochemistry, St. Louis MO, USA, under the supervision of Prof. Barry L. Marrs. In 1979, he was appointed Lecturer in Plant Biochemistry, and in 1981, he was promoted to become Associate Professor of Plant Biochemistry, at the Faculty of Sciences, University of Bologna. As a research fellow of the European Molecular Biology Organization (EMBO) in 1981, 1983 and 1991, he visited several European laboratories, namely: Department of Biochemistry and Microbiology, St. Andrews University, St. Andrews Scotland U.K.; Département de Biologie Cellulaire et Moléculaire, Centre National Recherche Scientifique (CNRS), Commissariat à l'Energie Atomique (CEA) Saclay Gif-sur-Yvette, France; Department of Microbiology, University of Göttingen, Göttingen, Germany, to investigate both the structure and the function of membrane redox-complexes in a variety of microbial genera. Zannoni's scientific interests now include bioenergetics and genomics of microbial remediation of metals and metalloids in planktonic cells and biofilms of *Rb. capsulatus* and *Pseudomonas pseudoalcaligenes*, molecular mechanisms of bacterial movement (chemo- and photo-taxis) and biofilm formation, alkane and naphthenic acid degradation by *Rhodococcus* spp., the use of microbial biofilms as electricity-producing systems, and, finally, bio-hydrogen anaerobic production by *Thermotoga*. Zannoni's pioneering work on hydrogen metabolism in *Rb. capsulatus* began in 1981 (European Community Solar Energy Research & Development). He is author and/or co-author of more than 130 publications in international research journals, and he has published several research as well as text-books for students. From 2004 to 2010, Prof. Zannoni has been the Head of the Department of Biology of the University of Bologna. He is presently acting as a Coordinator of the Master's degree in Molecular & Industrial Biotechnology at the Department of Pharmacy & Bio-Technology, University of Bologna – Alma Mater Studiorum, Italy. See his website <http://www.unibo.it/docenti/davide.zannoni> for further information.

Roberto De Philippis, Associate Professor of Microbial Biotechnology, received his Laurea degree in Chemistry from the University of Florence, Italy, in 1978; his thesis was on the chemical interactions between nucleic acids and amino acids, as studied by means of NMR and EPR techniques. During 1978–1981, he was a research scientist at the Research Center on Plastic Polymers at "Montedison" SpA, Milan, Italy; during 1981–1983 he was responsible for the scientific and technical aspects of Baker's yeast production at a Food Industry in Florence. During 1984–1990, he was a Research Fellow at the Institute of Agricultural and Technical Microbiology, University of Florence. During 1990–2001, he served as a Lecturer at the Department of Food and Microbiological Science and Technology, University of Florence. From 2001, he has been an Associate Professor of Microbial Biotechnology at the University of Florence, Department of Agrifood Production and Environmental Sciences. He is, at the same time, an Associate Researcher at the Institute of Chemistry of Organometallic Compounds, Italian National Research Council (ICCOM-CNR), Florence. His research activity is mainly concerned with the physiology and biochemistry of photosynthetic bacteria. In particular, Roberto is studying the physiology and the possible biotechnological exploitation of phototrophic microorganisms in the production of biopolymers of industrial interest or in processes related to the production of energy from renewable resources or for the treatment of polluted waters. He is also involved in studies on the formation of phototrophic biofilms on monuments or in the stabilization of desert soils by the use of phototrophic microorganisms. He has been hosted for his research by several Institutions in China, India, Israel, Mexico, Portugal and USA. He has published more than 80 scientific papers in international peer-reviewed journals, ten chapters in books, and has participated in more than 90 international and national Congresses. During 1999–2001, Roberto was Secretary/Treasurer, and currently, he is President-elect of the International Society for Applied Phycology. He is an Assistant Editor of the *Journal of Applied Phycology*. From 2010, he has been a Delegate for Italy in the IEA-HIA (International Energy Agency-Hydrogen Implementing Agreement) New Annex 21 "Bio-inspired and Biological Hydrogen". See the following website for further information: http://www.unifi.it/p-doc-2012-200001-D-3f2a3d29392930.html

Contributors

Francisco Gabriel Acién-Fernández
Department of Chemical Engineering, University of Almería, Almería, Spain
facien@ual.es

Alessandra Adessi
Department of Agrifood Production and Environmental Sciences, University of Florence, Florence, Italy
alessandra.adessi@unifi.it

Giacomo Antonioni
Department of Civil, Chemical, Environmental and Materials Engineering (DICAM), University of Bologna, Bologna, Italy
giacomo.antonioni3@unibo.it

Sara E. Blumer-Schuette
Department of Chemical and Biomolecular Engineering, North Carolina State University, Raleigh, NC, USA
rmkelly@ncsu.edu

Hermann Bothe
Botanical Institute, The University of Cologne, Cologne, Germany
aeb75@uni-koeln.de

Martina Cappelletti
Laboratory of General and Applied Microbiology, Department of Pharmacy and BioTechnology (FaBiT), University of Bologna, Bologna, Italy
martina.cappelletti2@unibo.it

Jonathan M. Conway
Department of Chemical and Biomolecular Engineering, North Carolina State University, Raleigh, NC, USA
rmkelly@ncsu.edu

Roberto De Philippis
Department of Agrifood Production and Environmental Sciences, University of Florence, Florence, Italy
Institute of Chemistry of Organometallic Compounds (ICCOM), CNR, Sesto Fiorentino, Florence, Italy
roberto.dephilippis@unifi.it

Alexandra Dubini
National Renewable Energy Laboratory, Golden, CO, USA
alexandra.dubini@nrel.gov

Carrie Eckert
National Renewable Energy Laboratory, Golden, CO, USA
carrie.eckert@nrel.gov

Ela Eroglu
School of Chemistry and Biochemistry, Centre for Strategic Nano-Fabrication, The University of Western Australia, Perth, Australia
ela.eroglu@uwa.edu.au

Inci Eroglu
Department of Chemical Engineering, Middle East Technical University, Ankara, Turkey
Department of Biotechnology, Middle East Technical University, Ankara, Turkey
ieroglu@metu.edu.tr

José Maria Fernández-Sevilla
Department of Chemical Engineering, University of Almería, Almería, Spain
jfernand@ual.es

Juan C. Fontecilla-Camps
Metalloproteins Unit, Institut de Biologie Structurale J.P. Ebel, Commissariat à l'Énergie Atomique, Centre National de la Recherche Scientifique, Université Joseph Fourier, UMR 5075, Grenoble, France
juan.fontecilla@ibs.fr

Dario Frascari
Department of Civil, Chemical, Environmental and Materials Engineering (DICAM), University of Bologna, Bologna, Italy
dario.frascari@unibo.it

Maria L. Ghirardi
National Renewable Energy Laboratory, Golden, CO, USA
Maria.Ghirardi@nrel.gov

Ufuk Gündüz
Department of Biotechnology, Middle East Technical University, Ankara, Turkey

Department of Biology, Middle East Technical University, Ankara, Turkey
ufukg@metu.edu.tr

Patrick C. Hallenbeck
Department of Microbiology and Immunology, University of Montreal, Montreal, QC, Canada
patrick.hallenbeck@umontreal.ca

Robert M. Kelly
Department of Chemical and Biomolecular Engineering, North Carolina State University, Raleigh, NC, USA
rmkelly@ncsu.edu

Paul W. King
National Renewable Energy Laboratory, Golden, CO, USA
Paul.King@nrel.gov

Sergey Kosourov
Institute of Basic Biological Problems, Russian Academy of Sciences, Pushchino, Moscow Region, Russia
sergks@rambler.ru

Pierre-Pol Liebgott
IRD, Aix-Marseille Université, CNRS/INSU, MIO, UM 110, Marseille Cedex 09, France
pierre-pol.liebgott@univ-amu.fr

Pin-Ching Maness
National Renewable Energy Laboratory, Golden, CO, USA
pinching.maness@nrel.gov

James B. McKinlay
Department of Biology, Indiana University, Bloomington, IN, USA
jmckinla@indiana.edu

Emilio Molina-Grima
Department of Chemical Engineering, University of Almería, Almería, Spain
emolina@ual.es

David W. Mulder
National Renewable Energy Laboratory, Golden, CO, USA
david.mulder@nrel.gov

William E. Newton
Department of Biochemistry, Virginia Polytechnic Institute and State University, Blacksburg, VA, USA
venewton@vt.edu

Paulo Oliveira
Bioengineering and Synthetic Microbiology Group, IBMC – Instituto de Biologia Molecular e Celular, Universidade do Porto, Porto, Portugal
paulo.oliveira@ibmc.up.pt

Bernard Ollivier
Aix-Marseille Université, Université du Sud Toulon-Var, CNRS/INSU, IRD, MIO, UM 110, Marseille Cedex 09, France
bernard.ollivier@univ-amu.fr

Ebru Özgür
MEMS Research and Application Center, Middle East Technical University, Ankara, Turkey
ebruozgur@gmail.com

Catarina C. Pacheco
Bioengineering and Synthetic Microbiology Group, IBMC – Instituto de Biologia Molecular e Celular, Universidade do Porto, Porto, Portugal
cclopes@ibmc.up.pt

Anne Postec
Aix-Marseille Université, Université du Sud Toulon-Var, CNRS/INSU, IRD, MIO, UM 110, Marseille Cedex 09, France

John M. Regan
Department of Civil and Environmental Engineering, The Pennsylvania State University, University Park, PA, USA
jregan@engr.psu.edu

Marc Rousset
CNRS, BIP, Marseille Cedex 20, France
rousset@imm.cnrs.fr

Paula Tamagnini
Bioengineering and Synthetic Microbiology Group, IBMC – Instituto de Biologia Molecular e Celular, Universidade do Porto, Porto, Portugal

Faculdade de Ciências, Departamento de Biologia, Universidade do Porto, Porto, Portugal
pmtamagn@ibmc.up.pt

Anatoly Tsygankov
Institute of Basic Biological Problems, Russian Academy of Sciences, Pushchino, Moscow Region, Russia
ttt-00@mail.ru

Anne Volbeda
Metalloproteins Unit, Institut de Biologie Structurale J.P. Ebel, Commissariat à l'Énergie Atomique, Centre National de la Recherche Scientifique, Université Joseph Fourier, UMR 5075, Grenoble, France
anne.volbeda@ibs.fr

Hengjing Yan
Institute for Collaborative Biotechnologies, Department of Chemistry and Biochemistry, University of California Santa Barbara, Santa Barbara, CA 93106, USA

Jianping Yu
National Renewable Energy Laboratory, Golden, CO, USA
jianping.yu@nrel.gov

Meral Yücel
Department of Biotechnology, Middle East Technical University, Ankara, Turkey

Department of Biology, Middle East Technical University, Ankara, Turkey
meraly@metu.edu.tr

Davide Zannoni
Department of Pharmacy and BioTechnology (FaBiT), University of Bologna, Bologna, Italy
davide.zannoni@unibo.it

Jeffrey V. Zurawski
Department of Chemical and Biomolecular Engineering, North Carolina State University, Raleigh, NC, USA
jvruraw@ncsu.edu

Author Index

Part I

General and Molecular Aspects of Bio-Hydrogen Generation

Chapter 1

Bioenergy from Microorganisms: An Overview

Patrick C. Hallenbeck*
Department of Microbiology and Immunology, University of Montreal, Montreal, Quebec H3C 3J7, Canada

Summary

Some of the challenges in meeting future energy needs and addressing climate change will need to be met through the development of bioenergy. The power and diversity of microbial metabolism, coupled with metabolic engineering and synthetic biology, can be used to produce a panoply of different biofuels from a variety of possible substrates. Here some of the underlying principles involved and an overview of the different production pathways under development are discussed. Critical factors in mature technologies and future directions in nascent technologies are reviewed.

*Author for correspondence, e-mail: patrick.hallenbeck@umontreal.ca

D. Zannoni and R. De Philippis (eds.), Microbial BioEnergy: Hydrogen Production, Advances in Photosynthesis and Respiration 38, pp. 3–21, DOI 10.1007/978-94-017-8554-9_1,

I. Climate Change and Future Energy Challenges

There can be little doubt that unprecedented climate change brought about by global warming due to excessive fossil fuel combustion is upon us. The recent increase in extreme weather events shows that noticeable changes in weather patterns are already here and thus this is not a problem for the future, distant or near, but one with which we must already cope. Although a variety of factors have increased atmospheric forcing over the millennia, including a considerable atmospheric CO_2 burden due to ancient prehistoric and early historic land clearing, this has greatly accelerated with industrialization and the development of fossil fuels as the primary energy source to drive this. At the same time, industrialization has created the modern world with greatly increased per capita GDP in the developed countries (OECD). Thus, at present energy use is intimately tied to per capita income levels while at the same time at the root of disastrous climate change effects. Of course, this dichotomy creates great resistance to changes in energy use policy as these are seen as directly affecting either the present high standard of living in the OECD, or the chance for developing countries to achieve a like life style.

A. The Scope of the Problem

It is certain that future energy demand (and use) will grow, inexorably driven by two compounding factors, population growth and growth in per capita energy usage. Growth in per capita energy usage is essentially due to income growth, as the "have nots" strive to achieve life style parity with the "haves". In fact, income growth and population growth have been considered the most powerful forces behind increasing energy demand. While world population has increased fourfold since 1900, over the same time period real income increased by a factor of 25 driving a 22.5-fold increase in energy consumption (BP 2011). This trend is obviously set to continue with world population widely believed to reach nine billion by 2050 (8.45 billion by 2035), fueled by an average annual growth rate close to 1 % (DOE International Energy Outlook 2011). As well, GDP is predicted to increase at an annual rate of 3.2 % at the same time (IEA World Energy Outlook 2010), effectively raising average incomes.

A very useful way to look at the component driving forces behind global carbon emission, broken down into four major factors, is given by the Kaya Identity (Yamaji et al. 1991):

$$CO_2 = P \times (GDP/P) \times (E/GDP) \times (CO_2/E) \quad (1.1)$$

Thus, total anthropogenic carbon emissions can be seen as a function of the total population (P), individual consumption (GDP/P, gross domestic product consumed per person), essentially individual income, the energy intensity of production (E/GDP), and the carbon intensity of energy use (CO_2/E).

However, in reality only two of these factors are amenable to manipulation to slow or stabilize total CO_2 emissions since changing the population growth rate has proven an intractable problem, and decreasing, or even stabilizing economic output (GDP) would directly impact average incomes, contrary to the economic policy of any government. The energy intensity of production can be decreased somewhat through efficiency measures, but this is relatively inflexible. A significant factor in energy intensity of production, at least on a regional level, is the

Abbreviations: CBP – Consolidated bioprocessing; DOE – Department of Energy; Fd – Ferredoxin; GDP – Gross Domestic Product; IEA – International Energy Agency; IPCC – Intergovernmental Panel on Climate Change; MEC – Microbial electrolysis cell; MFC – Microbial fuel cell; NFO – NADH:ferredoxin oxidoreductase; OECD – Organisation for Economic Cooperation and Development; PFL – Pyruvate:formate lyase; PFO – Pyruvate:ferredoxin oxidoreductase mix of industries/services that form the economic base. As OECD countries have switched to more of service economies, energy intensities in those countries steadily declined. However, this was not a really a savings in global carbon emissions, merely an "outsourcing" of emissions to countries like China as manufacturing to satisfy consumer appetites in OECD countries shifted overseas. Thus, although global emissions were rising at about 1 % up to 2000, since then global emissions have been increasing at ~3.2 % per year (Raupach et al. 2007). This means that the world is on track to surpass even the most pessimistic IPCC (Intergovernmental Panel on Climate Change) emissions scenario (WMO 2012).

One goal that has been adopted in principle, although the policy changes needed to put this in practice have not been made, is to prevent the mean global temperature increase from surpassing 2 °C above preindustrial levels. It is thought that this requires limiting atmospheric CO_2 levels to 450 ppm. Under present policies, basically BAU (Business As Usual), we will go well beyond this as demand for total energy, and consequently emissions, more than doubling by 2050. Indeed, we very well might have to confront a world where mean global temperatures have risen more than 4 °C, with largely unknowable, but assuredly drastic effects (New et al. 2011). The challenges to be met in response to the ever increasing demand for energy are enormous. This can be seen by regarding present energy producing infrastructure, "steel in the ground". Existing fossil fuel plants in themselves commit us to enough future carbon dioxide emissions to nearly reach the 450 ppm cut-off proposed by many (Davis et al. 2010; Hoffert 2010). In other words, in order to realistically reach a goal set by many, we will need to have a complete moratorium on the construction of new fossil fuel burning power plants. Obviously, this is a scenario that is completely unacceptable to all, especially the developing countries, where new economic growth is so tightly tied to bringing online more power. In fact, considerations of global energy equity suggest that they should not be asked to do so (see below).

In fact, any scenario for reducing the growth in carbon emissions requires the decarbonisation of energy production, i.e. reducing the CO_2/E term in Eq. 1.1. Thus, the introduction of carbon neutral fuels is required, and the more the better, considering the amount of new energy that will be needed under almost any future global energy scenario. It has been estimated that even in an optimistic scenario where there are substantial changes in energy intensity of production, maintaining economic growth while remaining at 450 ppm will require the introduction of 30 terawatts of carbon neutral fuel by 2050 (Hoffert 2010). Indeed, just getting to 1 terawatt of carbon neutral fuel has been called the "one terawatt challenge".

B. Global Energy Equity and Energy Justice

Another important factor in considering present and future energy production/consumption is energy equity. In 2011, the OECD countries accounted for only 18 % world's population, while non-OECD countries represented 82 %. Even though growth in carbon emissions is much faster in non-OECD as these countries strive to increase per capita income, and as they take on some of the carbon emissions out-sourced from OECD countries, they still only account for about a share of emissions that equals that of the OECD. In order to examine questions of global energy equity and energy justice one needs to look on a per capita basis at the regional emissions that are behind the global figures. In other words, energy usage is very unevenly distributed when measures such as per capita emissions and per capita energy consumption are examined. By these measures, countries like the USA (per capita energy usage 10.2 kW, per capita emissions, 5.5 t C/y) truly stand out compared with the poorest developing countries (per capita energy usage 0.11 kW, per capita emissions, 0.06 t C/y) (Raupach et al. 2007). Thus, one could argue that the OECD countries bear by far the largest burden in dealing with global warming and carbon emissions. This is especially true if one takes into account "legacy carbon emissions", since if one totals the emissions since the beginning of the industrial revolution, the OECD countries have contributed 77 % of the present excess atmospheric carbon dioxide. Thus, there are many questions surrounding future energy supplies and climate change which, although they should be framed in terms of scientific knowledge and informed projections, lie more in the realm of politics and public policy (Hallenbeck 2012a). In what follows we examine a small focused area in what is one of the major challenges presently facing humankind, how can microbes be used to make carbon neutral replacement fuels. Here, many recent scientific studies have shown that there is a great deal that can be done with either existing organisms, or modified ones, to potentially produce true sustainable, renewable fuels.

II. A Wide Variety of Biofuels

A. First Generation Biofuels

Biofuels production is already at large scale, primarily to supply the transportation sector, which relies almost exclusively (97 %) on the use of liquid fossil fuels. Moreover, this is an important target area since it is second only to the industrial sector in current and projected total fossil fuel consumption. In addition, while stationary power consumption is largely indifferent to the form of the energy carrier, mobile use requires a fuel source that can be stored on board at a sufficiently high energy density. Thus it is perceived to be critically important to partially or completely replace presently used carburants, gasoline or diesel, with a renewable energy carrier. Biofuel production has greatly increased worldwide, mostly driven by adopted government policies in the forms of incentives; subsidies and alternative fuel mandates. Thus, first generation biofuels (Table 1.1) are already being produced at large scale, with worldwide production of ethanol and biodiesel of 50 billion and 9 billion liters respectively in 2007. However, there are a number of problems with these first generation biofuels. First, it has become obvious that these first generation biofuels technologies are of questionable sustainability, and that in the long term it is untenable to produce biofuels in competition with food crops (Cassman and Liska 2007; Waldrop 2007; Scharlemann and Laurance 2008; Tollefson 2008; Tilman et al. 2009). Secondly, it can be questioned if the biofuels presently under commercial production, bioethanol and biodiesel, are ultimately the best biofuels since they have a number of undesirable characteristics (Keasling and Chou 2008). Thus, there is a need to go beyond these first generation biofuels, both in what substrates are used and, ultimately, in what products are made.

B. Beyond First Generation Biofuels

A number of methodologies, variably called synthetic biology or Metabolic engineering, provide powerful techniques that could be applied to various aspects of the biofuels problem (Fig. 1.1). These include genetic engineering, what are basically relatively simple changes made through simple gene knockouts, or small additions of heterologous DNA, to much more involved and complicated changes brought about by importing entire metabolic pathways, directed evolution, or genome shuffling. The production of a biofuel can be divided into two basic stages; substrate conversion to key metabolic intermediates, and conversion of these metabolic intermediates into a biofuel. Metabolic engineering could usefully impact each of these areas in several ways (Fig. 1.1).

1. Molecular Engineering for the Deconstruction of Lignocellulosic Biomass

To begin with, the production of large quantities of biofuels without seriously impacting the world food supply requires the ability to utilize biomass resources with their potentially substantial quantities of energy stored

Table 1.1. Comparison of different strategies for lignocellulosic deconstruction.

Process	Possible advantages	Possible disadvantages
Separate hydrolysis and fermentation	Reactor size and operating conditions easily optimized Enzyme specificity and efficiency can be adjusted to substrate Newly discovered or engineered enzymes easily incorporated	Costly enzyme production Enzymes need to be cloned from different organisms Two stage system required increasing operational costs and complexity
Native consolidated bioprocessing	Direct conversion of cellulose to biofuel possible Single stage process; simple facility, easy operation Avoidance of inhibition of cellulose degradation by monomers Uses existing metabolic machinery	Optimal temperatures for cellulose degradation and fermentation may be different Low rates and yields of useful products by native organism Low titers of active enzymes due to inefficient anaerobic growth
Engineered consolidated bioprocessing	Optimal cellulose degradative capacity in efficient fermenter Single stage process Cost-effective production of cellulases Designer cellulosomes can be fabricated	Need for complex Metabolic engineering, expression of multiple components May lack synergistic factors found in native organism

Adapted from Hallenbeck et al. (2011)

in lignocellulosic compounds (Sagar and Kartha 2007; Field et al. 2008; Martindale and Trewavas 2008; Tilman et al. 2009). At present, no organism is known that is both capable of effective lignocellulose deconstruction and the efficient conversion of the resulting five and six carbon sugars into the key metabolic intermediates found in the central metabolic pathways.

The capacity to directly degrade cellulosic materials into fermentable monomers, a process called lignocellulosic, occurs naturally in a few organisms, but Metabolic engineering could be used to transfer this ability to other organisms (Fig. 1.1). However, the capacity to use the wide variety of hexose and pentose sugars resulting from the hydrolysis of such complex substrates would also be required and this ability does not necessarily reside in organisms that carryout rapid high yielding fermentations. Thus, Metabolic engineering could also be applied to modify a strong fermenter such that it is able to use a wider range of substrates, or to increase the fermentative powers of an organism with an omnivorous appetite (Ghosh and Hallenbeck 2009, 2010, 2012).

Although various waste streams can and should be targeted for conversion to biofuels as these represent readily available substrates that are free or low cost and require treatment anyway, to achieve the scale needed for significant biofuels production the use of lignocellulosic containing biomass will be necessary. However, effective deconstruction of these materials has proven problematic despite some years of concerted effort due to the almost crystalline state of the cellulose component and the intractability of the lignin (Field et al. 2008; Tilman et al. 2009; Sims et al. 2010; Hallenbeck et al. 2011). Different strategies have been developed in response to this problem (Table 1.2) (Lynd et al. 2002; Hallenbeck et al. 2011). The most technologically advanced process, already near commercial scale, involves physical/chemical pretreatment followed by the addition of the various enzymes required for depolymerization to monomers. However, in reality this involves the addition of

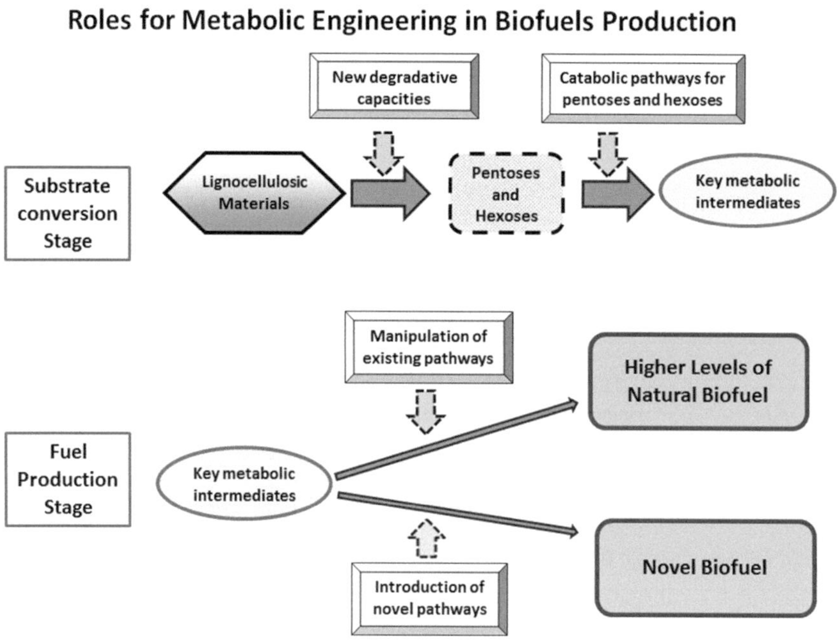

Fig. 1.1. Different possible roles for Metabolic engineering in biofuels production. There are a number of ways in which Metabolic engineering can be used to expand biofuels production going beyond first generation biofuels. First, Metabolic engineering can potentially be used in the substrate conversion stage to create organisms with newly abilities to degrade complex lignocellulosic substrates, by far the largest substrate pool available from non-food biomass. In addition, organisms with existing strong downstream capabilities can have their substrate range extended to include the capacity to use the pentoses and hexoses derived from lignocellulosic substrates. These strategies expand substrate conversion to key metabolic intermediates. Secondly, the conversion of these key intermediates to biofuels can be modified in two distinct ways. The production of a biofuel that is normally made by an organism can be augmented by changes in existing pathways, creating increased flux to the desired biofuel. This can be either through changes causing increased activity in the relevant pathway or by decreasing flux into non biofuel pathways. Another additional strategy is to engineer in novel enzymes and pathways which enable the organism to produce novel biofuels (Taken from Hallenbeck 2012b).

prodigious amounts of enzymes which must therefore be prepared at very large scale. The costly economics of this is presently the limiting factor in applying this strategy. Another possibility is to use the natural ability of some organisms to simultaneously carryout cellulose saccharification and fermentation. They are able to do this since they naturally produce the required enzymes in a very large extracellular macromolecular complex called a cellulosome that anchors them directly to the lignocellulosic substrate. This combination of all the required processes in one step is called consolidated bioprocessing (CBP), a technology which could potentially dramatically reduce the cost of biofuels production. Some of the advantages and disadvantages of natural CBP for biofuels production are given in Table 1.2. However, unfortunately, the known organisms that are capable of CBP are probably not suitable for industrial-scale biofuels production since the rates and yields of suitable products are too low. In addition, they are not known to produce the desired drop-in biofuels.

Although some of these problems might be rectified by Metabolic engineering, attention has turned towards bringing this capability to efficient fermenters using different Metabolic engineering strategies. This could lead to a process with a number of advantages over the processes already developed (Table 1.2). The approaches under study vary from attempts to express foreign cellulases to creating organ-

Table 1.2. Comparison of different strategies for lignocellulosic deconstruction.

Process	Possible advantages	Possible disadvantages
Separate hydrolysis and fermentation	Reactor size and operating conditions easily optimized Enzyme specificity and efficiency can be adjusted to substrate Newly discovered or engineered enzymes easily incorporated	Costly enzyme production Enzymes need to be cloned from different organisms Two stage system required increasing operational costs and complexity
Native consolidated bioprocessing	Direct conversion of cellulose to biofuel possible Single stage process; simple facility, easy operation Avoidance of inhibition of cellulose degradation by monomers Uses existing metabolic machinery	Optimal temperatures for cellulose degradation and fermentation may be different Low rates and yields of useful products by native organism Low titers of active enzymes due to inefficient anaerobic growth
Engineered consolidated bioprocessing	Optimal cellulose degradative capacity in efficient fermenter Single stage process Cost-effective production of cellulases Designer cellulosomes can be fabricated	Need for complex Metabolic engineering, expression of multiple components May lack synergistic factors found in native organism

Adapted from Hallenbeck et al. (2011)

Table 1.3. Characteristics of an ideal biofuels producer.

Required characteristics	
Yields	High >90 %
Productivity	High reduce capitol and energy inputs
Tolerance to product	High achieve high titers
Robustness	Tolerance to inhibitory substances in crude substrates
Ease of growth	Minimal, inexpensive growth factor requirement
Desirable characteristics	
Cell yield	Low, avoid disposal problem
GRAS (generally regarded as safe)	Avoid regulatory complications or need for post- production sterilization

Adapted from Hallenbeck et al. (2011)

isms with artificial minicellulosomes. The success of displaying a functional minicellulosome on the surface of an organism that already produces high titers of a biofuel would lead the way to achieving a truly industrially relevant CBP microorganism.

Lignocellulose deconstruction will of course give a complex mixture of five and six carbon sugars, principally glucose and xylose. For the successful fermentation of this complex mixture to a biofuel to be successful, the microbial strain must be capable of using the majority of the fermentable substrate available. However, the majority of the mainstream organisms presently used industrially cannot. These organisms meet most of the criteria for successful biofuel (ethanol) fermentation (Table 1.3), and much experience has already been gained with them on large scale fermentations. Thus, this is a fruitful area for the application of Metabolic engineering and some effort has already gone into creating derivatives of these strains that are capable of degrading the relevant pentoses and hexoses (Ghosh and Hallenbeck 2012).

2. Molecular Engineering for Biofuels Production

A large number of different biofuels are being investigated for use as mobile energy carriers (Atsumi and Liao 2008; Atsumi et al. 2008a, b, 2010; Connor and Liao 2008; Keasling and Chou 2008; Lee et al. 2008; Steen et al. 2008, 2010; Stephanopoulos 2008; Wackett 2008; Yan and Liao 2009; Dellomonaco et al. 2010). In these microbial based processes, monomers, ideally derived from deconstructed lignocellulosic biomass, are converted through central metabolic pathways to common metabolic intermediates which then can follow a number of metabolic pathways to produce a large variety of potentially useful compounds. This can be done either using traditional fermentation pathways, suitably modified to increase the rate and yield of biofuels production, or by tapping into pathways normally used for biosynthesis to channel metabolites into a variety of novel molecules of possible interest as biofuels (Fig. 1.2).

Alternative fuels that are currently being investigated include: bioethanol, biobutanol, longer-chain alcohols, biohydrogen, and fatty acid derivatives such as biodiesels and alkanes. Many of these potential fuel compounds have a number of advantages over bioethanol as an alternative fuel. These include having a higher energy density, closer to that of gasoline, being less corrosive, thus adapted to current infrastructure, being less volatile, and, finally, having the capability of being used in existing internal combustion engines without dilution. However, Metabolic engineering is necessary since microorganisms do not naturally produce these compounds in high quantities. A successful strategy will give a strain capable of the efficient and economical bioconversion of non-food feedstocks to biofuels at high rates and near stoichiometric yields.

How the first stage of a biofuels production process might be improved was discussed above. Once the pentoses and hexoses enter

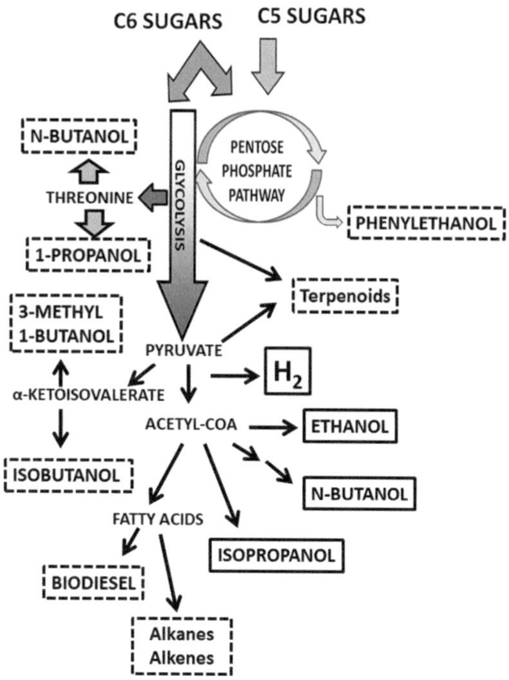

Fig. 1.2. A wide variety of possible biofuels can be made. The production of many different biofuels is possible using microbial metabolic capabilities. Some products can be made through traditional fermentation, with suitable modifications to increase yields and rates (*solid outlines*). In addition, existing biosynthetic pathways can be changed to make a variety of novel biofuels (*dashed outlines*). Pathways and products shown have already been experimentally demonstrated with *Escherichia coli* (Taken from Hallenbeck 2012b).

central metabolic pathways, key metabolic intermediates will be formed that could be used in one of two ways to generate molecules suitable for fuel use (Fig. 1.1). Metabolic engineering can be used to increase flux through an existing pathway thus increasing rates and yields of a naturally occurring metabolite that can serve as a biofuel; ethanol, butanol, and hydrogen. This can be achieved by blocking alternate pathways that channel metabolic flux into undesirable side products, or increasing the levels of key enzymes in the desired pathway.

Many studies have shown that novel biofuels can be made by the addition of foreign enzymes and pathways (Fig. 1.2). Thus, suitably modified *E. coli* has been shown to make isopropanol (Atsumi et al.

2008b), n-butanol (Atsumi and Liao 2008; Inui et al. 2008) or even isobutanol (Atsumi et al. 2010). The keto-acid pathways normally functioning in the biosynthesis of the amino acids threonine and norvaline have successfully been subverted for the production of a variety of butanol derivatives (Atsumi and Liao 2008; Connor and Liao 2008; Shen and Liao 2008). Likewise, terpenoid biosynthetic pathways can be successfully changed to produce a variety of compounds that could potentially be used as biofuels (Fortman et al. 2008; Rude and Schirmer 2009). The fatty acid biosynthesis pathways, normally used by the organism to make lipids using acetyl-CoA as precursor, can be adapted for the production of a variety of fatty acid ethyl esters (biodiesel) and alkanes and alkenes (Steen et al. 2010).

III. The Microbial Production of Hydrogen

Of all the possible biofuels, hydrogen has some unique advantages, but also has some drawbacks. On the positive side, hydrogen has the highest gravimetric energy density of any fuel. On the other hand, it has the lowest volumetric density, rasing serious challenges for storage, particularly for mobile uses. Hydrogen can be converted to mechanical energy with much greater efficiency since it can be used in fuel cells at about twice the efficiency of combustion in an internal combustion engine. While all biofuels are potentially carbon neutral since they are made from feedstocks produced by recent carbon dioxide fixation, hydrogen as a fuel offers some unique opportunities. For one thing, hydrogen can be produced by biophotolysis (see below), a carbon independent pathway. In common with other biofuels, hydrogen can be produced by other means using biomass derived substrates. However, unlike other biofuels with which CO_2 is emitted when they are combusted, with hydrogen production CO_2 is emitted during production, thus allowing easy capture and possible sequestration. Finally, combustion of hydrogen is cleaner than the combustion of other biofuels some of which emit pollutants that can be harmful to the health as well as the environment, for example combustion of ethanol can give significant amounts of acetaldehyde.

A. Modes of Biological Hydrogen Production

Hydrogen is an important metabolic intermediate or end product in a number of different microbial metabolisms. Consequently, there are a number of distinct ways in which hydrogen can be produced (Fig. 1.3). Each has its own particular advantages as well as challenges for practical application (Table 1.4). There are basically two different light dependent processes, one of which, biophotolysis, uses the captured solar energy to drive water-splitting photosynthesis and derive hydrogen from the high energy electrons created. The second approach, photofermentation, harnesses the capacity of bacterial-type photosynthesis to use the captured light energy to carry out what would otherwise be thermodynamically unfavorable hydrogen production from some substrates (organic acids). Dark fermentative hydrogen production uses the natural ability of some organisms to evolve hydrogen as an end product to rid themselves of excess electrons generated during anaerobic metabolism. Finally, microbial electrolysis uses the natural ability of some microbes to respire anaerobically with an external electrode as terminal electron acceptor. Addition of some voltage to the current thus generated allows hydrogen evolution at the cathode.

1. Light Driven Biohydrogen Production

A very large amount of free energy is available in the annual solar flux and two different biological processes are being studied for the conversion of captured solar energy to hydrogen (Hallenbeck and Benemann 2002; Ghirardi et al. 2009; Ghirardi and Mohanty 2010; Hallenbeck 2011).

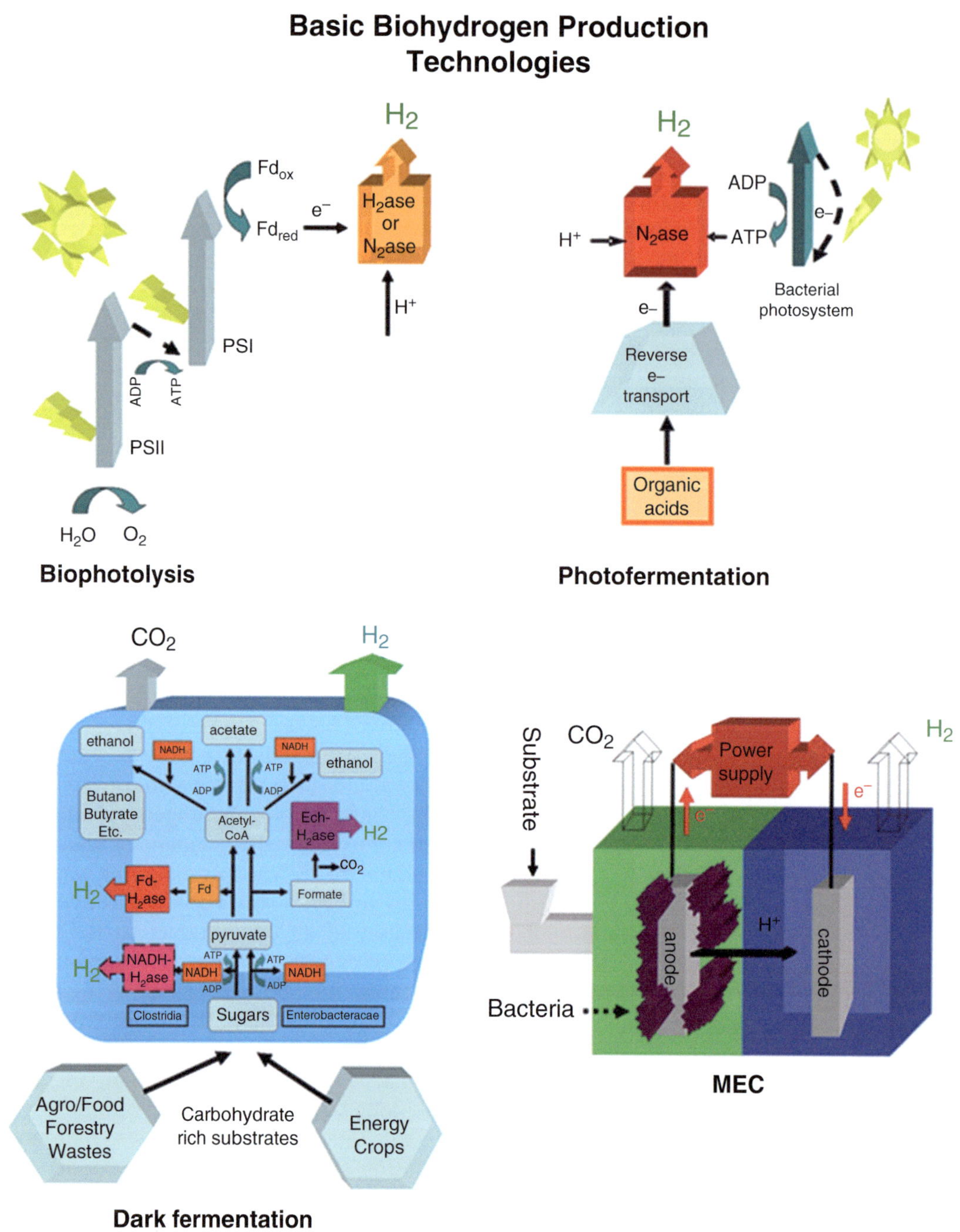

Fig. 1.3. Basic biological hydrogen production technologies. The four basic types of hydrogen producing systems are shown. There are two light-dependent processes, biophotolysis and photofermentation. These are carried out by two different types of organisms. Biophotolysis is possible with organisms, green algae and cyanobacteria, which have plant type photosynthesis and thus use water as a substrate. Photofermentation is carried out by photosynthetic bacteria which only possess a single photosystem and thus cannot split water, using organic compounds as electron donor instead. Dark fermentation involves different bacteria with different metabolic pathways which are capable of the anaerobic breakdown of organic compounds, in particular sugars, to produce hydrogen and a variety of side products, organic acids and alcohols. In MECs, the anode plays a critical role, accepting electrons coming from the anaerobic respiratory activity of microbes that are capable of interacting with external electron acceptors. Addition of complementary voltage allows hydrogen evolution at the electrode.

Table 1.4. Comparison of biohydrogen production technologies.

Method	Advantages	Disadvantages
Biophotolysis	Abundant substrate: water Simple products: H_2, O_2	Low light conversion efficiencies Oxygen sensitive H_2ase Expensive photobioreactors
Photofermentation	Complete conversion of organic acid wastes to H_2, CO_2 Possible waste treatment credits	Low light conversion efficiencies High energy demand by N_2ase Expensive photobioreactors
Dark fermentation	No direct solar input needed Variety of waste streams can be used Simple reactor technology	Incomplete substrate degradation Low H_2 yields
Microbial electrolysis	Complete conversion of organic compounds, sugars and acids, to H_2 and CO_2 Potential waste treatment credits	Low charge densities Expensive cathodes High energy input required for high rates and yields

a. Biophotolysis

In the most attractive system, biophotolysis, solar energy would be captured and used to decompose water, an abundant substrate, to hydrogen and oxygen. Actually, this type of process can be of two types, direct and indirect biophotolysis. In direct biophotolysis, electrons coming from the water splitting reaction are boosted by photosystem II (PSII) and photosystem I (PSI) to reduce ferredoxin which in turn directly reduces a hydrogen evolving enzyme. In indirect biophotolysis, some form of carbohydrate is produced from the reduced ferredoxin and this serves as a chemical energy carrier between the water-splitting photosynthetic reaction and the hydrogen producing reaction. Systems that are based on biophotolysis can potentially involve the basically incompatible reactions of simultaneous oxygen evolution while reducing protons with an oxygen sensitive enzyme. Thus, in indirect biophotolysis these two reactions can be separated in time and/or space. Two types of systems where the mechanism of hydrogen evolution is different have been investigated; hydrogen production by heterocystous cyanobacteria, and hydrogen production by green algae, principally sulphur-deprived cultures of the green alga, *Chlamydomonas*.

H_2 production by cyanobacteria. Two different types of cyanobacteria, prokaryotes capable of oxygenic photosynthesis, are known to evolve hydrogen. The enzyme responsible for the majority of cyanobacterial hydrogen production is nitrogenase, which in the absence of other reducible substrates continues to turnover, reducing protons to hydrogen in a relatively slow reaction ($6.4\ s^{-1}$) which also requires substantial energy input (2 ATP/e^-; 4 ATP/H_2). One group of cyanobacteria grow in filaments and differentiate specialized cells called heterocysts under conditions of nitrogen limitation (Kumar et al. 2010; Mariscal and Flores 2010). Heterocysts provide a microaerobic environment which allows the oxygen sensitive nitrogen fixation process to take place in the midst of environment supersaturated with oxygen. Heterocysts do not express photosystem II and therefore do not split water and evolve oxygen. They also cannot fix carbon dioxide since they lack the Calvin-Benson-Bassham cycle and therefore depend upon fixed carbon imported from neighboring vegetative cells. The imported sucrose is metabolized in the heterocyst through the oxidative pentose pathway (Summers et al. 1995; Lopez et al. 2010).

Since hydrogen production in the heterocysts depends upon sucrose which was produced in the adjacent vegetative cells, this is in fact indirect biophotolysis on a microscopic scale, which reduces the maximal theoretical conversion efficiency (see below). However, this system is inherently robust and hence has been extensively studied for over three and a half decades (Benemann and Weare 1974; Weissman and Benemann 1977). Light conversion efficiencies established early on, 0.4 % under laboratory conditions, 0.1 % under natural insolation, have not been much improved upon since (Tsygankov et al. 2002; Yoon et al. 2006; Sakurai and Masukawa 2007). There may be some room for improvement as theoretical efficiencies with this nitrogenase based system are around 4.6 % (Hallenbeck 2011). As found for other photosynthetic processes, part of the reduction in efficiency is thought to be due to the inefficient use of light energy at high light intensities. This might be improved by reducing the size of the photosynthetic antennae, allowing more efficient use of high light intensities by the culture.

Hydrogen can also be produced by unicellular cyanobacteria which obviously lack heterocysts. These cyanobacteria possess nitrogenase and are able to fix nitrogen without the protection afforded by the heterocyst. The problems with oxygen sensitivity are at least partly circumvented in nature through circadian transcriptional control which drives maximal transcription of photosynthesis during daylight, and maximal transcription of nitrogenase during darkness. Light-driven nitrogenase catalyzed hydrogen production by the unicellular *Cyanothece* has recently been demonstrated, although this was at low light intensities in the presence of glycerol, allowing consumption of oxygen through respiration and argon sparging removal of evolved oxygen (Min and Sherman 2010).

Of course these organisms could also be used in a true indirect biophotolysis process in which carbon is fixed in the first stage through oxygen-evolving photosynthesis creating reductant that can later be used in a second, anaerobic, hydrogen-producing stage. This separates in time and space the oxygen-sensitive proton reduction reaction from oxygen-producing photosynthesis. Such a process was recently demonstrated on an experimental level where the non-heterocystous *Plectonema boryanum* was cycled multiple times through an aerobic, nitrogen limited stage allowing for glycogen accumulation, and a second anaerobic, hydrogen producing stage (Huesemann et al. 2010). In another scenario, the unicellular cyanobacterium *Synechococcus*, was shown to convert biomass accumulated during first stage photosynthetic to hydrogen during a second stage dark fermentation with a 12 % efficiency (i.e. 1.44 moles H_2/mole hexose) (McNeely et al. 2010).

H_2 production by green algae. It has been recognized for over half a century that some species of green algae are capable of a short-lived burst of hydrogen production catalyzed by a FeFe-hydrogenase upon re-illumination of dark-adapted (anaerobic) cultures. Recently, sustained hydrogen production by illuminated cultures was demonstrated using two stages; an initial stage permitting photosynthesis and growth followed by sulphur deprivation which provides the anaerobic conditions necessary for sustained hydrogen production (Melis et al. 2000). Sulfur deprivation reduces photosystem II activity since under these conditions cells are unable to replace photodamaged D1 protein. At some point (the compensation point) the much lowered rate of oxygen evolution is less than the rate of respiratory oxygen consumption. The algal culture thus becomes anaerobic and hydrogenase is induced with hydrogen production lasting over a period of days. The electrons required for hydrogen evolution come from several different pathways, with only about 50 % coming directly from the water splitting action of photosystem II (direct biophotolysis). An additional 50 % of the electrons come from stored metabolites, such as starch, which were produced during the first stage of this two stage process.

Although a large number of studies have attempted to improve this system by examining various mutants and operational parameters, low light conversion efficiencies remain a critical limiting parameter. Low efficiencies are inherent in a process which is based on reducing PSII activity by 75–90 % and removing oxygen through respiration of substrate that could otherwise be used for hydrogen production. Therefore, attaining efficiencies within the realm of a practical system can only be achieved by moving away from the sulphur-deprived paradigm. One possible solution, using a hydrogenase that is (relatively) insensitive to oxygen inactivation, is apparent, although it is not obvious how to achieve this.

b. Photofermentation by Photosynthetic Bacteria

Photofermentation is another method that uses captured solar energy to drive hydrogen production, in this case from organic compounds, principally organic acids, by purple non-sulfur photosynthetic bacteria. There have been a large number of studies on photofermentative hydrogen production by a different organisms using a variety of substrates with investigations into the effects of a variety of factors including light intensity, nutrient regime, cell immobilization, etc., as compiled elsewhere (Li and Fang 2009; Adessi and De Philippis 2012). These bacteria, when grown photoheterotrophically under nitrogen limiting conditions which induce the hydrogen evolution catalyst, nitrogenase, can carry out the nearly stoichiometric conversion of various organic acids to hydrogen. The necessary energy inputs, ATP and high energy electrons, are generated through the action of bacterial photosynthesis. Hydrogen production under these conditions is thought to reflect the need for metabolic redox balance, with the necessary reoxidation of NADH coming from the hydrogen evolution process (Laguna et al. 2010; McKinlay and Harwood 2010). Thus, photofermentation has been used to demonstrate the conversion of a variety of substrates, usually organic acids, to hydrogen. Many studies have demonstrated that this process can use a variety of waste streams rich in these substrates, or others, such as the crude glycerol fraction derived from biodiesel manufacture (Sabourin-Provost and Hallenbeck 2009; Keskin and Hallenbeck 2012). Indeed, at present a great deal of work is examining the use of this process to convert the liquid products produced during dark hydrogen fermentations, thus increasing overall hydrogen yields (Keskin et al. 2011; Adessi et al. 2012; Keskin and Hallenbeck 2012).

Even though the substrate conversion yields are high, there are a number of drawbacks to this system that prevent practical application; volumetric hydrogen production rates are low, and, in common with other light-dependent systems, light conversion efficiencies are also low. Taken together both these factors mean that photobioreactors covering inordinately large surface areas would be required. In the future photofermentation could possibly be improved through several approaches that might increase rates, yields, or photosynthetic efficiencies. These strategies include; elimination of competing pathways; hydrogen consumption, polyhydroxybutyrate production, or carbon dioxide fixation, and substitution of hydrogenase for nitrogenase should decrease the photon requirement, bringing about higher conversion efficiencies. This strategy might also bring about increased volumetric rates of hydrogen production as well due to the much higher turnover rates of FeFe-hydrogenases ($6–12 \times 10^{-3}\,s^{-1}$) compared to nitrogenase ($6.4^{-1}\,s^{-1}$). As with other photosynthetic systems, increased light conversion efficiencies at high light intensities might be obtained by decreasing the photosynthetic antenna size.

c. Theoretical and Practical Limits to Light Conversion Efficiencies

The conversion efficiency of any process with solar energy as in input is a key parameter that directly affects its physical footprint. Theoretical conversion efficiencies should be based on total solar insolation of which only 45 % can be used by chlorophyll *a* containing organisms, green algae and cyanobacteria, with up to 70 % being usable by purple non-sulfur bacteria. A detailed discussion of theoretical conversion efficiencies in microbial hydrogen production is given elsewhere

(Hallenbeck 2011). One obvious difference is that systems based on nitrogenase have lower theoretical efficiencies than systems based on hydrogenase due to the extra energy (ATP) requirement of nitrogenase. Thus, biophotolysis by green algae can be predicted to have a maximum efficiency of 12.2 % and only 4.1 % for biophotolysis by heterocystous cyanobacteria. Photofermentation of organic substrates by photosynthetic bacteria can be predicted to have a maximal efficiency of 8.5 % (Hallenbeck 2011). Of course, in practice efficiencies will be much lower due to a several factors. One important factor is thought to be the photosynthetic antenna size. Photosynthetic cultures normally have antenna sizes adapted for efficient capture at low light intensities. At the high light intensities that would be encountered in an effective light conversion system, the excess energy is captured and then wasted as thermal energy or fluorescence, as much as 80–90 % at maximum light intensities. Mutants with reduced antenna size might therefore have higher efficiencies, but this remains to be fully demonstrated. An additional amount of energy is needed for cell growth, maintenance and any additional metabolic burdens, such as respiration in sulfur-deprived green algae.

2. Hydrogen Production by Dark Fermentation

Many different microbes have long been recognized to produce hydrogen during various types of anaerobic fermentations of carbohydrate-rich substrates. Other materials are poor substrates for fermentative hydrogen production; the fermentation of only a few amino acids gives hydrogen, and net hydrogen production from lipids is only possible at very low hydrogen partial pressures. Either pure substrates (usually glucose) or a variety of wastes have been used in studies on dark fermentative hydrogen production (Kapdan and Kargi 2006; Li and Fang 2007; Abo-Hashesh and Hallenbeck 2012). The use of various waste streams is desirable of course, but it requires either an omnivorous heterotrophic organism or a consortium of organisms with a wide range of catabolic activities.

The metabolic details of the hydrogen production are fairly well understood now (Fig. 1.4). This aspect has been extensively reviewed (Hallenbeck 2005, 2009, 2011, 2012c; Ghosh and Hallenbeck 2009). Basically, sugars are broken down through glycolysis to pyruvate, generating ATP and NADH. The fate of pyruvate is different depending upon the organism and metabolic pathway. Thus, a variety of enzymes and hydrogenases can potentially participate, but the net result is the production of a maximum of one mole of hydrogen per mole of pyruvate. In one pathway, pyruvate can be converted to formate and acetyl-CoA. The resulting formate can then be split to give hydrogen and CO_2 through several different membrane associated hydrogenases (Hallenbeck 2012c). Alternatively, pyruvate can be immediately oxidized to acetyl-CoA, giving a CO_2 and reducing ferredoxin. The reduced ferredoxin can drive hydrogen production by several different hydrogenases. In both cases, the acetyl-CoA that is made is used to form a variety of liquid fermentation products, ethanol, acetate, butanol, butyrate, acetone, etc., depending upon the organism, the redox state of the substrate and the need for NAD regeneration for cellular metabolism. The amount of NADH generated during glycolysis depends upon the oxidation state of the substrate, and thus drives the subsequent pattern and relative proportions of fermentation products made.

In order to produce more than 2 moles of hydrogen per mole of glucose (glucose catabolism gives two pyruvates), or glucose equivalent, additional hydrogen must be derived from the NADH made during glycolysis by oxidation of glyceraldehyde-3-phosphate (G-3-P). The stumbling block is that generation of hydrogen from NADH is thermodynamically unfavorable since the $NAD^+/NADH$ couple has an equilibrium midpoint potential ($E^{0\prime}$) of −320 mV whereas that of the H^+/H_2 couple is −420 mV. Therefore, under standard equilibrium conditions (1 atmosphere H_2) energy must be put into the system to make hydrogen ($\Delta G^{0\prime} = -nF\Delta E^{0\prime} = +19.3$ kJ/mol). All things being equal, these considerations would predict that hydrogen production from NADH should only proceed at low hydrogen partial pressures.

A variety of hydrogenases have been described in organisms belonging to the Firmicutes that might be proposed to function in producing hydrogen from NADH (Fig. 1.4). Since none of these are present in *Escherichia coli* and similar organisms, these are restricted to a maximum of 2 H_2/glucose. If all the excess NADH could be converted to hydrogen, the organisms that contain the relevant pathways would be able to produce 4 H_2/glucose. However, as noted above, this is unrealistic thermodynamically. A number of possible mechanisms exist (Fig. 1.4). NADH might directly reduce a specific hydrogenase or NADH might be used to produce reduced ferredoxin through the action of NADH ferredoxin oxidoreducatse. However, using NADH to generate H_2 may more likely than once thought as many organisms are thought to possess a bifurcating hydrogenase, an enzyme capable of coupling some of the free energy available in the oxidation of reduced ferredoxin to the oxidation of NADH thus compensating for the free energy needed to reduce hydrogenase with NADH (Schut and Adams 2009). The molecular details of this unique energy coupling mechanism are not presently known and the actual energetics will be determined by the prevailing hydrogen partial pressures and cellular concentrations of NAD^+, NADH, Fd_{ox}, and Fd_{red}.

Several major advances in dark fermentative hydrogen production have been recently made. Various types of immobilized systems have been developed, allowing the achievement of high volumetric rates of production. With this technology it is now possible to favor the development and maintenance of mixed communities degrading a variety of complex substrates and producing hydrogen under non-sterile conditions. Other advances have come from the application of Metabolic engineering which has shown that microbes can be manipulated into producing the maximum yields predicted from metabolic pathways (Abo-Hashesh et al. 2011). Additional strategies for further advances have been proposed (Hallenbeck et al. 2012). However, the bottom line is that presently achievable

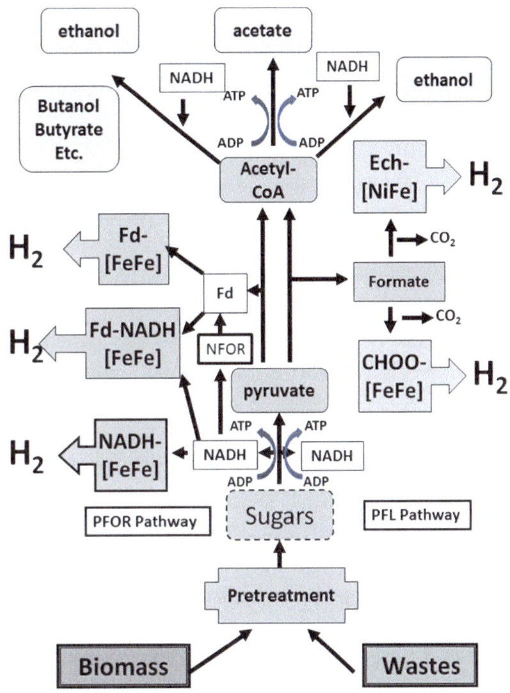

Fig. 1.4. Dark fermentative hydrogen producing pathways. In fermentations with hydrogen as one of the products, as in many other fermentations, glucose is broken down to pyruvate, generating ATP and NADH. Pyruvate is then converted to acetyl-CoA, and depending upon the organism, either formate, through the PFL pathway, or reduced ferredoxin and CO_2, through the PFO pathway. Formate can be converted to hydrogen and CO_2, by either the formate hydrogen lyase pathway which contains a [NiFe] hydrogenase (the Ech hydrogenase), or possibly in some other organisms another pathway which contains a formate dependent [FeFe] hydrogenase. NADH, generated during glycolysis, is oxidized through the production of various reduced carbon compounds, typically ethanol. A variety of [FeFe] hydrogenases can be used to reoxidize ferredoxin and produce hydrogen, including a ferredoxin-dependent H_2ase (Fd-[FeFe]). In some cases, NADH can also be used in hydrogen production, either by reducing ferredoxin (NFO), by directly reducing H_2ase (NADH-[FeFe]), or as a co-substrate with reduced ferredoxin (Fd-NADH-[FeFe]). Excess NADH is used to produce other reduced fermentation products. In both cases, acetyl-CoA can also be used to produce ATP. *Fd* ferredoxin, *NFO* NADH ferredoxin oxidoreductase, *PFL* pyruvate formate lyase, *PFO* pyruvate:ferredoxin oxidoreductase.

yields, at most 33 % of the theoretical 12 moles of hydrogen per mole of glucose, are unacceptable. These yields are not competitive with substrate conversion to other biofuels, which can already occur at 80–90 % yields. In addition, these low yields also mean the production of large amounts of side products that, at the scale necessary for significant production of a biofuel, would present an enormous waste disposal problem. Thus, the major challenge to the practical use of dark fermentation for biological hydrogen production is achieving acceptable yields. A number of strategies for overcoming this barrier have been recently suggested, including further Metabolic engineering, and the development of hybrid, two-stage systems that would convert the fermentation side products to methane or hydrogen (Hallenbeck and Ghosh 2009; Hallenbeck 2011; Hallenbeck et al. 2012). There are three distinct second stages for hybrid systems; anaerobic digestion of fermenter effluents to produce methane, photofermentation of the organic acids that are produced to hydrogen, or their conversion to hydrogen using microbial electrolysis cells.

3. *Microbial Electrolysis*

The last few years have seen the rapid development of a novel technique for producing hydrogen from a variety of substrates using what are called microbial electrolysis cells (MECs) (Logan et al. 2008; Geelhoed et al. 2010; Liu et al. 2010). These are based on microbial fuel cells (MFCs) which have been under investigation for decade. Both types of cells use microbes which can interact metabolically with an electrode. A variety of mechanisms are involved. Microbial metabolism degrades various organic compounds to CO_2, protons, and electrons with the electrode (anode) acting as an electron sink in a type of anaerobic respiration. The current that is generated can be used as a power source (MFC) or additional voltage can be added to drive hydrogen evolution at the cathode (MEC). This allows, for example, the conversion of acetate (−0.279 V) to hydrogen (−0.414 V) in a cathodic reaction against the thermodynamic gradient with the application of a relatively small voltage (−0.135 V). However, in practice, more voltage than this is required due to a number of physic-chemical and microbial factors.

MECs can be constructed in a variety of configurations and with different materials for the anodes and cathodes. Many studies have been carried out with two-chamber devices with ion permeable membranes separating the anodic and cathodic chambers. Although this configuration has a number of advantages, there are a number of problems with this approach. The separation of the two bulk liquids can lead to inhibitory pH changes with acidification of the anodic chamber and basification of the cathodic chamber. In addition, the membrane can contribute significantly to the overall resistance of the cell, thus creating a greater voltage requirement for hydrogen evolution.

Some of these problems can potentially be circumvented with single chamber MECs which have been recently shown to generate higher current densities than dual chamber configurations and give significantly higher hydrogen production rates (Call and Logan 2008; Hu et al. 2008). On the other hand, there may be several drawbacks to single chamber MECs which could decrease either hydrogen yields or coulombic efficiencies (Lee and Rittmann 2010). For example, increased methane production may come at the expense of produced hydrogen with this configuration, thus decreasing yields. In addition, microbes present at the anode might consume the produced hydrogen using the anode as electron sink, thus creating a futile cycle and decreasing efficiencies drastically (Lee and Rittmann 2010).

Thus, practical use of MECs will require that a number of challenges are addressed. These include the development of low cost, efficient electrode materials and the development of cell geometries and biocompatible buffers that reduce internal resistances. Given the rapid progress with this technology over the recent past, it might well reach the level required for practical application in a relatively short period of time, giving a useful technology for the conversion of various waste streams to hydrogen.

Acknowledgements

Biofuels research in my laboratory is supported by NSERC and FQRNT. I want to thank the many students, undergrad, graduate, and visiting, that have participated in the many on-going biofuels projects in my laboratory.

References

Abo-Hashesh M, Hallenbeck PC (2012) Fermentative hydrogen production. In: Hallenbeck PC (ed) Microbial technologies in advanced biofuels production. Springer, New York, pp 77–92

Abo-Hashesh M, Wang R, Hallenbeck PC (2011) Metabolic engineering in dark fermentative hydrogen production; theory and practice. Bioresour Technol 110:1–9

Adessi A, De Philippis R (2012) Hydrogen production: photofermentation. In: Hallenbeck PC (ed) Microbial technologies in advanced biofuels production. Springer, New York, pp 53–75

Adessi A, De Philippis R, Hallenbeck PC (2012) Combined systems for maximum substrate conversion. In: Hallenbeck PC (ed) Microbial technologies in advanced biofuels production. Springer, New York, pp 107–126

Atsumi S, Liao JC (2008) Directed evolution of *Methanococcus jannaschii* citramalate synthase for biosynthesis of 1-propanol and 1-butanol by *Escherichia coli*. Appl Environ Microbiol 74:7802–7808

Atsumi S, Cann AF, Connor MR, Shen CR, Smith KM, Brynildsen MP, Chou KJY, Hanai T, Liao JC (2008a) Metabolic engineering of *Escherichia coli* for 1-butanol production. Metab Eng 10:305–311

Atsumi S, Hanai T, Liao JC (2008b) Non-fermentative pathways for synthesis of branched-chain higher alcohols as biofuels. Nature 41:86–89

Atsumi S, Wu TY, Eckl EM, Hawkins SD, Buelter T, Liao JC (2010) Engineering the isobutanol biosynthetic pathway in *Escherichia coli* by comparison of three aldehyde reductase/alcohol dehydrogenase genes. Appl Microbiol Biotechnol 85:651–657

Benemann JR, Weare NM (1974) Hydrogen evolution by nitrogen-fixing *Anabaena cylindrica* cultures. Science 184:174–175

BP Energy Outlook 2030 (2011) http://www.bp.com/sectiongenericarticle800.do?categoryId=9037134&contentId=7068677. Accessed 1 Nov 2012

Call D, Logan BE (2008) Hydrogen production in a single chamber microbial electrolysis cell (MEC) lacking a membrane. Environ Sci Technol 42:3401–3406

Cassman KG, Liska AJ (2007) Food and fuel for all: realistic or foolish? Biofuels Bioprocess Bioref 1:18–23

Connor MR, Liao JC (2008) Engineering of an *Escherichia coli* strain for the production of 3-methyl-1-butanol. Appl Environ Microbiol 74:5769–5775

Davis SJ, Caldeira K, Matthews HD (2010) Future CO_2 emissions and climate change from existing energy infrastructure. Science 329:1330–1333

Dellomonaco C, Fava F, Gonzalez R (2010) The path to next generation biofuels: successes and challenges in the era of synthetic biology. Microb Cell Fact 9:3–18

DOE International Energy Outlook (2011) http://www.eia.gov/forecasts/ieo/. Accessed 1 Nov 2012

Field CB, Campbell JE, Lobell DB (2008) Biomass energy: the scale of the potential resource. Trends Ecol Evol 23:65–72

Fortman JL, Chhabra S, Mukhopadhyay A, Chou H, Lee TS, Steen E, Keasling JD (2008) Biofuel alternatives to ethanol: pumping the microbial well. Trends Biotechnol 26:375–381

Geelhoed JS, Hamelers HVM, Stams AJM (2010) Electricity-mediated biological hydrogen production. Curr Opin Microbiol 13:307–315

Ghirardi ML, Mohanty P (2010) Oxygenic hydrogen photoproduction – current status of the technology. Curr Sci 98:499–507

Ghirardi ML, Dubini A, Yu JP, Maness PC (2009) Photobiological hydrogen-producing systems. Chem Soc Rev 38:52–61

Ghosh D, Hallenbeck PC (2009) Fermentative hydrogen yields from different sugars by batch cultures of metabolically engineered *Escherichia coli* DJT135. Int J Hydrog Energy 34:7979–7982

Ghosh D, Hallenbeck PC (2010) Response surface methodology for process parameter optimization of hydrogen yield by the metabolically engineered strain *Escherichia coli* DJT135. Bioresour Technol 101:1820–1825

Ghosh D, Hallenbeck PC (2012) Advanced bioethanol production. In: Hallenbeck PC (ed) Microbial technologies in advanced biofuels production. Springer, New York, pp 165–181

Hallenbeck PC (2005) Fundamentals of the fermentative production of hydrogen. Water Sci Technol 52:21–29

Hallenbeck PC (2009) Fermentative hydrogen production: principles, progress, and prognosis. Int J Hydrog Energy 34:7379–7389

Hallenbeck PC (2011) Microbial paths to renewable hydrogen production. Biofuels 2:285–302
Hallenbeck PC (2012a) Biofuels, the larger context. In: Hallenbeck PC (ed) Microbial technologies in advanced biofuels production. Springer, New York, pp 3–12
Hallenbeck PC (2012b) The future of biofuels, biofuels of the future. In: Hallenbeck PC (ed) Microbial technologies in advanced biofuels production. Springer, New York, pp 261–268
Hallenbeck PC (2012c) Fundamentals of dark hydrogen fermentations; multiple pathways and enzymes. In: Azbared N, Levin D (eds) State of the art and progress in production of biohydrogen. Bentham Science Publishing, Oak Park, pp 94–111
Hallenbeck PC, Benemann JR (2002) Biological hydrogen production; fundamentals and limiting processes. Int J Hydrog Energy 27:1185–1193
Hallenbeck PC, Ghosh D (2009) Advances in fermentative biohydrogen production: the way forward? Trends Biotechnol 27:287–297
Hallenbeck PC, Ghosh D, Abo-Hashesh M, Wang R (2011) Metabolic engineering for enhanced biofuels production with emphasis on the biological production of hydrogen. In: James CT (ed) Advances in chemistry research, vol 6. Nova Publishers, New York, pp 125–154
Hallenbeck PC, Abo-Hashesh M, Ghosh D (2012) Strategies for improving biological hydrogen production. Bioresour Technol 110:1–9
Hoffert MI (2010) Farewell to fossil fuels? Science 329:1292–1293
Hu H, Fan Y, Liu H (2008) Hydrogen production using singlechamber membranefree microbial electrolysis cells. Water Res 42:4172–4178
Huesemann MH, Hausmann TS, Carter BM, Gerschler JJ, Benemann JR (2010) Hydrogen generation through indirect biophotolysis in batch cultures of the nonheterocystous nitrogen-fixing cyanobacterium *Plectonema boryanum*. Appl Biochem Biotechnol 162:208–220
IEA World Energy Outlook (2010) http://www.worldenergyoutlook.org/publications/weo-2010/. Accessed 1 Nov 2012
Inui M, Suda M, Kimura S, Yasuda K, Suzuki H, Toda H, Yamamoto S, Okino S, Suzuki N, Yukawa H (2008) Expression of *Clostridium acetobutylicum* butanol synthetic genes in *Escherichia coli*. Appl Microbiol Biotechnol 77:1305–1316
Kapdan IK, Kargi F (2006) Bio-hydrogen production from waste materials. Enzyme Microb Technol 38:569–582
Keasling JD, Chou H (2008) Metabolic engineering delivers next-generation biofuels. Nat Biotechnol 26:298–299
Keskin T, Hallenbeck PC (2012) Enhancement of biohydrogen production by two-stage systems: dark and photofermentation. In: Baskar C, Baskar S, Dhillon RS (eds) Biomass conversion. Springer, Berlin, pp 313–340
Keskin T, Abo-Hashesh M, Hallenbeck PC (2011) Photofermentative hydrogen production from wastes. Bioresour Technol 102:8557–8568
Kumar K, Mella-Herrera RA, Golden JW (2010) Cyanobacterial heterocysts. Cold Spring Harb Perspect Biol 2:a000315. doi:10.1101/cshperspect.a000315
Laguna R, Joshi GS, Dangel AW, Amanda K, Luther AK, Tabita FR (2010) Integrative control of carbon, nitrogen, hydrogen, and sulfur metabolism: the central role of the Calvin–Benson–Bassham cycle. In: Hallenbeck PC (ed) Recent advances in phototrophic prokaryotes. Springer, New York, pp 265–271
Lee HS, Rittmann BE (2010) Significance of biological hydrogen oxidation in a continuous single-chamber microbial electrolysis cell. Environ Sci Tech 44:948–954
Lee SK, Chou H, Ham TS, Lee TS, Keasling JD (2008) Metabolic engineering of microorganisms for biofuels production: from bugs to synthetic biology to fuels. Curr Opin Biotechnol 19:556–563
Li CL, Fang HHP (2007) Fermentative hydrogen production from wastewater and solid wastes by mixed cultures. Crit Rev Environ Sci Technol 37:1–39
Li RY, Fang HHP (2009) Heterotrophic photo fermentative hydrogen production. Crit Rev Environ Sci Technol 39:1081–1108
Liu H, Hu H, Chignell J, Fan Y (2010) Microbial electrolysis: novel technology for hydrogen production from biomass. Biofuels 1:129–142
Logan BE, Call D, Cheng S, Hamelers HVM, Sleutels THJA, Jeremiasse AW, Rozendal RA (2008) Microbial electrolysis cells for high yield hydrogen gas production from organic matter. Environ Sci Technol 42:8630–8640
Lopez-Igual R, Flores E, Herrero A (2010) Inactivation of a heterocyst-specific invertase indicates a principal role of sucrose catabolism in heterocysts of *Anabaena* sp. J Bacteriol 192:5526–5533
Lynd LR, Weimer PJ, van Zyl WH, Pretorius IS (2002) Microbial cellulose utilization: fundamentals and biotechnology. Microbiol Mol Biol Rev 66:506–577
Mariscal V, Flores F (2010) Multicellularity in a heterocyst-forming cyanobacterium: pathways for intercellular communication. In: Hallenbeck PC (ed) Recent advances in phototrophic prokaryotes. Springer, New York, pp 125–135
Martindale W, Trewavas A (2008) Fueling the 9 billion. Nat Biotechnol 26:1068–1069

McKinlay JB, Harwood CS (2010) Carbon dioxide fixation as a central redox cofactor recycling mechanism in bacteria. Proc Natl Acad Sci U S A 107:11669–11675

McNeely K, Xu Y, Bennette N, Bryant DA, Dismukes GC (2010) Redirecting reductant flux into hydrogen production via metabolic engineering of fermentative carbon metabolism in a cyanobacterium. Appl Environ Microbiol 76:5032–5038

Melis A, Zhang LP, Forestier M, Ghirardi ML, Seibert M (2000) Sustained photobiological hydrogen gas production upon reversible inactivation of oxygen evolution in the green alga *Chlamydomonas reinhardtii*. Plant Physiol 122:127–135

Min HT, Sherman LA (2010) Hydrogen production by the unicellular, diazotrophic cyanobacterium *Cyanothece* sp. strain ATCC 51142 under conditions of continuous light. Appl Environ Microbiol 76:4293–4301

New M, Liverman DA, Schroder H, Anderson K (2011) Four degrees and beyond: the potential for a global temperature increase of four degrees and its implications. Phil Trans R Soc A 369:6–19

Raupach MR, Marland G, Ciais P, Le Quèré C, Canadell JG, Klepper G (2007) Global and regional drivers of accelerating CO_2 emissions. Proc Natl Acad Sci U S A 104:10288–10293

Rude MA, Schirmer A (2009) New microbial fuels: a biotech perspective. Curr Opin Microbiol 12:274–281

Sabourin-Provost G, Hallenbeck PC (2009) High yield conversion of a crude glycerol fraction from biodiesel production to hydrogen by photofermentation. Bioresour Technol 100:3513–3517

Sagar AD, Kartha S (2007) Bioenergy and sustainable development? Annu Rev Environ Res 32:131–167

Sakurai H, Masukawa H (2007) Promoting R & D in photobiological hydrogen production utilizing mariculture-raised cyanobacteria. Mar Biotechnol 9:128–145

Scharlemann JPW, Laurance WF (2008) How green are biofuels? Science 319:43–44

Schut GJ, Adams MWW (2009) The iron-hydrogenase of *Thermotoga maritima* utilizes ferredoxin and NADH synergistically: a new perspective on anaerobic hydrogen production. J Bacteriol 191:4451–4457

Shen CR, Liao JC (2008) Metabolic engineering of *Escherichia coli* for 1-butanol and 1-propanol production via the keto-acid pathways. Metab Eng 10:312–320

Sims REH, Mabee W, Saddler JN, Taylor M (2010) An overview of second generation biofuel technologies. Bioresour Technol 101:1570–1580

Steen EJ, Chan R, Prasad N, Myers S, Petzold C, Redding A, Ouellet M, Keasling JD (2008) Metabolic engineering of *Saccharomyces cerevisiae* for the production of n-butanol. Microb Cell Fact 7:36. doi:10.1186/1475-2859-7-36

Steen EJ, Kang Y, Bokinsky G, Hu Z, Schirmer A, McClure A, del Cardayre SB, Keasling JD (2010) Microbial production of fatty-acid-derived fuels and chemicals from plant biomass. Nature 463:559–563

Stephanopoulos G (2008) Metabolic engineering: enabling technology for biofuels production. Metab Eng 10:293–294

Summers ML, Wallis JG, Campbell EL, Meeks JC (1995) Genetic evidence of a major role for glucose-6-phosphate-dehydrogenase in nitrogen-fixation and dark growth of the cyanobacterium *Nostoc* sp. strain ATCC-29133. J Bacteriol 177:6184–6194

Tilman D, Socolow R, Foley JA, Hill J, Larson E, Lynd L, Pacala S, Reilly J, Searchinger T, Somerville C, Williams R (2009) Beneficial biofuels: the food, energy, and environment trilemma. Science 325:270–271

Tollefson J (2008) Not your father's biofuels. Nature 451:880–883

Tsygankov AA, Fedorov AS, Kosourov SN, Rao KK (2002) Hydrogen production by cyanobacteria in an automated outdoor photobioreactor under aerobic conditions. Biotechnol Bioeng 80:777–783

Wackett LP (2008) Microbial-based motor fuels: science and technology. Microb Biotechnol 1:211–225

Waldrop MM (2007) Kill king corn. Nature 449:637

Weissman JC, Benemann JR (1977) Hydrogen production by nitrogen-starved cultures of *Anabaena cylindrica*. Appl Environ Microbiol 33:123–131

WMO IPCC (2012) Special report: emissions scenarios

Yamaji K, Matsuhashi R, Nagata Y, Kaya Y (1991) An integrated system for CO_2/energy/GNP analysis: case studies on economic measures for CO_2 reduction in Japan. In: Workshop on CO_2 reduction and removal: measures for the next century, International Institute for Applied Systems Analysis, Laxenburg, Austria, 19 Mar 1991

Yan Y, Liao JC (2009) Engineering metabolic systems for production of advanced fuels. J Ind Microbiol Biotechnol 36:471–479

Yoon JH, Hae Shin J, Kim M-S, Jun Sim S, Park TH (2006) Evaluation of conversion efficiency of light to hydrogen energy by *Anabaena variabilis*. Int J Hydrog Energy 31:721–727

Chapter 2

Structural Foundations for O_2 Sensitivity and O_2 Tolerance in [NiFe]-Hydrogenases

Anne Volbeda and Juan C. Fontecilla-Camps*
Metalloproteins Unit, Institut de Biologie Structurale J.P. Ebel, Commissariat à l'Énergie Atomique, Centre National de la Recherche Scientifique, Université Joseph Fourier, UMR 5075, 41 rue Jules Horowitz, 38027 Grenoble, France

Summary

Nature has evolved three different ways of metabolizing hydrogen, represented by the anaerobic [Fe]-, [FeFe]- and [NiFe]-hydrogenases. Structural and functional studies of these enzymes have unveiled the unusual composition of their active sites and characterized their catalytic mechanisms. From a biotechnological viewpoint, the most interesting hydrogenases are those that contain a [NiFe] moiety in their active sites. Some of these enzymes are O_2-resistant and can rapidly reductively recover from oxygen exposure whereas others are O_2-tolerant and can oxidize H_2 even at atmospheric oxygen levels. O_2-resistant [NiFeSe]-hydrogenases have one of the Cys ligands of the active site replaced by a SeCys and do not display the hard-to-reactivate "unready" state provoked by O_2. The reasons for this property might be related to the formation of O_2-derived Se-O bonds, which are weaker than

*Author for correspondence, e-mail: juan.fontecilla@ibs.fr

D. Zannoni and R. De Philippis (eds.), *Microbial BioEnergy: Hydrogen Production, Advances in Photosynthesis and Respiration 38, pp. 23–41*, DOI 10.1007/978-94-017-8554-9_2,

S-O bonds and, consequently, easier to break upon reduction. Conversely, membrane-bound O_2-tolerant hydrogenases have an unusual proximal (relative to the active site) $[Fe_4S_3]$ cluster coordinated by six Cys ligands. This cluster can rapidly send two successive electrons to the active site helping to reduce oxygen to water there. Some microorganisms posses more than one hydrogenase and use them in different ways. For instance, there are three well-characterized [NiFe]-hydrogenases in the model bacterium *Escherichia coli*. They are highly regulated and each one plays a specific role: microaerobic/anaerobic H_2 uptake, anaerobic H_2 evolution and, protection from O_2-induced damage, respectively. These enzymes are discussed in connection with the metabolic changes *E. coli* undergoes during its transit through the intestinal tract of the host. O_2-tolerant hydrogenases have been used to build bio-fuel cells that can function under air. Also, O_2-resistant [NiFeSe]-hydrogenases have been attached to TiO_2 particules for H_2 production from solar energy. Hydrogenase active sites have also served as a source of inspiration for the synthesis of organometallic catalysts.

I. Introduction

Many microorganisms use enzymes called hydrogenases to oxidize molecular hydrogen or reduce protons according to the reaction $H_2 \leftrightarrow 2H^+ + 2e^-$. Two major unrelated enzyme classes exist: the [FeFe]- and the [NiFe]-hydrogenases (Vignais and Billoud 2007; Fontecilla-Camps et al. 2007). The former are found in bacteria and some green algae, fungi and protozoa, whereas the latter are widespread in both bacteria and archaeans but absent from eukaryotes. A special class of enzymes called [Fe]-hydrogenases couples H_2 oxidation with the reduction of methenyl-tetrahydromethanopterin ($HC\text{-}H_4MPT^+$) without electron transfer to an external redox partner. When provided with the reaction products, H^+ + $H_2C\text{-}H^4MPT$, H_2 is evolved (Thauer et al. 2010). In general, [Fe]-hydrogenases, which are only found in archaeal species, are irreversibly inactivated by O_2. [FeFe]-hydrogenases are generally more active in proton reduction than [NiFe]- hydrogenases, which are more biased to H_2 oxidation. However, there are exceptions to this rule, such as the periplasmic [FeFe]-hydrogenase from *Desulfovibrio desulfuricans* which is an uptake enzyme (Nicolet et al. 1999). Proton reduction is physiologically important to eliminate excessive reducing power generated by photosynthetic and fermentative processes. Conversely, microorganisms can use the low-potential electrons generated by hydrogen oxidation for respiration with different terminal electron acceptors such as, for example, fumarate, nitrate, carbon dioxide, sulfur, sulfate, the heterodisulfide CoM-S-S-CoB (between coenzyme M and coenzyme B) in methanogenic archaeans and, in some exceptional cases, dioxygen. Hydrogen oxidation is also used to recycle H_2 generated by nitrogenases during N_2 reduction to ammonia by azototrophic bacteria.

[FeFe]-hydrogenases are generally O_2-sensitive but can be reactivated if they are progressively exposed to this gas, like is the case during aerobic purification. However, in the presence of excessive reducing power the enzyme metal centers can be irreversibly damaged due to the formation of radical oxygen species. In general, such sensitivity towards O_2 is a stumbling block for biotechnological applications. Conversely, many [NiFe]-hydrogenases are, at least in vitro, only reversibly inactivated by O_2 and some are present in aerobic organisms that can couple hydrogen oxidation to oxygen reduction. Because of these interesting properties we will now discuss this class of enzymes in more detail.

Abbreviations: EPR – Electron Paramagnetic Resonance spectroscopy; FTIR – Fourier Transform InfraRed spectroscopy; Pt – Platinum; TiO_2 – Titanium dioxide

II. [NiFe]-Hydrogenases

A. Types of [NiFe]-Hydrogenases

The simplest [NiFe]-hydrogenase consists of a large subunit containing the active Ni-Fe site and a small subunit typically harboring a proximal [Fe_4S_4], a medial [Fe_3S_4] and a distal [Fe_4S_4] cluster, which transfer electrons to and from the active site, located in the large subunit (Fig. 2.1a). Exceptions are the [NiFeSe]-hydrogenases that have a medial [Fe_4S_4] cluster and the O_2-tolerant enzymes that contain a modified proximal [Fe_4S_3] cluster (see below). [NiFe]-hydrogenases are either periplasmic or cytoplasmic H_2-uptake enzymes. Heterodimeric cytoplasmic enzymes typically function in the recycling of H_2 produced by microbial nitrogenases, as in cyanobacteria (Bothe et al. 2010, see also Chaps. 6 and 8). Heterodimeric periplasmic [NiFe]-hydrogenases have been extensively studied in sulfate-reducing bacteria (Fontecilla-Camps et al. 2007). They allow these organisms to use H_2 as an electron donor for the reduction of sulfate via a complex and still incompletely characterized electron transfer pathway, reviewed by Matias et al. (2005), that starts with a water-soluble cytochrome c_3 electron carrier. Based on sequence homologies (Vignais and Billoud 2007), all [NiFe]-hydrogenases have a common heterodimeric core that resembles the first structure of the enzyme from *Desulfivibrio gigas* published by Volbeda et al. (1995). In most hydrogenases this basic core forms part of larger protein complexes with different redox partners, including quinones in the cytoplasmic membrane and ferredoxin, NAD or NADP in the cytoplasm. In methanogenic archaea the quinones are replaced by methanophenazine and NAD is normally replaced by coenzyme F_{420} (8-hydroxy-5-deazaflavin), although there are also hyperthermophilic archaea that use NADP (Horch et al. 2012). In addition, methanogens have hydrogenases that are coupled with a heterodisulfide reductase to reduce the S-S bond between coenzymes M and B, produced in the last step of methanogenesis (Thauer et al. 2010).

Many of the multi-subunit complexes show striking homologies with NADH: ubiquinone oxidoreductase, also known as respiratory complex I. Although the latter does not have a Ni-Fe active site, the homology extends even to the basic heterodimeric hydrogenase core. The known structure and organization of both the hydrophilic and membrane-bound hydrophobic subunits of complex I has been used to construct homology-based models of several multisubunit hydrogenases (Efremov and Sazanov 2012). These include the so-called bidirectional hydrogenases, reviewed by Horch et al. (2012),

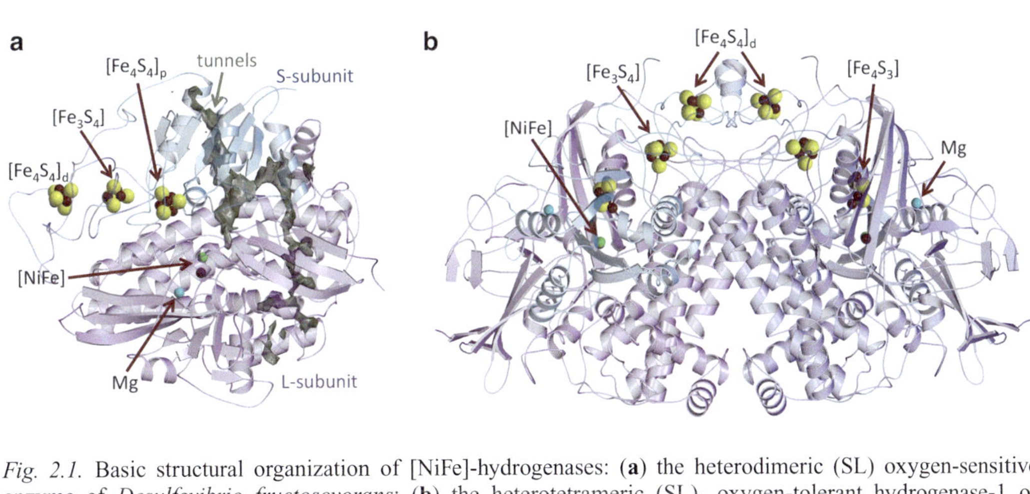

Fig. 2.1. Basic structural organization of [NiFe]-hydrogenases: (**a**) the heterodimeric (SL) oxygen-sensitive enzyme of *Desulfovibrio fructosovorans*; (**b**) the heterotetrameric $(SL)_2$ oxygen-tolerant hydrogenase-1 of *Escherichia coli.*

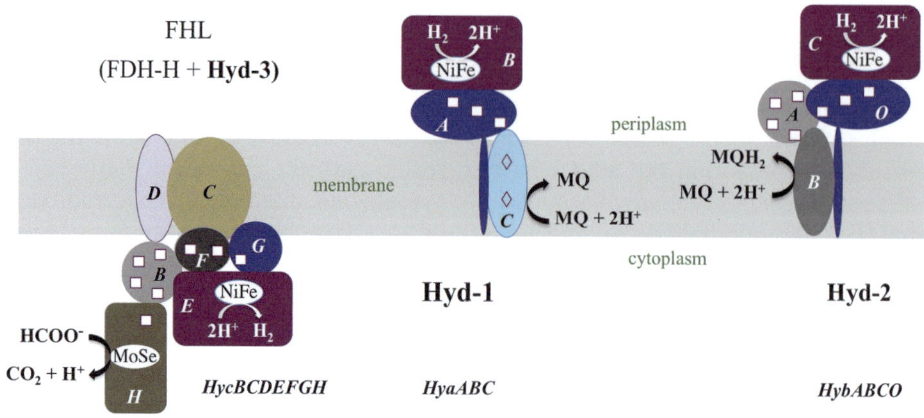

Fig. 2.2. Multisubunit complexes of the three *E. coli* [NiFe]-hydrogenases *Ec*Hyd-1, *Ec*Hyd-2 and *Ec*Hyd-3. The latter is part of the formate:hydrogenase lyase (FHL) complex which contains also formate dehydrogenase H (FDH-H). Electron-transferring iron-sulfur clusters are highlighted as squares, *b*-type hemes as diamonds and active sites as ellipsoids. MQ and MQH_2 are the oxidized and reduced forms of menaquinone. Subunits are labeled with capital letters, the corresponding genes are given in *italics* underneath.

which are found, for example, in the cytoplasm of photosynthetic bacteria. These enzymes typically consist of five different subunits in Bacteria and three in Archaea, and exchange electrons with either NAD(P) or coenzyme F_{420}, via a flavin-containing diaphorase subunit. The O_2-tolerant soluble [NiFe]-hydrogenase of *Ralstonia eutropha* (*Re*MBH) is also a member of this group of enzymes. Other multisubunit hydrogenases related to complex I are the proton pumping energy converting hydrogenases (Ech) found in the membrane fraction of Archaea like *Methanosarcina barkeri*, which use a ferredoxin as redox partner and consist of at least six subunits (Hedderich 2004), and *Ec*Hyd-3, which is a H_2 evolving hydrogenase of the enteric bacterium the *Escherichia coli*.

B. The [NiFe]-Hydrogenases of Escherichia coli

This enteric anaerobic bacterium has three well-studied multisubunit [NiFe]-hydrogenases (Fig. 2.2) called *Ec*Hyd-1, *Ec*Hyd-2 and *Ec*Hyd-3 (Pinske et al. 2012). Although these three membrane-bound hydrogenases have similar amino acid sequences, they are associated with different kinds of subunits (see below). Genomic annotation indicates the presence of a fourth hydrogenase although it has not been identified in the bacterium (Redwood et al. 2007). *Ec*Hyd-2 is a periplasmic membrane–bound hydrogenase, very active in hydrogen uptake (Dubini et al. 2002), that is expressed under microaerobic and anaerobic respiration. It transfers electrons resulting from H_2 oxidation to the (mena)quinone pool in the membrane, via its small subunit HybO, to the ferredoxin-like HybA subunit that contains four iron sulfur clusters and the intrinsic membrane subunit HybB devoid of metal (Dubini et al. 2002). These electrons are subsequently used in the reduction of fumarate to succinate on the cytoplasmic side of the membrane (Kröger et al. 2002). *Ec*Hyd-2 has a long anchoring α-helix, which corresponds to the C terminal segment of HybO. *Ec*Hyd-3 is part of the cytoplasmic, membrane–bound formate–hydrogenlyase complex consisting of seven different subunits, which catalyze the transformation of formate, a fermentation product, to CO_2 and H_2 (Leonhartsberger et al. 2002). This reaction prevents acidification of the bacterium cytoplasm and allows for hydrogen recycling. Whereas both *Ec*Hyd-2 and *Ec*Hyd-3 are very O_2-sensitive, *Ec*Hyd-1 is air-tolerant and oxidizes hydrogen at potentials significantly higher than those of

*Ec*Hyd-2 (Laurinavichene et al. 2002; Lukey et al. 2010). Amino acid sequence comparisons show that *Ec*Hyd-1 is related to membrane-bound hydrogenases (MBH) functioning in aerobic respiration in Knallgas bacteria. Like other enzymes from *E. coli*, *Ec*Hyd-1 is anchored to the membrane by a long trans–membrane α-helix from the small subunit and by an intrinsic membrane protein, cytochrome *b* (Dubini et al. 2002). The enzyme is a dimer of heterodimers (Volbeda et al. 2012), which are composed of a small and large subunit, i.e. a $(SL)_2$ dimer (Fig. 2.1b). Unlike MBH (but like *Ec*Hyd-3), *Ec*Hyd-1 is repressed by O_2 and highly expressed under fermentative growth (Pinske et al. 2012). Under these conditions the quinone pool is likely to be completely reduced, making a possible role of *Ec*Hyd-1 in anaerobic respiration difficult to rationalize. Indeed, experiments using *E. coli* mutants have shown that hydrogen produced by *Ec*Hyd-3 is mostly oxidized by *Ec*Hyd-2 and not by *Ec*Hyd-1 (Redwood et al. 2007). Consequently, *Ec*Hyd-1 and other O_2-tolerant enzymes could have a different function than respiration, such as defense against oxidative stress. This role has been proposed in the case of homoacetogenic bacteria living in termite guts (Boga and Brune 2003) and in the Fe(III)-reducing bacterium *Geobacter sulfurreducens* (Tremblay and Lovley 2012).

Most O_2-tolerant enzymes have significantly higher cluster redox potentials and much lower in vitro H_2 oxidation or production activities than the O_2-sensitive enzymes. High cluster potentials are probably beneficial for O_2 tolerance. In the following sections we will focus on the effects of O_2 on the structure and function of [NiFe]-hydrogenases, emphasizing the results obtained from crystallographic studies since 2001.

III. Structural Studies of O_2-Sensitive [NiFe]-Hydrogenases

Interpretation of the structural results obtained for [NiFe]-hydrogenases has been often complicated by the presence of mixtures of redox and protonation states in the crystals. This is due to the numerous redox states these enzymes display upon reduction from inactive oxidized to catalytically active species. In addition to O_2, molecules like H_2S, which can be present in significant amounts during the purification of enzymes from sulfate reducing bacteria, may react with the active site, producing another source of heterogeneity. Furthermore, the redox state of the structure may change depending on the X-ray dose used for collecting the crystallographic data. The different enzyme states have been extensively characterized by EPR and FTIR spectroscopic studies (De Lacey et al. 2007; Lubitz et al. 2007). Inactive oxidized states may be defined as unready and ready, which respectively give rise to EPR signals called Ni-A and Ni-B depending on whether they activate slowly or rapidly upon treatment with H_2. We obtained an almost pure Ni-B preparation from an active, anaerobically-purified periplasmic heterodimeric [NiFe]-hydrogenase of *Desulfovibrio (D.) fructosovorans* by exposing it to a mixture of 5 % H_2 and 95 % N_2 at pH 9.0 followed by exposure to a 100 % O_2 atmosphere at 0 °C. This sample was crystallized under air (Volbeda et al. 2005). The crystal structure showed the presence of spherical electron density between the Ni and Fe atoms of the active site that we assigned to a bridging hydroxide ligand (Fig. 2.3a). The crystallographic analysis of aerobically-purified enzyme in its "as-isolated" state showed a significantly more elongated electron density bridging the Ni and Fe ions (Fig. 2.3b). As aerobically purified enzyme is known to be mostly in the unready Ni-A state, we associated this observation with the presence of a peroxide species in this form. In addition, a small density feature close to a bridging cysteine thiol suggested its partial oxidation to a sulfenate (Volbeda et al. 2005). However, using data collected from a different crystal of "as isolated" enzyme, we observed spherical electron density bridging the Ni and Fe ions, as well as a small peak close to a bridging thiol. This crystal may have been overexposed to X-rays, as we also noticed the decarboxylation of several Asp and Glu

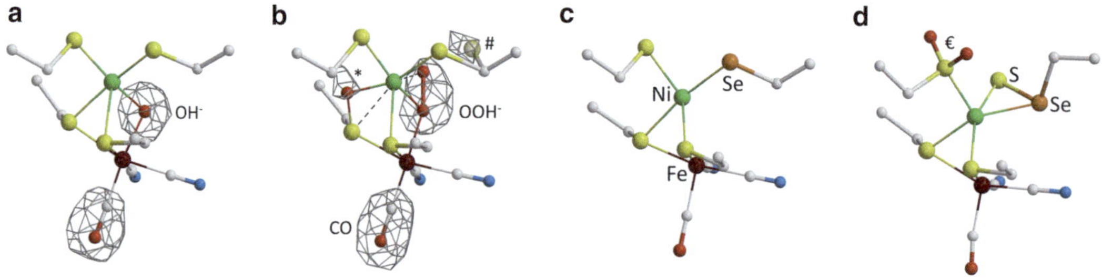

Fig. 2.3. Crystallographic models of the Ni-Fe(−Se) active site: (**a**) the ready Ni-B state in the *D. fructosovorans* enzyme; (**b**) the as-isolated, mainly unready mixture of the same enzyme; (**c**) the H_2-reduced [NiFeSe]-hydrogenase of *Desulfomicrobium baculatum*; (**d**) the major fraction of the as-isolated [NiFeSe]-hydrogenase of *D. vulgaris* Hildenborough. The gray grids in (**a**) and (**b**) depict an averaged omit electron density map, * denotes a partial oxidation of a Ni-Fe bridging thiolate ligand to a sulfenate, # an alternative conformation of a terminal Ni-bound thiolate and € a double oxidation of the other terminal Ni-bound thiolate to a sulfinate.

residues and the cleavage of a solvent-exposed disulfide bond (Volbeda et al. 2002). For the other crystals previously mentioned, there were no such signs of radiation damage.

The results described above may be explained as follows: active enzyme contains enough electrons in the active site and the [Fe-S] clusters to reduce O_2 completely, according to:

$$O_2 + 3H^+ + 4e^- \rightarrow H_2O + OH^- \quad (2.1)$$

Two of the required electrons could come from a bridging hydride bound to either Ni(III)-Fe(II) (Ni-C) or Ni(II)-Fe(II) (Ni-R) in active enzyme (Brecht et al. 2003; Fontecilla-Camps et al. 2007; Pandelia et al. 2010) and the remaining two could be provided by reduced [Fe-S] clusters. One water molecule escapes the active site whereas the other one remains trapped as a bound hydroxide. During aerobic purification the initially reduced enzyme will gradually oxidize. However, the redox potentials for the Ni(II)/Ni(III) and $[Fe_3S_4]^+/[Fe_3S_4]^0$ cluster couples are positive enough to provide two electrons for O_2 reduction at the increasingly higher potentials encountered during enzyme purification:

$$O_2 + H^+ + 2e^- \rightarrow OOH^- \quad (2.2)$$

The produced reactive peroxide species may oxidize thiolates to sulfenate (Forman et al. 2010), maybe after reduction of Ni(III) to Ni(II) because sulfenates are better ligands for the latter (Farmer et al. 1993):

$$Cys\text{-}S^- + OOH^- \rightarrow Cys\text{-}SO^- + OH^- \quad (2.3)$$

Reaction (2.3), which also produces bound hydroxide, is thermodynamically very favorable (Söderhjelm and Ryde 2006). Consequently, it must have a rather large kinetic barrier in order to explain the predominant detection of the less stable peroxide intermediate. Our observations with the crystal overexposed to X-rays suggest that sulfenates may be further reduced to water and thiolate by photoelectrons produced by this radiation. Similar active site modifications have been reported for *D. vulgaris* Miyazaki (Ogata et al. 2005) and *Allochromatium (A.) vinosum* [NiFe]-hydrogenase (Ogata et al. 2010) in their "as-isolated" state. In the first case, a complicated mixture was observed including a partially occupied bridging peroxide, a partial modification of both a Ni-Fe bridging and a terminally Ni-bound thiolate to a sulfenate, and possibly, an additional fraction containing an inorganic sulfur ligand (S^{2-} or HS^-). In the second case, a spherical Ni-Fe bridging electron density was observed, along with a partial modification of a bridging

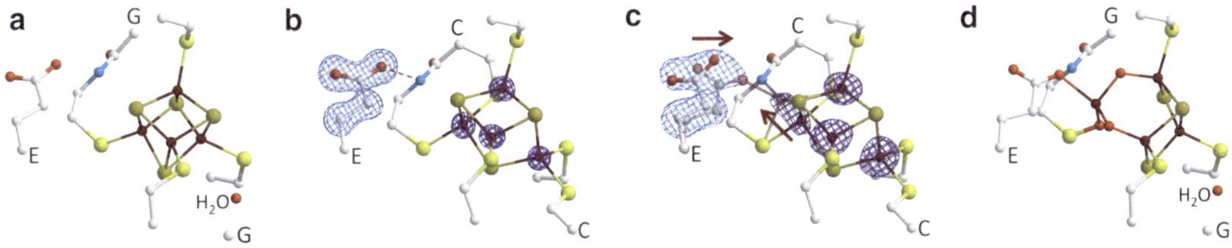

Fig. 2.4. Crystallographic models of the proximal iron-sulfur cluster: (**a**) [Fe_4S_4]-cluster in the *D. fructosovorans* enzyme, (**b**) [Fe_4S_3]-cluster in *E. coli* hydrogenase-1, observed in the H_2-reduced enzyme, (**c**) [Fe_4S_3]-cluster observed as a mixture of two states in as-isolated *E. coli* Hyd-1, (**d**) [$Fe_4S_3O_3$]-cluster observed in a fraction of the as-isolated *D. desulfuricans* ATCC 27,774 enzyme. Violet and blue grids denote anomalous difference and omit electron density maps.

thiolate to sulfenate. Using the data of our overexposed crystal of *D. fructosovorans* [NiFe]-hydrogenase and including quantum mechanical methods in the crystallographic refinement, Söderhjelm and Ryde (2006) obtained a similar result. This included a small fraction with a sulfenate modification of a bridging thiol (at a density peak that we had earlier attributed to noise) and an even smaller fraction with the Ni terminal thiolate ligand modified to sulfenate as observed by Ogata et al. (2005) in the *D. vulgaris* Miyazaki enzyme (although not exactly in the same conformation). In conclusion, although the exact identity of states like Ni-A and the one-electron more reduced unready Ni-SU form (Fig. 2.3b) is still debated, our interpretations seem to be compatible with all the discussed crystallographic, as well as with other experimental results, as previously reviewed by Fontecilla-Camps et al. (2007). Besides reacting at the active site, O_2 may also react with and presumably damage Fe-S clusters, as exemplified by the partial conversion of the proximal [Fe_4S_4] to a [$Fe_4S_3O_3$] cluster observed in the crystal structures of *D. desulfuricans* ATCC 27774 [NiFe]-hydrogenase (Matias et al. 2001) (Fig. 2.4d) and *D. vulgaris* Hildenborough [NiFeSe]-hydrogenase (Marques et al. 2010).

In theory there are at least three ways to decrease the oxygen sensitivity of a [NiFe]-hydrogenase: (A) to limit the access of O_2 to the active site, (B) to speed up the activation of oxidized states and (C) to avoid the formation of reactive oxygen species according to reactions (2.2) and (2.3) by keeping enough electrons available for the complete reduction of O_2 to water. Studies with mutants have shown that strategy A, which limits oxygen access through the tunnel connecting the active site to the protein exterior (Montet et al. 1997) may indeed explain the O_2-tolerance of H_2 sensors, also called regulatory hydrogenases (Buhrke et al. 2005; Duche et al. 2005). This possibility was predicted from sequence alignments with O_2-sensitive hydrogenases (Volbeda et al. 2002). However, the function of these sensors is to activate the synthesis of hydrogenases when hydrogen is present by interacting with a histidine protein kinase, which in turn, modulates the activity of a response regulator-transcription factor (Elsen et al. 2003; Buhrke et al. 2004). As hydrogenases they have very little activity. We will not review the elegant studies carried out with mutants of the tunnel and other regions using the O_2-sensitive periplasmic *D. fructosovorans* [NiFe]-hydrogenase as they are described in Chap. 3 of this book. Instead, we will next discuss those enzymes that are either naturally O_2-resistant or O_2-tolerant by using strategies B and C, respectively.

IV. Structural Studies of O_2-Resistant [NiFeSe]-Hydrogenases

In some [NiFe]-hydrogenases, one of the cysteine ligands of the Ni is naturally substituted by a seleno-cysteine (SeCys) and, in addition, the mesial [Fe_3S_4] is replaced by a

$[Fe_4S_4]$ cluster. In general, as recently reviewed by Baltazar et al. (2011), such [NiFeSe]-hydrogenases have much higher catalytic activity than the [NiFe] enzymes and they often appear to be less O_2-sensitive. The crystal structure of [NiFeSe]-hydrogenase from the sulfate reducing bacterium *Desulfomicrobium baculatum*, was reported by Garcin et al. (1999). It was both the first structure determined for this class of hydrogenase and one of the first, together with the structure reported by Higuchi et al. (1999) for *D. vulgaris* Miyazaki [NiFe]-hydrogenase, with a reduced active site, probably in the Ni-C state (Fig. 2.3c). Based on several sources (Brecht et al. 2003; Fontecilla-Camps et al. 2007; Pandelia et al. 2010), a hydride is postulated to bridge the Ni and Fe ions in the active Ni-C form, replacing the hydroxide found in the Ni-B state. More recently, Marques et al. (2010) have reported on the structure of the [NiFeSe]-hydrogenase from *D. vulgaris* Hildenborough. This structure contains a mixture of oxidized states and includes three different conformations for its SeCys residue. Quite unexpectedly, about 70 % of the structure appears to contain a doubly oxidized thiol (a sulfinate) and a persulfurated SeCys (Fig. 2.3d). Assuming it contains Ni(II), which is reasonable given the absence of EPR signals from oxidized [NiFeSe]-hydrogenase preparations, this highly oxidized active site will require no less than seven electrons and up to seven protons to be converted into the reduced Ni-C state (Fig. 2.3c): two electrons will be needed to reduce the Se-S bond, producing either H_2S or HS^-; four additional electrons will be required to reduce the two S-O bonds, producing two H_2O molecules, and one more electron must be employed, along with the oxidation of Ni(II) to Ni(III), to reduce a proton and generate the hydride. It is difficult to reconcile such a highly oxidized structure with the inherent O_2-resistance of [NiFeSe]-hydrogenases, because sulfinates are thermodynamically very stable hyperoxidized species that normally require dedicated enzymes for their reduction (Poole and Nelson 2008). Consequently, we conclude that additional studies, including structures of intermediate oxidation states, will be required to understand how the structure reported by Marques et al. (2010) could be generated and to determine whether it is easily activated. These studies should also shed light on the role of SeCys in activation and O_2 resistance. One possible reason for the SeCys/Cys substitution in these enzymes is the fact that Se-O bonds that might be formed upon air exposure are inherently weaker than S-O bonds (Parkin et al. 2008) and, consequently, easier to break.

V. Structural Studies of O_2-Tolerant Membrane-Bound [NiFe]-Hydrogenases

Amino acid sequence comparisons (Pandelia et al. 2012) have shown that a family of oxygen-tolerant hydrogenases has two supernumerary small subunit cysteine residues in the coordination sphere of the proximal cluster (Figs. 2.4a, b). These enzymes can oxidize hydrogen at the 21 % atmospheric oxygen level. As first shown in the case of *Aquifex aeolicus* hydrogenase 1 (*Aa*Hyd-1), the proximal cluster (PC) is involved in two one–electron redox process, involving PC1/PC2 (formal +1/+2) and PC2/PC3 (formal +2/+3) states. The higher potential PC2/PC3 redox transition does not change between pH 6.4 and 7.4. Other adaptations for their tolerance to oxygen are (i) a lower K_m for H_2 than the K_I for O_2 and (ii) the fact that all the metal centers have more positive potentials when compared to oxygen–sensitive NiFe hydrogenases. It has been shown that the superoxidized proximal cluster has more ferric character than standard clusters (Pandelia et al. 2011). This observation agrees with the proposition that the two oxidation steps of the proximal cluster correspond to 3Fe(II)-1Fe(III) $\rightarrow$ 2Fe(II)-2Fe(III) and 2Fe(II)-2Fe(III) $\rightarrow$ 1Fe(II)-3Fe(III) changes (Goris et al. 2011; Pandelia et al. 2011) in the PC1/PC2 and PC2/PC3 redox couples, respectively. It is noteworthy that these two redox couples are separated by a narrow potential difference of about 0.2 V (see also Roessler et al. 2012), compared to others, such as the one of high potential iron sulfur protein, where the unnatural super-reduction from $[Fe_4S_4]^{+2}$ to $[Fe_4S_4]^{+1}$ involves a much higher potential drop of 1 V relative to the +3/+2 redox couple (Heering et al. 1995). The structures of three oxygen-tolerant hydrogenases have been published including one solved by us (Fritsch et al. 2011b; Shomura et al. 2011; Volbeda et al. 2012). These studies have shown that the proximal cluster has an unusual structure where the supernumerary Cys19 bridges two iron ions and Cys120 terminally binds another one. Thus, these two cysteine residues replace a sulfide ligand. Site–directed mutagenesis has shown that the supernumerary Cys19 is crucial for oxygen tolerance whereas the supernumerary Cys120 plays a less important role (Lukey et al. 2011). The proximal cluster displays a remarkable plasticity undergoing a major conformational change when it goes from the PC2 to the PC3 state. This change involves the migration of one of the iron ions of the cluster towards the amide N of Cys20 forming a bond with it (Fig. 2.4c). In addition, this iron ion now binds the carboxylate group of a glutamate residue. An equivalent glutamate also binds the corresponding iron in oxygen-damaged proximal clusters (Fig. 2.4d). We have used our *Ec*Hyd-1 structure (Volbeda et al. 2012) to calculate and reproduce previously generated Mössbauer and EPR spectroscopic data using *Aa*Hyd-1. Our calculations show that the amide–N deprotonation, required to form the Fe-N bond, is mediated by the carboxylate group of the glutamic acid mentioned above. This residue is hydrogen-bonded to another glutamate residue, which is part of a proton transfer chain that normally operates by moving protons from the active site to the molecular surface (Fontecilla et al. 2007; Fdez Galván et al. 2008). However, when exposed to oxygen, the enzyme operates in the opposite direction by sending both protons and electrons to the active site. Under normal anaerobic conditions hydrogen uptake makes the cluster oscillate between PC1 and PC2, like in oxygen-sensitive hydrogenases. However, when the enzyme is exposed to O_2, and if the active site is in the Ni–C state with bound hydride, oxygen will be reduced to peroxide. In order to avoid subsequent oxidative damage this species has to be rapidly reduced to water. As mentioned above, this is mediated by the proximal cluster, which goes from PC1 to PC3 in two rapid successive one–electron reduction steps. Evidence for a water channel close to the active site has been also obtained from our structure. This channel is essential for evacuating the water generated upon oxygen reduction to the molecular surface. Our calculations show that the unique iron that forms a bond with the N amide atom of Cys20 is the one that gets oxidized from ferrous to ferric when the cluster goes from PC2 to PC3. Our conclusion is that if the active site stays in the Ni-B state and in the absence of H_2, there will not be electrons available to reduce the proximal cluster from the PC3 to the PC2 and PC1 state. Thus, both the Ni-B form and the superoxidized proximal cluster in the PC3 state protect the integrity of the hydrogenase when exposed to molecular oxygen. From a biotechnological standpoint these hydrogenases have potential applications in bio-fuel cells (see below). Conversely, and because of their more positive redox potentials relative to standard hydrogenases, these enzymes cannot be effectively used for hydrogen evolution.

*Ec*Hyd-1 is naturally bound to the periplamic side of the cytoplasmic membrane. It forms a dimer of heterodimers bringing the two distal clusters within 12 Å (Fig. 2.1b), a distance compatible with fast electron transfer (Page et al. 2003). So, it is possible to postulate that electrons generated at the active site of one monomer could be transferred to the active site of the other, i.e. the active site of one of the enzymes could help jumpstarting the other (Volbeda et al. 2012). This arrangement lowers the probability of the simultaneous oxygen-induced deactivation of the two hydrogenases in the dimer. Frielingsdorf et al. (2011) have proposed a trimeric arrangement for the heterodimers of the oxygen–tolerant *Re*MBH. We have modeled such a trimer and found that the distance between two distal clusters is too large to allow for efficient electron transfer. Furthermore, amino acid sequence comparisons for regions involved in monomer–monomer recognition indicate that they are well conserved in *Ec*Hyd-1 and *Re*MBH (not shown). This strongly suggests that *Re*MBH also forms a dimer of heterodimers. Furthermore, the same oligomeric state has been found in other hydrogenases, both of the O_2–sensitive and O_2–tolerant kinds, such as those from *Allochromatium vinosum* (Ogata et al. 2010) and *Hydrogenovibrio marinus* (Shomura et al. 2011), respectively.

VI. Regulation of Hydrogenase Expression and Activity: The Example of *Escherichia coli*

In order to elucidate the regulation and role of the three well-characterized H_2ases in *E. coli* one has to look into its fluctuating lifecycle from the moment it is ingested to the moment it is excreted by the host. Alexeeva et al. (2002) have put forward the concept of perceived aerobiosis that is defined as the extent to which the bacterium will use oxidative catabolism. At over 50 % aerobiosis, *E. coli* respires O_2 using low-affinity cytochrome *bo* oxidase. Conversely, below 40 % aerobiosis, the high-affinity cytochrome *bd-I* oxidase is expressed, upregulated by ArcA, the anoxic redox control regulator (Alexeeva et al. 2000). Under these conditions, cytochrome *bd-I* oxidase becomes the major terminal oxygen reductase and, thanks to its activity the intracellular oxygen tension is kept low enough to allow for, (i) pyruvate–formate lyase activity, which generates formate from pyruvate, and (ii) protection of the bacterium from oxidative damage induced by dyes (Alvarez et al. 2010). At lower oxygen concentrations, cytochrome *bd-I* oxidase expression is repressed by the fumarate-nitrate reduction regulator (FNR). FNR is a transcriptional regulator of respiratory pathway genes that becomes activated at 0.5 % O_2 when *E. coli* goes from microaerobic to anaerobic growth conditions (Becker et al. 1996). Anaerobic conditions cause the expression of an additional cytochrome oxidase called *bd-II,* which is co-regulated with the expression of the O_2-tolerant *Ec*Hyd-1 (Dassa et al. 1991). As *bd-I, bd-II* has high affinity for oxygen and is well suited to function in an anaerobic environment. In microorganisms such as *Azotobacter vinelandii*, which possess the highly oxygen–sensitive, nitrogen-reducing nitrogenase, high-affinity cytochrome oxidases afford protection against oxygen–induced damage (Poole and Hill 1997). The physiological role of *Ec*Hyd-1 has not been clearly determined. Most in vitro experiments are not well suited to clarify this point because one has to look at the natural environment where this bacterium grows in order to understand when and why the three different H_2ases are expressed. *Ec*Hyd-1 expression is upregulated under stressful conditions such as carbon and phosphate starvation, osmotic shock, and stationary phase conditions (Atlung et al. 1997) and both *Ec*Hyd-1 and *Ec*Hyd-3 are highly expressed under fermentative conditions, i.e., their expression is stimulated by formate (Brøndsted and Atlung 1994). These conditions are naturally found in the anoxic terminal segment of the gastro–intestinal tract of the host. The role of *Ec*Hyd-3 in recycling hydrogen and preventing acidification of the cytoplasm according to the reaction:

$$HCOO^- + H^+ \rightarrow CO_2 + H_2 \qquad (2.4)$$

is well established. But, in the case of *Ec*Hyd-1 it is not easy to explain why an oxygen-tolerant H_2ase is highly expressed under fermentative conditions when there is excess of reducing equivalents and electron acceptors are scarce (except for endogenously produced fumarate) (Pinske et al. 2012). Furthermore, because under these conditions the quinone pool should be fully reduced, it would make little sense to generate additional electrons from hydrogen oxidation. Conversely, the enzyme will very rapidly reduce any traces of oxygen present in the periplasmic space. This is so because, under these conditions, O_2 will constitute the only sink for H_2-generated electrons. As discussed above, our crystal structure (Fig. 2.1b) and electron transfer rate calculations favor direct oxygen reduction to water as the main activity of this enzyme when anaerobic *E. coli* is exposed to this gas, according to the Knallgas reaction:

$$O_2 + 2H_2 \rightarrow 2H_2O \qquad (2.5)$$

As long as there is H_2 being produced by *Ec*Hyd-3 from formate, *Ec*Hyd-1 will oxidize it and use the resulting electrons to reduce O_2 if any is present.

Several experiments have shown that *Ec*Hyd-1 cannot reduce low-potential artificial electron acceptors (Pinske et al. 2011). The enzyme, however, is capable of reducing nitroblue tetrazolium (NBT), a redox dye with E_0'=−80 mV (Pinske et al. 2012). This activity does not require the presence of the cognate membranous cytochrome *b*, indicating that the reduction is performed directly by the H_2ase. The catalytic bias of *Ec*Hyd-1 to hydrogen oxidation is related to its oxygen tolerance. Indeed, it has been shown that this enzyme has an overpotential of about +50 mV when compared to *Ec*Hyd-2 (Lukey et al. 2010). This overpotential also implies that the activity/inactivity switch of *Ec*Hyd-1 is shifted to higher potentials than in the case of *Ec*Hyd-2. Conversely, this over-potential prevents *Ec*Hyd-1 from being able to reduce protons or low–potential dyes (Lukey et al. 2010; Pinske et al. 2011).

The role of *Ec*Hyd-2 and the regulation of its expression are easier to rationalize. This enzyme, which resembles H_2ases from sulfate–reducing bacteria in terms of its catalytic properties, is expressed under microaerobic and anaerobic conditions. Its physiological role is hydrogen uptake (Dubini et al. 2002). *Ec*Hyd-2 can use fumarate as electron acceptor and its expression correlates with fumarate respiration (Pinske et al. 2012). Conversely, nitrate is a repressor of the expression of this enzyme. *Ec*Hyd-2 shows low benzyl viologen (BV) reduction activity in cell extracts. By comparison, *Ec*Hyd-3 is very effective in reducing this dye, which, although not biologically relevant, is related to the potentials at which this enzyme functions in proton reduction (Pinske et al. 2011).

VII. [NiFe]-Hydrogenase Maturation

The biosynthesis of the Ni-Fe active site is a complex energy-consuming and species-specific process. Moreover, when there are several hydrogenases in the same species, each of them has its own maturation machinery. In the extensively studied biosynthetic pathway of *Ec*Hyd-3 (Böck et al. 2006) at least ten gene products are involved. The cyanide precursor $H_2NC(O)P_i$ (carbamoylphosphate, here abbreviated CP) is produced by CP-synthetase from L-glutamine and bicarbonate in reactions 1–3 (Scheme 2.1), with concomitant consumption of two ATP molecules (Thoden et al. 1997). CP is converted by *HypF* to $H_2NC(O)$-AMP, also in an ATP-dependent reaction (reactions 4–5). After formation of a *HypEF* complex, of known structure (Shomura and Higuchi 2012), the H_2NCO group is transferred to the C-terminal cysteine of HypE (reaction 6). The resulting thiocarbamate is subsequently dehydrated in yet another ATP-dependent reaction to thiocyanate (reaction 7), followed by CN transfer to Fe bound to *HypCD* (reaction 8) in a putative *HypCDE* complex (Watanabe et al. 2007). Reactions 1–8 are repeated for the transfer of a second CN ligand to Fe, whereas CO is provided by a so far unknown donor (Bürstel et al. 2011) in reaction 9. The resulting $FeCO(CN)_2$ moiety is next transferred to *apo-pre-HycE* (reaction 10), followed by a SlyD-dependent Ni transfer (Chung and Zamble 2011; Kaluarachchi et al. 2012) from the GTPase *HypB* (reaction 11). Maturation is finished by the cleavage of a short C-terminal peptide of *pre-HycE* (reaction 12) by the endopeptidase *HycI*. The Ni insertion machinery further involves Ni transfer to *HypB* from the *HypA* carrier (reaction 15), which itself is charged with Ni

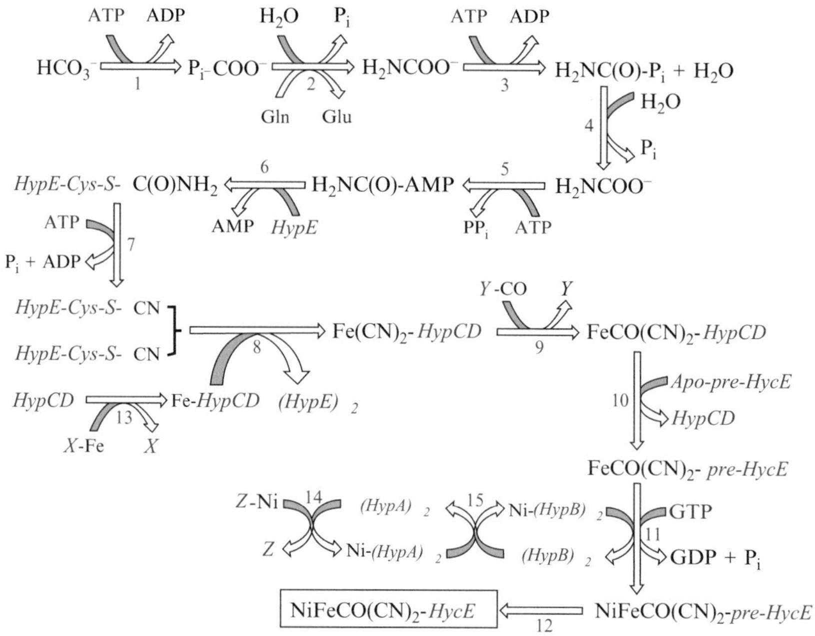

Scheme 2.1. Maturation of the Ni-Fe-containing HycE subunit of *Ec*Hyd-3. The enzymes/proteins involved, with reactions numbered 1–15, are carbamoylphosphate synthetase (1–3), *HypF* (4–7), *HypE* (7–8), *HypC and HypD* (8–10, 13), HypB and *SlyD* (11, 15), *HycI* (12) and *HypA* (14–15). *X*, *Y* and *Z* are unknown donors of Fe, CO and Ni, respectively.

by an unknown donor (reaction 14). The donor of Fe to *HypCD* (reaction 10) is also unknown. Taking all the reactions into account, at least eight ATP molecules and one GTP are required to complete the active site maturation. However, given the incomplete characterization of the pathways for metal transfer and production of the CO ligand, the actual energy requirements could be significantly higher.

Significant insight into the molecular aspects of the *Ec*Hyd-3 large subunit maturation has been provided by the X-ray structure determinations of *HypA*, *HypB*, *HypC*, *HypD*, *HypE*, *HypF*, *HycI* and *SlyD* (Watanabe et al. 2009; Xia et al. 2009; Gasper et al. 2006; Chan et al. 2012; Watanabe et al. 2007; Shomura et al. 2007; Rangarajan et al. 2008; Shomura and Higuchi 2012; Petkun et al. 2011; Kumarevel et al. 2009; Loew et al. 2010), but some details remain unclear. In addition, although *apo-pre-HycE* has been generally assumed to be devoid of metal, a recent report suggests that a mutant of *HybC*, the homologous unprocessed large subunit of *Ec*Hyd-2, may actually contain a labile $[Fe_4S_4]$ cluster at the active site position in the mature subunit (Soboh et al. 2012). If this were also the case for the native, unprocessed subunit, an additional step would be needed involving removal of the cluster, before incorporation of the Ni-Fe site.

An interesting aspect is the sensitivity of the maturation process towards oxygen. In the maturation of the O_2-tolerant *Re*MBH, extra gene products are involved that allow production of active enzyme under aerobic conditions (Fritsch et al. 2011a). The same applies to *Se*Hyd-5, the homologous O_2-tolerant hydrogenase-5 of *Salmonella enterica* serovar Typhimurium (Parkin et al. 2012). However, the latter organism also produces an O_2-tolerant Hyd-1 that, like the related *Ec*Hyd-1 enzyme, is only expressed under anoxic conditions. The maturation of these enzymes is most likely O_2-sensitive, because it does not involve gene products related to those used for *Re*MBH and *Se*Hyd-5 maturation, under air. In conclusion, aerobic production of O_2-tolerant [NiFe]-hydrogenases requires O_2-tolerant maturation.

VIII. Biotechnological Applications

Using electrochemistry, Vincent and collaborators (2005a, b, 2007) have studied mechanisms of catalysis, electron transfer, activation and inactivation, and defined important properties such as O_2 tolerance and CO resistance of H_2ases in physical terms. These enzymes are alternatives to noble metals for the production of hydrogen from solar energy (Jones et al. 2002). The latter are nonselective and can be poisoned by environmental pollutants whereas the enzymes are highly specific and relatively resistant. Like Pt, H_2ases produce hydrogen with minimal overpotential and are catalytically very efficient. For that reason, these enzymes are promising targets for developing new catalysts with biotechnological applications.

A. Membrane-Bound [NiFe]-Hydrogenase

The oxygen–tolerant hydrogenase from *Ralstonia metallidurans* CH34 and the fungal O_2-reductase laccase have been adsorbed to graphite electrodes to build an open bio–fuel cell that could generate electricity from 3 % hydrogen, under normal atmospheric conditions and in aqueous solution (Vincent et al. 2005a, 2006). This setup was shown to be capable of powering a wristwatch for several hours. Although the hydrogenase had to be activated after a few hours, this experiment has opened the possibility of powering electronic devices using low hydrogen concentrations in air. Because of the very high specificity of the enzymes, which is not the case of Pt that catalyzes both the anodic and cathodic sides of the reaction, no costly membrane is required in the fuel cell setup (Fig. 2.5a).

B. [NiFeSe]-Hydrogenase

Reisner et al. (2009) and Reisner and Armstrong (2011) have carried out a systematic study of enzyme efficiency by coupling colloidal semi-conductor TiO_2 nanoparticles with a synthetic ruthenium photosensitizer to different H_2ases. When a H_2ase is attached to an n-type semiconducting surface, rather than to a metallic or semi-metallic material like graphite, the direction of catalysis can be altered with a bias towards reduction reactions. This is convenient in the case of hydrogen production. The work by Reisner et al. is a proof of concept: the dye injects an electron into the conduction band of TiO_2 when exposed to visible light. This, in turn, oxidizes the dye and reduces TiO_2, which transfers electrons directly to the adsorbed H_2ase that reduces protons to molecular hydrogen. A sacrificial electron donor reduces the dye, closing the cycle (Fig. 2.5b). Reisner et al. (2009) and Reisner and Armstrong (2011) concluded that the most efficient available system included the [NiFeSe]-H_2ase from *Dm. baculatum* and the tris(bipyridyl) ruthenium photosensitizer RuP. The latter fulfills several requirements including (1) an absorption band in the visible spectrum, (2) stable attachment to TiO_2, (3) efficient charge separation and (4) long-term stability upon irradiation. Conversely, the choice of *Dm. baculatum* H_2ase was determined by several factors: (1) it has good hydrogen production activity; (2) it can be rapidly reactivated at low potentials after O_2–induced inactivation; (3) it can operate in the presence of about 1 % O_2 and (4), there is significant proton reduction even at 5 % H_2, which is usually inhibitory to H_2ases. However, these characteristics are not enough to render a H_2ase optimal for hydrogen production. A simple calculation indicates that the distal $[Fe_4S_4]$ cluster of *Dm. baculatum* H_2ase is surrounded by a negatively charged surface patch. Thus, the interaction between the enzyme and the TiO_2 particle is mostly controlled by localized polar interactions rather than overall electrostatic interactions. As a conclusion, the

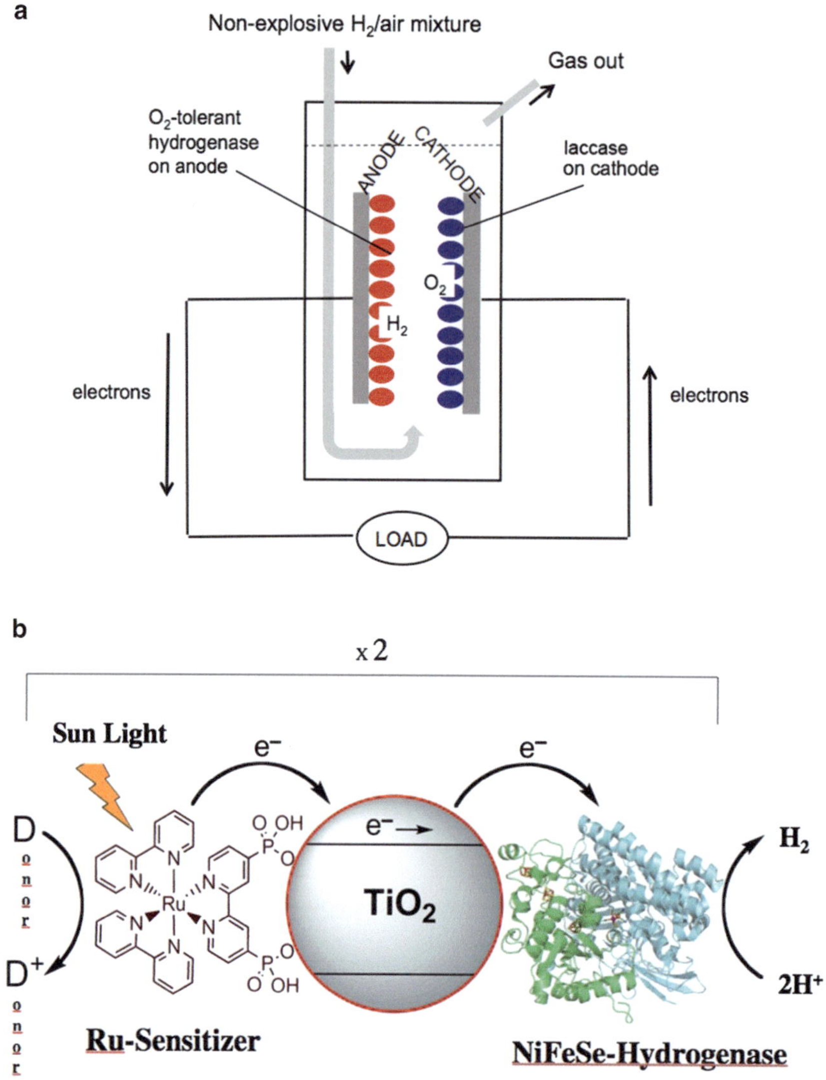

Fig. 2.5. (**a**) A bio-fuel cell comprising a graphite cathode modified with high potential fungal laccase and a graphite anode modified with the O_2-tolerant membrane-bound hydrogenase of *Ralstonia metallidurans* CH34 in aqueous electrolyte under an atmosphere of 3 % H_2 in air (Adapted from Vincent et al. 2005a); (**b**) Schematic representation of visible-light driven H_2 evolution with [NiFeSe]-H_2ase attached to RuP dye sensitized TiO_2 nanoparticles. Excitation by visible light in the presence of a sacrificial electron donor (Donor), causes RuP to inject an electron into the conduction band of the semi-conductor TiO_2. The electrons, which are transferred directly to the adsorbed [NiFeSe]-H_2ase, reduce H^+ from the buffered aqueous solution generating H_2. The three [Fe_4S_4]-clusters (indicated in the figure) form a "wire" responsible for electron transfer to and from the active site of the [NiFeSe]-H_2ase. The structure of the sensitiser RuP is also shown (*left side*). Two cycles are required to generate a H_2 molecule (Adapted from Reisner et al. 2009).

authors found that *Dm. baculatum* H_2ase has the very well suited property of being titaniaphilic. Taken together, these results should play an important role in the future design and assembly of robust H_2ase-nanoparticle devices including mesoporous 3D electrodes for enzyme–fuel cells or biosensors. The production of hydrogen at room temperature from neutral water without redox–mediators represents a significant step towards the development of an artificial system mimicking photosynthetic green algae. In a related approach, Lubner et al. (2010) have connected photosystem I (PS I) and an [FeFe]-H_2ase and have assayed electron transfer between the two components via light-induced H_2 generation.

C. Bio-inspired Artificial Hydrogen Catalysts

Helm et al. (2011) have reported on which may be the most efficient bio-inspired catalyst synthesized so far. It is the synthetic nickel complex, $[Ni(P(Ph)_2N(Ph)_2)](BF_4)_2$, $(P(Ph)_2N(Ph)) = 1,3,6$-triphenyl-1-aza-3,6-diphosphacycloheptane, which catalyzes the production of H_2 with protonated dimethylformamide as the proton donor. Turnover frequencies of 106,000 per second have been obtained in the presence of 1.2 M of water. This remarkably fast catalyst combines features of the two major types of H_2ases: a Ni ion ([NiFe]-H_2ase) and pendant amines that function as proton relays ([FeFe]-H_2ases). A computational study on a related compound suggests that proton transfers between the amine nitrogen and the nickel are favored relative to a direct nitrogen-to-nitrogen proton transfer (O'Hagan et al. 2011). This result supports our proposition that the bridgehead atom of the thiolate-containing small molecule at the [FeFe]-H_2ase active site is nitrogen (Nicolet et al. 2001).

IX. Conclusions

The structural studies of [NiFe]-H_2ases have shed considerable light on the catalytic mechanism of hydrogen uptake and proton reduction. Both processes are of biotechnological interest and are the subject of very active research. One major goal in this field is the coupling of solar energy to hydrogen production. The O_2-resistant [NiFeSe]-H_2ase has proven to be a very effective H_2 producer thanks to its very high affinity for TiO_2 particles. Another promising domain is the use of the O_2-tolerant enzymes in bio-fuel cells although the relative fragility of these molecules limits their application at present time. Maybe more importantly, the active sites of hydrogenases have inspired the synthesis of novel catalysts with very good performances. In addition, structural strategies such as the one employed by the $[Fe_4S_3]$ cluster of the O_2-tolerant [NiFe]-H_2ases, which is capable of two-electron redox chemistry, should also inspire new ways of designing synthetic catalysts for hydrogen oxidation.

Acknowledgements

The authors thank the Commissariat à l'Energie Atomique et aux Energies Alternatives (CEA) and the Centre National de la Recherche Scientifique (CNRS) for institutional funding and the Agence Nationale de la Recherche for several contracts concerning the subject of this chapter. Erwin Reisner is thanked for providing Fig. 2.5b.

References

Alexeeva S, de Kort B, Sawers G, Hellingwerf KJ, de Mattos JT (2000) Effects of limited aeration and of the ArcAB system on intermediary pyruvate catabolism in *Escherichia coli*. J Bacteriol 182:4934–4940

Alexeeva S, Hellingwerf KJ, de Mattos JT (2002) Quantitative assessment of oxygen availability: perceived aerobiosis and its effect on flux distribution in the respiratory chain of *Escherichia coli*. J Bacteriol 184:1402–1406

Alvarez AF, Malpica R, Contreras M, Escamilla E, Georgellis D (2010) Cytochrome *d* but not cytochrome *o* rescues the toluidine blue growth sensitivity of arc mutants of *Escherichia coli*. J Bacteriol 192:391–399

Atlung T, Knudsen K, Heerfordt L, Brøndsted L (1997) Effects of sigmaS and the transcriptional activator AppY on induction of the *Escherichia coli* hya and cbdAB-appA operons in response to carbon and phosphate starvation. J Bacteriol 179:2141–2146

Baltazar CSA, Marques MC, Soares CM, DeLacey AM, Pereira IAC, Matias PM (2011) Nickel–iron–selenium hydrogenases – an overview. Eur J Inorg Chem 2011:948–962

Becker S, Holighaus G, Gabrielczyk T, Unden G (1996) O_2 as the regulator signal for FNR-dependent gene regulation in *Escherichia coli*. J Bacteriol 178:4515–4521

Böck A, King PW, Blokesch M, Posewitz MC (2006) Maturation of hydrogenases. Adv Microb Physiol 51:1–71

Boga HI, Brune A (2003) Hydrogen-dependent oxygen reduction by homoacetogenic bacteria isolated from termite guts. Appl Environ Microbiol 69:779–786

Bothe H, Schmitz O, Yates MG, Newton WE (2010) Nitrogen fixation and hydrogen metabolism in cyanobacteria. Microbiol Mol Biol Rev 74:529–551

Brecht M, Van Gastel M, Buhrke T, Friedrich B, Lubitz W (2003) Direct detection of a hydrogen ligand in the [NiFe] center of the regulatory H_2-sensing hydrogenase from *Ralstonia eutropha* in its reduced state by HYSCORE and ENDOR spectroscopy. J Am Chem Soc 125:13075–13083

Brøndsted L, Atlung T (1994) Anaerobic regulation of the hydrogenase 1 (*hya*) operon of *Escherichia coli*. J Bacteriol 176:5423–5428

Buhrke T, Lenz O, Porthun A, Friedrich B (2004) The H_2-sensing complex of *Ralstonia eutropha*: interaction between a regulatory [NiFe] hydrogenase and a histidine protein kinase. Mol Microbiol 51:1677–1689

Buhrke T, Lenz O, Krauss N, Friedrich B (2005) Oxygen tolerance of the H_2-sensing [NiFe] hydrogenase from *Ralstonia eutropha* is based on limited access of oxygen to the active site. J Biol Chem 280:23791–23796

Bürstel I, Hummel P, Siebert E, Wisitruangsakul N, Zebger I, Friedrich B, Lenz O (2011) Probing the origin of the metabolic precursor of the CO ligand in the catalytic center of [NiFe] hydrogenase. J Biol Chem 286:44937–44944

Chan K-H, Li T, Wong C-O, Wong K-B (2012) Structural basis for GTP-dependent dimerization of hydrogenase maturation factor HypB. PLoS One 7:e30547

Chung KC, Zamble DB (2011) The Escherichia coli metal-binding chaperone SlyD interacts with the large subunit of [NiFe]-hydrogenase 3. FEBS Lett 585:291–294

Dassa J, Fsihi H, Marck C, Dion M, Kieffer-Bontemps M, Boquet PL (1991) A new oxygen-regulated operon in *Escherichia coli* comprises the genes for a putative third cytochrome oxidase and for pH 2.5 acid phosphatase (*appA*). Mol Gen Genet 229:341–352

De Lacey AL, Fernández VM, Rousset M, Cammack R (2007) Activation and inactivation of hydrogenase function and the catalytic cycle: spectroelectrochemical studies. Chem Rev 107:4304–4330

Dubini A, Pye RL, Jack RL, Palmer T, Sargent F (2002) How bacteria get energy from hydrogen: a genetic analysis of periplasmic hydrogen oxidation in *Escherichia coli*. Int J Hydrog Energy 27:1413–1420

Duché O, Elsen S, Cournac L, Colbeau A (2005) Enlarging the gas access channel to the active site renders the regulatory hydrogenase HupUV of *Rhodobacter capsulatus* O_2 sensitive without affecting its transductory activity. FEBS J 272:3899–3908

Efremov RG, Sazanov LA (2012) The coupling mechanism of respiratory complex I – a structural and evolutionary perspective. Biochim Biophys Acta 1817:1785–1795

Elsen S, Duché O, Colbeau A (2003) Interaction between the H_2 sensor HupUV and the histidine kinase HupT controls HupSL hydrogenase synthesis in *Rhodobacter capsulatus*. J Bacteriol 185:7111–7119

Farmer PJ, Reibenspies JH, Lindahl PA, Darensbourg M (1993) Effect of sulfur site modification on the redox potentials of derivatives of [N, N'-bis(2-mercaptoethyl)-1,5-diazocyclooctanato]nickel(II). J Am Chem Soc 115:4665–4674

Fdez Galván I, Volbeda A, Fontecilla-Camps JC, Field MJ (2008) A QM/MM study of proton transport pathways in a [NiFe] hydrogenase. Proteins 73:195–203

Fontecilla-Camps JC, Volbeda A, Cavazza C, Nicolet Y (2007) Structure/function relationships of [NiFe]- and [FeFe]-hydrogenases. Chem Rev 107:4273–4303

Forman HJ, Maiorino M, Ursini F (2010) Signaling functions of reactive oxygen species. Biochemistry 49:835–842

Frielingsdorf S, Schubert T, Pohlmann A, Lenz O, Friedrich B (2011) A trimeric supercomplex of the oxygen-tolerant membrane-bound [NiFe]-hydrogenase from *Ralstonia eutropha* H16. Biochemistry 50:10836–10843

Fritsch J, Lenz O, Friedrich B (2011a) The maturation factors HoxR and HoxT contribute to oxygen tolerance of membrane-bound [NiFe] hydrogenase in *Ralstonia eutropha* H16. J Bacteriol 193:2487–2497

Fritsch J, Scheerer P, Frielingsdorf S, Kroschinsky S, Friedrich B, Lenz O, Spahn CMT (2011b) The crystal structure of an oxygen tolerant hydrogenase uncovers a novel iron-sulphur centre. Nature 479:249–252

Garcin E, Vernede X, Hatchikian EC, Volbeda A, Frey M, Fontecilla-Camps JC (1999) The crystal structure of a reduced [NiFeSe] hydrogenase provides an image of the activated catalytic center. Structure 7:557–566

Gasper R, Scrima A, Wittinghofer A (2006) Structural insights into HypB, a GTP-binding protein that regulated metal binding. J Biol Chem 281:27492–27502

Goris T, Wait AF, Saggu M, Fritsch J, Heidary N, Stein M, Zebger I, Lendzian F, Armstrong FA, Friedrich B, Lenz O (2011) A unique iron-sulfur cluster is crucial for oxygen tolerance of a [NiFe]-hydrogenase. Nat Chem Biol 7:310–318

Hedderich R (2004) Energy-converting [NiFe] hydrogenases from archaea and extremophiles: ancestors of complex I. J Bioenerg Biomembr 36:65–75

Heering HA, Bulsink YBM, Hagen WR, Meyer TE (1995) Reversible super reduction of the cubane [4Fe-4S]$^{(3+,2+,1+)}$ in the high-potential iron-sulfur protein under nondenaturating conditions. Eur J Biochem 232:811–817

Helm ML, Stewart MP, Bullock RM, Rakowski DuBois M, DuBois DL (2011) A synthetic nickel electrocatalyst with a turnover frequency above 100,000 s^{-1} for H_2 production. Science 333:863–866

Higuchi Y, Ogata H, Miki K, Yasuoka N, Yagi T (1999) Removal of the bridging ligand atom at the Ni-Fe active site of [NiFe] hydrogenase upon reduction with H_2, as revealed by X-ray structure analysis at 1.4 Å resolution. Structure 7:549–556

Horch M, Lauterbach L, Lenz O, Hildebrandt P, Zebger I (2012) NAD(H)-coupled hydrogen cycling – structure-function relationships of bidirectional [NiFe] hydrogenases. FEBS Lett 586:545–556

Jones AK, Sillery E, Albracht SPJ, Armstrong FA (2002) Direct comparison of the electrocatalytic oxidation of hydrogen by an enzyme and a platinum catalyst. Chem Commun 8:866–867

Kaluarachchi H, Altenstein M, Sugumar SR, Balbach J, Zamble DB, Haupt C (2012) Nickel binding and [NiFe]-hydrogenase maturation by the metallchaperone SlyD with a single metal-binding site in *Escherichia coli*. J Mol Biol 417:28–35

Kröger A, Biel S, Simon J, Gross R, Unden G, Lancaster CRD (2002) Fumarate respiration of *Wolinella succinogenes*: enzymology, energetics and coupling mechanism. Biochim Biophys Acta 1553:23–38

Kumarevel T, Tanaka T, Bessho Y, Shinkai A, Yokoyama S (2009) Crystal structure of hydrogenase maturating endopeptidase HycI from *Escherichia coli*. Biochem Biophys Res Commun 389:310–314

Laurinavichene TV, Zorin NA, Tsygankov AA (2002) Effect of redox potential on activity of hydrogenase 1 and hydrogenase 2 in *Escherichia coli*. Arch Microbiol 178:437–442

Leonhartsberger S, Korsa I, Böck A (2002) The molecular biology of formate metabolism in enterobacteria. J Mol Microbiol Biotechnol 4:269–276

Loew C, Neumann P, Tidow H, Weininger U, Haupt C, Friedrich-Epler B, Scholz C, Stubbs MT, Balbach J (2010) Crystal structure determination and functional characterization of the metallochaperone SlyD from *Thermus thermophilus*. J Mol Biol 398:375–390

Lubitz W, Reijerse E, Van Gastel M (2007) [NiFe] and [FeFe] hydrogenases studied by advanced magnetic resonance techniques. Chem Rev 107:4331–4365

Lubner CE, Knörzer P, Silva PJ, Vincent KA, Happe T, Bryant DA, Golbeck JH (2010) Wiring an [FeFe]-hydrogenase with photosystem I for light-induced hydrogen production. Biochemistry 49:10264–10266

Lukey MJ, Parkin A, Roessler MM, Murphy BJ, Harmer J, Palmer T, Sargent F, Armstrong FA (2010) How *Escherichia coli* is equipped to oxidize hydrogen under different redox conditions. J Biol Chem 285:3928–3938

Lukey MJ, Roessler MM, Parkin A, Evans RM, Davies RA, Lenz O, Friedrich B, Sargent F, Armstrong FA (2011) Oxygen-tolerant [NiFe]-hydrogenases: the individual and collective importance of supernumerary cysteines at the proximal Fe-S cluster. J Am Chem Soc 133:16881–16892

Marques MC, Coelho R, De Lacey AL, Pereira IAC, Matias PM (2010) The three-dimensional structure of [NiFeSe] hydrogenase from *Desulfovibrio vulgaris* Hildenborough: a hydrogenase without a bridging ligand in the active site in its oxidised, "as-isolated" state. J Mol Biol 396:893–907

Matias PM, Soares CM, Saraiva LM, Coelho R, Morais J, Le Gall J, Carrondo MA (2001) [NiFe] hydrogenase from *Desulfovibrio desulfuricans* ATCC 27774: gene sequencing, three-dimensional structure determination and refinement at 1.8 Å and modelling studies of its interaction with the tetrahaem cytochrome c_3. J Biol Inorg Chem 6:63–81

Matias PM, Pereira IAC, Soares CM, Carrondo MA (2005) Sulphate respiration from hydrogen in *Desulfovibrio bacteria*: a structural biology overview. Prog Biophys Mol Biol 89:292–329

Montet Y, Amara P, Volbeda A, Vernede X, Hatchikian EC, Field MJ, Frey M, Fontecilla-Camps JC (1997) Gas access to the active site of Ni-Fe hydrogenases probed by X-ray crystallography and molecular dynamics. Nat Struct Biol 4:523–526

Nicolet Y, Piras C, Legrand P, Hatchikian CE, Fontecilla-Camps JC (1999) *Desulfovibrio desulfuricans* iron hydrogenase: the structure shows unusual coordination to an active site Fe binuclear center. Structure 7:13–23

Nicolet Y, De Lacey AL, Vernède X, Fernandez VM, Hatchikian EC, Fontecilla-Camps JC (2001) Crystallographic and FTIR spectroscopic evidence of changes in Fe coordination upon reduction of the active site of the Fe-only hydrogenase from *Desulfovibrio desulfuricans*. J Am Chem Soc 123:1596–1601

O'Hagan M, Shaw WJ, Raugei S, Chen S, Yang JY, Kilgore UJ, DuBois DL, Bullock RM (2011) Moving protons with pendant amines: proton mobility in a nickel catalyst for oxidation of hydrogen. J Am Chem Soc 133:14301–14312

Ogata H, Hirota S, Nakahara A, Komori H, Shibata N, Kato T, Kano K, Higuchi Y (2005) Activation process of [NiFe] hydrogenase elucidated by high-resolution X-Ray analyses: conversion of the ready to the unready state. Structure 13:1635–1642

Ogata H, Kellers P, Lubitz W (2010) The crystal structure of the [NiFe] hydrogenase from the photosynthetic bacterium *Allochromatium vinosum*: characterization of the oxidized enzyme Ni-a state. J Mol Biol 420:428–444

Page CC, Moser CC, Dutton ET (2003) Mechanism for electron transfer within and between proteins. Curr Opin Chem Biol 7:551–556

Pandelia ME, Ogata H, Lubitz W (2010) Intermediates in the catalytic cycle of [NiFe] hydrogenase: functional spectroscopy of the active site. ChemPhysChem 11:1127–1140

Pandelia ME, Nitschke W, Infossi P, Giudici-Orticoni MT, Bill E, Lubitz W (2011) Characterization of a unique [FeS] cluster in the electron transfer chain of the oxygen tolerant [NiFe] hydrogenase from *Aquifex aeolicus*. Proc Natl Acad Sci U S A 108:6097–6102

Pandelia ME, Lubitz W, Nitschke W (2012) Evolution and diversification of group 1 [NiFe] hydrogenases. Is there a phylogenetic marker for O_2-tolerance? Biochim Biophys Acta 1817:1565–1575

Parkin A, Goldet G, Cavazza C, Fontecilla-Camps JC, Armstrong FA (2008) The difference a Se makes? Oxygen-tolerant hydrogen production by the [NiFeSe]-hydrogenase from *Desulfomicrobium baculatum*. J Am Chem Soc 130:13410–13416

Parkin A, Bowman L, Roessler MM, Davies RA, Palmer T, Armstrong FA, Sargent FA (2012) How salmonella oxidises H_2 under aerobic conditions. FEBS Lett 586:536–544

Petkun S, Shi R, Li Y, Asinas A, Munger C, Zhang L, Waclawek M, Soboh B, Sawers RG, Cygler M (2011) Structure of hydrogenase maturation protein HypF with reaction intermediates shows two active sites. Structure 19:1773–1783

Pinske C, Krüger S, Soboh B, Ihling C, Kuhns M, Braussemann M, Jaroschinsky M, Sauer C, Sargent F, Sinz A, Sawers RG (2011) Efficient electron transfer from hydrogen to benzyl viologen by the [NiFe]-hydrogenases of *Escherichia coli* is dependent on the coexpression of the iron-sulfur-containing small subunit. Arch Microbiol 193:893–903

Pinske C, Jaroschinsky M, Sargent F, Sawers G (2012) Zymographic differentiation of NiFe]-hydrogenases 1, 2 and 3 of *Escherichia coli* K-12. BMC Microbiol 12:134. doi:10.1186/1471-2180-12-134

Poole RK, Hill S (1997) Respiratory protection of nitrogenase activity in *Azotobacter vinelandii*–roles of the terminal oxidases. Biosci Rep 17:303–317

Poole LB, Nelson KJ (2008) Discovering mechanisms of signaling-mediated cysteine oxidation. Curr Opin Chem Biol 12:18–24

Rangarajan ES, Asinas A, Proteau A, Munger C, Baardsnes J, Iannuzzi P, Matte A, Cygler M (2008) Structure of [NiFe] hydrogenase maturation protein HypE from *Escherichia coli* and its interaction with HypF. J Bacteriol 190:1447–1458

Redwood MD, Mikheenko IP, Sargent F, Macaskie L (2007) Dissecting the roles of *Escherichia coli* hydrogenases in biohydrogen production. FEMS Microbiol Lett 278:48–55

Reisner E, Armstrong FA (2011) A TiO_2 nanoparticle system for sacrificial solar H_2 production prepared by rational combination of hydrogenase with a ruthenium photosensitizer. Methods Mol Biol 743:107–117

Reisner E, Powell DJ, Cavazza C, Fontecilla-Camps JC, Armstrong FA (2009) Visible light-driven H_2 production by hydrogenases attached to dye-sensitized TiO_2 nanoparticles. J Am Chem Soc 131:18457–18466

Roessler MM, Evans RM, Davies RA, Harmer JR, Armstrong FA (2012) EPR spectroscopic studies of the Fe-S clusters in the O_2-tolerant [NiFe]hydrogenase Hyd-1 from *E. coli*, and characterization of the unique [4Fe-3S] cluster by HYSCORE. J Am Chem Soc 134:15581–15594

Shomura Y, Higuchi Y (2012) Structural basis for the reaction mechanism of S-carbamoylation of HypE by HypF in the maturation of [NiFe]-hydrogenases. J Biol Chem 287:28409–28419

Shomura Y, Komori H, Miyabe N, Tomiyama M, Shibata N, Higuchi Y (2007) Crystal structures of hydrogenase maturation protein HypE in the apo and ATP-bound forms. J Mol Biol 372: 1045–1054

Shomura Y, Yoon KS, Nishihara H, Higuchi Y (2011) Structural basis for [4Fe-3S] cluster in the oxygen-tolerant membrane-bound [NiFe]-hydrogenase. Nature 479:253–256

Soboh B, Kuhns M, Braussemann M, Waclawek M, Muhr E, Pierik AJ, Sawers RG (2012) Evidence for an oxygen-sensitive iron-sulfur cluster in an immature large subunit species of *Escherichia coli* [NiFe]-hydrogenase 2. Biochem Biophys Res Commun 424:158–163

Söderhjelm P, Ryde U (2006) Combined computational and crystallographic study of the oxidised states of [NiFe] hydrogenase. J Mol Struct (THEOCHEM) 770:199–219

Thauer RK, Kaster AK, Goenrich M, Schick M, Hiromoto T, Shima S (2010) Hydrogenases from methanogenic archaea, nickel, a novel cofactor and H_2 storage. Annu Rev Biochem 79:507–536

Thoden JB, Holden HM, Wesenberg G, Raushel FM, Rayment I (1997) Structure of carbamoyl phosphate synthetase: a journey of 96 Å from substrate to product. Biochemistry 36:6305–6316

Tremblay PL, Lovley DR (2012) Role of the NiFe hydrogenase Hya in oxidative stress defense in *Geobacter sulfurreducens*. J Bacteriol 194:2248–2253

Vignais MV, Billoud B (2007) Occurrence, classification and biological function of hydrogenases: an overview. Chem Rev 107:4206–4272

Vincent KA, Cracknell JA, Lenz O, Zebger I, Friedrich B, Armstrong FA (2005a) Electrocatalytic hydrogen oxidation by an enzyme at high carbon monoxide or oxygen levels. Proc Natl Acad Sci U S A 102:16951–16954

Vincent KA, Parkin A, Lenz O, Albracht SP, Fontecilla-Camps JC, Cammack R, Friedrich B, Armstrong FA (2005b) Electrochemical definitions of O_2 sensitivity and oxidative inactivation in hydrogenases. J Am Chem Soc 127:18179–18189

Vincent KA, Cracknell JA, Clark JR, Ludwig M, Lenz O, Friedrich B, Armstrong FA (2006) Electricity from low-level H_2 in still air – an ultimate test for an oxygen tolerant hydrogenase. Chem Commun 48:5033–5035

Vincent KA, Parkin A, Armstrong FA (2007) Investigating and exploiting the electrocatalytic properties of hydrogenases. Chem Rev 107:4366–4413

Volbeda A, Charon MH, Piras C, Hatchikian EC, Frey M, Fontecilla-Camps JC (1995) Crystal structure of the nickel-iron hydrogenase from *Desulfovibrio gigas*. Nature 373:580–587

Volbeda A, Montet Y, Vernède X, Hatchikian EC, Fontecilla-Camps JC (2002) High resolution crystallographic analysis of *Desulfovibrio fructosovorans* [NiFe] hydrogenase. Int J Hydrog Energy 27:1449–1461

Volbeda A, Martin L, Cavazza C, Matho M, Faber BW, Roseboom W, Albracht SPJ, Garcin E, Rousset M, Fontecilla-Camps JC (2005) Structural differences between the ready and unready oxidized states of [NiFe]-hydrogenases. J Biol Inorg Chem 10:239–249

Volbeda A, Amara P, Darnault C, Mouesca J-M, Parkin A, Roessler MM, Armstrong FA, Fontecilla-Camps JC (2012) X-ray crystallographic and computational studies of the O_2-tolerant [NiFe]-hydrogenase 1 from *Escherichia coli*. Proc Natl Acad Sci U S A 109:5305–5310

Watanabe S, Matsumi R, Arai T, Atomi H, Imanaka T, Miki K (2007) Crystal structures of HypC, HypD and HypE: insights into cyanation reaction by thiol redox signaling. Mol Cell 27:29–40

Watanabe S, Arai T, Matsumi R, Atomi H, Imanaka T, Miki K (2009) Crystal structure of HypA, a nickel-binding metallochaperone for [NiFe] hydrogenase maturation. J Mol Biol 394:448–459

Xia W, Li H, Sze K-H, Sun H (2009) Structure of a nickel chaperone, HypA, from *Helicobacter pylori* reveals two distinct metal binding sites. J Am Chem Soc 131:10031–10040

Chapter 3

Engineering Hydrogenases for H_2 Production: Bolts and Goals

Marc Rousset*
CNRS, BIP, 31 chemin Joseph Aiguier, 13402 Marseille Cedex 20, France

and

Pierre-Pol Liebgott
IRD, Aix-Marseille Université, CNRS/INSU, MIO, UM 110, 13288 Marseille Cedex 09, France

Summary

Hydrogenases are efficient biological catalysts of H_2 oxidation and production. Most of them are inhibited by oxygen, and a prerequisite for their use in biotechnological applications under air is to improve their oxygen tolerance. A few bacteria, however, contain hydrogenases that activate H_2 even in the presence of O_2. Intriguingly, molecular, kinetic and spectroscopic

*Author for correspondence, e-mail: rousset@imm.cnrs.fr

D. Zannoni and R. De Philippis (eds.), Microbial BioEnergy: Hydrogen Production, Advances in Photosynthesis and Respiration 38, pp. 43–77, DOI 10.1007/978-94-017-8554-9_3,

studies lead to assume that different mechanisms might be responsible for the resistance, depending on the enzyme type. In order to better understand the molecular bases of resistance to O_2 inhibition, this chapter focuses on the hydrogenases and their reaction with O_2 and examines the different strategies to lead to engineer kinetically efficient hydrogenases operating under aerobic conditions.

I. Introduction

Climate change, along with the rapid depletion of oil and gas reserves, prompt the world to turn to a search for a clean energy sources to provide the energy necessary for present and projected human activities. A variety of possible fuel sources are being examined at present. Among these, dihydrogen (H_2) has been identified as a clean and renewable energy carrier and is found to be one potential alternative to fossil fuel energy and has drawn a worldwide attention as a future energy source (Mason 2007). Interestingly, the pioneering notion of a "Dihydrogen Energy System" drew inspiration from the French science-fiction novel "The mysterious Island" by Jules Vernes (1874), where the idea of using H_2 as an energy carrier first appeared. However, over 90 % of the production of H_2 remains based upon steam reforming of hydrocarbons and coal gasification, which starts from fossil fuels and requires high temperature and pressure conditions. So, a sustainable, renewable supply of H_2 to power this economy is required. Alternative methods of H_2 generation include electrolysis of water and biological means. The development of new biotechnological processes, designed to meet the future energy demand, may take advantage of microbes that have been using H_2 from very early in the evolution of life (Perez-Arellano et al. 1998; Andersson and Kurland 1999). Many organisms including some *Bacteria*, *Archaea* and unicellular eukaryotes have an active H_2 metabolism, utilizing the cleavage of H_2 to gain energy, or H_2 production to release reducing power (Casalot and Rousset 2001). It has been estimated that these microorganisms produce or consume more than 200 million tons of H_2 per year (Richardson and Stewart 1990). These processes, carried out by hydrogenase occur via the inter-conversions between the molecular hydrogen and two protons plus two electrons ($H_2 \leftrightarrow 2H^+ + 2e^-$).

In this sense, an outlook in the production of hydrogen from water and light energy would be to use photosynthetic microorganisms, such as cyanobacteria and green algae (Antal et al. 2011; Carrieri et al. 2011). Green algae have many areas of potential improvements that often overlap with those of cyanobacteria and are currently recognized as better photobiological hydrogen producers from a demonstrated solar efficiency standpoint (Ghirardi et al. 2009). In principle, there is absorption of light in the form of photons by the photosystem which are going generate a strong oxidant that can oxidize water into protons, electrons/reducing equivalents and O_2. Thereafter, the electrons reduce protons to form H_2, carried out by hydrogenases (Fig. 3.1). In this sense, photosynthetic production of H_2 using water as a source of electrons and sunlight as the source of energy, driving proton reduction, is the most desirable process. Both water and light are available to an almost unlimited

---

Abbreviations: DFT – Density function theory; ENDOR – Electron nuclear double resonance; EPR – Electron paramagnetic resonance; EXAFS – Extended X-Ray absorption fine structure; Fd – Ferredoxin; FHL – Formate hydrogen lyase; FNR – Ferredoxin NADPH reductase; FTIR – Fourier Transform InfraRed spectroscopy; Hmd – H_2-forming methylenetetrahydromethanopterin dehydrogenase; MBH – Membrane-bound hydrogenase; PFV – Protein film voltammetry; RH – Regulatory hydrogenase; SH – Soluble hydrogenase; SHE – Standard hydrogen electrode; WT – Wild type;

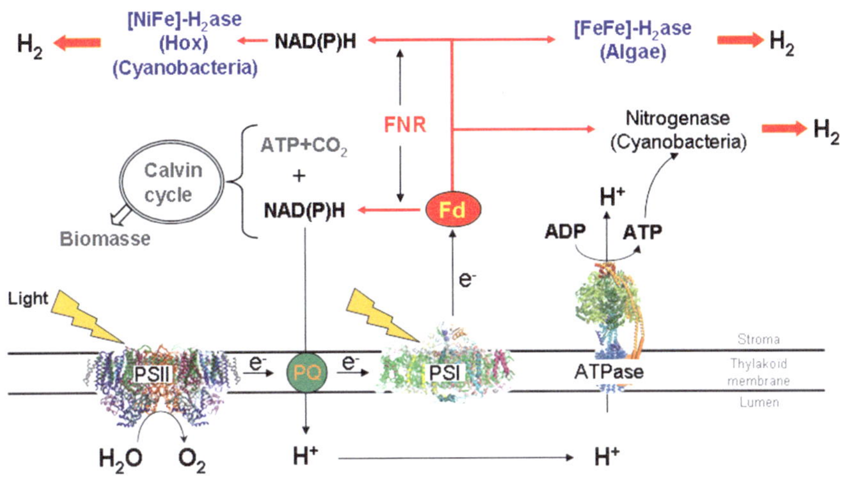

Fig. 3.1. H_2 production by cyanobacteria and algae. Water (*bold*) is oxidized to O_2 by photosystem II (*PSII*) and electrons are transferred to photosystem I (*PSI*) via the plastoquinone pool (*PQ*). Photosystem I transfers electrons to ferredoxin (*Fd*) which can donate electrons to [FeFe]-hydrogenase (H_2ase) in algae or to nitrogenase (N_2ase) in some cyanobacteria. Fd electrons can also be transferred to $NAD(P)^+$ by a ferredoxins NAD reductase (*FNR*). NAD(P)H can donate electrons to [NiFe]-hydrogenase in cyanobacteria.

extent and in addition to H_2 only O_2 is formed, whereas greenhouse gases are avoided. Moreover, photosynthetic production of H_2 is potentially very efficient in terms of energy conservation, since 10 % of the incident light energy can theoretically be recovered into H_2 (Prince and Kheshgi 2005). For example, an average sun-light flux of 46 Mwh/ha/day can be converted, with a 10 % yield, in 1,650 m^3 of H_2 per hectare per day, which represents 145 TOE (ton of oil equivalent) per hectare per year.

In addition to being used to the in vivo H_2 production, hydrogenases are known to be used in a variety of biotechnological applications including biofuel cells, biosensors, prevention against microbial-induced corrosion, and the generation and regeneration of NAD(P) cofactors. In the case of biofuel cell, the nature of the hydrogenases has inspired researchers worldwide to use them as biocatalysts predominantly to replace platinum electrode in hydrogen fuel cells (Fig. 3.2). Indeed, platinum is limited in availability and very expensive and therefore the use of hydrogenases would be good candidates to replace this precious metal in fuel cells. Moreover the hydrogenase-coated electrode confers greater fuel specificity and turnover rates than the platinum and it could be used to as an alternative to allow operation of biofuel cells at neutral pH and ambient temperatures, which are the conditions much more favorable for the handling of fuel cells (Ikeda and Kano 2001; Morozov et al. 2002).

Most applications of H_2-oxidizing and H_2-producing catalysts require them to function in air. For example, an O_2-stable H_2-production catalyst is an essential component for the photobiological production of H_2, a device that uses solar energy to split water into H_2 and O_2. Similarly, use of hydrogenases in biofuel cells requires them to remain active in the presence of O_2 since biofuel cells must work necessarily with O_2 (Figs. 3.1 and 3.2). However, the major barrier of developing an economically viable systems, is the oxygen sensitivity of the vast majority of hydrogenases because their active site react strongly with O_2 (Cournac et al. 2004; Léger et al. 2004; Oh et al. 2011).

At present, structure function relationship studies in hydrogenases have mainly remained in the basic research realm, aimed

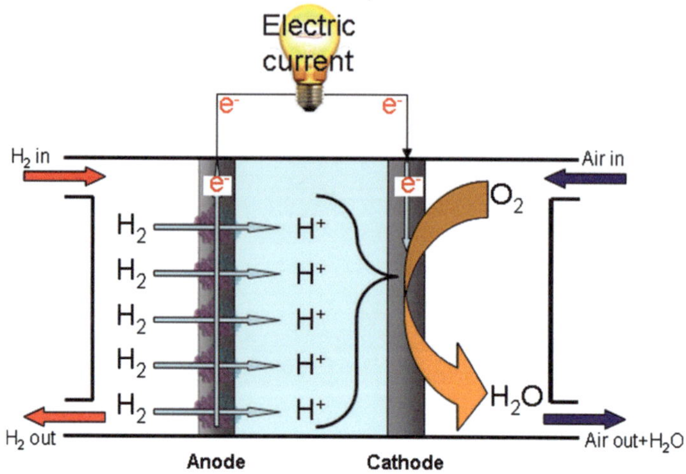

Fig. 3.2. Working principle of biofuel cells using [NiFe]-hydrogenases at the anode.

at understanding the enzyme catalytic mechanism (De Lacey et al. 2005). In order to obtain a sufficient level of enzyme efficiency and robustness for technological purposes, hydrogenases must be functionally optimized by improving efficiency H_2-oxidation or production while improving O_2 tolerance. These challenges can be approached through genetic engineering by two different strategies: at the cellular level, by metabolic engineering, it is possible to create favorable conditions to improve the H_2-production by deletion of hydrogen-uptake system (Liang et al. 2009), in inserting of non native hydrogenases more efficient and tolerant to O_2 (Wells et al. 2011) or to simply avoid O_2 exposure (Kruse et al. 2005; Rupprecht et al. 2006; Henstra et al. 2007). At the enzyme level, by protein engineering, the goal will be to improve hydrogenases so that they can outcompete other enzymes for substrate utilization, use of a thermodynamically more favorable substrate or become more O_2-tolerant. In this chapter, we will review hydrogenase structure-function relationship studies in which new properties of modified enzymes might serve as an inspiration source for rational optimization of hydrogenases for biotechnological processes.

In the first part, we will describe the different hydrogenases, and then we focus on these which are naturally O_2-tolerants. Thereafter, we will process the engineering approaches in three parts: (1) Improving H_2 production by heterologous expression of non-native hydrogenases or their over-expression. (2) Improving the electron transfer by increasing the substrate specificity or redirect redox intermediates. (3) Engineering protein to improve the tolerance of the hydrogenases to O_2.

II. Classification and Physiological Properties of Hydrogenases

Hydrogenases are metalloproteins which are involved in the metabolic machinery of a wide variety of microorganisms by catalyzing the reversible heterolytic splitting of dihydrogen according to the elementary reaction: $H_2 <==> H^- + H^+ <==> 2H^+ + 2e^-$.

Since 1931, when hydrogenases were described by Stephenson and Stickland, extensive research has been conducted in this area. Biochemical, spectroscopic, phylogenetic studies made possible to separate three groups of hydrogenases on the basis of the metal content of their active site: [Fe]-, [FeFe]- and [NiFe]-hydrogenase (Vignais et al. 2001; Cournac et al. 2004).

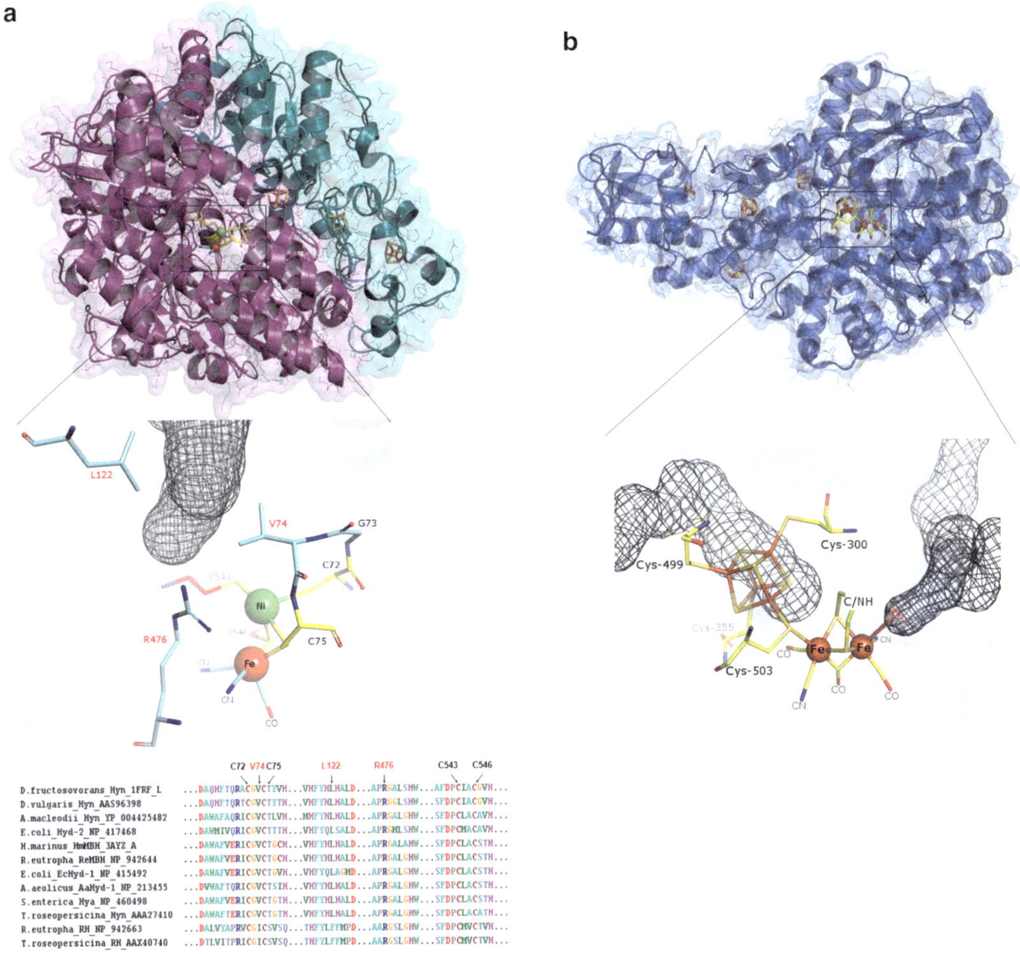

Fig. 3.3. Structure of [NiFe]-hydrogenase, [FeFe]-hydrogenase and [Fe]-hydrogenase. Panel (**a**) shows a structural model of the periplamsic [NiFe]-hydrogenase from *D. fructosovorans* (1YQW). The active site is buried in the large subunit (*purple*) whereas the small subunit contains three consecutive iron-sulfur clusters (*light blue*). Panel (**b**) shows the structure of [FeFe]-hydrogenase of *C. pasteurianum* (CpI: 3C8Y). All cofactors (active site and iron-sulfur) are localized in the same subunit (*blue*). The PyMOL Molecular Graphics System was used for visualization. The *lower part* of the figure shows the catalytic centers of three hydrogenases. The extremity of hydrophobic channels for gas diffusion were computed using Caver2.0 and appear as gray meshes for the panels **a** and **b**. Panel (**a**) residues, component the entrance gate of gases, are indicated in *red* and the multiple sequence alignment (ClustalW) of some [NiFe]-hydrogenase large subunits, is presented.

A. Generalities

The three classes of hydrogenases are evolutionarily unrelated but share similar nonprotein ligand assemblies at their active site that are not observed elsewhere in biology. They all contain a complex active-site cofactor that consists of at least one Fe atom coordinated by varying numbers of cysteine-S ligands and biologically unique carbon monoxide (CO) and in most cases additional cyanides (CN) ligands (Fig. 3.3). Therefore, this structure $Fe(CO)_2$ or Fe(CO-CN) likely represents the minimal cofactor making hydrogenase activity possible. These metal cofactors are synthesized in a coordinated post-translational process that involves up to nine hydrogenase-specific auxiliary proteins

(Böck et al. 2006; Lenz et al. 2010; Mulder et al. 2011). In the bimetallic hydrogenases the active-site cofactor is electronically coupled to FeS clusters that direct the electrons from the active site to the protein surface or from an external electron donor to the active site. Moreover, analysis of crystals structure from all three classes of hydrogenases revealed a network of hydrophobic cavities and channels (Fig. 3.3), or packing defects that form pathways connecting the active site to the surface of the enzyme (Volbeda et al. 1995; Montet et al. 1997; Nicolet et al. 1999, 2002; Fontecilla-Camps et al. 2007; Hiromoto et al. 2009; Mulder et al. 2011; Hong and Pachter 2012; Nicolet and Fontecilla-Camps 2012). Molecular dynamics simulations and xenon mapping of [NiFe]-hydrogenases show that these channels facilitate the diffusion of H_2 between the bulk of solvent and the active site (Montet et al. 1997; Cohen et al. 2005; Fontecilla-Camps et al. 2007; Leroux et al. 2008; Liebgott et al. 2010; Topin et al. 2012). In addition, the pathways can facilitate diffusion of small gas molecules such as CO and O_2 to access the active site, which in the case of CO leads to a reversible inhibition and in the case of O_2, to complete but non destructive inhibition.

The redox chemistry of hydrogenases is rich and involves many intermediate states, as beside catalyzing H_2 oxidation/production, they can also interact with gaseous molecules (i.e. CO, O_2) and become inhibited. It is thus important to gain an understanding into how hydrogenases catalyze H_2 production and oxidation, to determine the mechanisms by which they are inactivated under oxidizing conditions and how they may become re-activated. A combination of spectroscopic and electrochemical methods has provided structural knowledge on the oxidised and reduced forms of the enzyme, namely electron paramagnetic resonance (EPR) spectroscopy, Fourier transform infrared (FTIR) spectroscopy and protein film voltammetry (PFV).

EPR is a spectroscopic technique widely used to study the hydrogenase. This technique allows the detection of chemical species that have unpaired electrons designated as paramagnetic. A large number of molecules contain such paramagnetic atoms such as nickel at the active site, or the iron atoms of the iron-sulfur centers in [NiFe]-hydrogenases. Upon reduction or oxidation, the metal atoms of the prosthetic groups go through several redox states and in some of these states, they become paramagnetic and can unambiguously be identified by their corresponding EPR spectrum. Magnetic coupling between these atoms also provide information of the active or inactive states.

Another powerful method for monitoring reactions at the active sites of hydrogenases is infrared spectroscopy. This is an unusual technique to use for studying enzymes, but for hydrogenases it exploits the fact that CN^- and (particularly) CO are strong infrared-active vibrational oscillators and their stretching frequencies appear in a spectral window where the rest of the protein and water do not absorb.

Beside spectroscopic studies that provide thermodynamic information at equilibrium states, kinetic parameters have equally been studied using protein film voltametry (PFV), where the hydrogenase is adsorbed to an electrode and its activity directly measured by electron transfer through the electrode under oxidizing or reducing potentials during gas exposure (Inhibitors: CO, O_2 or substrate: H_2) (Vincent et al. 2007; Armstrong et al. 2009).

B. Classification

1. [Fe]-Hydrogenase (EC 1.12.98.2)

This type of enzyme was found only in a small group of methanogenic Archaea and has been described for the first time in *Methanothermobacter thermoautotrophicum* (Zirngibl et al. 1992). This enzyme catalyzes CO_2 reduction to methane using H_2 (Vignais and Billoud 2007). Based on the metal content of their active site, although in the past they were considered as metal-free hydrogenases, they have been recently designated as "iron-sulfur-cluster free hydrogenase" or simply [Fe]-dihydrogenases (Armstrong and Albracht 2005). This enzyme is also known as H_2-forming Methylenetetrahydromethanopterin Dehydrogenase (Hmd) (Corr and Murphy 2011). The structure of the active site and functional models have been reported in 2008 (Shima 2008; Hiromoto et al. 2009). This hydrogenase differs from the others hydrogenases not only by the primary and tertiary structures but also by the fact that the iron, required for enzyme activity is not redox active. Furthermore, the hydrogenase activity is rapidly lost under aerobic conditions and in presence of light (Lyon et al. 2004) which would make the isolation and the characterization very difficult.

2. [FeFe]-Hydrogenase (EC 1.12.7.2)

[FeFe]-hydrogenase have been found mainly in Gram positives and in eukaryotes, as well as in few anaerobic Gram negative (Atta and Meyer 2000; Horner et al. 2000, 2002). It should be emphasized that these are the only type of hydrogenase that is found in eukaryotes but that is absent in the *Archaea* domain (Cournac et al. 2004). In this sense, the hydrogenases present in eukaryotic microorganisms (Green algae) are [FeFe]-hydrogenases only. Generally, [FeFe]-hydrogenases are usually involved in H_2 production but they were also reported to function as an uptake hydrogenase. Indeed, the location of hydrogenases in the bacterial cell reflects the enzyme's function (Nicolet et al. 2000). For instance, the periplasmic *Desulfovibrio desulfuricans* [FeFe]-hydrogenase (DdH) is involved in dihydrogen uptake. Protons resulting from this dihydrogen oxidation create a gradient across the membrane that is thought to be coupled to ATP synthesis in the cytoplasm. *Clostridium pasteurianum* [FeFe]-hydrogenase I (CpI) is a cytoplasmic enzyme that accepts electrons from ferredoxin and generates dihydrogen with protons as electron acceptors.

The only [FeFe]-hydrogenase structures from anaerobic soil bacterium *Cl. pasteurianum* (CpI) (Peters et al. 1998) (pdb code 1FEH and 3C8Y) and sulfate reducing bacterium *D. desulfuricans* (DdH) (Nicolet et al. 1999) (pdb code 1HFE) revealed an unique active site metal cluster, termed as the H-cluster, where catalysis takes place. The H-cluster is composed of a binuclear $[2Fe]_H$ center bound to a $[4Fe\text{-}4S]_H$ subcluster by a bridging cystein, the [4Fe-4S] center is attached to the protein by four cystein ligands. The $[2Fe]_H$ center is coordinated by five diatomic CN^- and CO ligands, as well as a non protein dithiomethylamine ligand (Fig. 3.3b). Molecular masses of [FeFe]-hydrogenases can vary from 45 to 130 kDa according to the number of subunits. [FeFe]-hydrogenases are mainly monomeric and contains only one catalytic subunit, but they often comprise additional domains, which accommodate FeS clusters. For example in the [FeFe]-hydrogenase I from *Cl. pasteurianum* (CpI), three accessory [4Fe4S] clusters and one [2Fe2S] cluster are believed to transfer electrons between the electron donor or acceptor at the protein surface and the active site at the center of the protein (Nicolet et al. 2002; Nicolet and Fontecilla-Camps 2012). Contrarily, the simplest characterized [FeFe]-hydrogenases are observed in the green algae, including *Chlamydomonas reinhardtii*, *Chlorella fusca*, and *Scenedesmus obliquus*, which express enzymes consisting of only the H-cluster without FeS-cluster domains (Florin et al. 2001; Horner et al. 2002; Forestier et al. 2003). These proteins have been exploited more recently for biochemical and spectroscopic characterization because they lack the additional FeS clusters observed in most native [FeFe]-hydrogenases that may complicate the direct examination of the H-cluster (Kamp et al. 2008; Silakov et al. 2009; Stripp et al. 2009; Mulder et al. 2011).

Even though [FeFe]-hydrogenases may appear as the best suited for hydrogen production purposes, enzyme engineering studies for these enzymes are still poorly developed because of their great sensitivity to oxidative damage, which makes any biochemical characterization very uncertain.

a. Reaction with O_2 and CO

In most [FeFe]-hydrogenases, oxygen inhibits the enzyme, possibly by binding to the open coordination site on the distal Fe of the $[2Fe]_H$ center, then form a reactive oxygen species that destroys the $[4Fe4S]_H$ subcluster. This distal Fe would be equally the hydrogen binding site and also the site of reversible CO binding and inhibition (Stripp et al. 2009). The chemical nature of the oxygen species bound to the H-cluster after the exposure to O_2 is not known, but density function theory (DFT) calculations on inactivated states of the H-cluster have proposed a Fe^{II}-Fe^{II} oxidation state for the $[2Fe]_H$ center, with a possible OH group terminally bound to the distal Fe (Liu and Hu 2002).

Oxygen inactivation has been studied using protein film electrochemistry and it was shown that the rate of inhibition of [FeFe]-hydrogenases would be limited by two steps: (1) the diffusion of oxygen through the protein to the active site pocket, and (2) the binding of oxygen to the $[2Fe]_H$ subcluster (Armstrong et al. 2009).

Interestingly, the level of O_2 inhibition varies among [FeFe]-hydrogenases, with I_{50} values ranging from less than a few seconds for *Chlamydomonas reinhardtii* enzymes, to several minutes for the clostridial enzymes (Böck et al. 2006; Baffert et al. 2008). Algal hydrogenases, which lack the additional accessory cluster domain found in bacterial enzymes, are typically more sensitive to O_2 inhibition than are the enzymes isolated from bacteria. However, it is clear that this elevated sensitivity is not solely due to the lack of the N-terminal accessory cluster (Böck et al. 2006; Stripp et al. 2009).

3. [NiFe]-Hydrogenase (EC 1.12.2.1)

The most numerous and best studied class of hydrogenases have been the [NiFe]-hydrogenases. This type of enzyme was found in Bacteria and Archaea domains. The core enzyme consists of two subunits; the large subunit is approximately 60 kDa and houses the Ni-Fe-active site, whereas the small subunit, of approximately 35 kDa, which can be of variable size and harbors typically three iron-sulfur clusters (Fig. 3.3a): a distal [4Fe-4S] cluster at the surface of the protein and furthest from the active site; a medial [3Fe-4S] cluster; and a proximal Fe-S cluster, with variable properties, the closest to the active site. The large and small subunit exhibit sequence homologies to subunits of NADH:ubiquinone oxidoreductase (Complex I) (Volbeda et al. 2012). In certain enzymes, additional subunits enable the interaction of these clusters with physiological electron carriers such as quinones, pyridine nucleotides (NAD(P)H), ferredoxins, and cytochromes (Cournac et al. 2004). Crystal structure analysis of heterodimeric [NiFe]-hydrogenases from *Desulfovibrio* species (Volbeda et al. 1995, 2002; Higuchi et al. 1997; Matias et al. 2001) (Fig. 3.3a), and photosynthetic bacterium *Allochromatium vinosum* (Ogata et al. 2010), revealed that the Ni-Fe cofactor is deeply buried in the large subunit. The Ni is coordinated to the protein via four thiol groups from conserved cysteine residues; two of these are bridging ligands that coordinate both Fe and Ni (Volbeda et al. 1995, 2002; Higuchi et al. 1997; Matias et al. 2001). Fourier Transform Infrared spectroscopy revealed that the Fe coordination sphere also possesses three diatomic ligands: one CO and two CN molecules (Volbeda et al. 1996; Pierik et al. 1999). The sixth iron coordination position is assumed to be occupied by a bridging hydride between iron and nickel (Pardo et al. 2006; De Lacey et al. 2007). Hydrophobic cavities which channel the gas substrate between the protein surface and the active site (Fig. 3.3a) (Montet et al. 1997; Volbeda et al. 2002; Teixeira et al. 2006) as well as a proton-conducting channel (Léger et al. 2004) were identified inside the hydrogenase.

a. Classification

Based on their primary protein sequences the [NiFe]-hydrogenases have been categorized into four different groups (Table 3.1): Group

Table 3.1. Classification of [NiFe]-hydrogenases and physiological activities (Modified from Vignais and Billoud 2007).

Group	Function	Physiological activity	Microorganisms
I	Membrane-bound H_2 uptake hydrogenases	Energy conservation	*D. fructosovorans, A. aeolicus, R. eutropha* H16, *E. coli*
IIa	"Cyanobacterial" uptake hydrogenases	Energy conservation	*Anabaena variabilis, Nostoc* sp. PCC7120, *A. aeolicus*
IIb	Regulatory hydrogenase	H_2-sensing components in genetic regulation of hydrogenase expression	*R. eutropha, Thiocapsa roseopersicina*
IIIa	F420-reducing hydrogenases	Energy conservation	Methanogens only (*Methanothermobacter marburgensis*)
IIIb	NADP-reducing	Energy conservation/ Fermentation	*Thiobacillus denitrificans, Pyrococcus furiosus, Thermococcus kodakarensis*
IIIc	F420-non reducing	Energy conservation	*Geobacter sulfureducens, Geobacter metallidurans, Methanococcus voltae*
IIId	Bidirectional NAD(P)-reducing	Energy conservation Redox poising	*T. roseopersicina, Synechocystis* sp. PCC6308, *R. eutropha* H16, *D. fructosovorans*
IV	Membrane bound H_2 evolving hydrogenases	Energy conserving, membrane associated H_2 evolving hydrogenase	*P. furiosus, T. kodakarensis, Methanosarcina barkeri, E. coli*

I, H_2-uptake enzymes, localized in the bacterial or archaeal cell membrane, are primarily involved in H_2 oxidation; Group II, (a) cyanobacterial uptake [NiFe]-hydrogenases, whose location is cytoplasmic and which are involved in N_2 fixation and (b) H_2 regulatory hydrogenases or hydrogenase sensors, which detect the presence of H_2 in the environment and trigger a cascade of regulation controlling the synthesis of hydrogenases; Group III, cytoplasmic bidirectional enzymes, these water soluble multi-protein complexes are dependent on NAD(P)H or $NAD(P)^+$ as cofactors; Group IV, energy conserving H_2-evolving hydrogenases, these membrane-associated hydrogenases generate H_2 from reduced ferredoxin with the concomitant generation/utilization of an ion gradient (Vignais and Billoud 2007).

b. Reaction with O_2 and CO

It can be stated, as a general rule, that hydrogenases of either type are inhibited by O_2, but individual sensitivities can vary in a wide extent. Thus, [NiFe]-hydrogenases are considered to be more robust than [FeFe]-hydrogenases because they can be totally reactivated after inhibition by O_2. In the case of [NiFe]-hydrogenases, O_2 has also been shown to oxidize directly the bimetallic active site (van der Zwaan et al. 1990) but the difference with [FeFe]-hydrogenases is that [NiFe]-hydrogenases are not damaged by O_2 as they can be reactivated by reduction. X-ray diffraction studies showed that the main structural difference between oxidized and reduced states of the active site is that oxygen species bridge the metals in the oxidized state (Garcin et al. 1999; Carepo et al. 2002; Volbeda et al. 2005). Therefore, reductions of the oxidized states are triggered by the removal of the bridging oxygen species, which allows H_2 to bind to the active site and catalytic turnover. Studies of EPR revealed that oxidized enzyme may exist under two different states which have been called Ni-A and Ni-B (Fig. 3.4). Another very important feature, that differentiates Ni-A and Ni-B is their reactivation kinetics (Cammack et al. 1986). Ni-B is called the '*ready*' state because it quickly becomes active upon reduction, while Ni-A is called the '*unready*' state because it needs a long period of incubation under reducing conditions before becoming active. This is illustrated by PFV experiments conducted on the *Allochromatium vinosum* [NiFe]-hydrogenase. After inactivation by O_2, the enzyme is reactivated by reduction at low potential under H_2 (Lamle et al. 2004). For example at -208 mV (SHE) and pH 6, the reactivation process occurs in two phases: a fast (instantaneous) phase corresponding to the reactivation of the Ni-B state and a slower phase (several hundreds of seconds) assigned to Ni-A. FTIR-spectroelectrochemical studies of different hydrogenases indicated that one-electron reduction of Ni-A and Ni-B leads to two different states: Ni-A leads to the Ni-SU state, and Ni-B leads to Ni-SI (De Lacey et al. 2007). The enzyme in the Ni-SI state is active, whereas the Ni-SU state is still inactive (De Lacey et al. 2007). The rate-limiting step of the reactivation process is the a gradual and spontaneous conversion of the Ni-SU to the active Ni-SI state (Lamle et al. 2005) (Fig. 3.4).

As regards the structures in either oxidized state, this stir up still a matter of debate. Indeed, X-ray diffraction, (Volbeda et al. 1995, 2002) EPR (van der Zwaan et al. 1990), ENDOR (Carepo et al. 2002; van Gastel et al. 2006) and EXAFS (Davidson et al. 2000) data indicate that in both states an oxygen species is bridging nickel and iron atoms (De Lacey et al. 2007; Pandelia et al. 2010b). Most of the studies agree with the presence of a hydroxide in the Ni-B state (Black et al. 1994; Davidson et al. 2000; Stein et al. 2001; Stadler et al. 2002; Volbeda et al. 2005) whereas the nature of the oxygen species in Ni-A is more controversial. Indeed, either oxo (Carepo et al. 2002), hydroxo (Davidson et al. 2000; Stein et al. 2001; Stadler et al. 2002; Pandelia et al. 2010b) or peroxo (Lamle et al. 2005; Volbeda et al. 2005) species have been proposed. However, the "peroxo" hypothesis was not ruled out in recent DFT and spectroscopic studies (van Gastel et al. 2008; Pandelia et al. 2010b) but rather a hydroxo species. Moreover, an ENDOR study of a sample of [NiFe]-hydrogenase aerobically oxidized in $H_2{}^{17}O$, demonstrated that the bridging [NiFe]- ligand in Ni-A originates from the solvent water (Carepo et al. 2002). Thus, this will mean that the debate continues on the question of the Ni-A structure with the aim to understand differences between Ni-A and Ni-B states (Fig. 3.4).

Most [NiFe]-hydrogenases are inhibited by CO in a competitive manner (Teixeira et al. 1987; Léger et al. 2004). It reaches the active site using the same gas channel as H_2 and O_2 (Fig. 3.3a) (Liebgott et al. 2010). It binds weakly to the Ni ion at the active site (Albracht 1994; Stadler et al. 2002; De Lacey et al. 2007; Lubitz et al. 2007), only after reductive activation of the enzyme to the Ni-SI forms, presumably when the bridging oxygen species had been removed from the active site. No binding of CO occurs when the enzyme is in the inactive states Ni-A, Ni-B, and Ni-SU. In addition, CO-inhibition blocks electron and proton transfer at the active site, although reduction at the proximal [4Fe-4S] cluster is detected (Stadler et al. 2002). The kinetics of CO inhibition has been studied by PFV experiments and the kinetics of CO binding was fast, about 10^8 s^{-1}/M. Thus, it has been shown that diffusion was the rate-limiting step of CO-inhibition. As a result, this inhibitor has been used to probe gas diffusion in hydrogenases (Leroux et al. 2008; Liebgott et al. 2010).

c. Specific Characteristics of the Naturally Occurring O_2-Tolerant [NiFe]-Hydrogenases

[NiFe] hydrogenases are considered to be more robust than [FeFe] hydrogenases because they can be totally reactivated after inhibition by O_2. Moreover, there are even a few examples in nature of relatively O_2-tolerant [NiFe] hydrogenases. Thus, the [NiFe] hydrogenase can be classified as either 'standard O_2-sensitive' or 'O_2-tolerant' based on their ability to function in the presence of O_2. O_2-tolerance defines one hydrogenase that retains some activity in the presence of O_2. The level of residual activity can vary depending on the enzyme, but it

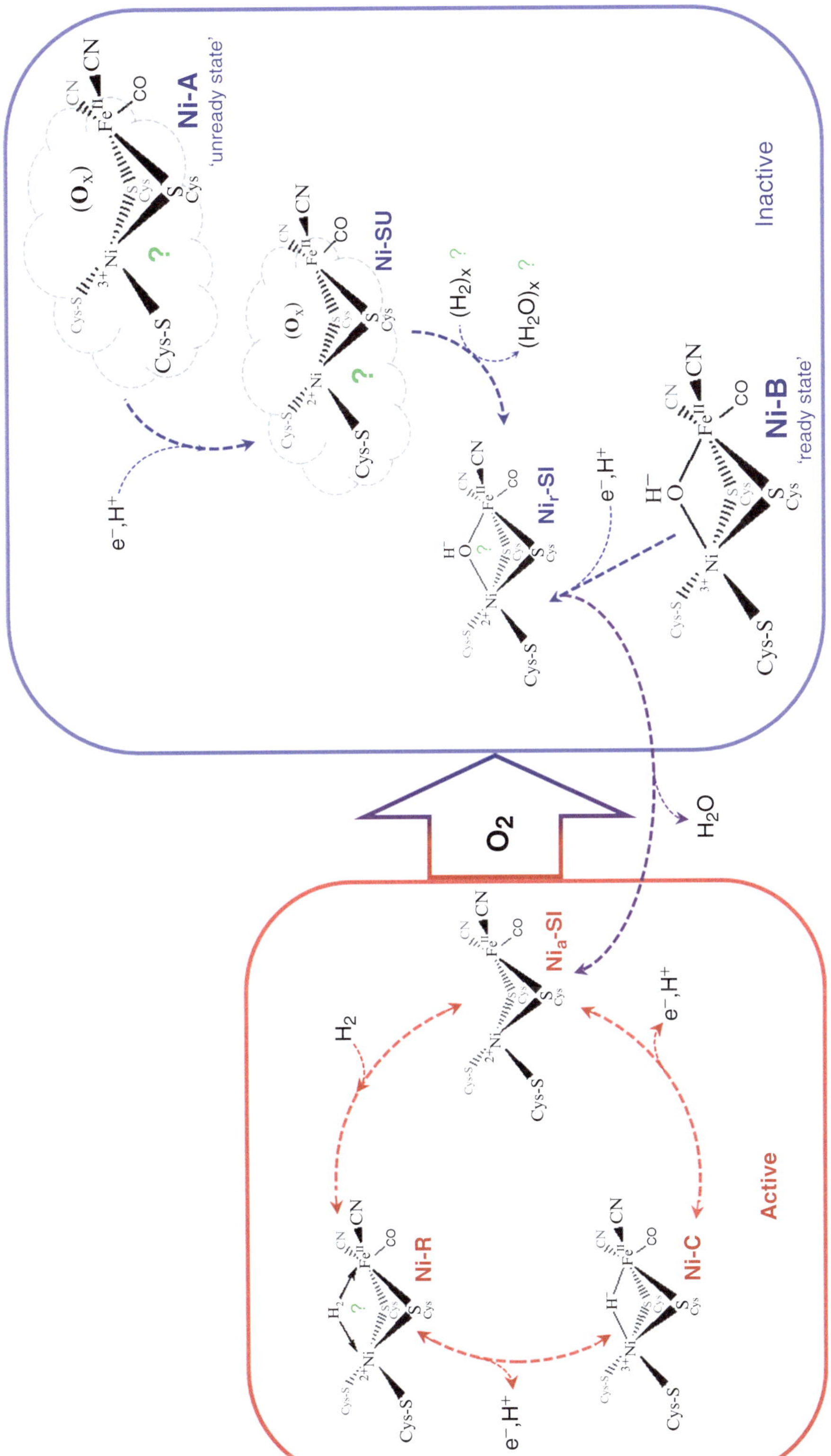

Fig. 3.4. Proposed reaction scheme for the standard O_2 sensitive [NiFe]-hydrogenase from *D. fructosovorans* based on electrochemical titrations, including both active and inactive intermediates. The nomenclature is based on that initially adopted to designate the EPR paramagnetic states (Ni-A, NiB, NiC) and the "Ni-S" designates an EPR silent state. In Ni-A and Ni-SU states, the binding oxygenic species presented by (O_x) has not been conclusively determined. There are multiple forms of Ni-C and Ni-R and these have been suggested to differ in terms of the redox state of the nearest (proximal) iron–sulfur cluster and/or protonation of the cysteinyl sulfur atoms.

should be remembered that trace amounts (a few μM) of O_2 readily inhibit standard O_2-sensitive [NiFe] hydrogenases.

The fact that a large number of microbes are able to use H_2 as the sole energy source in the course of aerobic respiration indicates that some [NiFe] hydrogenases afford a specific protection against detrimental effects of O_2 (Tremblay and Lovley 2012). This aerobic H_2 oxidation occurs in phylogenetically diverse groups of prokaryotes such as the nitrogen-fixing bacterium *Bradyrhizobium japonicum* (Kaneko et al. 2002), the photosynthetic proteobacterium *Rhodobacter capsulatus* (Strnad et al. 2010), the hyperthermophilic bacterium *Aquifex aeolicus* (Deckert et al. 1998) and the well-studied proteobacterium *Ralstonia eutropha* H16 that contains even three indigenous O_2-tolerant hydrogenases (Schwartz et al. 2003; Pohlmann et al. 2006; Lenz et al. 2010; Bürstel et al. 2011).

R. eutropha H16 is a bacterium that lives in soil and water and it is one of organisms able to grow chemolithoautotrophically using hydrogen as the sole energy source and dioxygen as terminal electron acceptor. This is called the Knallgas (detonating gas) reaction. The three distinct O_2-tolerant [NiFe] hydrogenases, that each serve unique physiological roles, are: a bidirectional cytoplasmic Soluble Hydrogenase (*Re*SH) able to generate reducing equivalents by reducing NAD^+ at the expense of hydrogen (Group III), a Regulatory Hydrogenase (*Re*RH) which acts in a signal transduction cascade to control transcription of hydrogenase genes (Group II) and a Membrane-Bound Hydrogenase (*Re*MBH) coupled to the respiratory chain (Group I) (Burgdorf et al. 2005). Interestingly, these O_2-tolerant hydrogenases are usually less active than standard O_2-sensitive enzymes. Their H_2 oxidation activities are reduced by a factor of about 5 for the *Re*SH (Ludwig et al. 2009a), about 50 for the *Re*MBH and about 500 for the *Re*RH (Vignais and Billoud 2007). The H_2 production activities are usually considerably weak, especially because of the strong inhibitory effect of H_2 (Goldet et al. 2008). However, these O_2-tolerant enzymes represent precious inspiration sources for the study of the molecular bases of O_2 inhibition.

Three strategies seem to have been developed by *R. eutropha* H16 to allow its [NiFe] hydrogenases to be catalytically active in the presence of dioxygen.

In the case of the SH (Group III), the O_2 resistance was assumed to be due to the presence of extra CN ligands at the active site (Happe et al. 2000) that might be incorporated by a specific maturation protein HypX (Bleijlevens et al. 2004). Deletion of *hypX* led to a lower O_2 resistance of the *Re*SH enzyme (Bleijlevens et al. 2004) while the activities of *Re*MBH (Buhrke and Friedrich 1998) and of the regulatory hydrogenase (*Re*RH) remained unaffected (Buhrke et al. 2001). However, the presence of a Ni-Bound cyanide under native conditions has been recently ruled out (Horch et al. 2010). In this context, the supply of low-potential electrons from the oxidation of NAD(P)H appears to play a major role in preserving catalytic activity under aerobic conditions in vivo. However, despite considerable efforts and promising insights (Horch et al. 2012), the structural and mechanistic basis for this property has still to be resolved. Bidirectional cytoplasmic soluble hydrogenases are of particular interest for biotechnological applications as they are suited for light driven hydrogen production in vivo (Prince and Kheshgi 2005) and the regeneration of NAD(P)H in biocatalytic processes (Okura et al. 1990; Ratzka et al. 2011). Such applications are particularly promising as some of these enzymes are oxygen tolerant in contrast to most other hydrogenases (Happe et al. 2000; Horch et al. 2012).

The *Re*RH and related enzymes adopted apparently another strategy consisting in reducing the gas channel size at the level of the interface with the active site cavity. At the end of the hydrophobic channel, near the active site, two hydrophobic residues, usually valine and leucine that are conserved in O_2-sensitive hydrogenases, are replaced by larger residues, respectively isoleucine and phenylalanine, in the O_2-tolerant hydrogen-sensors

(Volbeda et al. 2002) (Fig. 3.3a). It has therefore been suggested that increasing the bulk of residues occupying these two positions may act like a molecular sieve, reducing the channel diameter at that point, thereby preventing efficient dioxygen access to the active site. This hypothesis was supported by two experiments in which the bulky amino-acids from *Re*RH were substituted by valine and leucine. In both cases, determination of the inactivation kinetics in the presence of dioxygen revealed that the mutated enzymes were inactivated after prolonged incubation and required a reductive activation to reach the maximum activity (Burgdorf et al. 2005; Duche et al. 2005). Even though the mutated enzymes became more sensitive to dioxygen than the wild-type, it should be noted that they retain a significant level of activity after prolonged dioxygen exposure, and therefore still belong to the O_2-tolerant group of hydrogenases.

Membrane bound hydrogenases of the group I represent the best studied group of O_2-tolerant hydrogenases and has recently gained extensive attention, due to their potential biotechnological importance. Thus, the O_2 tolerance of MBH (group I) has been recently discovered through the hydrogenase crystal structures of *R. eutropha* H16 (*Re*MBH, pdb accession number 3RGW) (Fritsch et al. 2011), *E. coli* (*Ec*Hyd-1, pdb 3UQY, 3USC and 3USE) (Volbeda et al. 2012), and *Hydrogenovibrio marinus* (*Hm*MBH, pdb 3AYX, 3AYY, 3AYZ) (Shomura et al. 2011) and spectroscopic data obtained from hydrogenases of *A. aeolicus* (*Aa*Hyd-1) (Pandelia et al. 2011) and *Re*MBH (Fritsch et al. 2011; Goris et al. 2011; Lukey et al. 2011). Therefore, the recent crystallographic structures from three members of the Group I revealed that there are no significant differences between the [NiFe] catalytic centres of standard or O_2-tolerance hydrogenases. The presence of the usual nickel signatures detected by EPR indicates that the chemistry at the active site is identical to that catalyzed by standard hydrogenases.

Remarkably, there are two additional cysteine residues in the close vicinity of the proximal FeS cluster that are absent in O_2-sensitive standard [NiFe] hydrogenases which lead to the construction of a unique proximal [4Fe-3S] cluster with in-total six coordinating cysteine residues (Fig. 3.5). These two additional cysteines are fully conserved in the O_2-tolerant hydrogenase (Pandelia et al. 2010a). The sulfur atom of an extra cysteine replaces one of the inorganic sulfides and thus becomes an intrinsic cluster ligand, whereas the second cysteine terminally coordinates one of the Fe atoms (Fig. 3.5). This particular structure studied by EPR in its reduced or oxidized form in *Hm*MBH, *Aa*Hyd-1 and *Re*MBH, indicate that the [4Fe-3S] cluster is stable in three oxidation states. These redox-dependent structural changes promoted by the surplus of cysteine coordination, give the potential to the proximal cluster to theoretically deliver two electrons for the O_2-reduction instead of one (Goris et al. 2011; Lukey et al. 2011). Thus when O_2-tolerant [NiFe]-hydrogenases are attacked by O_2, they might fully reduce O_2 to water, thereby avoiding the production of reactive oxygen species that would damage or block the active site. As a result, it is currently assumed that the electron deficiency during O_2 attack might be responsible for the formation of the Ni-A inactive form ('unready' state) (Ogata et al. 2009, 2010). O_2-tolerant hydrogenases would escape the 'unready' state by forming only the 'ready' conformation (Ni-B) (in which O_2 has been fully reduced), which re-activates very easily to re-join the catalytic cycle (Armstrong et al. 2009; Lenz et al. 2010). However, in the case of *Aa*Hyd-1, a weak Ni-A signal has been reported to appear after O_2 exposure (Guiral et al. 2006), while only Ni-B was detected in a recent study (Pandelia et al. 2010b). The *Ec*Hyd-1 exhibits a Ni-A signal upon aerobic isolation but this signal is then barely detectable when the enzyme is exposed to O_2 after activation (Lukey et al. 2010).

In PFV experiments, the activity of the *Re*MBH and *Aa*Hyd-1 recover extremely

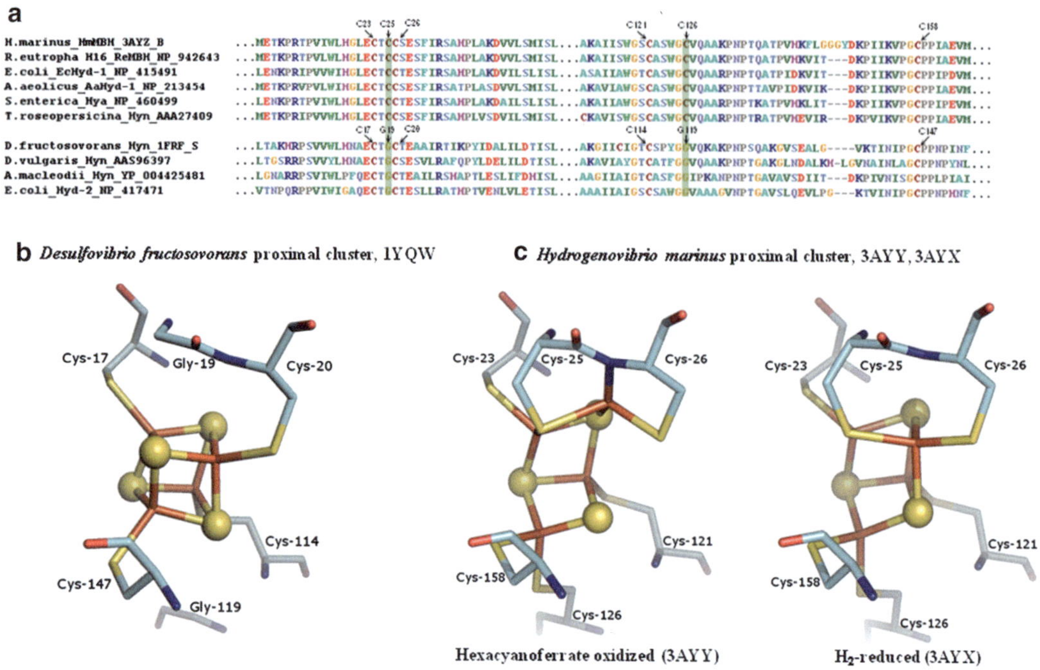

Fig. 3.5. Multiple sequence alignment and structural comparison of the proximal clusters between O_2-sensitive and O_2-tolerant hydrogenases. Panel (**a**) shows the multiple sequence alignment comparing the cysteines which ligate the proximal cluster in O_2-sensitive [NiFe]-hydrogenases and O_2-tolerant enzymes. Panels (**b**) and (**c**) shows the structural comparison between the proximal FeS clusters of *D. fructosovorans* (O_2-sensitive hydrogenase, 1YQW) and *H. marinus* (O_2-tolerant hydrogenase, 3AYY and 3AYX). Panel (**c**) Details of the structural changes associated with 'super-oxidation' of the proximal cluster in membrane bound O_2-tolerant [NiFe]-hydrogenases (Adapted from Parkin and Sargent 2012).

fast after O_2 exposure (Armstrong et al. 2009; Pandelia et al. 2010b). The O_2-tolerance properties of these enzymes are therefore likely to be due to a fast reactivation rate, as shown in recent electrochemical studies (Armstrong et al. 2009; Liebgott et al. 2010; Pandelia et al. 2010b) and by the lack of the Ni-A signal after aerobic inactivation.

The exploitation of these hydrogenases and their molecular determinants is a major challenge for a broad range of biotechnological applications. Indeed, microorganisms harboring optimized hydrogenases may play a major role in H_2 generation for fuels. Bio-fuel cells and biosensors also represent an important potential application of these hydrogenases as immobilized enzymes. At last, these hydrogenases might also allow searches in the metabolic engineering in the aim to improve the H_2 production or in protein engineering to mimic molecular determinants responsible to the tolerance towards of O_2.

III. Maturation of Hydrogenases: Specific and Complex Process

In metal-containing enzymes, complex active sites generally require specific machineries for their synthesis and assembly. Indeed, assembly of the active hydrogenase involves sophisticated biological processes, such as careful co-ordination of cofactor biosynthesis and insertion, subunit recruitment, and protein target processes (Vignais et al. 2001; Paschos et al. 2002; Böck et al. 2006).

A. [FeFe]-Hydrogenase Maturation: Protein Machinery

In order to be catalytically active after its synthesis, the [FeFe]-hydrogenase polypeptide encoded by the *hydA* gene has to incorporate the H-cluster and, when required, accessory [Fe-S] clusters. This post-translational process is extraordinarily complex as it involves a number of difficult reactions including: (i) the synthesis of CO, CN and the dithiolate bridging ligand; (ii) the assembly of the di-iron active site sub-cluster; (iii) its incorporation into the enzyme already containing the [4Fe-4S] component of the H-cluster and (iv) the assembly and transfer of the accessory FeS clusters.

The [FeFe]-hydrogenase maturation protein machinery was initially discovered in the eukaryotic green alga *Chlamydomonas reinhardtii* incapable of H_2 production (Posewitz et al. 2004). The disruption of either the *hydEF* or *hydG* (Hyd machinery) gene resulted in a mutant that proved to be unable to produce hydrogen, even though full-length hydrogenase accumulated. Genes encoding for HydE, HydF, and HydG are present in all organisms capable of synthesizing an active [FeFe]-hydrogenase (HydA) (Meyer 2007). Thus, it can be concluded that HydEF and HydG provide the minimal protein machinery necessary for the synthesis and assembly of the H-cluster. Whether other proteins are required for an optimal maturation process has not been demonstrated so far. Moreover, several reports of heterologous expression of active [FeFe]-hydrogenases have demonstrated that the Hyd machinery from one organism can be successfully used for the maturation of an enzyme from another. For example, expression of an active HydA1 enzyme from *C. reinhardtii* or *Scenedesmus obliquus* (green algae) has been shown to be possible using *Clostridium acetobutycicum*, another [FeFe]-hydrogenase synthesizing organism, as the expression host (Girbal et al. 2005). Further evidence for the lack of selectivity of the Hyd machinery came from the observation that co-expression of HydE, HydF and HydG from the bacterium *Cl. acetobutylicum* with various algal and bacterial [FeFe]-hydrogenases in *E. coli* resulted in purified enzymes with specific activities that were not very different from those of their counterparts from native sources (Böck et al. 2006). Finally, the bacterium *Shewanella oneidensis* proved to be an efficient system for the expression and maturation of HydA1 from *C. reinhardtii* (Sybirna et al. 2008).

B. [NiFe]-Hydrogenase Maturation: Protein Machinery

At present, cloning [NiFe]-hydrogenases is still very difficult and the progresses realised recently remain very limited (Burgdorf et al. 2005; Ludwig et al. 2009b). Even though [NiFe]-hydrogenase operons are highly conserved and exhibit a high degree of similarity, each maturation system is specific to the corresponding structural subunits, probably because of tight protein-protein interactions occurring during processing (Leach et al. 2007). Indeed, the complex architecture of the active site of [NiFe]-hydrogenases with their diatomic ligand (CN and CO) requires a specific and complex maturation system. There are two main groups of genes responsible for maturation, which are differentiated by their resultant phenotypes. The first group of genes is mainly located on the same transcription unit as the structural genes. Disruption of this group of genes specifically impairs the processing or activity of the hydrogenase encoded *in cis* in the operon, without affecting the maturation of other hydrogenases. The maturation processes mediated by the products of this family of accessory genes cannot be complemented *in trans* by homologous genes from the other hydrogenase operons, regardless of the degree of similarity (Sauter et al. 1992; Menon et al. 1994; Bernhard et al. 1996). This specific barrier is one of the key reasons for the failure of the active hydrogenase production in heterologous hosts. The second group is another set of the *hyp* ('p' for pleiotropic) genes which encode proteins, involved in the insertion of Ni, Fe, CO and CN into the active site (Jacobi et al. 1992; Maier et al. 1996; Wolf et al. 1998; Böck et al. 2006; Mulder et al. 2011; Petkun et al. 2011). Mutations of these genes affect the synthesis and activity of all the hydrogenase isoenzymes. However, the functions of this set of genes can be complemented *in trans* by heterologous genes (Chaudhuri and Krasna 1990).

C. O_2-Tolerant [NiFe]-Hydrogenase Maturation: Protein Machinery

It would be particularly fruitful to take advantage of the properties of the O_2-tolerant hydrogenases by cloning their corresponding genes into organisms of biotechnological interests. Although [NiFe]-hydrogenase exhibit reversible inhibition by oxygen, the sensitivity of hydrogen production in presence of O_2 is a multifaceted problem, since hydrogenase transcription, and likely maturation and assembly, might be also inhibited by exposure to atmospheric oxygen (Soboh et al. 2012). However, among O_2-tolerant hydrogenases, some are synthesized solely under aerobic conditions (Lukey et al. 2011; Tremblay and Lovley 2012). *E. coli* could be one of the most informative model systems for understanding the biosynthesis of O_2-tolerant enzymes because the bacterium produces both O_2-tolerant MBH (Hyd-1) and standard O_2-sensitive (Hyd-2) hydrogenases. Among the three types of O_2-tolerant hydrogenases, only the MBH family appears to require specific maturation proteins, necessary for the synthesis of the unique [4Fe-3S] proximal cluster (Figs. 3.5 and 3.6). Hyd-1 is produced from an operon of six genes, *hyaABCDEF*, where HyaA is the small subunit, HyaB is the large subunit, HyaC is a cytochrome that anchors the [NiFe]-hydrogenase to the membrane and the HyaD is the specific protease required for large subunit maturation-terminal processing. There are therefore two extra genes *hyaE* and *hyaF* that are not required for the assembly of standard O_2-sensitive hydrogenases (Hyd-2 and others) and are apparently only involved in the assembly of O_2-tolerant respiratory enzymes (Schubert et al. 2007).

In *R. eutropha* H16 the HyaE homolog is HoxO, which has been shown to interact with the small subunit during biosynthesis and is essential for MBH activity in that organism (Schubert et al. 2007) (Fig. 3.6). In *Rhizobium leguminosarum* the HyaE homolog, HupG, was shown to be only required for small subunit maturation under aerobic conditions (Manyani et al. 2005), and in *E. coli*, which only expresses Hyd-1 under anaerobic conditions, the *hyaE* gene was dispensable for Hyd-1 biosynthesis (Dubini and Sargent 2003). Similarly, the HyaF (HoxQ in *R. eutropha* H16 and HupH in *R. leguminosarum*) is also absolutely required for MBH activity. HyaF interacts with HyaE to form a complex together with the small subunit during assembly (Schubert et al. 2007). Genetic two-hybrid studies suggested that HyaE interacted strongly with HyaA, the small subunit, (Dubini and Sargent 2003). It is possible, therefore, that HyaE-like proteins have a role to play in assembly of the [4Fe-3S] cluster. Shomura et al. (2011) suggested that additional negative charges around the proximal [4Fe-3S] cluster in the final structure may be important for its stabilization. Interestingly, HyaE has a perfect thioredoxin-like fold (Parish et al. 2008), except that in the HyaE protein, acid residues (two aspartates and two glutamates) are found at the same position as the redox-active cysteines, originally present in true thioredoxins (Parish et al. 2008). Thus, it is possible that this negatively charged region mediates protein-protein interactions with the small subunit, though it can also be considered that it might be involved in protecting the proximal cluster until the large subunit has docked correctly with its small subunit partner.

IV. Enzyme and Metabolic Engineering to Improve H_2 Production

Genetic modifications of hydrogen metabolism or hydrogenases, can be very promising strategies to achieve an efficient H_2 production system or to improve hydrogenases as biocatalysts. During the last 5 years, many reviews dealt with this topic and summarized the scientific and technological hurdle encountered (Ghirardi et al. 2007; Germer et al. 2009; Brentner et al.

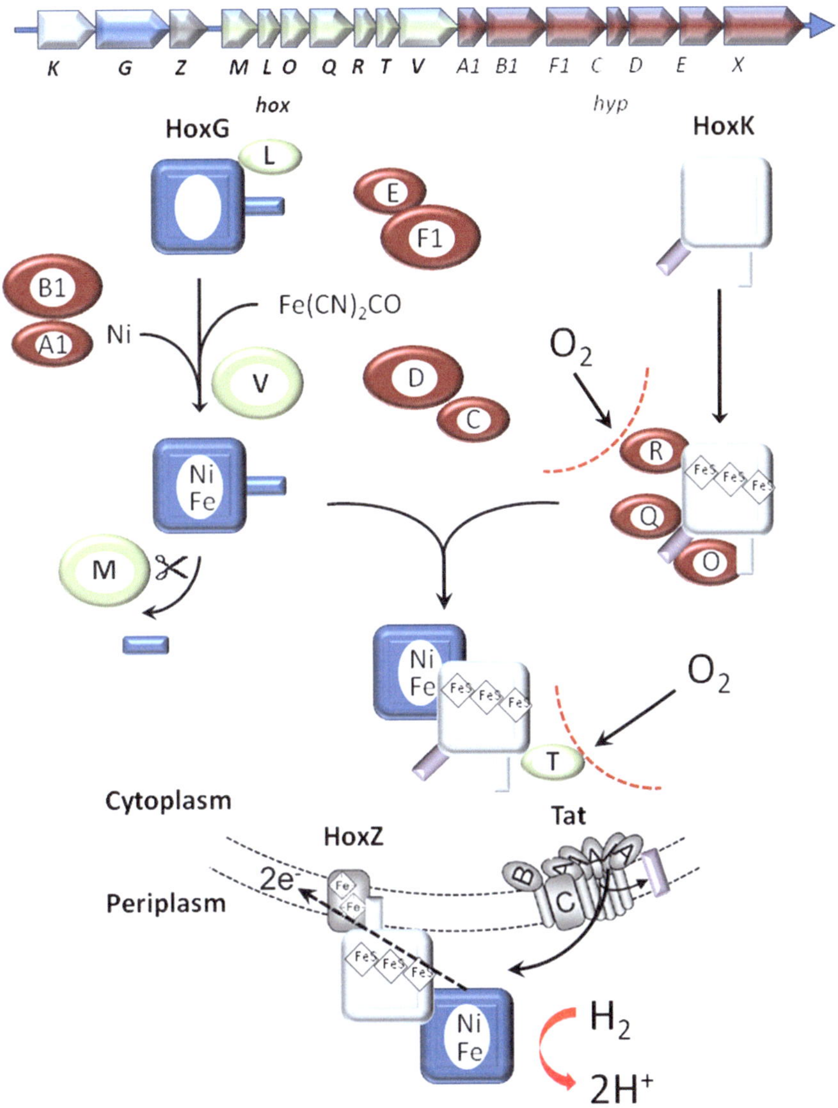

Fig. 3.6. Schematic maturation process of the [NiFe] hydrogenase. The *R. Eutropha* MBH gene cluster is presented at the *top*. *hoxK* encodes the small subunit and *hoxG* encodes the large subunit. The genes in *red* and *green* encode the accessory proteins, the role of which in the assembly of the active is presented (Adapted from Fritsch et al. 2011).

2010; McKinlay and Harwood 2010; Abo-Hashesh et al. 2011; Hallenbeck et al. 2012). In this section, we will discuss the different processes of engineering used to generate significant improvement in the production of hydrogen or hydrogenases, it will be divided in three points: (A) the heterologous expression and overexpression of interesting hydrogenases; (B) enhancing the efficiency of H_2 production by redirecting the flow of reducing equivalents toward hydrogenases, and (C) increasing the O_2 tolerance of hydrogenase by enzyme engineering.

A. Heterologous-Expression and Overexpression of Hydrogenases

In order to characterize their structure-function properties in greater detail, and to use hydrogenases for biotechnological applications, reliable methods for rapid, high-yield expression and purification are required. Owing to genetic manipulations, the purification of recombinant hydrogenases is greatly facilitated by the use of affinity tags, such as His or StrepII-tag, inserted at the N- or C-terminus (Kim et al. 2012). One approach that has been adopted in order to enhance the amount of enzymes or promote the H_2 production, is the engineering of a stable hydrogenase either through its production in a heterologous host or through overexpression (English et al. 2009; Abo-Hashesh et al. 2011). The interest of heterologous expression systems is not to be demonstrated but is still limited to [FeFe] hydrogenases (Böck et al. 2006; Nicolet and Fontecilla-Camps 2012).

1. [FeFe]-Hydrogenase

In recent years, several groups have developed different strategies for the expression of recombinant [FeFe]-hydrogenases (English et al. 2009) (Table 3.2). These include using as hosts organisms bacteria expressing naturally native [FeFe]-hydrogenase (Girbal et al. 2005; Sybirna et al. 2008), or using *E. coli* for heterologous expression in a more common way. At present, in order to obtain large amounts of [FeFe]-hydrogenases for biochemical and biophysical studies, the studies are focused on the high yield heterologous production mainly in *E. coli* (Kuchenreuther et al. 2010; Yacoby et al. 2012). On the other hand, the overexpression of [FeFe]-hydrogenase is also an important factor for the improvement of H_2 production. Indeed, the most efficient hydrogen-producing enzymes are [FeFe]-hydrogenases, which can have an activity about 10–100 times higher than that of [NiFe]-hydrogenases. As a result, overexpressing (HydA) in *Cl. paraputrificum* M-21 (Morimoto et al. 2005) improved the H_2 yield from 1.4 to 2.4 mol H_2 per mol of glucose. Jo et al. (2009) also reported that *Cl. tyrobutyricum* JM1 showed an improved H_2 yield (1.8 mol H_2/mol glucose) compared to the parental strain (1.2 mol H_2/mol glucose), when HydA was overexpressed. Regarding metabolic engineering, the inactivation of *ack*, which encodes acetate kinase of *Cl. tyrobutyricum* (Liu et al. 2006) increased the H_2 production yield by 1.5-fold compared to the wild-type strain.

2. [NiFe]-Hydrogenase

The main interests in producing [NiFe]-hydrogenases in heterologous hosts are to improve hydrogen production by focusing on (a) the heterologous expression of bidirectional [NiFe]-hydrogenases of the group III and (b) to take advantage of the O_2 tolerance properties of some [NiFe]-hydrogenases (Carrieri et al. 2011) (Table 3.2). However, as discussed above, the development of heterologous expression systems for the biosynthesis and molecular engineering of [NiFe]-hydrogenases is challenging due to the complexity and the high specificity of the maturation process. There are several documented examples of non-functional heterologous expression, which have only recently been reported for a limited number of organisms. For example, heterologous expression of [NiFe]-hydrogenases from *Rhodococcus opacus*, *Desulfovibrio vulgaris*, and *Synechocystis* sp. PCC6803 all resulted in the production of non-functional hydrogenases (Voordouw et al. 1987; Grzeszik et al. 1997; Maeda et al. 2007). For this reason, expression and purification of [NiFe]-hydrogenase for structural and in vitro studies are most often accomplished through the development of plasmid -based expression in homologous strains or closely related species as expression hosts (Rousset et al. 1998; Burgdorf et al. 2005; Ludwig et al. 2009b).

E. coli is an ideal microorganism commonly used in genetic engineering due to its well-characterized genome, well known

Table 3.2. Expression of [FeFe]- and [NiFe]-hydrogenases in different systems and their activities.

[FeFe]-hydrogenase genes	Expression host	H_2 evolution activity (μmol $H_2.min^{-1}.mg^{-1}$)	Bibliography
Chlamydomonas reinhardtii	*Escherichia coli*	0.4	Posewitz et al. (2004)
Clostridium acetobutylicum	*C. acetobutylicum*	10	Girbal et al. (2005)
Chlamydomonas reinhardtii	*C. acetobutylicum*	760	Girbal et al. (2005)
Scenedesmus obliquus	*C. acetobutylicum*	633	Girbal et al. (2005)
Clostridium paraputificum M-21	*C. paraputrificum*	1.7 fold more	Morimoto et al. (2005)
Clostridium acetobutylicum	*E. coli*	75.2	King et al. (2006)
Chlamydomonas reinhardtii	*S. oneidensis*	700	Sybirna et al. (2008)
Clostridium tyrobutyricum	*C. tyrobutyricum*	1.7 fold more	Jo et al. (2009)
Enterobacter aerogenes	*Enterobacter aerogenes*	1.95 fold more	Zhao et al. (2010)
Chlamydomonas reinhardtii	*E. coli*	641	Kuchenreuther et al. (2010)
Clostridium pasteurianum	*E. coli*	1,087	Kuchenreuther et al. (2010)
Chlamydomonas reinhardtii	*E. coli*	1,000	Yacoby et al. (2012)
[NiFe]-hydrogenase genes	**Expression host**	**H_2 uptake/evolution activity, MV-linked (μmol $H_2.min^{-1}.mg^{-1}$)**	**Bibliography**
Desulfovibrio gigas (*Group I*)	*D. fructosovorans*	Nd	Rousset et al. (1998)
Desulfovibrio fructosovorans (*Group I*)	*D. fructosovorans*	700 (up.)	Rousset et al. (1998)
Rhodococcus opacus (*Group III*)	*Ralstonia eutropha*	7.8 (up.)	Porthun et al. (2002)
Ralstonia eutropha (*Group II*)	*E. coli*	Nd	Posewitz et al. (2004)
Ralstonia eutropha (*Group II*)	*Ralstonia eutropha*	1.8 (up.)	Buhrke et al. (2005)
Ralstonia eutropha (*Group I*)	*Pseudomonas stutzeri*	19 (up.)	Lenz et al. (2005)
Synechocystis sp. PCC6803	*E. coli*	Nd	Maeda et al. (2007)
Pyrococcus furiosus (*Group III*)	*E. coli*	89 (evol.)	Sun et al. (2010)
Hydrogenovibrio marinus (*Group I*)	*E. coli*	0.06 (evol.)	Kim et al. (2011)
Alteromonas macleodii (*Group I*)	*Synechococcus elongatus*	$\sim 1.10^{-6}$ (evol.)	Weyman et al. (2011)
Thiocapsa roseopersicina (*Group I*)	*Synechococcus elongatus*	$\sim 1.10^{-6}$ (evol.)	Weyman et al. (2011)
Alteromonas macleodii (*Group I*)	*E. coli*	$\sim 4.10^{-3}$ (evol.)	Weyman et al. (2011)
Thiocapsa roseopersicina (*Group I*)	*E. coli*	$\sim 4.10^{-4}$ (evol.)	Weyman et al. (2011)
Klebsiella oxytoca (*Group IV*)	*K. oxytoca*	11 (evol.)	Bai et al. (2012)

metabolism, and its ability to utilize a wide range of carbon sources including hexoses and pentoses. In addition, *E. coli* has the potential advantages over at least some other microorganisms to exhibit a rapid growth, has simple nutritional requirements and also harbors four [NiFe]-hydrogenases. Two hydrogenases, one O_2-tolerant hydrogenase (Hyd-1) and one standard O_2-sensitive hydrogenase (Hyd-2) encoded by the *hya* and *hyb* operons respectively, are involved in periplasmic hydrogen uptake. Two others, hydrogenase 3 and 4, are part of cytoplasmically oriented hydrogenase complexes. Hydrogenase 3, encoded by the *hyc* operon, produces hydrogen from formate as a part of the formate hydrogen lyase complex (FHL-1), which is active in hydrogen production during mixed-acid fermentation at acidic pHs (Vignais et al. 2001). Hydrogenase 4, encoded by the *hyf* operon, appears to be cryptic under normal circumstances (Self et al. 2004).

Thus, *E. coli* has a set of specific maturation proteins, which might be used for the maturation of heterologous enzymes (Porthun et al. 2002; Maroti et al. 2003, 2009). Although *E. coli* is perhaps the most useful organism as a target for metabolic engineering, the lack of any NADH-dependent hydrogenases (Group III) is one major hurdle for the engineering of hydrogen metabolism in this organism. Indeed, these bidirectional cytoplasmic hydrogenases function reversibly in their physiological setting, coupling hydrogen uptake or emission, to oxidation and reduction of cellular coenzymes, such as NAD(P)H (Cournac et al. 2004). So, in *E. coli*, NAD(P)H generated through carbon metabolism cannot be used directly for H_2 production (Schmitz et al. 2002). Therefore, expressing a heterologous NAD(P)H-dependent hydrogenase is one of the main goals for increasing the maximum H_2 yield in *E. coli*.

a. Heterologous Expression of Bidirectional [NiFe]-Hydrogenases (Group III) in *E. coli*

The [NiFe]-hydrogenase of the cyanobacterium *Synechocystis* sp. PCC6803 is a well-studied representative of the bidirectional [NiFe]-hydrogenases from Group III. Therefore, the maturation process has been reconstituted to allow functional expression of this hydrogenase in *E. coli* (Maeda et al. 2007; Wells et al. 2011; Zheng et al. 2012). As a result, the introduction of the bidirectional [NiFe]-hydrogenase of *Synechocystis* sp. PCC6803 has altered the whole metabolism for hydrogen production in *E. coli*. Firstly, the hydrogen productivity was enhanced up to 41-fold in comparison with *Synechocystis* sp. PCC6803 (Maeda et al. 2007). Secondly the expression of this hydrogenase also showed a distinct H_2 production pathway than the one initially presents in *E. coli* (Wells et al. 2011). Finally, this heterologous expression has suppressed the transcription of native uptake [NiFe]-hydrogenases (Hya and Hyb) (Maeda et al. 2007; Zheng et al. 2012), increasing significantly the production of H_2. Similarly, the successful expression in *E. coli* of a recombinant cytoplasmic, NADP-dependent hydrogenase from *Pyrococcus furiosus*, an anaerobic hyperthermophile increased significantly the production of H_2. Remarkably, the native *E. coli* maturation machinery was able to generate a functional hydrogenase when transformed with only the genes encoding the hydrogenase structural subunits and the C-terminal protease (Sun et al. 2010).

b. Heterologous Expression of O_2-Tolerant [NiFe]-Hydrogenase (Group I)

The biotechnological goal of algal and cyanobacterial hydrogen production is to divert the reducing equivalents away from normal growth functions and to redirect them toward hydrogenases. All the enzymatic components for hydrogen production from water splitting and sunlight are present in cyanobacteria. As discussed above, hydrogenases are inactivated by molecular oxygen, which represent a major technological hurdle for hydrogen production from cyanobacteria at high solar efficiencies. Thus, strategies for overcoming this barrier include heterologous expression of a more oxygen-tolerant hydrogenase in cyanobacteria. In this connection, a system of heterologous expression has been developed to express O_2-tolerant [NiFe]-H_2ase belonging to group I from the bacteria *Alteromonas macleodii* and *Thiocapsa roseopersicina* (Vargas et al. 2011) in the cyanobacterium *Synechococcus elongatus* (Maroti et al. 2009); (Vargas et al. 2011). The cloned enzymes were active, indicating that it is possible to express hydrogenases in the cyanobacteria.

Recently, Kim et al. (2011) have succeeded in performing a heterologous expression of oxygen-tolerant *Hydrogenovibrio marinus* [NiFe]-hydrogenase in *E. coli*. Interestingly, recombinant *H. marinus* [NiFe]-hydrogenase produced of sevenfold to ninefold more hydrogen than did *E. coli* [NiFe]-hydrogenase (Hyd-1) in a gaseous environment containing 5–10 % (v/v) oxygen. Likewise, the same team (Kim et al. 2012) has improved the H_2 production (1.3 fold more) of this recombinant *H. marinus* [NiFe]-hydrogenase in co-expressing it with the proteorhodopsin under light conditions.

B. Substrate Selectivity, Competition and Linking

When expressed in vivo, hydrogenases interact with electron carriers which are generally at the junction of numerous redox reactions (respiration, CO_2 fixation, assimilation, etc.). A key point is that these reactions that appear as competitors for biotechnological purposes are often essential for cell survival or development. This explains in part the difficulty and the slow progress in biohydrogen research. One proposed research direction, to overcome this kind of limitation, would be to increase the amount of specific redox partners of hydrogenases. In this sense, a recent study has showed that the overexpression of ferredoxin-NADPH-reductase (FNR) coupled to that of ferredoxin (Fd) and a [FeFe]-hydrogenase, increased the H_2 production when NADPH was added in the medium (Weyman et al. 2011).

Another strategy to favor H_2 production would be to carry out an expression of synthetic or chimerical enzymes based on native hydrogenase fused to the electron transfer subunit (Kontur et al. 2012). In the case of algal [FeFe]-hydrogenases for instance, the redox partner is Fd (Fig. 3.1), which is also involved in photosynthetic carbon fixation via NADPH production by FNR. Indeed, under the anaerobic conditions that support hydrogen production, there is a significant loss of photosynthetic electrons toward NADPH production supplied by ferredoxin:NADP$^+$ oxidoreductase (FNR). No algal [FeFe]-hydrogenase structure is available yet, but Horner et al. (2002) modeled algal hydrogenase structure and charge distribution, and identified a set of aminoacids likely to participate in electrostatic interaction with algal Fd. Mutagenesis experiments conducted at the (putative) Fd binding sites in hydrogenase (and also at the Fd binding site in FNR) could be a way to modify relative affinities of these enzymes for their substrate and ultimately tune these affinities for an optimal ratio between photosynthetic and hydrogen-producing capabilities. Some results illustrated the feasibility of utilizing directly attached redox partners for H_2 production in vivo (Agapakis et al. 2010). Indeed, it has been reported that H_2 production via Fd-dependent hydrogenase can be improved by manipulating the interaction between hydrogenase and Fd via protein surface engineering in *E. coli*. Some chimerical enzymatic complexes have been expressed in *E. coli*, in which a ferredoxin and heterologous [FeFe] hydrogenase were either immobilized in a modular protein scaffold or directly attached to each other via an amino acid linker. H_2 production from cells containing these complexes showed 3-fold and 4.4-fold increases in H_2 production, respectively, over cells containing separate versions of the same proteins (Agapakis et al. 2010). However, yields from these complexes were relatively low (<0.1 mol H_2/mol glucose). More recently, a complex consisting of the Fd fused to the [FeFe]-hydrogenase HydA from *C. reinhardtii* (green algae), was shown to improve the light-dependent H_2 production in vitro when incubated with purified PSI (Yacoby et al. 2011). This algal Fd-HydA fusion prevents the competition between FNR and HydA that both exhibit affinity with Fd. Moreover, Fd-HydA fusion improved HydA function in several respects. First, the specific activities were up to sixfold higher than for the native HydA. Second, the fusion successfully insulates its internal Fd electrons, because only 10 % of the electrons are lost with external competitors such as FNR. Third, the fusion was able to overcome the limitation caused by FNR, as more than 60 % of photosynthetic electrons were diverted to hydrogen production, compared to less than 10 % for nonfused HydA (Yacoby et al. 2011).

Another tempting approach to favor H_2 production would be to tightly connect a specific electron carrier or a photosystem with hydrogenase, making the electrons flow directly from the photosystem to the hydrogenase or from the hydrogenase to the electron acceptor, avoiding competition with the bulk of electron carriers. In a study, Ihara et al. (2006) engineered a 'hard-wired' protein complex consisting of a hydrogenase and a photosystem. They designed an artificial fusion protein composed of the membrane-bound [NiFe]-hydrogenase from *R. eutropha* H16 and the peripheral photosystem I (PSI) subunit PsaE (involved in the docking of Fd to the PSI) of the cyanobacterium *Thermosynechococcus elongatus*. The resulting hydrogenase-PsaE fusion protein when associated with PsaE-free PSI spontaneously formed a complex which showed light-driven hydrogen production at a rate of 0.58 μmol H_2/mg chlorophyll/h. The complex retained accessibility to the native electron acceptor Fd, which is necessary for autotrophic growth of these cells. But unfortunately, the activity was totally suppressed in the presence of the physiological PSI partners, Fd and FNR. In an attempt to establish a H_2 photoproduction system in which the activity is not interrupted by Fd and FNR, the same group introduced a chimeric protein of PsaE and cytochrome c3 (cytc3) from *D. vulgaris* into the cyanobacterium *Synechocystis* sp. PCC6803 (Ihara et al. 2006). The covalent adduct of cytc3 and PsaE assembled with PsaE-free PSI and formed a complex which was still able to reduce Fd for photosynthesis (approximately 20 % of the original activity). Interestingly, this complex was able to drive hydrogen production when coupled with hydrogenase from *D. vulgaris* even in the presence of Fd and FNR, although the rate was limited (around 0.30 μmol H_2/mg chlorophyll/h). These results suggest, however, that this type of complex may eventually be modified to produce H_2 in vivo. More recently, *Cl. acetobutylicum* [FeFe]-hydrogenase was fused with *Synechococcus* sp. PCC 7002 PSI via a 1,6-hexanedithiol molecular wire (Lubner et al. 2010). However, while this complex also catalyzed light-dependent H_2 production in vitro, the abiotic nature of the wire make the system difficult to implement in vivo. Another system in which a fused PSI-[NiFe]-hydrogenase complex was immobilized on a gold electrode (Krassen et al. 2009) produced H_2 at the equivalent of 3 mmol H_2/mg chlorophyll/h (Bürstel et al. 2011), it then was suggested that this system is analogous to the in vivo situation where membrane-bound PSI receives electrons from the photosynthetic electron transport chain (Bürstel et al. 2011).

C. Protein Engineering to Improve the O_2-Tolerance of Hydrogenases

To develop a viable H_2 technology, hydrogenases should work in presence of O_2. Indeed, the direct biological photoproduction of hydrogen at the expense of water oxidation will unavoidably lead to a certain exposure to O_2. Similarly, the bio-fuel cells operate necessarily with O_2 to realize the oxidation of H_2 by hydrogenases. Thus, improving hydrogenase oxygen resistance is then a major challenge for a broad range of biotechnological applications such as hydrogen photoproduction, bio-fuel cells, and biosensors. As said above, there are currently several projects in progress with the single objective of identifying and expressing natural O_2-tolerant [NiFe]-hydrogenases in model host organisms in order to over-express the hydrogenases of interest or to improve H_2 production.

Another strategy is to carry out molecular engineering studies directly on the hydrogenases by directed mutagenesis for instance. Significant interest surrounds molecular engineering studies aimed at achieving hydrogenases with low levels of sensitivity to O_2 (Bürstel et al. 2011), and standard O_2-sensitive [NiFe]-hydrogenases are often the target of these studies because they are reversibly inhibited by oxygen unlike [FeFe]-hydrogenases. One major difficulty, however, lies in the complexity of the maturation process associated with the low production of native [NiFe]-hydrogenases. One of the most studied is the periplasmic O_2-sensitive [NiFe]-hydrogenase from *Desulfovibrio fructosovorans* (Rousset et al. 1998). This enzyme is soluble, highly produced, genetically accessible and it can be crystallized without much difficulties. All the mutants of interests were fully characterized, at the structural, spectroscopic and kinetic levels. Studies of this enzyme, at the molecular level, enabled to determine the mechanisms of electron transfer (Dementin et al. 2006, 2011), of proton transfer (Léger et al. 2004) and to increase the tolerance towards O_2 (Dementin et al. 2009; Liebgott et al. 2010; Dementin et al. 2011). Research strategies to increase the tolerance towards oxygen have been primarily inspired by some key features observed in the O_2-tolerant hydrogenase from *R. eutropha* H16 (Bleijlevens et al. 2004; Burgdorf et al. 2005; Fritsch et al. 2011). Thus, two general strategies have been followed: (1) slowing down the oxygen diffusion along the gas channel and (2) changing the reactivity of oxygen with the active site.

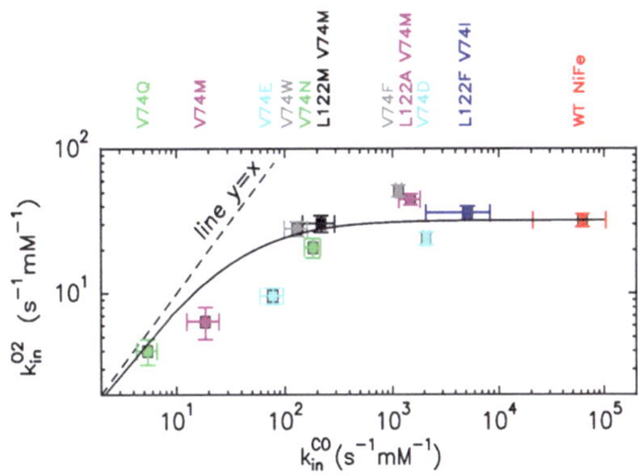

Fig. 3.7. Rate of inhibition by O_2 ($k_{in}^{O_2}$) plotted against the rate of binding of CO (k_{in}^{CO}). The *dashed line* depicts $y=x$ for which O_2-inhibition rate would be a linear function of diffusion rate; the *plain line* is the best fit to equation: $1/k_{in}^{O_2} = 1/k_{in}^{CO} + 1/k_{in}^{O_2}$ max, with $k_{in}^{O_2}$ max = 32 s^{-1} $mM(O_2)^{-1}$. *Error bars* represent either the deviation from the average of three to five independent determinations or the estimated error introduced from the extrapolation to 40 °C (Adapted from Liebgott et al. 2010).

1. Slowing Down of the Oxygen Diffusion Along the Gas Channel

Firstly, it is important to understand how the structure of the tunnel in hydrogenases determines the diffusion rate and possibly the selectivity of the enzymes with respect to substrates and inhibitors of similar sizes. From a multiscale simulation approach, associated with the comparison between the sequences and biochemical properties of homologous [NiFe] hydrogenases, it has been proposed that diffusion in these enzymes is controlled by two gates, which guard the entrance of the active site (Fig. 3.3a), and may determine the accessibility of the active site and therefore the resistance to O_2. One is located between residues 74 and 476 and the other between residues 74 and 122 (Fig. 3.3a) (Volbeda and Fontecilla-Camps 2004; Kim et al. 2012; Topin et al. 2012) (We use *D. fructosovorans* amino acids numbering throughout). This hypothesis has been erratically supported by the results of mutagenesis studies. Both *R. eutropha* H16 and *R. capsulatus* RH (regulatory hydrogenase) become oxygen sensitive when the two conserved bulky amino-acids, Phe122 and Ile74, are replaced with smaller residues, Leu and Val respectively, which are conserved in "standard," O_2-sensitive [NiFe] hydrogenases, such as those from *D. fructosovorans* and *A. vinosum* (Burgdorf et al. 2005; Duche et al. 2005).

Conversely, inspired by the RH, substitutions of the Val74 and Leu122 with isoleucine and phenylalanine, respectively, were carried out in the [NiFe]-hydrogenase from *D. fructosovorans* (Fig. 3.7). Surprisingly, these substitutions did not improve O_2 tolerance and did not significantly modify the catalytic properties of the enzyme under anoxic conditions (Dementin et al. 2009). Consequently, the residue bulkiness at these positions was not the only parameter affecting O_2 tolerance. Indeed, the existence of two control points in different locations might explain why the reduction in the experimental diffusion rate does not simply correlate with the width of the main gas channel. Moreover, the orientation or chemical nature of the side chain are also crucial to slow down the diffusion rate and the inhibition rate by O_2 (Leroux et al. 2008).

a. Slowing Down of the Diffusion Rate

In order to explore the respective roles of the bulk, hydrophobicity, charge and polarity, a number of single and double mutants at positions 122 and/or 74 have been constructed and purified. Val74 was replaced with aspartate (D), asparagine (N), tryptophan (W), glutamate (E), glutamine (Q), isoleucine (I), and phenylalanine (F) while Leu122 was replaced with phenylalanine, alanine (A) and methionine (M) (Liebgott et al. 2010). The mutants were screened using PFV to estimate different parameters such as the rate of inhibition by CO ($k_{in}CO$; the limiting step is the diffusion rate) and the rate of O_2-inhibition ($k_{in}O_2$; the limiting step is the reaction rate) (Leroux et al. 2008; Liebgott et al. 2010) (Fig. 3.7).

The effect of the mutations that keeps the side chain hydrophobic indicates a simple correlation between bulk and diffusion rate. Compared to the wild-type (WT), the diffusion rate is reduced by about two orders of magnitude for the V74F and about three orders of magnitude for the V74W. The increase of the molecular volume (calculated according to Ref. Häckel et al. 1999) of the amino-acid side chains lining the channel has a strong influence on the diffusion rate. In this set of experiments, the effect observed is only due the steric hindrance that shrivels the tunnel (Fig. 3.7).

In the case of polar amino-acids, two levels of impact were observed on diffusion, involving both bulk and charge. The influence of changing the size and polarity of the residue at position 74 was analyzed by comparing the V74E, V74D, V74Q and V74N mutants two at a time. Increasing the length of the side chain by one carbon (V74D to V74E, or V74N to V74Q) slows the diffusion rate about 30-fold. The magnitude of this effect lies within the same range as that observed with hydrophobic residues. Nevertheless, charge also matters as the substitution V74 to D74 slows the diffusion by a factor 40, while the molecular volume of aspartate is 40 % smaller than that of valine (Häckel et al. 1999). The polarity has an even stronger impact as the substitution V74 to Q74 reduces the diffusion by about four orders of magnitude, while the molecular volumes of these two amino acids are quite similar (Häckel et al. 1999). Within the polar amino acids, replacing a carboxylic acid with an amide, keeping the Van der Waals volume constant (V74E to V74Q, or V74D to V74N), slows diffusion by a factor of about 12. The two contributions, size and polarity, are independent of each other, and therefore the combination of the two (V74D to V74Q) decreases the rate of diffusion by more than two orders of magnitude. The electrostatic interaction of the amino acid side chains lining the tunnel is therefore very unfavourable to CO diffusion (Fig. 3.7). This could be due to a direct interaction of the polar group with the gas molecule or the Arg476 residue (residue strictly conserved within [NiFe]-hydrogenases, which lead to complications in actual experimental mutations), or to the stabilization of a water molecule that would be part of the barrier to ligand entry, as observed in certain myoglobin mutants (Nienhaus et al. 2003).

b. Slowing Down of the Inhibition Rate by O_2

In [NiFe] hydrogenase, the rate of inhibition by CO is about four orders of magnitude faster than the rate of aerobic inhibition (3.10^4 s^{-1}/M) (Léger et al. 2004; Liebgott et al. 2010). Considering that CO and O_2 diffuse within the protein at about the same rate, this observation implies that the rate of inhibition by O_2 is limited by the reaction at the active site. Mutations such as V74N or V74W decrease the rate of intramolecular diffusion by blocking the tunnel, but this has no effect on the overall reaction with O_2 because the diffusion process does not limit the inhibition rate (Fig. 3.7). However, other mutations decrease the rate of diffusion in such a large extent (three orders of magnitude for V74E or four orders of magnitude for V74Q) that this step becomes limiting because is slower than the reaction of O_2 at the active site, thus decreasing the overall rate of inhibition by O_2.

2. Changing of the Reactivity of Oxygen with the Active Site

Contrary to the *Re*RH, the O_2 tolerance of both MBH and SH is performed by a peculiar reaction of the active site with O_2. In the case of MBH the presence of a unique proximal [4Fe-3S] cluster is thought to be responsible for the tolerance of these hydrogenases towards O_2. The complexity and specificity of the maturation process of this unique [4Fe-3S] cluster will make the realization of variant enzymes, mimicking the capacities of O_2-tolerant [NiFe] Hydrogenases (MBH), more difficult. However, drawing inspiration from *Re*SH various mutations have been carried out. Amino acids like cysteine and methionine have a high affinity for oxygen, which results in a strong reactivity with reactive oxygen species and they are also known to participate in oxidative stress responses and protection in several proteins (Kim et al. 2001; Stadtman et al. 2002; Stadtman 2004, 2006; Reddie and Carroll 2008). With this knowledge in mind, several mutants at position 74 and 122 have been performed to replace valine by methionine or cysteine (Dementin et al. 2009).

a. Introduction of Methionines in the Gas Channel Near the Active Site

In order to decrease the sensitivity of the [NiFe] hydrogenase from *D. fructosovorans*, the effects of replacing Val74 and Leu122 with methionines have been tested. Indeed, methionines placed at the entrance of the active site cavity at positions V74M and L122M may protect the Ni-Fe site from oxidation, either by reacting or at least by interacting with the oxygen species present at the active site under oxidizing conditions (Volbeda et al. 2005). The goal of that substitution was twofold: slowing diffusion and modifying the reactivity with O_2.

At diffusion rate level, the molecular volume of methionine is 30 % larger than that of valine and about the same as leucine (Häckel et al. 1999). The diffusion rate in the case of the V74M-L122M is decreased by more than two orders of magnitude and even three orders of magnitude for the V74M (Fig. 3.7), which goes far beyond the expected effect of the volume increase on the diffusion. The interaction of gases with methionine is therefore stronger than a simple steric obstruction. Indeed, another interesting property of the V74M-L122M mutant was detected by PFV: unlike the WT enzyme, it partly reactivates in the presence of H_2 even under very oxidizing conditions (Dementin et al. 2009) (Fig. 3.8). The V74M mutant also reactivates under H_2 at high potential but to a lesser extent (Dementin et al. 2009). This process is slow and has a small amplitude, but it is significant because under the very oxidizing conditions used in the experiments, one would expect nothing but the inactivation of the enzyme (Vincent et al. 2007; Leroux et al. 2008) (Fig. 3.8).

Crystallographic and spectroscopic studies showed that methionines are not modified in the oxidized enzymes but that they interact with the active site by modifying its environment. FTIR studies determined that the mutants were inactivated more slowly and reactivated more rapidly than the native enzyme. This slower inactivation is attributed to a reduced active site accessibility that is due to partial tunnel obstruction by the mutations (Leroux et al. 2008). However the faster activation necessarily involves a quicker removal of the bound oxygen species. This was assumed to involve methionine that would stabilize the rearrangement of the oxygen species that is necessary to allow its protonation, facilitating its escape from the oxidized enzyme. As a result, the phenotype of the V74M and V74M-L122M mutants is not a consequence of a modification of the structure of the active site, but rather reveals subtle changes in the kinetics of the reaction with O_2.

These different studies clearly showed that the V74M-L122M and V74M mutated hydrogenases became O_2 tolerant, since these mutant hydrogenases continued to operate in the presence of 150 μM of O_2, which is close to the O_2 concentration of

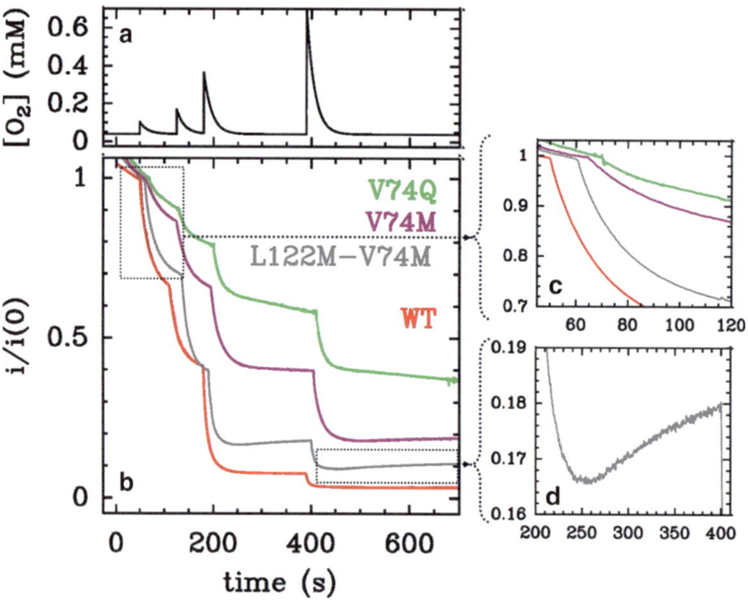

Fig. 3.8. The inhibition by O_2 of *D. fructosovorans* [Ni-Fe]-hydrogenase selected mutants. (**a**) The change in dioxygen concentration plotted against time, reconstructed from the amount of O_2 injected and the time constant of the exponential decay; the latter is calculated from fitting the change in current. (**b**) The plain lines show the change in [Ni-Fe]-hydrogenase activity (current *i* normalized by its value *i*(0) just before the inhibitor is added); (**c**, **d**) Enlarged views of the data in **b**, showing the decreases in current after the first exposure to O_2 (**c**) and the partial reactivation of the L122M-V74M mutant (**d**). E=+200 mV versus SHE, T=40 °C, pH 7, electrode rotation rate w=2 kr.p.m (Adapted from Liebgott et al. 2010).

200 µM in air-equilibrated solutions (Dementin et al. 2009).

b. Introduction of a Cysteine in the Gas Channel Near the Active Site

As described above, the O_2-tolerance of membrane-bound [NiFe]-hydrogenases, results from the fact that they only convert into the Ni-B state under O_2. Moreover, the reactivation rate of this inactive state is greater than that of the same species in O_2-sensitive enzymes. In order to test the reaction of a thiol function with O_2, the valine 74 in the wild type enzyme, has been exchanged with a cysteine (Liebgott et al. 2011). The obtained mutant has showed an activity during several minutes under oxygenated atmosphere. Further FTIR and PFV experiments showed that the inhibition of V74C by oxygen led to the formation of Ni-B state only, unlike the wild-type enzyme (Fig. 3.9). Moreover, the Ni-B state of V74C reactivated 17–25 times faster than the same species in the wild type (Fig. 3.9). Two other point mutations, at the level of valine 74, have been constructed to further address the role of cysteine 74 in modulating oxygen reactivity. The V74S and V74N mutations have showed also a significant increase of the rate of reactivation after inactivation under aerobic conditions (~ 4 and 11 times faster than wild type), although these rates are smaller than that of V74C (Fig. 3.10) Thus, the V74C mutant exhibits similar features typical of the naturally-occurring oxygen-tolerant hydrogenases (MBH). However, in the resistant *Ec*MBH, *Re*MBH and *Aa*MBH enzymes, it has been hypothesized that an additional source of electrons supplied to the active site would be responsible for the conversion into the Ni-B species only. V74C cannot supply electrons but from a combination of biochemical, spectroscopic (EPR, FTIR), X-ray and PFV studies, the authors proposed that V74C accelerates the electron transfer from the FeS clusters to the active site. This may promote (i) complete reduction of O_2 into

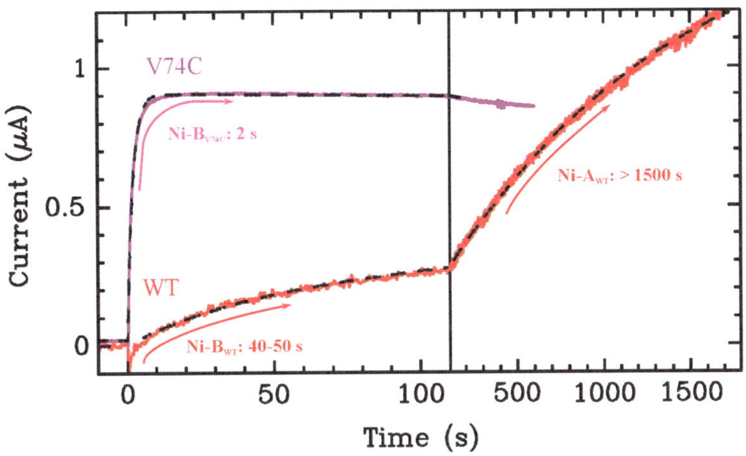

Fig. 3.9. Reactivation of the WT (*red) and V74C (purple) enzymes, at-90* mV, pH 5.5, 1 atm of H_2, after aerobic inactivation at + 215 mV, under 1 atm of Argon. The WT reactivates in two phases (NiB and NiA or 'ready' and 'unready' states respectively). In contrast, the reactivation of the V74C mutant is essentially monophasic and has a time constant of reactivation 20 times faster than for WT (Adapted from Liebgott et al. 2011).

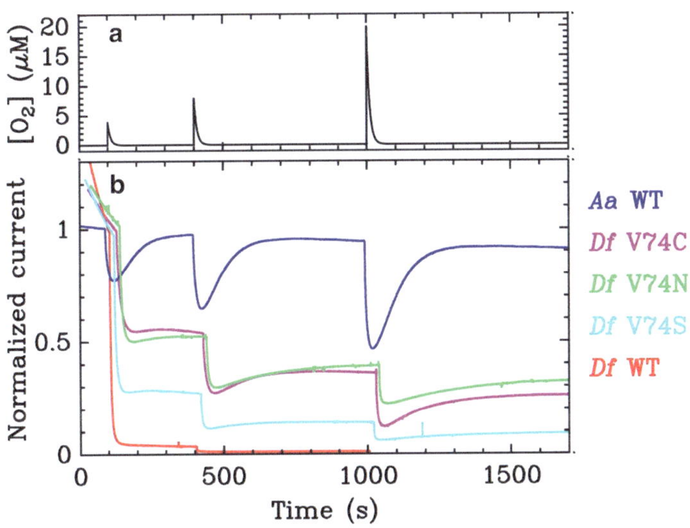

Fig. 3.10. Aerobic inhibition of *D. fructosovorans* WT (*red*), V74C, (*purple*), V74S (*cyan*), and V74N (*green)* and *A. aeolicus* MBH (*dark blue*) hydrogenases adsorbed at a rotating electrode poised at +140 mV, 40 °C, pH 5.5, 1 atm H_2. (**a**) Oxygen concentration in the cell during experiment. The three peaks correspond to injections of 100, 200, and 500 μL of an air-saturated solution (**b**) Effect of O_2 on enzyme's activity. The traces were recorded after a 200 s (WT) or 500 s (V74C) poise at +140 mV (Adapted from Liebgott et al. 2011).

H_2O and OH^-, which favors the formation of the Ni-B state and (ii) fast reactivation of this state (Dementin et al. 2011). Thus, with the V74M and V74M-L122M mutants (Dementin et al. 2009), the V74C mutant provides another evidence that it is possible to improve the oxygen tolerance of hydrogenases, which is a prerequisite for their biotechnological use in air. Interestingly, the V74C mutant was not designed to copy the sequence of the oxygen resistant enzymes, these results are totally unexpected but they suggest new strategies for conceiving oxygen-tolerant enzymes.

The main outcome of these studies is the demonstration that it is possible to induce

O_2-tolerance in a [NiFe] hydrogenase, in such a way that modified enzymes are active in the presence of O_2 at concentrations close to that in air-equilibrated solutions. The fact that the substitutions tested are located in conserved regions makes it possible to engineer [NiFe]-hydrogenases in a wide range of organisms, as heterologous expression of [NiFe]-hydrogenase is still difficult. This achievement opens the way for future development in the field of biological hydrogen production or utilization.

V. Conclusion

Hydrogenase engineering is a very difficult issue not only because of the complexity of these enzymes but also because of the vital processes in which they are involved. However, in spite of the difficulties, research is very active in several countries because of the potential spin offs that might participate in the development of a new hydrogen energy economy. The enzyme bias, substrate specificity and oxygen resistance are the main domains in which some progress have already been made, opening the way towards future applications. But other issues, like heterologous expression of [NiFe]-hydrogenases that would facilitate molecular research and organism engineering or deciphering the catalytic mechanism that would allow the development of biomimetic catalysts, are also the subjects of intense research and will contribute to biohydrogen implementation.

References

Abo-Hashesh M, Wang R, Hallenbeck PC (2011) Metabolic engineering in dark fermentative hydrogen production; theory and practice. Bioresour Technol 102:8414–8422

Agapakis C, Ducat D, Boyle P, Wintermute E, Way J, Silver P (2010) Insulation of a synthetic hydrogen metabolism circuit in bacteria. J Biol Eng 4:3

Albracht SP (1994) Nickel hydrogenases: in search of the active site. Biochim Biophys Acta 1188:167–204

Andersson SGE, Kurland CG (1999) Origins of mitochondria and hydrogenosomes. Curr Opin Microbiol 2:535–541

Antal T, Krendeleva T, Rubin A (2011) Acclimation of green algae to sulfur deficiency: underlying mechanisms and application for hydrogen production. Appl Microbiol Biotechnol 89:3–15

Armstrong FA, Albracht SP (2005) [NiFe]-hydrogenases: spectroscopic and electrochemical definition of reactions and intermediates. Philos Transact A Math Phys Eng Sci 363:937–954

Armstrong FA, Belsey NA, Cracknell JA, Goldet G, Parkin A, Reisner E, Vincent KA, Wait AF (2009) Dynamic electrochemical investigations of hydrogen oxidation and production by enzymes and implications for future technology. Chem Soc Rev 38:36–51

Atta M, Meyer J (2000) Characterization of the gene encoding the [Fe]-hydrogenase from *Megasphaera elsdenii*. Biochim Biophys Acta 1476:368–371

Baffert C, Demuez M, Cournac L, Burlat B, Guigliarelli B, Bertrand P, Girbal L, Leger C (2008) Hydrogen-activating enzymes: activity does not correlate with oxygen sensitivity. Angew Chem Int Ed Engl 47:2052–2054

Bai LP, Wu XB, Jing LJ, Liu J, Long MN (2012) Hydrogen production by overexpression of hydrogenase subunit in oxygen-tolerant Klebsiella oxytoca. Int J Hydrogen Energy 37:13227–13233

Bernhard M, Schwartz E, Rietdorf J, Friedrich B (1996) The *Alcaligenes eutrophus* membrane-bound hydrogenase gene locus encodes functions involved in maturation and electron transport coupling. J Bacteriol 178:4522–4529

Black LK, Fu C, Maier RJ (1994) Sequences and characterization of *hupU* and *hupV* genes of *Bradyrhizobium japonicum* encoding a possible nickel-sensing complex involved in hydrogenase expression. J Bacteriol 176:7102–7106

Bleijlevens B, Buhrke T, van der Linden E, Friedrich B, Albracht SP (2004) The auxiliary protein HypX provides oxygen tolerance to the soluble [NiFe]-hydrogenase of *Ralstonia eutropha* H16 by way of a cyanide ligand to nickel. J Biol Chem 279:46686–46691

Böck A, King PW, Blokesch M, Posewitz MC (2006) Maturation of hydrogenases. Adv Microb Physiol 51:1–71

Brentner LB, Peccia J, Zimmerman JB (2010) Challenges in developing biohydrogen as a sustainable energy source: implications for a research agenda. Environ Sci Technol 44:2243–2254

Buhrke T, Friedrich B (1998) hoxX (hypX) is a functional member of the *Alcaligenes eutrophus hyp* gene cluster. Arch Microbiol 170:460–463

Buhrke T, Bleijlevens B, Albracht SP, Friedrich B (2001) Involvement of *hyp* gene products in maturation of the H(2)-sensing [NiFe] hydrogenase of *Ralstonia eutropha*. J Bacteriol 183:7087–7093

Buhrke T, Lenz O, Krauss N, Friedrich B (2005) Oxygen tolerance of the H2-sensing [NiFe] hydrogenase from Ralstonia eutropha H16 is based on limited access of oxygen to the active site. J Biol Chem 280:23791–23796

Burgdorf T, Lenz O, Buhrke T, van der Linden E, Jones AK, Albracht SP, Friedrich B (2005) [NiFe]-hydrogenases of *Ralstonia eutropha* H16: modular enzymes for oxygen-tolerant biological hydrogen oxidation. J Mol Microbiol Biotechnol 10:181–196

Bürstel I, Hummel P, Siebert E, Wisitruangsakul N, Zebger I, Friedrich B, Lenz O (2011) Probing the origin of the metabolic precursor of the CO ligand in the catalytic center of [NiFe] hydrogenase. J Biol Chem 286:44937–44944

Cammack R, Fernandez VM, Schneider K (1986) Activation and active sites of nickel-containing hydrogenases. Biochimie 68:85–91

Carepo M, Tierney DL, Brondino CD, Yang TC, Pamplona A, Telser J, Moura I, Moura JJ, Hoffman BM (2002) ^{17}O ENDOR detection of a solvent-derived Ni-(OH(x))-Fe bridge that is lost upon activation of the hydrogenase from *Desulfovibrio gigas*. J Am Chem Soc 124:281–286

Carrieri D, Wawrousek K, Eckert C, Yu J, Maness P-C (2011) The role of the bidirectional hydrogenase in cyanobacteria. Bioresour Technol 102:8368–8377

Casalot L, Rousset M (2001) Maturation of the [NiFe] hydrogenases. Trends Microbiol 9:228–237

Chaudhuri A, Krasna AI (1990) Restoration of hydrogenase activity in hydrogenase-negative strains of *Escherichia coli* by cloned DNA fragments from *Chromatium vinosum* and *Proteus vulgaris*. J Gen Microbiol 136:1153–1160

Cohen J, Kim K, King P, Seibert M, Schulten K (2005) Finding gas diffusion pathways in proteins: application to O_2 and H_2 transport in CpI [FeFe]-hydrogenase and the role of packing defects. Structure 13:1321–1329

Corr MJ, Murphy JA (2011) Evolution in the understanding of [Fe]-hydrogenase. Chem Soc Rev 40:2279–2292

Cournac L, Guedeney G, Peltier G, Vignais PM (2004) Sustained photoevolution of molecular hydrogen in a mutant of *Synechocystis* sp. strain PCC 6803 deficient in the type I NADPH-dehydrogenase complex. J Bacteriol 186:1737–1746

Davidson G, Choudhury SB, Gu Z, Bose K, Roseboom W, Albracht SP, Maroney MJ (2000) Structural examination of the nickel site in chromatium vinosum hydrogenase: redox state oscillations and structural changes accompanying reductive activation and CO binding. Biochemistry 39:7468–7479

De Lacey AL, Fernandez VM, Rousset M (2005) Native and mutant nickel-iron hydrogenases: unravelling structure and function. Coord Chem Rev 249:1596–1608

De Lacey AL, Fernandez VM, Rousset M, Cammack R (2007) Activation and inactivation of hydrogenase function and the catalytic cycle: spectroelectrochemical studies. Chem Rev 107:4304–4330

Deckert G, Warren PV, Gaasterland T, Gaasterland T, Young WG, Lenox AL, Graham DE, Overbeek R, Snead MA, Keller M, Aujay M, Huber R, Feldman RA, Short JM, Olsen GJ, Swanson RV (1998) The complete genome of the hyperthermophilic bacterium *Aquifex aeolicus*. Nature 392:353–358

Dementin S, Belle V, Bertrand P, Guigliarelli B, Adryanczyk-Perrier G, De Lacey AL, Fernandez VM, Rousset M, Leger C (2006) Changing the ligation of the distal [4Fe4S] cluster in NiFe hydrogenase impairs inter- and intramolecular electron transfers. J Am Chem Soc 128:5209–5218

Dementin S, Leroux F, Cournac L, de Lacey AL, Volbeda A, Léger C, Burlat B, Martinez N, Champ S, Martin L, Sanganas O, Haumann M, Fernández VM, Guigliarelli B, Fontecilla-Camps JC, Rousset M (2009) Introduction of methionines in the gas channel makes [NiFe] hydrogenase aero-tolerant. J Am Chem Soc 131:10156–10164

Dementin S, Burlat B, Fourmond V, Leroux F, Liebgott PP, Abou Hamdan A, Léger C, Rousset M, Guigliarelli B, Bertrand P (2011) Rates of intra- and intermolecular electron transfers in hydrogenase deduced from steady-state activity measurements. J Am Chem Soc 133:10211–10221

Dubini A, Sargent F (2003) Assembly of Tat-dependent [NiFe] hydrogenases: identification of precursor-binding accessory proteins. FEBS Lett 549:141–146

Duche O, Elsen S, Cournac L, Colbeau A (2005) Enlarging the gas access channel to the active site renders the regulatory hydrogenase HupUV of *Rhodobacter capsulatus* O_2 sensitive without affecting its transductory activity. FEBS J 272:3899–3908

English CM, Eckert C, Brown K, Seibert M, King PW (2009) Recombinant and in vitro expression systems for hydrogenases: new frontiers in basic and applied studies for biological and synthetic H_2 production. Dalton Trans 45:9970–9978

Florin L, Tsokoglou A, Happe T (2001) A novel type of iron hydrogenase in the green alga *Scenedesmus obliquus* is linked to the photosynthetic electron transport chain. J Biol Chem 276:6125–6132

Fontecilla-Camps JC, Volbeda A, Cavazza C, Nicolet Y (2007) Structure/function relationships of [NiFe]- and [FeFe]-hydrogenases. Chem Rev 107:4273–4303

Forestier M, King P, Zhang L, Posewitz M, Schwarzer S, Happe T, Ghirardi ML, Seibert M (2003) Expression of two [Fe]-hydrogenases in *Chlamydomonas reinhardtii* under anaerobic conditions. Eur J Biochem 270:2750–2758

Fritsch J, Scheerer P, Frielingsdorf S, Kroschinsky S, Friedrich B, Lenz O, Spahn CMT (2011) The crystal structure of an oxygen-tolerant hydrogenase uncovers a novel iron-sulphur centre. Nature 479:249–252

Garcin E, Vernede X, Hatchikian EC, Volbeda A, Frey M, Fontecilla-Camps JC (1999) The crystal structure of a reduced [NiFeSe] hydrogenase provides an image of the activated catalytic center. Structure 7:557–566

Germer F, Zebger I, Saggu M, Lendzian F, Schulz R, Appel J (2009) Overexpression, isolation, and spectroscopic characterization of the bidirectional [NiFe] hydrogenase from *Synechocystis* sp. PCC 6803. J Biol Chem 284:36462–36472

Ghirardi ML, Posewitz MC, Maness PC, Dubini A, Yu J, Seibert M (2007) Hydrogenases and hydrogen photoproduction in oxygenic photosynthetic organisms. Annu Rev Plant Biol 58:71–91

Ghirardi ML, Dubini A, Yu J, Maness PC (2009) Photobiological hydrogen-producing systems. Chem Soc Rev 38:52–61

Girbal L, von Abendroth G, Winkler M, Benton PM, Meynial-Salles I, Croux C, Peters JW, Happe T, Soucaille P (2005) Homologous and heterologous overexpression in *Clostridium acetobutylicum* and characterization of purified clostridial and algal Fe-only hydrogenases with high specific activities. Appl Environ Microbiol 71:2777–2781

Goldet G, Wait AF, Cracknell JA, Vincent KA, Ludwig M, Lenz O, Friedrich B, Armstrong FA (2008) Hydrogen production under aerobic conditions by membrane-bound hydrogenases from *Ralstonia* species. J Am Chem Soc 130:11106–11113

Goris T, Wait AF, Saggu M, Fritsch J, Heidary N, Stein M, Zebger I, Lendzian F, Armstrong FA, Friedrich B, Lenz O (2011) A unique iron-sulfur cluster is crucial for oxygen tolerance of a [NiFe]-hydrogenase. Nat Chem Biol 7:310–318

Grzeszik C, Lubbers M, Reh M, Schlegel HG (1997) Genes encoding the NAD-reducing hydrogenase of *Rhodococcus opacus* MR11. Microbiology 143:1271–1286

Guiral M, Tron P, Belle V, Aubert C, Léger C, Guigliarelli B, Giudici-Orticoni M-T (2006) Hyperthermostable and oxygen resistant hydrogenases from a hyperthermophilic bacterium *Aquifex aeolicus*: physicochemical properties. Int J Hydrog Energy 31:1424–1431

Häckel M, Hinz H-J, Hedwig GR (1999) Partial molar volumes of proteins: amino acid side-chain contributions derived from the partial molar volumes of some tripeptides over the temperature range 10–90 °C. Biophys Chem 82:35–50

Hallenbeck PC, Abo-Hashesh M, Ghosh D (2012) Strategies for improving biological hydrogen production. Bioresour Technol 110:1–9

Happe RP, Roseboom W, Egert G, Friedrich CG, Massanz C, Friedrich B, Albracht SP (2000) Unusual FTIR and EPR properties of the H_2-activating site of the cytoplasmic NAD-reducing hydrogenase from *Ralstonia eutropha*. FEBS Lett 466:259–263

Henstra AM, Sipma J, Rinzema A, Stams AJM (2007) Microbiology of synthesis gas fermentation for biofuel production. Curr Opin Biotechnol 18:200–206

Higuchi Y, Yagi T, Yasuoka N (1997) Unusual ligand structure in Ni-Fe active center and an additional Mg site in hydrogenase revealed by high resolution X-ray structure analysis. Structure 5:1671–1680

Hiromoto T, Ataka K, Pilak O, Vogt S, Stagni MS, Meyer-Klaucke W, Warkentin E, Thauer RK, Shima S, Ermler U (2009) The crystal structure of C176A mutated [Fe]-hydrogenase suggests an acyl-iron ligation in the active site iron complex. FEBS Lett 583:585–590

Hong G, Pachter R (2012) Inhibition of biocatalysis in [FeFe] hydrogenase by oxygen: molecular dynamics and density functional theory calculations. ACS Chem Biol 7:1268–1275

Horch M, Lauterbach L, Saggu M, Hildebrandt P, Lendzian F, Bittl R, Lenz O, Zebger I (2010) Probing the active site of an O2-tolerant NAD+–reducing [NiFe]-hydrogenase from Ralstonia eutropha H16 by in situ EPR and FTIR spectroscopy. Angew Chem Int Ed Engl 49:8026–8029

Horch M, Lauterbach L, Lenz O, Hildebrandt P, Zebger I (2012) NAD(H)-coupled hydrogen cycling-structure-function relationships of bidirectional [NiFe] hydrogenases. FEBS Lett 586:545–556

Horner DS, Foster PG, Embley TM (2000) Iron hydrogenases and the evolution of anaerobic eukaryotes. Mol Biol Evol 17:1695–1709

Horner DS, Heil B, Happe T, Embley TM (2002) Iron hydrogenases–ancient enzymes in modern eukaryotes. Trends Biochem Sci 27:148–153

Ihara M, Nishihara H, Yoon KS, Lenz O, Friedrich B, Nakamoto H, Kojima K, Honma D, Kamachi T, Okura I (2006) Light-driven hydrogen production by a hybrid complex of a [NiFe]-hydrogenase and the cyanobacterial photosystem I. Photochem Photobiol 82:676–682

Ikeda T, Kano K (2001) An electrochemical approach to the studies of biological redox reactions and their applications to biosensors, bioreactors, and biofuel cells. J Biosci Bioeng 92:9–18

Jacobi A, Rossmann R, Bock A (1992) The *hyp* operon gene products are required for the maturation of

catalytically active hydrogenase isoenzymes in *Escherichia coli*. Arch Microbiol 158:444–451
Jo JH, Lee DS, Kim J, Park JM (2009) Effect of initial glucose concentrations on carbon material and energy balances in hydrogen-producing Clostridium tyrobutyricum JM1. J Microbiol Biotechnol 19:291–298
Kamp C, Silakov A, Winkler M, Reijerse EJ, Lubitz W, Happe T (2008) Isolation and first EPR characterization of the [FeFe]-hydrogenases from green algae. Biochim Biophys Acta 1777:410–416
Kaneko T, Nakamura Y, Sato S, Minamisawa K, Uchiumi T, Sasamoto S, Watanabe A, Idesawa K, Iriguchi M, Kawashima K, Kohara M, Matsumoto M, Shimpo S, Tsuruoka H, Wada T, Yamada M, Tabata S (2002) Complete genomic sequence of nitrogen-fixing symbiotic bacterium *Bradyrhizobium japonicum* USDA110. DNA Res 9:189–197
Kim YH, Berry AH, Spencer DS, Stites WE (2001) Comparing the effect on protein stability of methionine oxidation versus mutagenesis: steps toward engineering oxidative resistance in proteins. Protein Eng 14:343–347
Kim JY, Jo BH, Cha HJ (2011) Production of biohydrogen by heterologous expression of oxygen-tolerant Hydrogenovibrio marinus [NiFe]-hydrogenase in Escherichia coli. J Biotechnol 155(3):312–319
Kim S, Lu D, Park S, Wang G (2012) Production of hydrogenases as biocatalysts. Int J Hydrogen Energy. http://dx.doi.org/10.1016/j.ijhydene.2012.03.033
King PW, Posewitz MC, Ghirardi ML, Seibert M (2006) Functional studies of [FeFe] hydrogenase maturation in an Escherichia coli biosynthetic system. J Bacteriol 188:2163–2172
Kontur WS, Noguera DR, Donohue TJ (2012) Maximizing reductant flow into microbial H_2 production. Curr Opin Biotechnol 23:382–389
Krassen H, Schwarze A, Friedrich B, Ataka K, Lenz O, Heberle J (2009) Photosynthetic hydrogen production by a hybrid complex of photosystem I and [NiFe]-hydrogenase. ACS Nano 3:4055–4061
Kruse O, Rupprecht J, Bader KP, Thomas-Hall S, Schenk PM, Finazzi G, Hankamer B (2005) Improved photobiological H_2 production in engineered green algal cells. J Biol Chem 280:34170–34177
Kuchenreuther JM, Grady-Smith CS, Bingham AS, George SJ, Cramer SP, Swartz JR (2010) High-yield expression of heterologous [FeFe] hydrogenases in *Escherichia coli*. PLoS One 5:e15491
Lamle SE, Albracht SP, Armstrong FA (2004) Electrochemical potential-step investigations of the aerobic interconversions of [NiFe]-hydrogenase from *Allochromatium vinosum*: insights into the puzzling difference between unready and ready oxidized inactive states. J Am Chem Soc 126:14899–14909
Lamle SE, Albracht SP, Armstrong FA (2005) The mechanism of activation of a [NiFe]-hydrogenase by electrons, hydrogen, and carbon monoxide. J Am Chem Soc 127:6595–6604
Leach MR, Zhang JW, Zamble DB (2007) The role of complex formation between the *Escherichia coli* hydrogenase accessory factors HypB and SlyD. J Biol Chem 282:16177–16186
Léger C, Dementin S, Bertrand P, Rousset M, Guigliarelli B (2004) Inhibition and aerobic inactivation kinetics of *Desulfovibrio fructosovorans* [NiFe] hydrogenase studied by protein film voltammetry. J Am Chem Soc 126:12162–12172
Lenz O, Gleiche A, Strack A, Friedrich B (2005) Requirements for heterologous production of a complex metalloenzyme: the membrane-bound [NiFe] hydrogenase. J Bacteriol 187:6590–6595
Lenz O, Ludwig M, Schubert T, Bürstel I, Ganskow S, Goris T, Schwarze A, Friedrich B (2010) H_2 conversion in the presence of O_2 as performed by the membrane-bound [NiFe]-hydrogenase of *Ralstonia eutropha*. ChemPhysChem 11:1107–1119
Leroux F, Dementin S, Burlat B et al (2008) Experimental approaches to kinetics of gas diffusion in hydrogenase. Proc Natl Acad Sci U S A 105:11188–11193
Liang Y, Wu X, Gan L, Xu H, Hu Z, Long M (2009) Increased biological hydrogen production by deletion of hydrogen-uptake system in photosynthetic bacteria. Microbiol Res 164:674–679
Liebgott PP, Leroux F, Burlat B, Dementin S, Baffert C, Lautier T, Fourmond V, Ceccaldi P, Cavazza C, Meynial-Salles I, Soucaille P, Fontecilla-Camps JC, Guigliarelli B, Bertrand P, Rousset M, Léger C (2010) Relating diffusion along the substrate tunnel and oxygen sensitivity in hydrogenase. Nat Chem Biol 6:63–70
Liebgott PP, De Lacey AL, Burlat B, Cournac L, Richaud P, Brugna M, Fernandez VM, Guigliarelli B, Rousset M, Léger C, Dementin S (2011) Original design of an oxygen-tolerant [NiFe] hydrogenase: major effect of a valine-to-cysteine mutation near the active site. J Am Chem Soc 133:986–997
Liu ZP, Hu P (2002) A density functional theory study on the active center of Fe-only hydrogenase: characterization and electronic structure of the redox states. J Am Chem Soc 124:5175–5182
Liu X, Zhu Y, Yang ST (2006) Construction and characterization of ack deleted mutant of *Clostridium tyrobutyricum* for enhanced butyric acid and hydrogen production. Biotechnol Prog 22:1265–1275
Lubitz W, Reijerse E, van Gastel M (2007) [NiFe] and [FeFe] hydrogenases studied by advanced magnetic resonance techniques. Chem Rev 107:4331–4365
Lubner CE, Knoïzer P, Silva PJN, Vincent KA, Happe T, Bryant DA, Golbeck JH (2010) Wiring an

[FeFe]-hydrogenase with photosystem I for light-induced hydrogen production. Biochemistry 49: 10264–10266

Ludwig M, Cracknell JA, Vincent KA, Armstrong FA, Lenz O (2009a) Oxygen-tolerant H_2 oxidation by membrane-bound [NiFe] hydrogenases of *Ralstonia* species. Coping with low level H_2 in air. J Biol Chem 284:465–477

Ludwig M, Schubert T, Zebger I, Wisitruangsakul N, Saggu M, Strack A, Lenz O, Hildebrandt P, Friedrich B (2009b) Concerted action of two novel auxiliary proteins in assembly of the active site in a membrane-bound [NiFe] hydrogenase. J Biol Chem 284: 2159–2168

Lukey MJ, Parkin A, Roessler MM, Murphy BJ, Harmer J, Palmer T, Sargent F, Armstrong FA (2010) How *Escherichia coli* is equipped to oxidize hydrogen under different redox conditions. J Biol Chem 285:3928–3938

Lukey MJ, Roessler MM, Parkin A, Evans RM, Davies RA, Lenz O, Friedrich B, Sargent F, Armstrong FA (2011) Oxygen-tolerant [NiFe]-hydrogenases: the individual and collective importance of supernumerary cysteines at the proximal Fe-S cluster. J Am Chem Soc 133:16881–16892

Lyon EJ, Shima S, Buurman G, Chowdhuri S, Batschauer A, Steinbach K, Thauer RK (2004) UV-A/blue-light inactivation of the 'metal-free' hydrogenase (Hmd) from methanogenic archaea. Eur J Biochem 271:195–204

Maeda T, Sanchez-Torres V, Wood TK (2007) Enhanced hydrogen production from glucose by metabolically engineered *Escherichia coli*. Appl Microbiol Biotechnol 77:879–890

Maier T, Binder U, Bock A (1996) Analysis of the *hydA* locus of *Escherichia coli*: two genes (*hydN* and *hypF*) involved in formate and hydrogen metabolism. Arch Microbiol 165:333–341

Manyani H, Rey L, Palacios JM, Imperial J, Ruiz-Argueso T (2005) Gene products of the *hupGHIJ* operon are involved in maturation of the iron-sulfur subunit of the [NiFe] hydrogenase from *Rhizobium leguminosarum* bv. *viciae*. J Bacteriol 187:7018–7026

Maroti G, Fodor BD, Rakhely G, Kovacs AT, Arvani S, Kovacs KL (2003) Accessory proteins functioning selectively and pleiotropically in the biosynthesis of [NiFe] hydrogenases in *Thiocapsa roseopersicina*. Eur J Biochem 270:2218–2227

Maroti G, Tong Y, Yooseph S, Baden-Tillson H, Smith HO, Kovàcs KL, Frazier M, Venter JC, Xu Q (2009) Discovery of [NiFe] hydrogenase genes in metagenomic DNA: cloning and heterologous expression in *Thiocapsa roseopersicina*. Appl Environ Microbiol 75:5821–5830

Mason JE (2007) World energy analysis: H_2 now or later? Energy Policy 35:1315–1329

Matias PM, Soares CM, Saraiva LM, Coelho R, Morais J, Le Gall J, Carrondo MA (2001) [NiFe] hydrogenase from *Desulfovibrio desulfuricans* ATCC 27774: gene sequencing, three-dimensional structure determination and refinement at 1.8 A and modelling studies of its interaction with the tetrahaem cytochrome c3. J Biol Inorg Chem 6:63–81

McKinlay JB, Harwood CS (2010) Photobiological production of hydrogen gas as a biofuel. Curr Opin Biotechnol 21:244–251

Menon NK, Chatelus CY, Dervartanian M, Wendt JC, Shanmugam KT, Peck HD Jr, Przybyla AE (1994) Cloning, sequencing, and mutational analysis of the *hyb* operon encoding *Escherichia coli* hydrogenase 2. J Bacteriol 176:4416–4423

Meyer J (2007) [FeFe] hydrogenases and their evolution: a genomic perspective. Cell Mol Life Sci 64:1063–1084

Montet Y, Amara P, Volbeda A, Vernede X, Hatchikian EC, Field MJ, Frey M, Fontecilla-Camps JC (1997) Gas access to the active site of Ni-Fe hydrogenases probed by X-ray crystallography and molecular dynamics. Nat Struct Biol 4:523–526

Morimoto K, Kimura T, Sakka K, Ohmiya K (2005) Overexpression of a hydrogenase gene in *Clostridium paraputrificum* to enhance hydrogen gas production. FEMS Microbiol Lett 246:229–234

Morozov SV, Karyakina EE, Zorin NA, Varfolomeyev SD, Cosnier S, Karyakin AA (2002) Direct and electrically wired bioelectrocatalysis by hydrogenase from *Thiocapsa roseopersicina*. Bioelectrochemistry 55:169–171

Mulder DW, Shepard EM, Meuser JE, Joshi N, King PW, Posewitz MC, Broderick JB, Peters JW (2011) Insights into [FeFe]-hydrogenase structure, mechanism, and maturation. Structure 19:1038–1052

Nicolet Y, Fontecilla-Camps JC (2012) Structure-function relationships in [FeFe]-hydrogenase active site maturation. J Biol Chem 287:13532–13540

Nicolet Y, Piras C, Legrand P, Hatchikian CE, Fontecilla-Camps JC (1999) *Desulfovibrio desulfuricans* iron hydrogenase: the structure shows unusual coordination to an active site Fe binuclear center. Structure 7:13–23

Nicolet Y, Lemon BJ, Fontecilla-Camps JC, Peters JW (2000) A novel FeS cluster in Fe-only hydrogenases. Trends Biochem Sci 25:138–143

Nicolet Y, Cavazza C, Fontecilla-Camps JC (2002) Fe-only hydrogenases: structure, function and evolution. J Inorg Biochem 91:1–8

Nienhaus K, Deng P, Kriegl JM, Nienhaus GU (2003) Structural dynamics of myoglobin: effect of internal

cavities on ligand migration and binding Biochem 42:9647–9658
Ogata H, Lubitz W, Higuchi Y (2009) [NiFe] hydrogenases: structural and spectroscopic studies of the reaction mechanism. Dalton Trans 37:7577–7587
Ogata H, Kellers P, Lubitz W (2010) The crystal structure of the [NiFe] hydrogenase from the photosynthetic bacterium *Allochromatium vinosum*: characterization of the oxidized enzyme (Ni-A State). J Mol Biol 402:428–444
Oh Y-K, Raj SM, Jung GY, Park S (2011) Current status of the metabolic engineering of microorganisms for biohydrogen production. Bioresour Technol 102:8357–8367
Okura I, Otsuka K, Nakada N, Hasumi F (1990) Regeneration of NADH and ketone hydrogenation by hydrogen with the combination of hydrogenase and alcohol dehydrogenase. Appl Biochem Biotechnol 24–25:425–430
Pandelia M-E, Fourmond V, Tron-Infossi P, Lojou E, Bertrand P, Léger C, Giudici-Orticoni M-T, Lubitz W (2010a) Membrane-bound hydrogenase I from the hyperthermophilic bacterium *Aquifex aeolicus*: enzyme activation, redox intermediates and oxygen tolerance. J Am Chem Soc 132:6991–7004
Pandelia M-E, Ogata H, Lubitz W (2010b) Intermediates in the catalytic cycle of [NiFe] hydrogenase: functional spectroscopy of the active site. ChemPhysChem 11:1127–1140
Pandelia M-E, Nitschke W, Infossi P, M-Trs G-O, Bill E, Lubitz W (2011) Characterization of a unique [FeS] cluster in the electron transfer chain of the oxygen tolerant [NiFe] hydrogenase from *Aquifex aeolicus*. Proc Natl Acad Sci U S A 108:6097–6102
Pardo A, De Lacey AL, Fernandez VM, Fan HJ, Fan Y, Hall MB (2006) Density functional study of the catalytic cycle of nickel-iron [NiFe] hydrogenases and the involvement of high-spin nickel(II). J Biol Inorg Chem 11:286–306
Parish D, Benach J, Liu G, Singarapu KK, Xiao R, Acton T, Su M, Bansal S, Prestegard JH, Hunt J, Montelione GT, Szyperski T (2008) Protein chaperones Q8ZP25_SALTY from *Salmonella typhimurium* and HYAE_ECOLI from *Escherichia coli* exhibit thioredoxin-like structures despite lack of canonical thioredoxin active site sequence motif. J Struct Funct Genomics 9:41–49
Parkin A, Sargent F (2012) The how and whys of aerobic hydrogen metabolism. Curr Opin Chem Biol 16:26–34
Paschos A, Bauer A, Zimmermann A, Zehelein E, Bock A (2002) HypF, a carbamoyl phosphate-converting enzyme involved in [NiFe] hydrogenase maturation. J Biol Chem 277:49945–49951
Perez-Arellano JL, Martin T, Lopez-Novoa JM, Sanchez ML, Montero A, Jimenez A (1998) BN 52021 (a platelet activating factor-receptor antagonist) decreases alveolar macrophage-mediated lung injury in experimental extrinsic allergic alveolitis. Mediators Inflamm 7:201–210
Peters JW, Lanzilotta WN, Lemon BJ, Seefeldt LC (1998) X-ray crystal structure of the Fe-only hydrogenase (CpI) from *Clostridium pasteurianum* to 1.8 angstrom resolution. Science 282:1853–1858
Petkun S, Shi R, Li Y, Asinas A, Munger C, Zhang L, Waclawek M, Soboh B, Sawers RG, Cygler M (2011) Structure of hydrogenase maturation protein HypF with reaction intermediates shows two active sites. Structure 19:1773–1783
Pierik AJ, Roseboom W, Happe RP, Bagley KA, Albracht SP (1999) Carbon monoxide and cyanide as intrinsic ligands to iron in the active site of [NiFe]-hydrogenases. $NiFe(CN)_2CO$, biology's way to activate H_2. J Biol Chem 274:3331–3337
Pohlmann A, Fricke WF, Reinecke F, Kusian B, Liesegang H, Cramm R, Eitinger T, Ewering C, Pötter M, Schwartz E, Strittmatter A, Voss I, Gottschalk G, Steinbüchel A, Friedrich B, Bowien B (2006) Genome sequence of the bioplastic-producing Knallgas bacterium *Ralstonia eutropha* H16. Nat Biotechnol 24:1257–1262
Porthun A, Bernhard M, Friedrich B (2002) Expression of a functional NAD-reducing [NiFe] hydrogenase from the gram-positive *Rhodococcus opacus* in the gram-negative *Ralstonia eutropha*. Arch Microbiol 177:159–166
Posewitz MC, Smolinski SL, Kanakagiri S, Melis A, Seibert M, Ghirardi ML (2004) Hydrogen photoproduction is attenuated by disruption of an isoamylase gene in *Chlamydomonas reinhardtii*. Plant Cell 16:2151–2163
Prince RC, Kheshgi HS (2005) The photobiological production of hydrogen: potential efficiency and effectiveness as a renewable fuel. Crit Rev Microbiol 31:19–31
Ratzka J, Lauterbach L, Lenz O, Ansorge-Schumacher MB (2011) Systematic evaluation of the dihydrogen-oxidising and NAD+-reducing soluble [NiFe]-hydrogenase from *Ralstonia eutropha* H16 as a cofactor regeneration catalyst. Biocatal Biotransfor 29:246–252
Reddie KG, Carroll KS (2008) Expanding the functional diversity of proteins through cysteine oxidation. Curr Opin Chem Biol 12:746–754
Richardson AJ, Stewart CS (1990) Hydrogen transfer between *Neocallimastix frontalis* and *Selenomonas ruminantium* grown in mixed culture. In: Bélaich JP, Bruschi M, Garcia JL (eds) Microbiology and biochemistry of strict anaerobes involved in interspe-

cies hydrogen transfer. Plenum Publishing Corporation, New York, pp 463–466

Rousset M, Montet Y, Guigliarelli B, Forget N, Asso M, Bertrand P, Fontecilla-Camps JC, Hatchikian EC (1998) [3Fe-4S] to [4Fe-4S] cluster conversion in *Desulfovibrio fructosovorans* [NiFe] hydrogenase by site-directed mutagenesis. Proc Natl Acad Sci U S A 95:11625–11630

Rupprecht J, Hankamer B, Mussgnug JH, Ananyev G, Dismukes C, Kruse O (2006) Perspectives and advances of biological H_2 production in microorganisms. Appl Microbiol Biotechnol 72:442–449

Sauter M, Bohm R, Bock A (1992) Mutational analysis of the operon (*hyc*) determining hydrogenase 3 formation in *Escherichia coli*. Mol Microbiol 6:1523–1532

Schmitz O, Boison G, Salzmann H, Bothe H, Schutz K, Wang SH, Happe T (2002) HoxE–a subunit specific for the pentameric bidirectional hydrogenase complex (HoxEFUYH) of cyanobacteria. Biochim Biophys Acta 1554:66–74

Schubert T, Lenz O, Krause E, Volkmer R, Friedrich B (2007) Chaperones specific for the membrane-bound [NiFe]-hydrogenase interact with the Tat signal peptide of the small subunit precursor in *Ralstonia eutropha* H16. Mol Microbiol 66:453–467

Schwartz E, Henne A, Cramm R, Eitinger T, Friedrich B, Gottschalk G (2003) Complete nucleotide sequence of pHG1: a *Ralstonia eutropha* H16 megaplasmid encoding key enzymes of H(2)-based lithoautotrophy and anaerobiosis. J Mol Biol 332:369–383

Self WT, Hasona A, Shanmugam KT (2004) Expression and regulation of a silent operon, *hyf*, coding for hydrogenase 4 isoenzyme in *Escherichia coli*. J Bacteriol 186:580–587

Shima S (2008) Structure of [Fe]-hydrogenase and the convergent evolution of the active site of hydrogenases. Seikagaku 80:846–849

Shomura Y, Yoon K-S, Nishihara H, Higuchi Y (2011) Structural basis for a [4Fe-3S] cluster in the oxygen-tolerant membrane-bound [NiFe]-hydrogenase. Nature 479:253–256

Silakov A, Wenk B, Reijerse E, Albracht SP, Lubitz W (2009) Spin distribution of the H-cluster in the H(ox)-CO state of the [FeFe] hydrogenase from *Desulfovibrio desulfuricans*: HYSCORE and ENDOR study of (14)N and (13)C nuclear interactions. J Biol Inorg Chem 14:301–313

Soboh B, Kuhns M, Braussemann M, Waclawek M, Muhr E, Pierik AJ, Sawers RG (2012) Evidence for an oxygen-sensitive iron-sulfur cluster in an immature large subunit species of *Escherichia coli* [NiFe]-hydrogenase 2. Biochem Biophys Res Commun 424:158–163

Stadler C, de Lacey AL, Montet Y, Volbeda A, Fontecilla-Camps JC, Conesa JC, Fernandez VM (2002) Density functional calculations for modeling the active site of nickel-iron hydrogenases. 2. Predictions for the unready and ready states and the corresponding activation processes. Inorg Chem 41:4424–4434

Stadtman ER (2004) Cyclic oxidation and reduction of methionine residues of proteins in antioxidant defense and cellular regulation. Arch Biochem Biophys 423:2–5

Stadtman ER (2006) Protein oxidation and aging. Free Radic Res 40:1250–1258

Stadtman ER, Moskovitz J, Berlett BS, Levine RL (2002) Cyclic oxidation and reduction of protein methionine residues is an important antioxidant mechanism. Mol Cell Biochem 234–235:3–9

Stein M, van Lenthe E, Baerends EJ, Lubitz W (2001) Relativistic DFT calculations of the paramagnetic intermediates of [NiFe] hydrogenase. Implications for the enzymatic mechanism. J Am Chem Soc 123:5839–5840

Stripp ST, Goldet G, Brandmayr C, Sanganas O, Vincent KA, Haumann M, Armstrong FA, Happe T (2009) How oxygen attacks [FeFe] hydrogenases from photosynthetic organisms. Proc Natl Acad Sci U S A 106:17331–17336

Strnad H, Lapidus A, Paces J, Ulbrich P, Vlcek C, Paces V, Haselkorn R (2010) Complete genome sequence of the photosynthetic purple nonsulfur bacterium *Rhodobacter capsulatus* SB 1003. J Bacteriol 192:3545–3546

Sun J, Hopkins RC, Jenney FE Jr, McTernan PM, Adams MWW (2010) Heterologous expression and maturation of an NADP-dependent [NiFe]-hydrogenase: a key enzyme in biofuel production. PLoS One 5:e10526

Sybirna K, Antoine T, Lindberg P, Fourmond V, Rousset M, Mejean V, Bottin H (2008) *Shewanella oneidensis*: a new and efficient system for expression and maturation of heterologous [Fe-Fe] hydrogenase from *Chlamydomonas reinhardtii*. BMC Biotechnol 8:73

Teixeira M, Fauque G, Moura I, Lespinat PA, Berlier Y, Prickril B, Peck HD Jr, Xavier AV, Le Gall J, Moura JJ (1987) Nickel-[iron-sulfur]-selenium-containing hydrogenases from *Desulfovibrio baculatus* (DSM 1743). Redox centers and catalytic properties. Eur J Biochem 167:47–58

Teixeira VH, Baptista AM, Soares CM (2006) Pathways of H_2 toward the active site of [NiFe]-hydrogenase. Biophys J 91:2035–2045

Topin J, Rousset M, Antonczak S, Golebiowski J (2012) Kinetics and thermodynamics of gas diffusion in a NiFe hydrogenase. Proteins 80:677–682

Tremblay P-L, Lovley DR (2012) Role of the NiFe hydrogenase Hya in oxidative stress defense in *Geobacter sulfurreducens*. J Bacteriol 194:2248–2253

van der Zwaan JW, Coremans JM, Bouwens EC, Albracht SP (1990) Effect of $^{17}O_2$ and ^{13}CO on EPR spectra of nickel in hydrogenase from *Chromatium vinosum*. Biochim Biophys Acta 1041:101–110

van Gastel M, Stein M, Brecht M, Schroder O, Lendzian F, Bittl R, Ogata H, Higuchi Y, Lubitz W (2006) A single-crystal ENDOR and density functional theory study of the oxidized states of the [NiFe] hydrogenase from *Desulfovibrio vulgaris* Miyazaki F. J Biol Inorg Chem 11:41–51

van Gastel M, Shaw JL, Blake AJ, Flores M, Schroder M, McMaster J, Lubitz W (2008) Electronic structure of a binuclear nickel complex of relevance to [NiFe] hydrogenase. Inorg Chem 47:11688–11697

Vargas WA, Weyman PD, Tong Y, Smith HO, Xu Q (2011) [NiFe] hydrogenase from *Alteromonas macleodii* with unusual stability in the presence of oxygen and high temperature. Appl Environ Microbiol 77:1990–1998

Vignais PM, Billoud B (2007) Occurrence, classification, and biological function of hydrogenases: an overview. Chem Rev 107:4206–4272

Vignais PM, Billoud B, Meyer J (2001) Classification and phylogeny of hydrogenases. FEMS Microbiol Rev 25:455–501

Vincent KA, Parkin A, Armstrong FA (2007) Investigating and exploiting the electrocatalytic properties of hydrogenases. Chem Rev 107:4366–4413

Volbeda A, Fontecilla-Camps JC (2004) Crystallographic evidence for a CO/CO(2) tunnel gating mechanism in the bifunctional carbon monoxide dehydrogenase/acetyl coenzyme A synthase from *Moorella thermoacetica*. J Biol Inorg Chem 9:525–532

Volbeda A, Charon MH, Piras C, Hatchikian EC, Frey M, Fontecilla-Camps JC (1995) Crystal structure of the nickel-iron hydrogenase from *Desulfovibrio gigas*. Nature 373:580–587

Volbeda A, Garcin E, Piras C, De Lacey AL, Fernandez VM, Hatchikian CE, Frey M, Fontecilla-Camps JC (1996) Structure of the [NiFe] hydrogenase active site: evidence for biologically uncommon Fe ligands. J Am Chem Soc 118:12989–12996

Volbeda A, Montet Y, Vernede X, Hatchikian CE, Fontecilla-Camps JC (2002) High-resolution crystallographic analysis of *Desulfovibrio fructosovorans* [NiFe] hydrogenase. Int J Hydrog Energy 27:1449–1461

Volbeda A, Martin L, Cavazza C, Matho M, Faber BW, Roseboom W, Albracht SP, Garcin E, Rousset M, Fontecilla-Camps JC (2005) Structural differences between the ready and unready oxidized states of [NiFe] hydrogenases. J Biol Inorg Chem 10:239–249

Volbeda A, Fontecilla-Camps JC, Sazanov L (2012) The evolutionary relationship between complex I and [NiFe]-hydrogenase. In: Sazanov L (ed) A structural perspective on respiratory Complex I: structure and function of NADH:ubiquinone oxidoreductase. Springer, Dordrecht, pp 109–121

Voordouw G, Hagen WR, Kruse-Wolters KM, van Berkel-Arts A, Veeger C (1987) Purification and characterization of *Desulfovibrio vulgaris* (Hildenborough) hydrogenase expressed in *Escherichia coli*. Eur J Biochem 162:31–36

Wells MA, Mercer J, Mott RA, Pereira-Medrano AG, Burja AM, Radianingtyas H, Wright PC (2011) Engineering a non-native hydrogen production pathway into *Escherichia coli* via a cyanobacterial [NiFe] hydrogenase. Metab Eng 13:445–453

Weyman PD, Vargas WA, Tong Y, Yu J, Maness P-C, Smith HO, Xu Q (2011) Heterologous expression of *Alteromonas macleodii* and *Thiocapsa roseopersicina* [NiFe] hydrogenases in *Synechococcus elongatus*. PLoS One 6:e20126

Wolf I, Buhrke T, Dernedde J, Pohlmann A, Friedrich B (1998) Duplication of *hyp* genes involved in maturation of [NiFe] hydrogenases in *Alcaligenes eutrophus* H16. Arch Microbiol 170:451–459

Yacoby I, Pochekailov S, Toporik H, Ghirardi ML, King PW, Zhang S (2011) Photosynthetic electron partitioning between [FeFe]-hydrogenase and ferredoxin:NADP+-oxidoreductase (FNR) enzymes in vitro. Proc Natl Acad Sci U S A 108:9396–9401

Yacoby I, Tegler LT, Pochekailov S, Zhang S, King PW (2012) Optimized expression and purification for high-activity preparations of algal [FeFe]-hydrogenase. PLoS One 7:e35886

Zhao J, Song W, Cheng J, Zhang C (2010) Cloning of Enterobacter aerogenes fh1A gene and overexpression of hydrogen production. Wei Sheng Wu Xue Bao 50:736–742

Zheng H, Zhang C, Lu Y, Jiang P-X, Xing X-H (2012) Alteration of anaerobic metabolism in *Escherichia coli* for enhanced hydrogen production by heterologous expression of hydrogenase genes originating from *Synechocystis* sp. Biochem Eng J 60:81–86

Zirngibl C, Van Dongen W, Schworer B, Von Bunau R, Richter M, Klein A, Thauer RK (1992) H_2-forming methylenetetrahydromethanopterin dehydrogenase, a novel type of hydrogenase without iron-sulfur clusters in methanogenic archaea. Eur J Biochem 208:511–520

Chapter 4

H_2 Production Using Cyanobacteria/Cyanobacterial Hydrogenases: From Classical to Synthetic Biology Approaches

Catarina C. Pacheco*, Paulo Oliveira*
Bioengineering and Synthetic Microbiology Group, IBMC – Instituto de Biologia Molecular e Celular, Universidade do Porto, Rua do Campo Alegre 823, 4150-180 Porto, Portugal

and

Paula Tamagnini**
Bioengineering and Synthetic Microbiology Group, IBMC - Instituto de Biologia Molecular e Celular, Universidade do Porto, Rua do Campo Alegre 823, 4150-180 Porto, Portugal

Faculdade de Ciências, Departamento de Biologia, Universidade do Porto, Rua do Campo Alegre, Edifício FC4, 4169-007 Porto, Portugal

Summary

The simple nutritional requirements of cyanobacteria, the availability of molecular tools and genome sequences, as well as the recent genome scale-models of *Synechocystis* sp. PCC 6803 make these photosynthetic prokaryotes attractive platforms for the production of added-value

*These Authors contributed equally to this work.

**Author for correspondence, e-mail: pmtamagn@ibmc.up.pt

D. Zannoni and R. De Philippis (eds.), Microbial BioEnergy: Hydrogen Production, Advances in Photosynthesis and Respiration 38, pp. 79–99, DOI 10.1007/978-94-017-8554-9_4,

compounds, namely hydrogen. Naturally, these organisms may contain up to three types of enzymes directly involved in hydrogen metabolism: one or more nitrogenases that evolve H_2 concomitantly with N_2 fixation, an uptake hydrogenase that recycles the H_2 released by the nitrogenase, and a bidirectional hydrogenase. Significant contributions from studies of genetic engineering, transcriptional regulation and maturation, and assessment of enzymatic activities in response to various environmental cues led to gaining further control of the mechanisms by which one can engineer cyanobacteria for H_2 production. In this chapter, the classical approaches of screening natural communities for improved activities, the manipulation of native enzymes and/or growth conditions, and the expression of heterologous hydrogenases into or from cyanobacteria will be summarized. Moreover, the recent synthetic biology approaches pursuing the use of cyanobacteria as photoautotrophic chassis, as well as the construction and characterization of synthetic parts and devices aiming at improving H_2 production will be reviewed. However, the generated knowledge and the molecular/synthetic tools developed in this field represent a valuable contribution not only to improve biohydrogen production but also to further drive the technology related to other biofuels and various industrial applications.

I. Introduction

Besides sharing the basic cellular features of other Bacteria, cyanobacteria possess unique and diagnostic characteristics. Distinctively, cyanobacteria are the only organisms ever to evolve coupled photosystems that harvest electrons from water and produce oxygen as a byproduct (Knoll 2008). They are photosynthetic prokaryotes typically possessing the ability to synthesize chlorophyll *a* (Whitton and Potts 2000) although four distinctive lineages produce alternative chlorophyll pigments (Swingley et al. 2008).

In terms of Earth history, cyanobacteria occupy a privileged position among organisms. As primary producers they play a significant role in Earth's carbon cycle; as nitrogen fixers, they also figure prominently in the nitrogen cycle (Knoll 2008). Moreover, they also loom large in our planet's redox history (Knoll 2008). One of the major changes on Earth, the introduction of oxygen into the atmosphere is widely accepted to be attributed to the photosynthetic activity of cyanobacteria. Cyanobacterial ecological plasticity is remarkable and their long evolutionary history is possibly related to their success in modern habitats. They are mostly found in aquatic, but also in many terrestrial environments (Whitton and Potts 2000). They can even be found growing near the limits for life in the dry deserts of Antarctica or in many thermal springs (Whitton and Potts 2000). Symbiotic interactions between cyanobacteria and other organisms are surprisingly diverse (Costa 2004): in these associations, cyanobacteria provide different hosts with fixed carbon, with the product of nitrogen fixation or with both products (Costa 2004). The secret that certainly contributes for the cyanobacterial easy adaptation to numerous ecological niches and makes them the driving force for shaping nearly every ecosystem on Earth is definitely their physiological flexibility.

Cyanobacteria are the only diazotrophs that perform oxygenic photosynthesis, and since the enzymatic complex carrying out nitrogen fixation – the nitrogenase – is extremely sensitive to oxygen, they present different strategies to cope with this incompatibility. Although these strategies are usually divided in spatial (N_2 fixation in specialized cells – the heterocysts) or temporal (N_2 fixation confined to the dark period) (Herrero et al. 2001), there is a range of mechanisms far more complex (for details see Berman-Frank et al. 2003). Nitrogen fixing cyanobacteria contain the so-called conventional nitrogenase with Mo and Fe in the active site, but some heterocystous strains

Abbreviations: OCD – Oxygen Consuming Device; SB – Synthetic Biology e.g. *Anabaena variabilis* may also contain “alternative” nitrogenases (Thiel 1993; Bothe et al. 2010). Under Mo-deprived conditions the MoFe nitrogenase is replaced by a V nitrogenase, or in the absence of this element by a Fe only-nitrogenase (the existence of this enzyme has not yet been established in cyanobacteria). The reduction of atmospheric nitrogen to ammonium is always accompanied by the formation of molecular hydrogen as a byproduct. It has been reported that alternative nitrogenases are better H_2 producers compared with the MoFe nitrogenases but very little work has been performed on this subject. For an overview see Bothe et al. (2010).

Closely related to the ability to fix nitrogen, cyanobacteria possess a distinct H_2 metabolism. The H_2 produced by nitrogenase is rapidly consumed by an uptake hydrogenase, an enzyme that has been found in all nitrogen fixing cyanobacteria examined so far. In 2006, Ludwig et al. (2006) reported the existence of a N_2-fixing strain, *Synechococcus* sp. BG 043511, naturally lacking an uptake hydrogenase and this was believed to be the first exception to the rule. However, recent work carried out by Skizim and co-workers has shown that this strain, now referred to as *Cyanothece* sp. Miami BG 043511, does in fact possess not only the genes but also a functional hydrogenase (Skizim et al. 2012). Additionally, cyanobacteria may contain a bidirectional hydrogenase, an enzyme that is generally present in non-nitrogen fixing strains. However, it is absent in *Gloeobacter violaceus* PCC 7421, a cyanobacterium that possesses a number of unique characteristics such as the absence of thylakoids (Nakamura et al. 2003).

In summary, cyanobacteria may contain up to three types of enzymes directly involved in hydrogen metabolism: one or more nitrogenases that evolve H_2 concomitantly with dinitrogen fixation, an uptake hydrogenase that recycles the H_2 released by the nitrogenase(s), and a bidirectional hydrogenase. In these organisms, H_2 is naturally produced by the nitrogenase or in specific conditions (e.g. dark anaerobic) by the bidirectional hydrogenase. Therefore, given what is known in terms of H_2 metabolism, minimal nutritional requirements, and physiology, cyanobacteria emerge as prime candidates for H_2 production. In this chapter we focus on the efforts to manipulate their native enzymes and/or environmental conditions, as well as in the recent synthetic biology approaches to improve H_2 production.

II. Transcriptional Regulation and Maturation of Cyanobacterial Hydrogenases

Transcriptional regulation of cyanobacterial hydrogenases has deserved some attention over time, and significant advances have been accomplished in this field. In an attempt to assess the role of hydrogenases in cell fitness and physiological flexibility, several different strains have been grown (or subjected) to numerous environmental conditions, and the understanding of which cues trigger an up- or down-regulation of the hydrogenases genes transcription is fairly understood. It is not our intention to extensively review this topic here (for more details, the reader is suggested to see Oliveira and Lindblad 2009; Bothe et al. 2010). Nevertheless, given the fact that several contributions have been made in recent years it is worth referring in a bit more detail the work carried out on the characterization of transcription factors and transcription networks.

On one hand, the cyanobacterial uptake hydrogenase seems to be under the control of NtcA in unicellular (Oliveira et al. 2004), filamentous (Leitão et al. 2005; Ferreira et al. 2007) and heterocystous (Weyman et al. 2008; Holmqvist et al. 2009) strains. NtcA operates global nitrogen regulation in cyanobacteria, and these observations support the notion that part of the hydrogen metabolism is clearly related with nitrogen metabolism, namely overlapping with the process of nitrogen fixation.

On the other hand, cyanobacterial AbrB-like regulators have been shown to play a key role in the bidirectional hydrogenase genes transcription. Cyanobacterial AbrB-like proteins (also known as CalA and CalB, CyAbrB clade A and clade B, or AbrB1 and AbrB2) present a certain degree of sequence homology to the *Bacillus subtilis* AbrB, which is a transition state regulator and involved in spore formation (Phillips and Strauch 2002). However, despite the sequence homology, the cyanobacterial AbrB-like proteins do not seem to be involved in the regulation of the same set of genes as in *B. subtilis*. Instead, the cyanobacterial AbrB-like regulator AbrB2 (CalB – Sll0822) was shown to have a broad range of regulatory points: (i) it works in parallel with NtcA to achieve flexible regulation of the nitrogen uptake system in *Synechocystis* sp. PCC 6803 (Ishii and Hihara 2008), (ii) AbrB2 is also directly involved in the modulation of low-CO_2-induced gene expression (Lieman-Hurwitz et al. 2009), and (iii) it regulates the promoter activity of functional antisense RNA of an operon that plays a crucial role in photoprotection of photosystem II under low carbon conditions (Eisenhut et al. 2012). Moreover, AbrB2 was demonstrated to interact with itself (Sato et al. 2007), suggesting to work as an oligomer, and most importantly to function as an autorepressor (Dutheil et al. 2012), inhibiting as well the bidirectional hydrogenase operon transcription in *Synechocystis* sp. PCC 6803 (Ishii and Hihara 2008; Dutheil et al. 2012). Furthermore, AbrB1 (CalA – Sll0359) has been equally implicated in the regulation of the hydrogenase operon in *Synechocystis* sp. PCC 6803, suggested to work as an activator (Oliveira and Lindblad 2008). The regulatory network operating on the fine tuning of the hydrogenase operon expression in *Synechocystis* sp. PCC 6803 becomes more complicated with three additional pieces of evidence: firstly, AbrB1 and AbrB2 have the capacity to interact with each other, as assessed by yeast two hybrid interactions (Sato et al. 2007) and his-tag pull down assays (Yamauchi et al. 2011), indicating that a regulatory balance may be achieved by protein-protein interactions; secondly, the AbrB-like proteins were shown to be modified on a post-translational level (Shalev-Malul et al. 2008), which introduces another level of possible regulation; and finally, a third transcription factor has also been shown to positively regulate the expression of this operon, namely LexA (Gutekunst et al. 2005), which can also be modified on a post-translational level (Oliveira and Lindblad 2011).

The action of AbrB-like proteins on the transcriptional regulation of the genes involved in hydrogen metabolism in cyanobacteria is not limited to the bidirectional hydrogenase. Recent work by Holmqvist and co-workers showed that AbrB1 binds specifically to the promoter region of *hupSL*, genes encoding the uptake hydrogenase in *Nostoc punctiforme* ATCC 29133 (Holmqvist et al. 2011), suggesting a possible active role in controlling their expression. *Anabaena* sp. PCC 7120 AbrB1 (Alr0946) was found to regulate negatively the hydrogenase accessory gene *hypC* (Agervald et al. 2010). Finally, in the filamentous non-heterocystous cyanobacterium *Lyngbya majuscula* CCAP 1446/4, NtcA was shown to interact specifically with the *hyp* genes promoter (Ferreira et al. 2007), indicating a possible co-regulation of *hyp* and *hup* genes via NtcA in response to various sources and amounts of nitrogen in the medium.

Hydrogenase accessory proteins, like HypC mentioned above are chaperones that assemble the various components of the hydrogenase NiFe active site and coordinate it with the enzyme's amino acid structure. A final maturation step consists of a proteolytic cleavage of the hydrogenase large subunit C-terminus, which can be regarded as a control checkpoint since it occurs only when all the active site components have been properly introduced. This step precedes the assembly between the hydrogenase small and large subunits, rendering a functional enzyme. The literature on this topic is quite scarce in respect to cyanobacteria and most of what is known regarding maturation of hydrogenases derives from work carried out in e.g. *Escherichia coli*. Considering that the hydrogenase accessory genes in *E. coli* share a high degree of homology with the cyanobacterial ones, it was generally assumed that the Hyp proteins in cyanobacteria would play a similar role. It was the work of Hoffmann and co-workers that finally set this question, when they were able to show convincingly that the hydrogenase accessory proteins in *Synechocystis* sp. PCC 6803 have a direct involvement in the bidirectional hydrogenase maturation (Hoffmann et al. 2006). In filamentous heterocystous strains the only work available addressing the role of the hydrogenase accessory genes on the enzyme's maturation was done by Lindberg and co-workers, assessing the function of *hupW* in *Anabaena* sp. PCC 7120 (Lindberg et al. 2012). In this article, *hupW* is presented as the specific protease that completes the uptake hydrogenase maturation, leading to a hydrogen evolving phenotype (see below). In summary, the main driving forces for studying transcriptional regulation of cyanobacterial hydrogenases and the complex process of hydrogenase maturation have been (i) to improve the understanding of the physiological relevance of the hydrogen metabolism in cyanobacteria, (ii) to unravel novel overlapping points with other metabolic pathways, and (iii) to gain further control on the mechanisms by which one can engineer cyanobacteria for H_2 production.

III. Strategies to Improve Cyanobacterial H_2 Production

Over the years, several strategies have been adopted in an ultimate effort to improve cyanobacterial H_2 production. In the present review, we divide these efforts into "classical" and "synthetic biology" approaches. The first includes conventional work carried out in different domains of cell and molecular biology, ranging from thorough selections of cyanobacterial H_2 producers to simple genetic manipulations of key enzymes involved in the hydrogen metabolism. The latter represents a modern approach, where standardized and interchangeable modules are routinely used to assist rational metabolic engineering methodologies.

A. Classical Approaches

1. Screening for Strains/Enzymes with Improved Activities

To work with widely used and well characterized cyanobacterial model organisms has clear advantages regarding the study of the hydrogen metabolism. Genetic tools for their manipulation are usually well established (enabling the production of both knock-outs and knock-ins) and the amount of physiological and genetic data available supports a reasonably good understanding of the organism's overall metabolism. However, it is evident that cyanobacterial strains yet to be described may have different and more attractive performances regarding hydrogen evolution than the standard reference cyanobacteria. Therefore, serious efforts have been put together to screen both existing culture collections and environmental samples to search for more efficient H_2 producers. The work carried out by Allahverdiyeva et al. (2010) represents a good example of such approach: they describe a screen of 400 cyanobacterial isolates from the Baltic Sea and Finnish lakes and conclude that roughly half of the tested strains produce detectable amounts of H_2. Interestingly, ten of the evaluated strains evolved similar or up to four times as much hydrogen as uptake hydrogenase deletion mutants created in various laboratories and specifically engineered to evolve higher amounts of H_2 (Allahverdiyeva et al. 2010). These newly described strains represent excellent candidates for further characterization and future genetic engineering. Cyanobacterial strains isolated from diverse environments, including terrestrial, freshwater and marine intertidal settings, have also been the subject of recent studies aiming at quantitatively comparing their hydrogenase activities under non-nitrogen fixing conditions (Kothari et al. 2012). The authors selected these particular environments because most of the hydrogen evolving cyanobacteria described so far are originally from freshwater bodies, making their study unique in respect to origin, morphology, taxonomy and phylogeny. The main conclusions withdrawn from the work are that strains isolated from freshwater and intertidal settings had a high incidence of hydrogen production (concurrent with the presence of *hoxH*, encoding the bidirectional hydrogenase large subunit), while all terrestrial isolates were negative for both traits. Moreover, and most interestingly, some novel strains displayed rates of hydrogen evolution several fold higher than those previously reported, making them potentially interesting for biohydrogen production (Kothari et al. 2012).

In addition, cyanobacterial strains that have been once the subject of studies regarding various metabolic processes (e.g. photosynthesis, and in particular nitrogen fixation) are now evaluated in terms of their H_2 evolution capacity. It is worth mentioning the work done with *Cyanothece* strains (Bandyopadhyay et al. 2010, 2011; Min and Sherman 2010a, b). *Cyanothece* is a genus of unicellular nitrogen-fixing cyanobacteria that have the remarkable capacity of performing both oxygenic photosynthesis (evolving molecular oxygen) and nitrogen-fixation (a highly oxygen sensitive reaction) in the same cell. The ability to fix N_2 and produce H_2 was assessed in six *Cyanothece* strains: ATCC 51142, PCC 7424, PCC 7425, PCC 7822, PCC 8801 and PCC 8802 (Bandyopadhyay et al. 2011). *Cyanothece* sp. ATCC 51142 showed the highest nitrogenase activity as well as H_2 production, followed by *Cyanothece* sp. PCC 8802 and PCC 8801. From all the strains evaluated in this work, *Cyanothece* sp. PCC 7425 was the only one unable to fix N_2 (and produce H_2) under aerobic conditions and with lower nitrogenase activity (Bandyopadhyay et al. 2011). More extensive work has been carried for *Cyanothece* sp. ATCC 51142. This cyanobacterium was shown to produce H_2 either through the bidirectional hydrogenase or the nitrogenase; in the latter case the yield was 30- to 60-fold higher (Min and Sherman 2010a). H_2 evolution by the hydrogenase was shown to be dependent on the electron transport by photosystem II whereas the activity of nitrogenase relies on photosystem I and respiration. Furthermore, it was demonstrated that H_2 production and N_2 fixation occurred at high rates even under continuous light conditions (when *hupSL* transcript levels are low). In addition, the results indicated that low-oxygen conditions (argon sparging) favor H_2 production and N_2 fixation by the nitrogenase (Min and Sherman 2010a). Also in 2010, Bandyopadhyay and co-workers reported that the same *Cyanothece* strain (ATCC 51142) could generate high levels of H_2 under N_2 fixing and natural aerobic conditions. The production yield was significantly enhanced in cultures supplemented with glycerol or high CO_2 concentrations, in which the excess of carbon source functions as a signal for enhanced nitrogenase activity (Bandyopadhyay et al. 2010). H_2 evolution was also evaluated in *Cyanothece* sp. Miami BG 043511 and production by the hydrogenase occurred in dark anoxic conditions, using the reducing power derived from fermentation, while H_2 production by the nitrogenase occurred in light, relying on photosystem I (Skizim et al. 2012). Further work performed by Min and Sherman (2010b) showed that in a *Cyanothece* sp. PCC 7822 deletion mutant (Δ*nifK*) defective in N_2 fixation, cells were unable to grow in absence of combined nitrogen and H_2 production was nearly abolished in air or low-oxygen conditions.

Cyanobacterial strains with outstanding physiological characteristics and selected for particular growth traits are also being target of thorough studies (Taton et al. 2012). The combination of their natural capacities and increased environmental fitness with the amenability to genetic modifications/manipulations represents an attractive and complementary way of moving forward towards a more sustainable cyanobacterial H_2 producer.

Technologies for high-throughput sequencing are developing fast and respective costs are dropping. Therefore, its application in metagenomics has been tremendous and it has enabled the acquisition of information at a rate never imagined before. Making use of this powerful technique, an alternative approach to screen strains for their natural ability to evolve molecular hydrogen has emerged, consisting instead of searching for novel hydrogenase genes, which may encode enzymes with a higher oxygen tolerance and/or a higher H_2 evolving activity. The focus is therefore concentrated on finding novel genes rather than on the organism itself, since maintaining and growing a microorganism under laboratory conditions may proof difficult. The global ocean sampling metagenomic database (Venter et al. 2004; Rusch et al. 2007; Yooseph et al. 2007) represents one of the most adequate starting points for such quest. Barz and co-workers (2010) have taken the first steps into this matter by searching the database for all the described families of hydrogenases. In parallel, the authors have also investigated DNA isolated from samples taken from the North Atlantic, Mediterranean Sea, North Sea, Baltic Sea and two fresh water lakes for the presence of genes encoding the bidirectional hydrogenase. Surprisingly, this study shows that hydrogenases are quite abundant in marine environments and that marine surface waters could be a source of oxygen-resistant uptake hydrogenases. Furthermore, this approach helps clarifying the primary function of hydrogenases based on their ecological distribution (Barz et al. 2010). Certainly in a near future, new metagenomic studies looking into unexplored environments will be presented and novel hydrogenase genes will be uncovered with possible improved features. The challenging venture will be to take that genetic information into a cyanobacterium, wire the functional hydrogenase with the organism's metabolism and enhance the organism performance in terms of H_2 production rates and sustainability. As a matter of fact, similar approaches have been done in which cyanobacteria are used to express a heterologous hydrogenase (see below), the main difference here being that the foreign hydrogenases usually originate from well-known organisms, grown under laboratory conditions and with various limitations.

2. Tuning Physiological/Environmental Conditions

Another point that has attracted considerable attention over the years regarding the improvement of cyanobacterial H_2 production is the environment in which the cells are grown (or kept) in. When media compositions were initially established, maximization of the cells capacity to evolve molecular hydrogen was not top priority. Therefore, extensive work has been carried out to find the best conditions for higher and more sustainable production rates and total amounts. Ananyev et al. (2008) and Carrieri et al. (2010, 2011) in *Arthrospira* (*Spirulina*) *maxima* CS-328 and Burrows et al. (2008, 2009, 2011) in *Synechocystis* sp. PCC 6803 constitute a few examples of such work.

The effort that is being carried out in this field is in close association to the developments achieved in bioreactor design. It is not our goal to review what is being done in this area (for good and recent reviews on the topic, the reader is suggested to read Rupprecht et al. 2006; Kumar et al. 2011; Show et al. 2011), but it is impossible to miss out the multiple and alternative methods implemented in the most varied scales to optimize cyanobacterial H_2 production.

3. Knockout Mutants/Protein Engineering

Once the hydrogen metabolism became described in cyanobacteria and hydrogenase activities determined and characterized, identifying the genes coding for the hydrogenases structural and accessory proteins became the necessary following step. The availability of genetic tools to manipulate cyanobacteria turned out to be extremely useful to further characterize these genes and to determine their specific role in the hydrogen metabolism. The gene coding for the uptake hydrogenase large subunit has been a clear candidate for knockout constructions in various cyanobacterial strains, namely in *Anabaena variabilis* ATCC 29413 (Happe et al. 2000), *Nostoc punctiforme* ATCC 29133 (Lindberg et al. 2002), *Anabaena* sp. PCC 7120 (Masukawa et al. 2002), *Nostoc* sp. PCC 7422 (Yoshino et al. 2007), *Anabaena siamensis* TISTR 8012 (Khetkorn et al. 2012). In all these studies, the knockout organism was able to evolve significant higher amounts of H_2 than the respective wild-type, as a direct consequence of the organism's incapacity to recycle the molecular hydrogen released as byproduct of nitrogen fixation. Moreover, a proteomic study identified the specific metabolic processes used by the *N. punctiforme* mutant NHM5 to maintain the high rate of nitrogen fixation despite the absence of the uptake hydrogenase (Ekman et al. 2011). Interestingly, all strains used in the studies mentioned above are invariably filamentous, heterocyst-forming cyanobacteria. In addition, a *hupW* (gene putatively coding for the uptake hydrogenase large subunit protease) deletion mutant in *Anabaena* sp. PCC 7120 was recently described (Lindberg et al. 2012). In this study, it was found that molecular hydrogen accumulates in comparable amounts as for the *hupL* knockout (Lindberg et al. 2012), suggesting that nitrogenase based H_2 evolution can also be attained by silencing the uptake hydrogenase maturation system.

It is quite apparent at this point that using nitrogenase as the H_2 producing enzyme in cyanobacteria represents a fairly common practice to accumulate desirable amounts of H_2. However, the nitrogenase is a very energy demanding enzymatic complex (it is estimated that 16 ATP molecules are necessary to fix one molecule of N_2 into two of NH_3) and H_2 production simply results as a byproduct (one molecule released per 16 ATP molecules). Therefore, engineering the cyanobacterial nitrogenase has deserved some focus with the clear goal of converting it into a more efficient H_2 producing device. Nevertheless, it should be noted that the nitrogenase is a quite complex multimeric enzymatic system, requiring metal coordination and incorporation into the protein's active site and insertion of FeS clusters that mediate the transfer of electrons between ferredoxin and the active site for N_2 reduction. Both of these processes depend on several well-known chaperones, but also on yet to be described mechanisms. Thus, the room for engineering cyanobacterial nitrogenases (or any other nitrogenase for that matter) is somewhat limited. Regardless of the difficulties, Weyman et al. (2010) created modifications on an alternative Mo-nitrogenase (Nif2) of *Anabaena variabilis*, which is expressed in vegetative cells grown with fructose under strictly anaerobic conditions. A V75I substitution in the α-subunit rendered an enzyme with greatly impaired acetylene reduction and reduced levels of $^{15}N_2$ fixation, but with a fourfold higher H_2 production in the presence of N_2 compared with the wild-type and similar H_2 production rates as the wild-type enzyme in an argon atmosphere (Weyman et al. 2010). The authors concluded that the amino acid substitution did not change the ability of the mutant enzyme to reduce substrates, but instead simply increased the selectivity for substrates. A more comprehensive study was carried out by Masukawa and co-workers, targeting the nitrogenase active site of *Anabaena* sp. PCC 7120. Their efforts were concentrated on six nitrogenase amino acid residues (Q193, H197, Y236, R284, S285, F388) predicted to be located within 5 Å of the metal MoFe active center, aiming at directing electron flow selectively towards proton reduction (Masukawa et al. 2010). The ability to fix N_2 was moderately or severely impaired in nearly all NifD variant strains examined, a necessary compromise if one aims to redistribute electrons toward proton reduction. In terms of H_2 accumulation during long-term incubations under physiologically relevant conditions, 11 selected *Anabaena* sp. PCC 7120 NifD variants (namely Q193S, Q193H, Q193L, Q193K, H197T, H197F, H197Q, Y236T, R284T, R284H and F388H) greatly exceeded the H_2 produced by the reference strains (ΔNifHΔHupL, ΔHupL, AnNifHΔHupL). Among the NifD variants, R284H presented the best performance, since the H_2 accumulated under N_2 after 1 week was 87 % of what was observed for the reference strains under an argon atmosphere (Masukawa et al. 2010). These encouraging results highlight the potential of using some of these nitrogenase variants as parental strains for additional engineering steps in a concerted effort to improve photobiological H_2 production. Still regarding the use of nitrogenase as the key factor for H_2 production in cyanobacteria, but moving away from the protein engineering approach, an alternative strategy has been attempted. This involved the deletion of the homocitrate synthase genes (*nifV1* and *nifV2*) in *Anabaena* sp. PCC 7120 (Masukawa et al. 2007). The catalytic site of dinitrogenase normally binds the FeMo cofactor to which homocitrate is ligated. However, mutating *nifV* in various microorganisms resulted in significantly different nitrogenase *in vivo* activities. By creating Δ*nifV1*, Δ*nifV2* and Δ*nifV1*Δ*nifV2* mutants in the Δ*hupL Anabaena* sp. PCC 7120 strain diazotrophic growth rates were decreased moderately to severely depending on the mutant in comparison to the rates of the parental strain (Masukawa et al. 2007). The simple fact that Δ*nifV1*Δ*nifV2* mutant cells could grow under combined nitrogen depleted conditions indicates that the nitrogenase of this strain can fix N_2 in the absence of homocitrate. Interestingly, hydrogen production was found to be higher and more sustainable under air in the Δ*nifV1*mutant than in any other evaluated strain, suggesting this can be a useful strategy for improving photobiological H_2 production by cyanobacteria (Masukawa et al. 2007).

The potential of the cyanobacterial bidirectional hydrogenase to produce molecular hydrogen has also been extensively explored. *Synechocystis* sp. PCC 6803, as one of the most well studied cyanobacterial strains, has been a clear candidate for metabolic engineering. According to Gutthann et al. (2007), cells of *Synechocystis* sp. PCC 6803 growing under normal conditions have four main electron sinks competing with each other, namely the Calvin-cycle (and respective inorganic carbon concentrating mechanisms), nitrate, oxygen and protons. Gutthann's suggestion was brought up based on various observations and some follow below. The *ndhB* gene codes for a key subunit of all NDH-1 (Type I NADPH-dehydrogenase) complexes, and consequently a knockout mutant of this gene leads to a strain devoid of any of such complexes. A Δ*ndhB* mutant (M55) has been created (Ogawa 1991) and one of its most remarkable features has to do with its slow growth rates when compared to the wild-type. The phenotype can be explained by an impairment of the inorganic carbon concentrating mechanisms and by a reduced carbon fixation capacity (Ogawa 1991; Ohkawa et al. 2000). This M55 mutant has been evaluated in terms of H_2 production ability and it was found to evolve significant amounts of H_2 (Cournac et al. 2002, 2004). M55 is specifically impaired in NADPH oxidation and to avoid an over-reducing state of the cell it is believed that the bidirectional hydrogenase releases the electron excess by evolving H_2. Alternatively, Baebprasert et al. (2011) have targeted the nitrate assimilation pathway by creating *Synechocystis* sp. PCC 6803 strains lacking functional nitrate reductase (Δ*narB*) or nitrite reductase (Δ*nirA*) or both, and tested for their ability to produce hydrogen. In this work, it is reported that all engineered strains present a higher capacity of hydrogen production as well as higher hydrogenase activity than those of wild-type, and the best hydrogen producer was found to be the double mutant. The reduction of nitrate to ammonia before its incorporation to amino acids requires two sequential reactions carried out by nitrate reductase and nitrite reductase, demanding a large amount of reducing power. Therefore, by eliminating such electron sink the cell undergoes an unbalanced redox state which can be attenuated by the reduction of protons via bidirectional hydrogenase. Furthermore, Gutthann et al. (2007) made use of various oxidases deletion mutants (cytochrome *c* oxidase – *ctaI*, alternative cytochrome *c* oxidase – *ctaII*, quinol oxidase – *cyd*) that are deficient in extracting electrons from cytochrome *c* or from plastoquinone to reduce O_2 to water, to assess several aspects of the hydrogen metabolism in *Synechocystis* sp. PCC 6803. The study shows that the deletion of quinol oxidase in combination with one of the two cytochrome *c* oxidases causes a prolonged H_2 production phase in the light and a higher maximal amount, supporting their suggestion that O_2 is an important electron sink that competes with protons for reductant. All these pieces of evidence seem to support the idea that the bidirectional hydrogenase in the unicellular cyanobacterium *Synechocystis* sp. PCC 6803 functions as a valve for low-potential electrons generated during the light reaction of photosynthesis, thus preventing a slowing down of electron transport (Appel et al. 2000; Carrieri et al. 2011; Pinto et al. 2012). Therefore, understanding the physiological role of the cyanobacterial bidirectional hydrogenase will definitely lead to a more rational design of metabolic pathways, envisioning engineered strains with improved hydrogen production capacity.

Alternative approaches have been adopted to improve cyanobacterial hydrogen production via bidirectional hydrogenase, including genetic engineering of fermentative pathways. Within this line, McNeely et al. (2010) report their efforts on redirecting reductant into hydrogen production using the unicellular *Synechococcus* sp. PCC 7002. This cyanobacterium has a quite distinct dark anaerobic metabolism when compared to other cyanobacteria, since it is capable of yielding up to five fermentation products (McNeely et al. 2010). Lactate was found to be the product excreted in higher amounts, and therefore an *ldhA* (coding for D-lactate dehydrogenase) knockout mutant was created. Among several phenotypic traits, Δ*ldhA* cells presented an altered fermentative flux in comparison to wild-type cells, which reflected on an up to fivefold increase in hydrogen production (McNeely et al. 2010). In addition, the same lab has also focused on the pyruvate:ferredoxin oxidoreductase (NifJ) of *Synechococcus* sp. PCC 7002, which reduces ferredoxin during fermentation of pyruvate to acetyl-coenzyme A (McNeely et al. 2011). Interestingly, and even though Δ*nifJ* cells presented a higher *in vitro* hydrogenase activity, levels of H_2 produced were 1.3-fold lower than those accumulated by wild-type cells (McNeely et al. 2011). Regardless of the lower performance of Δ*nifJ* cells in terms of H_2 production, this study contributed significantly to a better understanding of the bidirectional hydrogenase role in *Synechococcus* sp. PCC 7002: it serves as a valve to reoxidize excess NADH formed during glycolysis, in addition to oxidizing reductant generated from ferredoxin, produced by the pyruvate:ferredoxin oxidoreductase (McNeely et al. 2011).

Despite all the efforts to find or produce an attractive and sustainable cyanobacterial H_2 producer, it became increasingly evident as new reports were available that different research groups with different laboratory routines grow these potentially interesting organisms in the most varied ways (e.g. light quality, intensity and regimen, media composition) and present H_2 evolution rates and amounts in multiple units and manners. Trying to compare activities and to assess which strain presents the best performance in terms of H_2 evolution represents therefore a really difficult challenge. Lopes Pinto et al. (2002) took on their hands the hard task to screen through the literature and present a uniform landscape of the cyanobacterial H_2 producers and respective H_2 evolving capacities. Alternatively, others opted to study various cyanobacterial strains (including both wild-types and H_2 evolving mutants), grown under the same culturing conditions and evaluated their H_2 evolving capacities, presenting a more comparable picture (Schütz et al. 2004).

4. Expression of Heterologous Hydrogenases in Cyanobacteria

The first work reporting the introduction and expression of a heterologous hydrogenase into a cyanobacterium, the unicellular non-N_2-fixing *Synechococcus elongatus* PCC 7942, was published in 2000 by Asada et al. These authors cloned the gene encoding the monomeric Fe hydrogenase I (*cpI*) from *Clostridium pasteurianum* downstream a strong native promoter and exchanged the Shine-Dalgarno sequence. After anaerobic adaption, the transformed *Synechococcus elongatus* PCC 7942 cells exhibited a significantly higher H_2 evolution (threefold increase) compared to the wild-type (Table 4.1).

Berto et al. (2011) reported to have expressed an active monomeric Fe hydrogenase, from the unicellular green alga *Chlamydomonas reinhardtii*, in another

Table 4.1. Expression of heterologous hydrogenases in cyanobacteria/use of cyanobacteria as a photoautothrophic chassis.

Cyanobacterial host	Fe-hydrogenase		NiFe-hydrogenase		Promoter	Site/plasmid	Protein expression	Protein activity	References
	Structural genes inserted	Accessory genes inserted	Structural genes inserted	Accessory genes inserted					
Synechococcus elongatus PCC 7942	*cpI* from *Clostridium pasteurianum*	None			Strong native promoter and exchanged Shine-Dalgarno sequence	Plasmid pKE4-9	Yes	Yes	Asada et al. (2000)
	hydA from *Clostridium acetobutylicum*	*hydEF*, *hydG* exogenous ferredoxins			P_{lac} – IPTG inducible (*hydA* and ferredoxins) P_{psbA1} – constitutive (*hydEF*, *hydG*)	Neutral site NS3 (*hydA*) Neutral site NS1 (*hydEF*, *hydG*)	Yes	Yes	Ducat et al. (2011b)
Synechococcus elongatus Δ*hoxYH* – strain PW416			*hynS and hynL* from *Alteromonas macleodii*	*orf1*, *cyt*, *orf2*, *hynD*, *hupH* and *hypCABDFE*	P_{trc} – IPTG inducible	Neutral site NS1	Yes	Yes	Weyman et al. (2011)
			hynSL from *Thiocapsa roseopersicina*	All accessory genes from *T. roseopersicina*		Neutral site NS2	Yes	No	
				+11 accessory genes from *A. macleodii*			Yes	Yes	
Synechocystis sp. PCC 6803	*hydA1* from *Chlamydomonas reinhardtii* (without the region encoding the N-terminal transit peptide)	None			P_{psbA2} – light inducible P_{ziaA} – Zn^{2+} inducible	Genome, between *sll1865* and *sll1864*	Yes	Yes	Berto et al. (2011)
Nostoc sp. PCC 7120	*hydA* and *hydB* (modified), from *Shewanella oneidensis* MR-1	S03922 *hydG*, S03924, *hydE, hydF*			P_{hetN} – heterocyst specific		Yes	Yes, in N_2 fixing conditions only	Gärtner et al. (2012)

unicellular non-N_2-fixing cyanobacterium *Synechocystis* sp. PCC 6803, in the absence of the native maturation system but already truncated at the N-terminal (Table 4.1). The authors suggested that the cyanobacterium hydrogenase accessory proteins could account for the folding/maturation of the algal hydrogenase, and that the low activity levels observed were consistent with what was expected for a heterologous enzyme working outside its specific environment (the enzyme is cloroplastidial in the alga) or that the cyanobacterial maturation machinery could impair the activity of the enzyme. To further prove their hypothesis they suggest expressing the *C. reinhardtii* hydrogenase in *Synechocystis* sp. PCC 6803 lacking the different *hyp* genes. In addition, *Synechocystis* sp. PCC 6803 mutant lacking the native hydrogenase could be used (see Pinto et al. 2012).

The genes encoding the oxygen tolerant and thermostable NiFe hydrogenase from *Alteromonas macleodii* Deep ecotype, as well as the HynSL from *Thiocapsa roseopersicina* were expressed in *Synechococcus elongatus* PCC 7942 hydrogenase knockout mutant (PW416 strain), using an IPTG-inducible promoter (Weyman et al. 2011). The *hynSL* genes from *A. macleodii*, together with 11 adjacent genes, were introduced into the previously identified neutral site 1, NS1 (Andersson et al. 2000). The *hynSL* genes of *T. roseopersicina* were introduced in neutral site 2 (NS2, Andersson et al. 2000) also via homologous recombination, together with different sets of accessory genes (Table 4.1). While the authors could demonstrate the presence of an active HynSL from *A. macleodii* in strain PW416, the one from *T. roseopersicina* was not active. However, this could be altered by adding the 11 accessory genes from *A. macleodii* (Weyman et al. 2011). Furthermore, the introduction of the bidirectional hydrogenase of *T. roseopersicina* (*hoxEFUYH*), that harbors similarities with the native cyanobacterial hydrogenase, was attempted. However, its co-expression with *T. roseopersicina* or *A. macleodii's* accessory genes was not sufficient to obtain an active enzyme in strain PW416 (Weyman et al. 2011).

Recently, and for the first time, a heterologous Fe hydrogenase, from *Shewanella oneidensis* MR-1, was successfully expressed in the heterocysts of a filamentous cyanobacterium *Anabaena* sp. PCC 7120 using the specific promoter P_{hetN} (Table 4.1). The spatial separation between photosynthesis and H_2 production overcomes the sensibility of the hydrogenase to oxygen by expressing it in a microaerobic compartment (Gärtner et al. 2012).

5. Expression of a Cyanobacterial Hydrogenase in Escherichia coli

The genes encoding the bidirectional hydrogenase (*hoxEFUYH*) from *Synechocystis* sp. PCC 6803 were cloned into *E. coli* and, the expression of the cyanobacterial hydrogenase lead to a threefold increase in H_2 production (Maeda et al. 2007). An optimization of the growth medium, replacing glucose by fructose, galactose or maltose resulted in 20 % increase in H_2 yield. Moreover, a time course evaluation of the H_2 production revealed that, after 18 h, there was over 41 times more H_2 production in the mutant compared with the wild-type. However, Maeda et al. (2007) also showed that in *E. coli* cells expressing the bidirectional hydrogenase from *Synechocystis* sp. PCC 6803, H_2 production is sustained by the native hydrogenase 3 while the HoxEFUYH inhibits H_2 uptake by hydrogenases 1 and 2.

B. Synthetic Biology Approaches

The emerging field of Synthetic Biology (SB) offers novel perspectives for the production of added-value compounds, namely biofuels. In this context, and due to their minimal nutritional requirements and metabolic plasticity, cyanobacteria could constitute exceptional photoautotrophic chassis for the production of hydrogen (Angermayr et al. 2009; Ducat et al. 2011a; Heidorn et al. 2011; Lindblad et al. 2012; Pinto et al. 2012). For this purpose, a wide toolbox containing well characterized standardized biological parts and devices should be developed.

In this section the recent advances in this field will be presented.

1. Cyanobacteria as Photoautotrophic Chassis

In SB applications parts, devices, and circuits must eventually be introduced into a host cell that is designated as chassis. The chassis should be based on a well-known organism with a sequenced genome, for which plenty of information is available (transcriptomic, proteomic and metabolomic) allowing the development of models to predict its behavior. Its genetic manipulation should also be easy and, preferably, the chassis should have a streamline genome making device implementation more effective and predictable (Etc Group 2007; O'Malley et al. 2008). The accomplishment of SB-based strategies for the production of biofuels, namely hydrogen, has been successfully explored using *Saccharomyces cerevisiae* and *E. coli* as chassis (Waks and Silver 2009; Agapakis et al. 2010). Still, there is a need to develop more/more predictable basic SB tools and to explore other chassis, namely photoautotrophic ones. In this context, cyanobacteria emerge as promising candidates due to their ability to use solar energy and CO_2 as energy and carbon sources respectively, thrive in different environments and metabolic plasticity. In addition, cyanobacteria have higher growth rates compared to plants and the molecular tools for their genetic manipulation are available, being easier to engineer than algae (Ducat et al. 2011a; Heidorn et al. 2011; Lindblad et al. 2012). The unicellular *Synechocystis* sp. PCC 6803 is the best studied cyanobacterial strain and its genome was the first to be sequenced among photosynthetic organisms (Kaneko et al. 1996). Moreover, the vast amount of data available allowed the construction and validation of genome-scale metabolic models – e.g. *i*Syn811 – that are powerful tools to develop a robust photoautotrophic chassis, and to predict changes when synthetic modules are introduced (Fu 2009; Montagud et al. 2010, 2011; Yoshikawa et al. 2011). Recently, a *Synechocystis* sp. PCC 6803 deletion mutant (Δ*hoxYH*), lacking an active bidirectional hydrogenase was produced and extensively characterized to be used as chassis for the introduction of heterologous hydrogenases/hydrogen producing devices (Pinto et al. 2012). The authors also developed vectors compatible with the BioBrick system that allow removing redundant genes and/or introducing synthetic parts into *Synechocystis* sp. PCC 6803 genome (Pinto et al. 2012). For the introduction of synthetic parts/devices, it is important to identify neutral sites i.e. genomic loci that can be disrupted without affecting cellular viability or causing any distinguishable phenotype (Clerico et al. 2007). The identification/validation of these neutral sites increases the chassis functionality since it permits the sequential integration of different/more complex devices into the genome. To date, four chromosomal loci have been used to introduce foreign DNA into *Synechocystis* sp. PCC 6803 (Williams 1988; Burnap et al. 1994; Aoki et al. 1995, 2011) but they have not been fully characterized. In another unicellular cyanobacterium *Synechococcus elongatus* sp. PCC 7942, three neutral sites (NS1, NS2 and NS3) have also been used for the integration of foreign DNA (Andersson et al. 2000; Clerico et al. 2007; Niederholtmeyer et al. 2010; Ducat et al. 2011b) but, similarly to *Synechocystis* sp. PCC 6803, they remain largely uncharacterized.

One of the current challenges in SB is the development of streamline genomes since lower genome complexity will facilitate chassis engineering, insulation of the introduced synthetic devices from the chassis regulatory network, and render systems behavior more predictable. A 15 % reduction of the *E. coli* genome led to unanticipated benefits such as higher electroporation efficiency and increased genome stability (Pósfai et al. 2006). In cyanobacteria, a blueprint for genome reduction of *Synechococcus elongatus* PCC 7942 has already been constructed (Delaye et al. 2011). In this case, the essential and non-essential genes were identified using a combination of methods:

unusual G+C content, unusual phylogenetic similarity and/or a small number of the highly iterated palindrome 1 (HIP1) plus unusual codon usage. This work will facilitate engineering *Synechococcus elongatus* PCC 7942, for which fewer tools are available but has higher growth rates and a smaller genome than *Synechocystis* sp. PCC 6803 (2.7 Mb compared to 3.6 Mb). At present, *Synechocystis* sp. PCC 6803 and *Synechococcus elongatus* PCC 7942 are among the best cyanobacterial candidates to be used as photoautotrophic chassis in SB applications.

2. Parts and Devices for H_2 Production

Synthetic Biology is also based on the premise that if one is able to learn how each modular component works and understands the interaction between parts, then is able to manipulate and recombine them (assembling them into devices and circuits) to design novel pathways with useful purposes. This implies the construction and characterization of standardized interchangeable biological parts with defined functions, that often need to be customized (e.g. codon optimized) to be fully functional in a given chassis. The development of standards and databases such as the BioBricks and the Registry of Biological Parts (Registry of Standard Biological Parts) facilitated the sharing of parts between researchers and led to the construction of numerous synthetic devices with applications in different areas such as pharmaceuticals and biofuels (Ro et al. 2006; Waks and Silver 2009; Agapakis et al. 2010). Although the quantitative control (predictable behaviour) of biological systems using a SB approach is possible, it still remains an iterative process. In different organisms, including cyanobacteria, an effort has been made in the characterization of promoters, ribosomal binding sites, terminators and regulatory elements, as well as in the development of BioBrick compatible plasmids (Shetty et al. 2008; Boyle and Silver 2009; Purnick and Weiss 2009; Huang et al. 2010). Still, there is the need to increase the repertoire of available parts, including orthogonal ones that should have the major advantage of insulating the synthetic devices and circuits from the genetic network of the chassis.

a. Hydrogen Producing Devices

Using a similar strategy as in the classical molecular biology approaches (see above), a gene encoding the monomeric Fe hydrogenase – HydA – from *Clostridium acetobutylicum* was expressed in the cyanobacterium *Synechococcus elongatus* PCC 7942 (Ducat et al. 2011b). The main difference here is that the authors made use of genes commercially synthesized, codon optimized, and acceptable for use in *E. coli* and other organisms, with all constructs in the BioBrick format (Ducat et al. 2011b). The genes encoding the hydrogenase maturation factors HydEF and HydG were cloned separately and placed under the *Synechococcus elongatus* PCC 7942 *psbA1* constitutive promoter, and then combined into a cassette that was subsequently integrated in the previously defined genomic neutral site NS1 (Clerico et al. 2007). The gene encoding HydA was inserted separately, downstream the IPTG-inducible promoter P_{lac}, and integrated into the neutral site NS3 (Niederholtmeyer et al. 2010) (see Table 4.1). Ducat et al. (2011b) demonstrated that the *in vivo* hydrogenase activity is connected to the light-dependent reactions of the electron transport chain, and that under anoxic conditions the heterologous enzyme is capable of supporting light-dependent hydrogen evolution at a rate 500-fold greater than that supported by the endogenous NiFe bidirectional hydrogenase. Moreover, to facilitate electron transfer to HydA and further increase hydrogen production, genes encoding heterologous ferredoxins (that constitute the strongest pair together with HydA) were inserted into *Synechococcus elongatus* PCC 7942 under an IPTG-inducible promoter (Ducat et al. 2011b). The expression of the *C. acetobutylicum* ferredoxin increased the rate of hydrogen evolution by twofold, and additional experiments demonstrate that the addition of supplemental ferredoxins or optimization of ferredoxin-hydrogenase interactions can both increase the flux of electrons towards hydrogenase as well as rewire the redox pathway.

In another SB approach, a single synthetic operon containing 12 genes, five encoding the structural subunits (*hoxEFUYH*) and seven the maturation factors (*hypA$_1$B$_1$CDEF*, *hoxW*), was constructed to express the *Synechocystis* sp. PCC 6803 bidirectional hydrogenase in *E. coli* (Wells et al. 2011). The 11.7 kb sequence was codon optimized for the host and placed under the control of a T7 promoter, originating the pSynHox construct. The authors clearly demonstrated the production of hydrogen in an *E. coli* strain without the native hydrogenases, and it was also shown that the hydrogen output could be increased when formate production was abolished, reinforcing the hypothesis that hydrogen production is coupled with the NADH/NADPH pools. However, the relative low levels of the recombinant enzyme compared to the native *Synechocystis* sp. PCC 6803 host indicate that there is still room for improvement, e.g. coupling the hydrogenase with electron donors. Concerning the maturation machinery, the authors concluded that to express the bidirectional hydrogenase of *Synechocystis* sp. PCC 6803 in *E. coli*, only *hypA1* and *hoxW* are essential since the host can complement the deletion of all the other maturation factors.

b. Oxygen Consuming Devices

The photobiological production of H_2 is severely compromised by the presence of O_2 due to the sensitivity of the H_2-evolving enzymes: hydrogenases and nitrogenases (Tamagnini et al. 2007; Bothe et al. 2010). As oxygenic phototrophs, cyanobacteria have developed spatial- or time-based strategies to separate the O_2-evolving from the O_2-sensitive processes. In parallel with the search for O_2-tolerant enzymes, the development of synthetic Oxygen Consuming Devices (OCDs) capable of reducing intracellular oxygen is highly desirable. The DESHARKY bioinformatic tool (Rodrigo et al. 2008), was used to identify several proteins, native and heterologous to *Synechocystis* sp. PCC 6803, that were subsequently selected to produce OCDs. These OCDs can be introduced into a chassis together with efficient hydrogenases/hydrogen producing devices to improved H_2 production or used for other industrial applications that require low intracellular O_2 pressures.

IV. Conclusions and Future Perspectives

The use of photosynthetic organisms, such as algae and cyanobacteria, is a valuable option for H_2/biofuels production. However, several issues should be addressed to achieve economical relevant results. Some are considered in this chapter, but others like solar energy conversion efficiencies (see for e.g. Masukawa et al. 2012) and the development of adequate bioreactors are equally important. For cyanobacterial H_2 production both native and heterologous enzymes have been used. Nitrogenases have a high energy demand and hydrogen is only produced as a by-product of nitrogen fixation. Thus, the search or engineering of nitrogenases that preferably will function as "hydrogenases" is a possibility. Regarding approaches using hydrogenases, Fe enzymes are generally more active and have less complex maturation systems than the NiFe hydrogenases, consequently they have been preferably used for heterologous expression. A better understanding of the regulation and maturation processes of the NiFe hydrogenases, notably of the small subunits is also necessary. To different extents, all H_2 evolving enzymes are quite sensitive to oxygen and the search/engineering of oxygen tolerant enzymes is being actively pursued. In the Synthetic Biology field, cyanobacteria emerge as prominent candidates to be used as photoautotrophic chassis for the accommodation of highly efficient hydrogenases/hydrogen producing devices. Protein fusions (e.g. hydrogenase and ferredoxin), the use of linkers and scaffolds would probably improve H_2 production. Nevertheless, efforts should be made to circumvent the impairment of H_2 production by oxygen. Development of synthetic oxygen consuming devices coupled with O_2 sensors, or compartmentalization of the process (in certain cells/cell types) are two of the possibilities. Interestingly, heterocystous cyanobacteria have microaerobic compartments – the heterocysts – where oxygen sensitive processes like N_2 fixation can occur. Moreover, the generation of genome scale models will help to re-direct the metabolism towards an efficient H_2 production.

Acknowledgements

This work was funded by FEDER funds through the Operational Competitiveness Programme – COMPETE and by National Funds through FCT – Fundação para a Ciência e a Tecnologia under the projects FCOMP-01-0124-FEDER-022718 (PEst-C/SAU/LA0002/2011), FCOMP-01-0124-FEDER-009003 (PTDC/BIA-MIC/100370/2008), and SFRH/BPD/64095/2009, SFRH/BPD/74894/2010. The research leading to these results has received funding from the European Union Seventh Framework Programme (FP7/2007-2013) under grant agreement number 308518 (CyanoFactory).

References

Agapakis CM, Ducat DC, Boyle PM, Wintermute EH, Way JC, Silver PA (2010) Insulation of a synthetic hydrogen metabolism circuit in bacteria. J Biol Eng 4:3

Agervald Å, Zhang X, Stensjö K, Devine E, Lindblad P (2010) CalA, a cyanobacterial AbrB protein, interacts with the upstream region of *hypC* and acts as a repressor of its transcription in the cyanobacterium *Nostoc* sp. strain PCC 7120. Appl Environ Microbiol 76:880–890

Allahverdiyeva Y, Leino H, Saari L, Fewer DP, Shunmugam S, Sivonen K, Aro E-M (2010) Screening for biohydrogen production by cyanobacteria isolated from the Baltic Sea and Finnish lakes. Int J Hydrog Energy 35:1117–1127

Ananyev G, Carrieri D, Dismukes GC (2008) Optimization of metabolic capacity and flux through environmental cues to maximize hydrogen production by the cyanobacterium "*Arthrospira* (*Spirulina*) *maxima*". Appl Environ Microbiol 74:6102–6113

Andersson CR, Tsinoremas NF, Shelton J, Lebedeva NV, Yarrow J, Min H, Golden SS (2000) Application of bioluminescence to the study of circadian rhythms in cyanobacteria. Methods Enzymol 305:527–542

Angermayr SA, Hellingwerf KJ, Lindblad P, de Mattos MJ (2009) Energy biotechnology with cyanobacteria. Curr Opin Biotechnol 20:257–263

Aoki S, Kondo T, Ishiura M (1995) Circadian expression of the *dnaK* gene in the cyanobacterium *Synechocystis* sp. strain PCC 6803. J Bacteriol 177:5606–5611

Aoki R, Goto T, Fujita Y (2011) A heme oxygenase isoform is essential for aerobic growth in the cyanobacterium *Synechocystis* sp. PCC 6803: modes of differential operation of two isoforms/enzymes to adapt to low oxygen environments in cyanobacteria. Plant Cell Physiol 52:1744–1756

Appel J, Phunpruch S, Steinmüller K, Schulz R (2000) The bidirectional hydrogenase of *Synechocystis* sp. PCC 6803 works as an electron valve during photosynthesis. Arch Microbiol 173:333–338

Asada Y, Koike Y, Schnackenberg J, Miyake M, Uemura I, Miyake J (2000) Heterologous expression of clostridial hydrogenase in the cyanobacterium *Synechococcus* PCC 7942. Biochim Biophys Acta Gene Struct Expr 1490:269–278

Baebprasert W, Jantaro S, Khetkorn W, Lindblad P, Incharoensakdi A (2011) Increased H_2 production in the cyanobacterium *Synechocystis* sp. strain PCC 6803 by redirecting the electron supply via genetic engineering of the nitrate assimilation pathway. Metab Eng 13:610–616

Bandyopadhyay A, Stöckel J, Min HT, Sherman LA, Pakrasi HB (2010) High rates of photobiological H_2 production by a cyanobacterium under aerobic conditions. Nat Commun 1:139

Bandyopadhyay A, Elvitigala T, Welsh E, Stöckel J, Liberton M, Min H, Sherman LA, Pakrasi HB (2011) Novel metabolic attributes of the genus *Cyanothece*, comprising a group of unicellular nitrogen-fixing cyanobacteria. mBio 2:e00214

Barz M, Beimgraben C, Staller T, Germer F, Opitz F, Marquardt C, Schwarz C, Gutekunst K, Vanselow KH, Schmitz R, LaRoche J, Schulz R, Appel J (2010) Distribution analysis of hydrogenases in surface waters of marine and freshwater environments. PLoS One 5:e13846

Berman-Frank I, Lundgren P, Falkowski P (2003) Nitrogen fixation and photosynthetic oxygen evolution in cyanobacteria. Res Microbiol 154:157–164

Berto P, D'Adamo S, Bergantino E, Vallese F, Giacometti GM, Costantini P (2011) The cyanobacterium *Synechocystis* sp. PCC 6803 is able to express an active [FeFe]-hydrogenase without additional maturation proteins. Biochem Biophys Res Commun 405:678–683

Bothe H, Schmitz O, Yates MG, Newton WE (2010) Nitrogen fixation and hydrogen metabolism in cyanobacteria. Microbiol Mol Biol Rev 74:529–551

Boyle PM, Silver PA (2009) Harnessing nature's toolbox: regulatory elements for synthetic biology. J R Soc Interface 6:535–546

Burnap RL, Qian M, Shen JR, Inoue Y, Sherman LA (1994) Role of disulfide linkage and putative intermolecular binding residues in the stability and binding of the extrinsic manganese-stabilizing protein to the photosystem II reaction center. Biochemistry 33:13712–13718

Burrows EH, Chaplen FWR, Ely RL (2008) Optimization of media nutrient composition for increased photofermentative hydrogen production by *Synechocystis* sp. PCC 6803. Int J Hydrog Energy 33:6092–6099

Burrows EH, Wong W-K, Fern X, Chaplen FWR, Ely RL (2009) Optimization of pH and nitrogen for enhanced hydrogen production by *Synechocystis* sp. PCC 6803 via statistical and machine learning methods. Biotechnol Prog 25:1009–1017

Burrows EH, Chaplen FWR, Ely RL (2011) Effects of selected electron transport chain inhibitors on 24h hydrogen production by *Synechocystis* sp. PCC 6803. Bioresour Technol 102:3062–3070

Carrieri D, Momot D, Brasg IA, Ananyev G, Lenz O, Bryant DA, Dismukes GC (2010) Boosting autofermentation rates and product yields with sodium stress cycling: application to production of renewable fuels by cyanobacteria. Appl Environ Microbiol 76:6455–6462

Carrieri D, Ananyev G, Lenz O, Bryant DA, Dismukes GC (2011) Contribution of a sodium ion gradient to energy conservation during fermentation in the cyanobacterium *Arthrospira* (*Spirulina*) *maxima* CS-328. Appl Environ Microbiol 77:7185–7194

Clerico EM, Ditty JL, Golden SS (2007) Specialized techniques for site-directed mutagenesis in cyanobacteria. In: Rosato E (ed) Circadian rhythms: methods and protocols. Humana Press, Totowa, pp 155–171

Costa JL (2004) The $tRNA^{Leu}$ (UAA) intron of cyanobacteria – towards understanding a genetic marker. Ph.D. thesis, Uppsala University, Uppsala

Cournac L, Mus F, Bernard L, Guedeney G, Vignais P, Peltier G (2002) Limiting steps of hydrogen production in *Chlamydomonas reinhardtii* and *Synechocystis* PCC 6803 as analysed by light-induced gas exchange transients. Int J Hydrog Energy 27:1229–1237

Cournac L, Guedeney G, Peltier G, Vignais PM (2004) Sustained photoevolution of molecular hydrogen in a mutant of *Synechocystis* sp. strain PCC 6803 deficient in the type I NADPH-dehydrogenase complex. J Bacteriol 186:1737–1746

Delaye L, González-Domenech CM, Garcillán-Barcia MP, Peretó J, de la Cruz F, Moya A (2011) Blueprint for a minimal photoautotrophic cell: conserved and variable genes in *Synechococcus elongatus* PCC 7942. BMC Genomics 12:25

Ducat DC, Way JC, Silver PA (2011a) Engineering cyanobacteria to generate high-value products. Trends Biotechnol 29:95–103

Ducat DC, Sachdeva G, Silver PA (2011b) Rewiring hydrogenase-dependent redox circuits in cyanobacteria. Proc Natl Acad Sci U S A 108:3941–3946

Dutheil J, Saenkham P, Sakr S, Leplat C, Ortega-Ramos M, Bottin H, Cournac L, Cassier-Chauvat C, Chauvat F (2012) The AbrB2 autorepressor, expressed from an atypical promoter, represses the hydrogenase operon to regulate hydrogen production in *Synechocystis* strain PCC6803. J Bacteriol 194:5423–5433

Eisenhut M, Georg J, Klähn S, Sakurai I, Mustila H, Zhang P, Hess WR, Aro E-M (2012) The antisense RNA As1_flv4 in the cyanobacterium *Synechocystis* sp. PCC 6803 prevents premature expression of the *flv4-2* operon upon shift in inorganic carbon supply. J Biol Chem 287:33153–33162

Ekman M, Ow SY, Holmqvist M, Zhang XH, van Wagenen J, Wright PC, Stensjö K (2011) Metabolic adaptations in a H_2 producing heterocyst-forming cyanobacterium: potentials and implications for biological engineering. J Proteome Res 10:1772–1784

Etc Group (2007) Extreme genetic engineering – an introduction to synthetic biology. http://www.etcgroup.org/sites/www.etcgroup.org/files/publication/602/01/synbioreportweb.pdf. 23 Oct 2012

Ferreira D, Leitão E, Sjöholm J, Oliveira P, Lindblad P, Moradas-Ferreira P, Tamagnini P (2007) Transcription and regulation of the hydrogenase(s) accessory genes, *hypFCDEAB*, in the cyanobacterium *Lyngbya majuscula* CCAP 1446/4. Arch Microbiol 188:609–617

Fu P (2009) Genome-scale modeling of *Synechocystis* sp. PCC 6803 and prediction of pathway insertion. J Chem Technol Biotechnol 84:473–483

Gärtner K, Lechno-Yossef S, Cornish AJ, Wolk CP, Hegg EL (2012) Expression of *Shewanella oneidensis* MR-1 [FeFe]-hydrogenase genes in *Anabaena* sp. strain PCC 7120. Appl Environ Microbiol 78:8579–8586

Gutekunst K, Phunpruch S, Schwarz C, Schuchardt S, Schulz-Friedrich R, Appel J (2005) LexA regulates the bidirectional hydrogenase in the cyanobacterium *Synechocystis* sp. PCC 6803 as a transcription activator. Mol Microbiol 58:810–823

Gutthann F, Egert M, Marques A, Appel J (2007) Inhibition of respiration and nitrate assimilation enhances photohydrogen evolution under low oxygen concentrations in *Synechocystis* sp. PCC 6803. Biochim Biophys Acta Bioenerg 1767:161–169

Happe T, Schütz K, Böhme H (2000) Transcriptional and mutational analysis of the uptake hydrogenase of the filamentous cyanobacterium *Anabaena variabilis* ATCC 29413. J Bacteriol 182:1624–1631

Heidorn T, Camsund D, Huang H-H, Lindberg P, Oliveira P, Stensjö K, Lindblad P (2011) Synthetic biology in cyanobacteria: engineering and analyzing novel functions. Methods Enzymol 497:539–579

Herrero A, Muro-Pastor A, Flores E (2001) Nitrogen control in cyanobacteria. J Bacteriol 183:411–425

Hoffmann D, Gutekunst K, Klissenbauer M, Schulz-Friedrich R, Appel J (2006) Mutagenesis of hydrogenase accessory genes of *Synechocystis* sp. PCC 6803 – additional homologues of *hypA* and *hypB* are not active in hydrogenase maturation. FEBS J 273:4516–4527

Holmqvist M, Stensjö K, Oliveira P, Lindberg P, Lindblad P (2009) Characterization of the *hupSL* promoter activity in *Nostoc punctiforme* ATCC 29133. BMC Microbiol 9:54

Holmqvist M, Lindberg P, Agervald Å, Stensjö K, Lindblad P (2011) Transcript analysis of the extended *hyp*-operon in the cyanobacteria *Nostoc* sp. strain PCC 7120 and *Nostoc punctiforme* ATCC 29133. BMC Res Notes 4:186

Huang HH, Camsund D, Lindblad P, Heidorn T (2010) Design and characterization of molecular tools for a Synthetic Biology approach towards developing cyanobacterial biotechnology. Nucleic Acids Res 38:2577–2593

Ishii A, Hihara Y (2008) An AbrB-like transcriptional regulator, Sll0822, is essential for the activation of nitrogen-regulated genes in *Synechocystis* sp. PCC 6803. Plant Physiol 148:660–670

Kaneko T, Sato S, Kotani H, Tanaka A, Asamizu E, Nakamura Y, Miyajima N, Hirosawa M, Sugiura M, Sasamoto S, Kimura T, Hosouchi T, Matsuno A, Muraki A, Nakazaki N, Naruo K, Okumura S, Shimpo S, Takeuchi C, Wada T, Watanabe A, Yamada M, Yasuda M, Tabata S (1996) Sequence analysis of the genome of the unicellular cyanobacterium *Synechocystis* sp. strain PCC 6803. II. Sequence determination of the entire genome and assignment of potential protein-coding regions. DNA Res 3:109–136

Khetkorn W, Lindblad P, Incharoensakdi A (2012) Inactivation of uptake hydrogenase leads to enhanced and sustained hydrogen production with high nitrogenase activity under high light exposure in the cyanobacterium *Anabaena siamensis* TISTR 8012. J Biol Eng 6:19

Knoll AH (2008) Cyanobacteria an earth history. In: Herrero A, Flores E (eds) The cyanobacteria: molecular biology, genomics and evolution. Caister Academic Press, Norfolk, pp 1–19

Kothari A, Potrafka R, Garcia-Pichel F (2012) Diversity in hydrogen evolution from bidirectional hydrogenases in cyanobacteria from terrestrial, freshwater and marine intertidal environments. J Biotechnol 162:105–114

Kumar K, Dasgupta CN, Nayak B, Lindblad P, Das D (2011) Development of suitable photobioreactors for CO_2 sequestration addressing global warming using green algae and cyanobacteria. Bioresour Technol 102:4945–4953

Leitão E, Oxelfelt F, Oliveira P, Moradas-Ferreira P, Tamagnini P (2005) Analysis of the *hupSL* operon of the nonheterocystous cyanobacterium *Lyngbya majuscula* CCAP 1446/4: regulation of transcription and expression under a light-dark regimen. Appl Environ Microbiol 71:4567–4576

Lieman-Hurwitz J, Haimovich M, Shalev-Malul G, Ishii A, Hihara Y, Gaathon A, Lebendiker M, Kaplan A (2009) A cyanobacterial AbrB-like protein affects the apparent photosynthetic affinity for CO_2 by modulating low-CO_2-induced gene expression. Environ Microbiol 11:927–936

Lindberg P, Schütz K, Happe T, Lindblad P (2002) A hydrogen-producing, hydrogenase-free mutant strain of *Nostoc punctiforme* ATCC 29133. Int J Hydrog Energy 27:1291–1296

Lindberg P, Devine E, Stensjö K, Lindblad P (2012) HupW protease specifically required for processing of the catalytic subunit of the uptake hydrogenase in the cyanobacterium *Nostoc* sp. strain PCC 7120. Appl Environ Microbiol 78:273–276

Lindblad P, Lindberg P, Oliveira P, Stensjö K, Heidorn T (2012) Design, engineering, and construction of photosynthetic microbial cell factories for renewable solar fuel production. Ambio 41:163–168

Lopes Pinto FA, Troshina O, Lindblad P (2002) A brief look at three decades of research on cyanobac-

terial hydrogen evolution. Int J Hydrog Energy 27:1209–1215
Ludwig M, Schulz-Friedrich R, Appel J (2006) Occurrence of hydrogenases in cyanobacteria and anoxygenic photosynthetic bacteria: implications for the phylogenetic origin of cyanobacterial and algal hydrogenases. J Mol Evol 63:758–768
Maeda T, Vardar G, Self W, Wood T (2007) Inhibition of hydrogen uptake in *Escherichia coli* by expressing the hydrogenase from the cyanobacterium *Synechocystis* sp. PCC 6803. BMC Biotechnol 7:25
Masukawa H, Mochimaru M, Sakurai H (2002) Disruption of the uptake hydrogenase gene, but not of the bidirectional hydrogenase gene, leads to enhanced photobiological hydrogen production by the nitrogen-fixing cyanobacterium *Anabaena* sp. PCC 7120. Appl Microbiol Biotechnol 58:618–624
Masukawa H, Inoue K, Sakurai H (2007) Effects of disruption of homocitrate synthase genes on *Nostoc* sp. strain PCC 7120 photobiological hydrogen production and nitrogenase. Appl Environ Microbiol 73:7562–7570
Masukawa H, Inoue K, Sakurai H, Wolk CP, Hausinger RP (2010) Site-directed mutagenesis of the *Anabaena* sp. strain PCC 7120 nitrogenase active site to increase photobiological hydrogen production. Appl Environ Microbiol 76:6741–6750
Masukawa H, Kitashima M, Inoue K, Sakurai H, Hausinger RP (2012) Genetic engineering of cyanobacteria to enhance biohydrogen production from sunlight and water. Ambio 41:169–173
McNeely K, Xu Y, Bennette N, Bryant DA, Dismukes GC (2010) Redirecting reductant flux into hydrogen production via metabolic engineering of fermentative carbon metabolism in a cyanobacterium. Appl Environ Microbiol 76:5032–5038
McNeely K, Xu Y, Ananyev G, Bennette N, Bryant DA, Dismukes GC (2011) *Synechococcus* sp. strain PCC 7002 *nifJ* mutant lacking pyruvate: ferredoxin oxidoreductase. Appl Environ Microbiol 77: 2435–2444
Min H, Sherman LA (2010a) Hydrogen production by the unicellular, diazotrophic cyanobacterium *Cyanothece* sp. strain ATCC 51142 under conditions of continuous light. Appl Environ Microbiol 76:4293–4301
Min H, Sherman LA (2010b) Genetic transformation and mutagenesis via single-stranded DNA in the unicellular, diazotrophic cyanobacteria of the genus *Cyanothece*. Appl Environ Microbiol 76:7641–7645
Montagud A, Navarro E, Fernandez de Cordoba P, Urchueguia JF, Patil KR (2010) Reconstruction and analysis of genome-scale metabolic model of a photosynthetic bacterium. BMC Syst Biol 4:156
Montagud A, Zelezniak A, Navarro E, de Cordoba PF, Urchueguia JF, Patil KR (2011) Flux coupling and transcriptional regulation within the metabolic network of the photosynthetic bacterium *Synechocystis* sp. PCC 6803. Biotechnol J 6:330–342
Nakamura Y, Kaneko T, Sato S, Mimuro M, Miyashita H, Tsuchiya T, Sasamoto S, Watanabe A, Kawashima K, Kishida Y, Kiyokawa C, Kohara M, Matsumoto M, Matsuno A, Nakazaki N, Shimpo S, Takeuchi C, Yamada M, Tabata S (2003) Complete genome structure of *Gloeobacter violaceus* PCC 7421, a cyanobacterium that lacks thylakoids. DNA Res 10:137–145
Niederholtmeyer H, Wolfstädter BT, Savage DF, Silver PA, Way JC (2010) Engineering cyanobacteria to synthesize and export hydrophilic products. Appl Environ Microbiol 76:3462–3466
O'Malley MA, Powell A, Davies JF, Calvert J (2008) Knowledge-making distinctions in synthetic biology. Bioessays 30:57–65
Ogawa T (1991) A gene homologous to the subunit-2 gene of NADH dehydrogenase is essential to inorganic carbon transport of *Synechocystis* PCC 6803. Proc Natl Acad Sci U S A 88:4275–4279
Ohkawa H, Price GD, Badger MR, Ogawa T (2000) Mutation of *ndh* genes leads to inhibition of CO_2 uptake rather than HCO_3^- uptake in *Synechocystis* sp. strain PCC 6803. J Bacteriol 182:2591–2596
Oliveira P, Lindblad P (2008) An AbrB-like protein regulates the expression of the bidirectional hydrogenase in *Synechocystis* sp. strain PCC 6803. J Bacteriol 190:1011–1019
Oliveira P, Lindblad P (2009) Transcriptional regulation of the cyanobacterial bidirectional Hox-hydrogenase. Dalton Trans 45:9990–9996
Oliveira P, Lindblad P (2011) Novel insights into the regulation of LexA in the cyanobacterium *Synechocystis* sp. strain PCC 6803. J Bacteriol 193:3804–3814
Oliveira P, Leitao E, Tamagnini P, Moradas-Ferreira P, Oxelfelt F (2004) Characterization and transcriptional analysis of *hupSLW* in *Gloeothece* sp. ATCC 27152: an uptake hydrogenase from a unicellular cyanobacterium. Microbiology 150: 3647–3655
Phillips ZE, Strauch MA (2002) *Bacillus subtilis* sporulation and stationary phase gene expression. Cell Mol Life Sci 59:392–402
Pinto F, van Elburg KA, Pacheco CC, Lopo M, Noirel J, Montagud A, Urchueguía JF, Wright PC, Tamagnini P (2012) Construction of a chassis for hydrogen production: physiological and molecular characterization of a *Synechocystis* sp. PCC 6803 mutant lacking a functional bidirectional hydrogenase. Microbiology 158:448–464

Pósfai G, Plunkett G III, Fehér T, Frisch D, Keil GM, Umenhoffer K, Kolisnychenko V, Stahl B, Sharma SS, de Arruda M, Burland V, Harcum SW, Blattner FR (2006) Emergent properties of reduced-genome *Escherichia coli*. Science 312:1044–1046

Purnick PE, Weiss R (2009) The second wave of synthetic biology: from modules to systems. Nat Rev Mol Cell Biol 10:410–422

Registry of Standard Biological Parts. http://partsregistry.org/Main_Page. 24 Oct 2012

Ro D-K, Paradise EM, Ouellet M, Fisher KJ, Newman KL, Ndungu JM, Ho KA, Eachus RA, Ham TS, Kirby J, Chang MCY, Withers ST, Shiba Y, Sarpong R, Keasling JD (2006) Production of the antimalarial drug precursor artemisinic acid in engineered yeast. Nature 440:940–943

Rodrigo G, Carrera J, Prather KJ, Jaramillo A (2008) DESHARKY: automatic design of metabolic pathways for optimal cell growth. Bioinformatics 24: 2554–2556

Rupprecht J, Hankamer B, Mussgnug JH, Ananyev G, Dismukes C, Kruse O (2006) Perspectives and advances of biological H_2 production in microorganisms. Appl Microbiol Biotechnol 72:442–449

Rusch DB, Halpern AL, Sutton G, Heidelberg KB, Williamson S, Yooseph S, Wu D, Eisen JA, Hoffman JM, Remington K, Beeson K, Tran B, Smith H, Baden-Tillson H, Stewart C, Thorpe J, Freeman J, Andrews-Pfannkoch C, Venter JE, Li K, Kravitz S, Heidelberg JF, Utterback T, Rogers Y-H, Falcón LI, Souza V, Bonilla-Rosso G, Eguiarte LE, Karl DM, Sathyendranath S, Platt T, Bermingham E, Gallardo V, Tamayo-Castillo G, Ferrari MR, Strausberg RL, Nealson K, Friedman R, Frazier M, Venter JC (2007) The *Sorcerer II* global ocean sampling expedition: northwest Atlantic through eastern tropical pacific. PLoS Biol 5:e77

Sato S, Shimoda Y, Muraki A, Kohara M, Nakamura Y, Tabata S (2007) A large-scale protein–protein interaction analysis in *Synechocystis* sp. PCC 6803. DNA Res 14:207–216

Schütz K, Happe T, Troshina O, Lindblad P, Leitao E, Oliveira P, Tamagnini P (2004) Cyanobacterial H_2 production – a comparative analysis. Planta 218:350–359

Shalev-Malul G, Lieman-Hurwitz J, Viner-Mozzini Y, Sukenik A, Gaathon A, Lebendiker M, Kaplan A (2008) An AbrB-like protein might be involved in the regulation of cylindrospermopsin production by *Aphanizomenon ovalisporum*. Environ Microbiol 10:988–999

Shetty RP, Endy D, Knight TF Jr (2008) Engineering BioBrick vectors from BioBrick parts. J Biol Eng 2:5

Show K-Y, Lee D-J, Chang J-S (2011) Bioreactor and process design for biohydrogen production. Bioresour Technol 102:8524–8533

Skizim NJ, Ananyev GM, Krishnan A, Dismukes GC (2012) Metabolic pathways for photobiological hydrogen production by nitrogenase- and hydrogenase-containing unicellular cyanobacteria *Cyanothece*. J Biol Chem 287:2777–2786

Swingley WD, Blankenship RE, Raymond J (2008) Insights into cyanobacterial evolution from comparative genomics. In: Herrero A, Flores E (eds) The cyanobacteria: molecular biology, genomics and evolution. Caister Academic Press, Norflok, pp 21–43

Tamagnini P, Leitao E, Oliveira P, Ferreira D, Pinto F, Harris DJ, Heidorn T, Lindblad P (2007) Cyanobacterial hydrogenases: diversity, regulation and applications. FEMS Microbiol Rev 31:692–720

Taton A, Lis E, Adin DM, Dong G, Cookson S, Kay SA, Golden SS, Golden JW (2012) Gene transfer in *Leptolyngbya* sp. strain BL0902, a cyanobacterium suitable for production of biomass and bioproducts. PLoS One 7:e30901

Thiel T (1993) Characterization of genes for an alternative nitrogenase in the cyanobacterium *Anabaena variabilis*. J Bacteriol 175:6276–6286

Venter JC, Remington K, Heidelberg JF, Halpern AL, Rusch D, Eisen JA, Wu D, Paulsen I, Nelson KE, Nelson W, Fouts DE, Levy S, Knap AH, Lomas MW, Nealson K, White O, Peterson J, Hoffman J, Parsons R, Baden-Tillson H, Pfannkoch C, Rogers Y-H, Smith HO (2004) Environmental genome shotgun sequencing of the Sargasso Sea. Science 304:66–74

Waks Z, Silver PA (2009) Engineering a synthetic dual-organism system for hydrogen production. Appl Environ Microbiol 75:1867–1875

Wells MA, Mercer J, Mott RA, Pereira-Medrano AG, Burja AM, Radianingtyas H, Wright PC (2011) Engineering a non-native hydrogen production pathway into *Escherichia coli* via a cyanobacterial [NiFe] hydrogenase. Metab Eng 13:445–453

Weyman P, Pratte B, Thiel T (2008) Transcription of *hupSL* in *Anabaena variabilis* ATCC 29413 is regulated by NtcA and not by hydrogen. Appl Environ Microbiol 74:2103–2110

Weyman PD, Pratte B, Thiel T (2010) Hydrogen production in nitrogenase mutants in *Anabaena variabilis*. FEMS Microbiol Lett 304:55–61

Weyman PD, Vargas WA, Tong YK, Yu JP, Maness PC, Smith HO, Xu Q (2011) Heterologous expression of *Alteromonas macleodii* and *Thiocapsa roseopersicina* [NiFe] hydrogenases in *Synechococcus elongatus*. PLoS One 6:e20126

Whitton B, Potts M (2000) Introduction to the cyanobacteria. In: Whitton B, Potts M (eds) The ecology of cyanobacteria – their diversity in time and space. Kluwer, Dordrecht, pp 1–11

Williams JGK (1988) Construction of specific mutations in photosystem II photosynthetic reaction center by genetic engineering methods in *Synechocystis* 6803. Methods Enzymol 167:766–778

Yamauchi Y, Kaniya Y, Kaneko Y, Hihara Y (2011) Physiological roles of the cyAbrB transcriptional regulator pair Sll0822 and Sll0359 in *Synechocystis* sp. strain PCC 6803. J Bacteriol 193: 3702–3709

Yooseph S, Sutton G, Rusch DB, Halpern AL, Williamson SJ, Remington K, Eisen JA, Heidelberg KB, Manning G, Li W, Jaroszewski L, Cieplak P, Miller CS, Li H, Mashiyama ST, Joachimiak MP, van Belle C, Chandonia J-M, Soergel DA, Zhai Y, Natarajan K, Lee S, Raphael BJ, Bafna V, Friedman R, Brenner SE, Godzik A, Eisenberg D, Dixon JE, Taylor SS, Strausberg RL, Frazier M, Venter JC (2007) The *Sorcerer II* global ocean sampling expedition: expanding the universe of protein families. PLoS Biol 5:e16

Yoshikawa K, Kojima Y, Nakajima T, Furusawa C, Hirasawa T, Shimizu H (2011) Reconstruction and verification of a genome-scale metabolic model for *Synechocystis* sp. PCC 6803. Appl Microbiol Biotechnol 92:347–358

Yoshino F, Ikeda H, Masukawa H, Sakurai H (2007) High photobiological hydrogen production activity of a *Nostoc* sp. PCC 7422 uptake hydrogenase-deficient mutant with high nitrogenase activity. Mar Biotechnol 9:101–111

Chapter 5

Hydrogen Production by Water Biophotolysis

Maria L. Ghirardi*, Paul W. King, David W. Mulder, Carrie Eckert, Alexandra Dubini, Pin-Ching Maness, and Jianping Yu
National Renewable Energy Laboratory, 15013 Denver West Parkway, Golden, CO 80401, USA

Summary

The use of microalgae for production of hydrogen gas from water photolysis has been studied for many years, but its commercialization is still limited by multiple challenges. Most of the barriers to commercialization are attributed to the existence of biological regulatory mechanisms that, under anaerobic conditions, quench the absorbed light energy, down-regulate linear electron transfer, inactivate the H_2-producing enzyme, and compete for electrons with the hydrogenase. Consequently, the conversion efficiency of absorbed photons into H_2 is significantly lower than its estimated potential of 12–13 %. However, extensive research continues towards addressing these barriers by either trying to understand and circumvent intracellular regulatory mechanisms at the enzyme and metabolic level or by developing biological systems that achieve prolonged H_2 production albeit under lower than 12–13 % solar conversion efficiency. This chapter describes the metabolic pathways involved in

*Author for correspondence, e-mail: maria.ghirardi@nrel.gov

D. Zannoni and R. De Philippis (eds.), Microbial BioEnergy: Hydrogen Production, Advances in Photosynthesis and Respiration 38, pp. 101–135, DOI 10.1007/978-94-017-8554-9_5,

biological H_2 photoproduction from water photolysis, the attributes of the two hydrogenases, [FeFe] and [NiFe], that catalyze biological H_2 production, and highlights research related to addressing the barriers described above. These highlights include: (a) recent advances in improving our understanding of the O_2 inactivation mechanism in different classes of hydrogenases; (b) progress made in preventing competitive pathways from diverting electrons from H_2 photoproduction; and (c) new developments in bypassing the non-dissipated proton gradient from down-regulating photosynthetic electron transfer. As an example of a major success story, we mention the generation of truncated-antenna mutants in *Chlamydomonas* and *Synechocystis* that address the inherent low-light saturation of photosynthesis. In addition, we highlight the rationale and progress towards coupling biological hydrogenases to non-biological, photochemical charge-separation as a means to bypass the barriers of photobiological systems.

I. Introduction

A. Photosynthetic Pathways for H_2 Production

The production of H_2 by oxygenic photosynthetic organisms (microalgae and cyanobacteria) is fueled by low potential reductant extracted from either water or stored sugars by the photosynthetic apparatus, and it is mediated by the hydrogenase enzymes. A nitrogenase-mediated H_2-production pathway is also found in certain cyanobacteria (Tsygankov 2007; Ghirardi et al. 2009; Bandyopadhyay et al. 2010; Bothe et al. 2011) but will not be further addressed in this chapter. As discussed in several recent reviews (Posewitz et al. 2008; Ghirardi et al. 2009; Hemschemeier and Happe 2011) and shown in Fig. 5.1, two major light-dependent, linear electron flow (LEF) pathways generate photosynthetic reductant: (a) *the Photosystem II (PSII)-dependent pathway*, in which electrons are extracted from water and transferred through PSII, the cytochrome (cyt) b_6f complex, and Photosystem I (PSI) to ferredoxin (FDX); and (b) *the PSII-independent pathway*, by which electrons released during glycolytic degradation of stored starch enter the photosynthetic electron transport chain at the level of plastoquinone (PQ), (possibly involving an NAD(P)H/PQ oxidoreductase enzyme) (Rumeau et al. 2005; Desplats et al. 2009) and are further transferred to cyt b_6/f, PSI, and FDX. In green algae, reduced FDX, generated by either of the two pathways, provides reductant to [FeFe]-hydrogenases (Happe and Naber 1993; Winkler et al. 2009), HYDA1 and HYDA2, that catalyze production of H_2. In cyanobacteria, an additional electron transfer (ET) step from reduced FDX to $NADP^+$ produces NADPH, the direct electron donor to [NiFe]-hydrogenases (Boison et al. 1998; Massanz et al. 1998; Antal et al. 2006). The presence of subunits in the cyanobacterial hydrogenases that are homologous to those present in the respiratory complex I suggest that these hydrogenases could be functionally connected to the respiratory pathway

Abbreviations: ATP – Adenosine triphosphate; CCCP – Carbonyl cyanide m-chloro phenyl hydrazone; CEF – Cyclic electron flow; CRR1 – Copper response regulator 1; DCIP – Dichlorophenol indophenol; DCMU – (3-(3,4-dichlorophenyl)-1,1-dimethylurea); DHG – Dehydroglycine; EPR – Electron paramagnetic resonance; ET – Electron transfer; ETR – Electron transport rate; FCCP – Carbonylcyanide *p*-fluoromethoxyphenylhydrazone; FDX – Ferredoxin; FNR – Ferredoxin/NADP oxido-reductase; FTIR – Fourier transform infrared spectroscopy; ISC – Iron-sulfur cluster; LEF – Linear electron flow; LHC – Light-harvesting complex; MBH – Membrane-bound hydrogenase; MWNT – Multi-walled carbon nanotubes; NAD(P) – Nicotinamide adenine (phosphate) dinucleotide; NPQ – Non-photochemical quenching; OEC – Oxygen-evolving complex; OCP – Orange carotenoid protein; PFR – Pyruvate/ferredoxin reductase; PQ – Plastoquinone; PSI – Photosystem I; PSII – Photosystem II; PTOX – Plastoquinone oxidase; SAM – *S*-adenosyl methionine; SAXS – Small angle X-ray scattering; SWNT – Single walled carbon nanotubes; WT – Wild-type

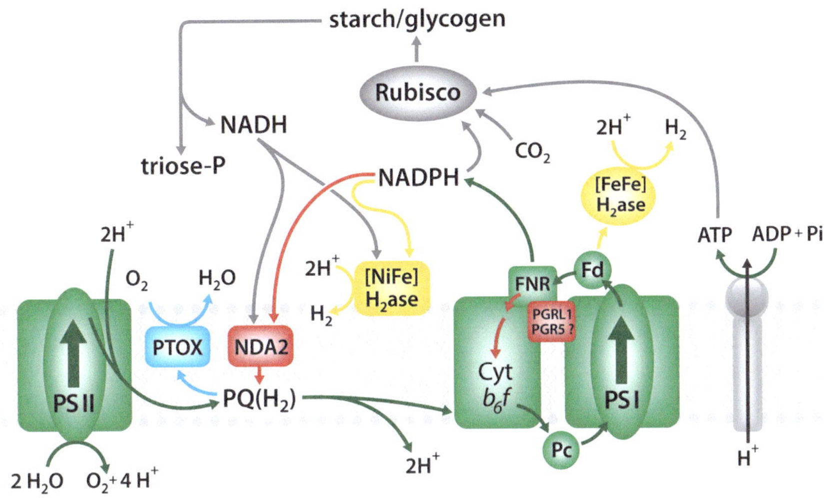

Fig. 5.1. Electron transfer pathways in photosynthetic microbes. Two ET pathways are indicated: (a) the PSII-dependent pathway (*in green*), transferring electrons linearly (*LEF*) from water oxidation by PSII to either the [FeFe]-hydrogenase (green algae) or the NAD(P)H-dependent [NiFe]-hydrogenase (cyanobacteria); and (b) the PSII-independent pathway (*in grey*), in which reductant (NADH) released from starch or glycogen degradation is used to reduce the PQ pool, in a process mediated by NDA2; this is followed by further electron-transfer steps through the cyt b_6f complex and PSI, and from there to FDX and [FeFe]-hydrogenase in green algae. In cyanobacteria, NADH is directly linked to the [NiFe]-hydrogenase. The *arrows* indicate ET reactions, proton translocation, and ATP/CO_2 utilization. PQ-dependent and independent CEF pathways are shown in *red*. Notice that FNR is placed in close contact with PSI, PGRL1, PGR5, and cyt b_6f based on biochemical evidence for formation of a supercomplex under growth conditions that favor CEF (see text); the mechanism for FNR-dependent ET within this supercomplex is not clear, and it is not known whether $NADP^+$ acts as an ET mediator in the process. The protein PTOX is shown in *blue*, and it is able to catalyze the oxidization of the PQ pool to complete a water cycle, which competes with photochemical reduction of the cyt b_6f complex.

(Schmitz et al. 1995; Appel and Schulz 1996), although no evidence for such a function exists (see Sect. II.B for more details). Similarly, a plastoquinone oxidase (PTOX) present in the thylakoids of green algae has been proposed to reduce O_2 molecules directly (Cournac et al. 2002; McDonald et al. 2011). Finally, green algae and cyanobacteria are also capable of fermentatively generating H_2 from extra or intracellular substrates (Brand et al. 1989; Mus et al. 2007; Hemschemeier and Happe 2011) albeit at much lower rates when compared to the light-dependent reactions (Meuser et al. 2012). It is important to note that the photosystems generate a single electron per absorbed photon but are catalytically coupled to enzymes that catalyze multiple electron reactions such as FNR, hydrogenase, and the O_2-evolving Mn-cluster system. Whether this difference impacts the efficiencies of ET and enzyme catalysis, for example through losses of electrons among the various competing reactions, is not known.

The evidence for direct ET from FDX to the HYDA1 hydrogenase in *Chlamydomonas* was first demonstrated in the early 80s (Roessler and Lien 1984). The steady-state kinetics of the ET reaction between FDX and algal hydrogenase have been partially characterized (Roessler and Lien 1984; Winkler et al. 2009), and the reaction was modeled as a direct transfer from the FDX iron-sulfur cluster to the [FeFe]-hydrogenase H-cluster (Florin et al. 2001; Chang et al. 2007; Long et al. 2008, 2009). More recently, Winkler et al. (2009) were able to reconstitute *in vitro* ET through PSI, FDX, and HYDA1, using ascorbate/DCIP as the electron donor to plastocyanin.

Although most of the photosynthetic reductant is transferred from FDX to HYDA1 in *Chlamydomonas*, Meuser et al. (2012) reported that both HYDA1 and HYDA2 are capable of accepting electrons from the photosynthetic ET chain (and thus from FDX), by investigating the H_2-production properties of three *Chlamydomonas* mutants lacking either HYDA1, HYDA2, or both enzymes. They observed that the *ΔHYDA2* mutant had about 70 % of H_2 photoproduction of the WT, while the *ΔHYD1* strain was able to maintain only 30 % of the rates of H_2 photoproduction compared to the wild-type (WT). Interestingly, both partial mutants had similar rates of fermentative H_2 production as the wild-type, suggesting that the latter is limited by factors other than hydrogenase activity. The double knockout mutant, *ΔHYDA1ΔHYDA2*, completely lacked all hydrogenase activity. The predominance of HYDA1 activity in *Chlamydomonas* was confirmed by RNAi experiments, where inhibition of the *HYDA1* gene led to almost total loss of hydrogenase activity in cell extracts (measured by methyl viologen (MV) reduction), while inhibition of *HYDA2* hardly affected the total hydrogenase activity (Godman et al. 2010).

In cyanobacteria, the bidirectional [NiFe]-hydrogenase has been demonstrated to be linked to NAD(P)H through its diaphorase components (see Sect. II.B); H_2 photoproduction occurs for a few seconds upon illumination, followed by H_2 uptake as O_2 accumulates.

It is important to point out that, besides generating low potential reductant, the photosynthetic electron transport chain is responsible for establishing a proton gradient across the thylakoid membrane which drives the synthesis of ATP by the ATP synthase. This is achieved additionally by means of cyclic electron flow (CEF) around PSI involving FDX and either the PQ pool or the cyt b_6f complex (Finazzi et al. 1999; Cournac et al. 2002; Alric 2010) as depicted in Fig. 5.1. CEF is relevant in green algae grown under high illumination and nutrient limiting conditions (Allen 2003; Eberhard et al. 2008) where it helps maintain optimal ATP/NADPH ratios for CO_2 fixation (Forti et al. 2003; Shikanai 2007; Cardol et al. 2009), but it is perhaps less significant in cyanobacteria (Bernat et al. 2009). The occurrence of CEF involves additional components such as the proteins PGRL1 and PGR5 (Tolleter et al. 2011), each having been first identified as having a role in CEF in *Arabidopsis thaliana* (DalCorso et al. 2008), and ssl0352 (Battchikova et al. 2011) in *Synechocystis* 6803. The activation of the CEF pathway is mediated by imbalances in the redox state of the PQ pool. An increase in the ratio of reduced/oxidized PQ, such as is observed either under PSII excitation (Allen 1992; Finazzi et al. 1999) or under anaerobic conditions due to electrons released from glycolytic starch degradation and lack of O_2 consumption by the plastoquinone oxidase (Alric 2010) triggers the activity of the STT7 kinase. This enzyme phosphorylates specific subunits of the light-harvesting complex II (LHCII), inducing its dissociation from PSII in the grana, translocation to the stroma lamellae, and re-association with PSI (Lemeille et al. 2010). This process is known as *state transitions*. Recently, this response has been shown to be accompanied by the formation of supercomplexes between PSI, FNR, PGRL1, PETO (a subunit of the cyt b_6/f complex), LHCI, and LHCII in *Chlamydomonas* (Iwai et al. 2010). FNR was shown to be tightly associated with PSI in the CEF supercomplex (see also Fig. 5.1), suggesting a new kinetic model for oxidation of NADPH and direct reduction of cyt b_6f and PSI. The presence of this supercomplex has been correlated to an increase in CEF in *Chlamydomonas* (Tolleter et al. 2011).

A second intracellular mechanism that regulates the rates of ET is non-photochemical quenching (NPQ) (Li et al. 2009). Under high light intensity, when photophosphorylation cannot keep up with the rate of ET, protons over-accumulate in the lumen causing its acidification (Papageorgiou et al. 2007) increasing the probability of formation of triplet chlorophyll, and subsequent generation of reactive oxygen species (ROS). As a response to the acidification of the lumen, the zeaxanthin/violoxanthin system is activated (however, see (Lambrev et al. 2012) for a different proposed mechanism for photoprotection). The zeaxanthin/violoxanthin system is responsible for de-epoxidation of the carotenoid violoxanthin bound to the Lhcbm1 subunit of PSII in green algae (Elrad et al. 2002) to form zeaxanthin, which is capable of quenching the absorbed excitation at the PSII light harvesting antenna, thus decreasing the amount of excitation energy that drives charge separation. In cyanobacteria, excess light energy is detected by the orange carotenoid protein (OCP) (Kirilovsky and Kerfeld 2012) that triggers NPQ through an unknown mechanism (Jahns and Holzwarth 2012). Both state transitions and NPQ limit the supply of reductant to the hydrogenase, and it has been suggested that inactivation of CEF and NPQ would result in organisms with higher rates of H_2 photoproduction.

A physiological method to sustain H_2 photoproduction by the green alga *Chlamydomonas reinhardtii* was developed by Melis et al. (2000). The method relies on the partial inactivation of photosynthetic O_2 evolution induced by the removal of sulfate from the growth medium. Under sulfur-deprived conditions, sealed cultures sequentially over-accumulate starch and inactivate PSII, become anaerobic, induce hydrogenase gene expression, and degrade Rubisco (Zhang et al. 2002). These changes result in the continuous production of H_2 gas for a period of about 3 days. Sulfur deprivation causes major metabolic changes in *Chlamydomonas* (Melis and Happe 2001; Winkler et al. 2002; Kosourov et al. 2003; Ghysels and Franck 2010) including the activation of NPQ and cyclic phosphorylation (Kruse et al. 2005; Johnson and Alric 2012); consequently, the overall rates of H_2 production by the cultures can never achieve levels close to the 13 % light conversion efficiency and are limited by the relative lower amount of active PSII centers. Indeed, the highest reported *incident light conversion efficiency*, 0.87 %, was achieved with sulfur-deprived, WT *Chlamydomonas* cultures immobilized in alginate (Kosourov and Seibert 2008).

The contribution of the two light-dependent pathways to H_2 production by sulfur-deprived algae has been extensively studied in the last few years. It has become apparent that the contribution from each pathway varies depending on the cultivation mode and prevailing growth conditions (Fouchard et al. 2005; Ghirardi et al. 2009; Ghysels and Franck 2010). For instance, the PSII-inhibitor DCMU completely eliminates H_2 photoproduction when added to photoautotrophic or photoheterotrophic cultures that have been anaerobically induced and illuminated for a few minutes (Gfeller and Gibbs 1984; Brand et al. 1989). However, by measuring the effect of DCMU addition throughout the sulfur deprivation process, it was observed that its effect was higher in the early stages of H_2 production (about 80 % inhibition) (Kosourov et al. 2003; Fouchard et al. 2005), while the inhibition level was lower when DCMU was added in the later stages (Laurinavichene et al. 2004). Moreover, Fouchard et al. (2005) reported complete inhibition of H_2 production if DCMU was added before starch accumulation and H_2 photoproduction occurred. The role of starch in H_2-production by sulfur-deprived cells has been extensively examined, and its involvement can be summarized as: (a) transcriptional activation of hydrogenase genes (Posewitz et al. 2004b) (b) removal of photosynthetically-evolved O_2 by serving as a substrate for respiration; and (c) source of reductant to the hydrogenase through the PSII-independent pathway.

Sulfur deprivation is not the only method to induce anaerobiosis/H_2 production in *Chlamydomonas*. Alternative methods that partially and temporarily inactivate PSII such as photoinhibition (Markov et al. 2006), nitrogen deprivation (Philipps et al. 2012), D1 temperature-sensitive mutations (Mazor et al. 2012), and the use of an inducible psbD-based chloroplast gene expression system (Surzycki et al. 2007) are capable of yielding sustained H_2 photoproduction alone or in combination with sulfur deprivation (Torzillo et al. 2009). However, so far, sulfur deprivation still yields the best rates of H_2 photoproduction, particularly as

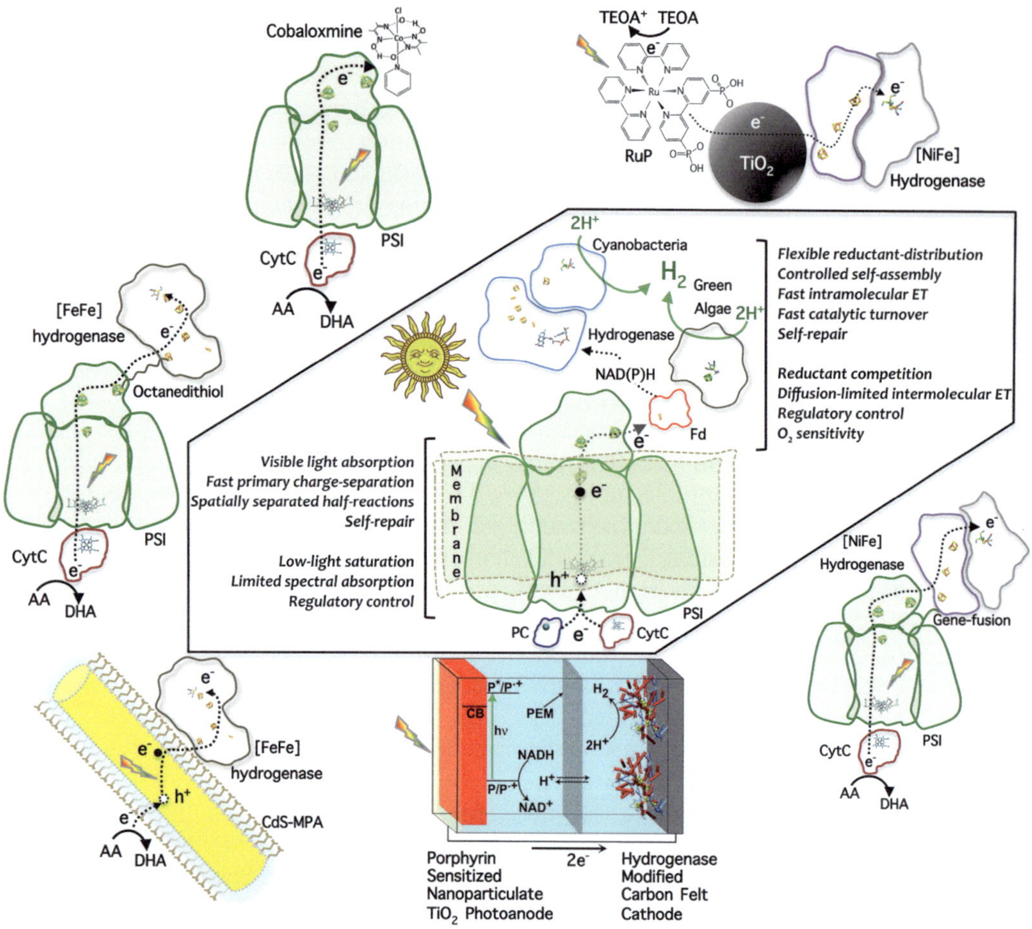

Fig. 5.2. Bio-inspired, photobiohybrid complexes and devices directly couple the catalytic power of enzymes and catalysts with light-harvesting by natural photosystems or artificial nanoparticles and photoelectrochemical cells for solar H_2 production. Photobiohybrids address some of the limitations found in model organisms being studied for solar conversion (e.g. light harvesting, diffusion-controlled ET), and how designs that are directed towards single catalytic process can improve turnover and quantum yields. Examples of photobiohybrids discussed in the text are shown clockwise from the *left*: a PSI-CytC fusion chemically linked to a [FeFe]-hydrogenase by octanedithiol (labeled) (Lubner et al. 2011); cobaloxime adsorbed onto PSI (Utschig et al. 2011); [NiFeSe]-hydrogenase adsorbed to particulate, dye-sensitized TiO_2 (Reisner et al. 2009b); [NiFe]-hydrogenase genetically fused to the *psaE* subunit to create a PSI-[NiFe]-hydrogenase complex (Ihara et al. 2006a, b; Krassen et al. 2009); [FeFe]-hydrogenase integrated into a dye-sensitized photoelectrochemical cell (Reprinted with permission, Hambourger et al. 2008); and [FeFe]-hydrogenase adsorbed to mercapto-propionic acid capped CdS nanorods (Brown et al. 2012). *AA* ascorbic acid, *CytC* cytochrome C, *CdS* cadmium sulfide nanorod, *DHA* dehydroascorbate, *ET* electron-transfer, *MPA* mercaptopropionic acid, *PC* plastocyanin, *RuP* ruthenium bipyridine phosphonic acid, *TEOA* triethanolamine.

other metabolic and reactor engineering barriers (see Sect. IV) are being addressed.

B. Photobiohybrid Systems

Photobiohybrids are defined here as devices that result from the integration of biocatalysts (hydrogenases) with light-harvesting nanomaterials (particle-based) or photoelectrochemical components (electrode-based) – see Fig. 5.2. Research in this area has benefitted from advances in material synthesis for improved, tunable control of band-gap energy (e.g., light absorption), band-edge redox levels (Chen et al. 2010; Peng 2010) and surface functionalization (Gaponik et al. 2002), as well as from a better understanding of the properties that control self-assembled layers on both nanoparticles and electrodes (Badia et al. 2000; Zhang et al. 2011). As a result, the knowledge gained from integrating enzymes with photochemical nanomaterials and electrodes in device architectures has helped to provide guidance and inspiration for practical designs of fully artificial molecular systems (Gust et al. 2000, 2009; Lubitz et al. 2008; Allakhverdiev et al. 2009; Navarro et al. 2010; Park and Holt 2010; Reisner 2011; Wang et al. 2011). The scope of this short review on photobiohybrids will address the reductive side of the biophotolytic process and will not address the recent progress made on the immobilization of PSII water oxidation catalysts on electrodes (Badura et al. 2008; Terasaki et al. 2008; Kato et al. 2012). Although the measured turnover and rates are below those obtained *in vivo*, these efforts are very promising steps towards developing tunable, electrochemical systems for investigating the mechanisms of the PSII-OEC water oxidation reaction.

Hydrogenases and other energy-converting enzymes that function in direct photoconversion with the primary reactions of photosynthesis are ideal catalysts, operating at or near the thermodynamic potential of the specific half-reaction with high turnover (Cracknell et al. 2008; Armstrong et al. 2009; Armstrong and Hirst 2011). For example, algal and bacterial [FeFe]-hydrogenases catalyze H_2 evolution at k_{cat} values of up to 10^4 s^{-1} (Madden et al. 2012), with prolonged stabilities when immobilized on electrode surfaces (Alonso-Lomillo et al. 2007; Bae et al. 2008; Hambourger et al. 2008). Moreover, [NiFeSe]-hydrogenases can evolve H_2 at high turnover (k_{cat} ~10^3 s^{-1}) and retain activity in the presence of O_2 (Vincent et al. 2006; Reisner et al. 2009a; Baltazar et al. 2011). Because hydrogenase turnover is fast, it is possible to match them with the high light-harvesting capacity (k_{Abs}>>solar flux) and broad spectral range of artificial photochemical materials (e.g., semiconductor nanomaterials). In principle, charge-transfer efficiencies and H_2 production rates can approach or even surpass those of photosynthetic organisms (Bolton et al. 1985; Blankenship et al. 2011) but at the cost of a self-repair process. Therefore, a significant challenge to artificial technologies is to develop designs that attain a careful balance of light absorption, conversion, and catalysis to avoid damaging and costly side-reactions. One example to address this in a particle-based approach strategy is to use heterostructured materials as artificial photosystems (i.e., PSI/PSII) that spatially confine electrons from holes, limiting rates of internal recombination (Dukovic et al. 2008; Amirav and Alivisatos 2010). The 2D-spatial confinement structurally mimics charge-separation across membranes in photosynthetic reaction centers and provides a design strategy towards improving photoconversion.

Photoconversion in biological systems requires the concerted, sequential ET between reaction centers, electron-carriers, and enzymes that is mediated through protein-protein complexes, as described in Sect. I.A. The matching of molecular shape, hydrophobicity, and surface electrostatics helps to drive formation of the biological ET complexes (Chang et al. 2007; Long et al. 2008, 2009; Winkler et al. 2009). These design principles can be used to guide engineering of the molecular interface that mediates the integration of molecules into functional photobiohybrids. In both particle and electrode systems, the surface chemistry determines the stability and conductivity of the interface, which in turn greatly affects external ET process (Armstrong et al. 1997; Zhang et al. 2002) and ultimately the efficiencies of photocatalysis. Several reports have demonstrated the successful integration of hydrogenases with photochemical light-harvesting molecules (Greene et al. 2012; Zadvornyy et al. 2012) and on conducting electrodes (Alonso-Lomillo et al. 2007; Hambourger et al. 2008; Brown et al. 2010, 2012; Woolerton et al. 2012), in some cases based on electrostatically guided assembly. One of the first examples of light-driven H_2 production by a photobiohybrid complex was in the 1980s with [FeFe]-hydrogenase and nanoparticulate TiO_2 (Cuendet et al. 1986). Later efforts integrated [NiFe]-hydrogenases with anatase TiO_2 (Pedroni et al. 1996; Selvaggi et al. 1999) and CdS (Shumilin et al. 1992). More recently, nanoparticulate TiO_2 has been used as a conducting material to couple light-capture and charge-separation by dye molecules to catalytic H_2 production by [NiFeSe]-hydrogenases (Reisner et al. 2009a, b). Assembly relied on interactions between the enzyme and surface groups (i.e. hydroxyls or sulfur vacancies). As a result, molecular orientations are not specifically optimized for ET. There have been significant efforts in chemical passivation of nanocrystal surfaces through the use of ligands that provide solvent exposed functional groups for assembly (Ai et al. 2007). Controlled adsorption of [FeFe]-hydrogenase on nanocrystalline materials has been achieved with mercapto-propionic acid-(MPA)-capped CdTe nanocrystals and bacterial [FeFe]-hydrogenases (Brown et al. 2010). The MPA ligand chemi-adsorbs via the S-group to CdTe surfaces, with the solvent-exposed carboxyl group of the ligand, promoting electrostatically controlled self-assembly with positively charged surfaces on the hydrogenase. Many hydrogenases, including the [FeFe]-hydrogenase, possess positively charged regions that mediate formation of an ET complex with electron donor molecules such as FDX (Moulis and Davasse 1995; Peters 1999; Chang et al. 2007; Long et al. 2009). Due to the high surface charge density, molecular complexes that are formed between the nanoparticle and hydrogenase possess strong binding kinetics suggesting formation of a metastable complex.

Once adsorbed, the light-conversion efficiencies in particle-based photobiohybrids can show a strong dependence on the molecular compositions (Brown et al. 2010, 2012). As the hydrogenase and nanoparticles self-assemble in solution, the resulting molecular distributions are comprised of an ensemble of complexes differing in stoichiometric ratio, which can be modeled based on a Poisson distribution. External ET in photobiohybrids, like in photosystems, occurs through the sequential process of single photoexcitation events where the macroscopic ET rate is directly proportional to incident flux. When coupled to multi-electron catalysis, subsequent allocation of electrons among multiple bound catalysts lowers the conversion efficiencies, presumably due to singly reduced states of the hydrogenase competing with back ET to the nanoparticle. Thus, each catalyst-nanoparticle ratio distribution contributes to a broad range of conversion efficiencies (Brown et al. 2010, 2012), and controlling the molecular stoichiometries in particle systems might support higher photoconversion efficiencies.

As catalysts in photoelectrochemical devices for solar H_2 production, hydrogenases have been immobilized on conducting electrodes in dye-sensitized photoelectrochemical cells and biofuel cells (Hambourger et al. 2008; Krishnan and Armstrong 2012). Charge balance was maintained by regenerating the ground state of the photoanode with an electrolyte or a reforming reaction (Hambourger et al. 2007, 2009). The conductive electrode can be composed of metal-based (e.g. TiO_2, or gold) or carbon-based materials (e.g. glassy carbon, pyrolytic graphite edge, carbon felt or carbon cloth) (Lojou et al. 2008; Joo et al. 2009; Gutierrez-Sanchez et al. 2011; Krassen et al. 2011; Morra et al. 2011). Nanostructured electrodes have also been used, for example single- or multi-walled carbon nanotubes (SWNT and MWNT, respectively) or TiO_2 nanorods (Bae et al. 2006; Alonso-Lomillo et al. 2007) on conductive supports. [NiFe]-hydrogenase adsorbed onto MWNT electrodes under an applied electric field have been covalently attached by peptide cross-linking chemistry (Guldi et al. 2005) resulting in high H_2 oxidation current densities with long-term stability (Alonso-Lomillo et al. 2007; Gutierrez-Sanchez et al. 2011). [FeFe]-hydrogenases and SWNTs have been shown to self-assemble into charge-transfer complexes observed by alterations in SWNT photoluminescence and Raman spectra (Blackburn et al. 2008; McDonald et al. 2008). These demonstrate that efficient electron exchange is possible with the small intrinsic bias between the SWNTs and hydrogenase. Thus, these can be used as highly efficient conductors for coupling redox catalysts to electrodes in solar devices.

An alternative strategy for developing photobiohybrids is to directly link hydrogenases, or synthetic H_2 catalysts to the light harvesting power of PSI since PSI has a quantum yield of 100 % for photon absorption to primary charge-separation, a long-live photoexcited state (~50 ms), and the photochemical potential (−580 mV vs. NHE) required for proton reduction (Lubner et al. 2011). The resulting photocatalytic complexes in turn directly couple the donor side of PSI to a catalyst, which in concept would prevent competition with other PSI acceptors. Strategies for directly linking PSI with hydrogenases include the use of alkane dithiols to chemically wire the PSI F_B [4Fe-4S] cluster to the distal [4Fe-4S] cluster of the *C. acetobutylicum* [FeFe]-hydrogenase. The molecular wire facilitates direct intermolecular ET from PSI to the hydrogenase for catalytic H_2 production under illumination (Lubner et al. 2010). An alternative strategy was to create a genetic fusion of the MBH [NiFe]-hydrogenase of *Ralstonia eutropha* to the PsaE subunit of PSI, which was used *in vitro* to reconstitute a functional hydrogenase-PSI complex. The fusion to PsaE allowed for direct ET to the hydrogenase and resulted in light-driven H_2 production both in solution (Ihara et al. 2006b) and when immobilized on a Au-electrode (Krassen et al. 2009). PSI has also been shown to function in photochemical catalysis in complexes with artificial H_2 production catalysts. When PSI was mixed with cobaloxime, the molecules self-assembled and catalyzed H_2 photoproduction (Utschig et al. 2011). Each of these examples confirms the versatility of PSI in driving photocatalysis and provides models for elucidating the engineering principles that control solar conversion towards improving photocatalytic efficiencies.

II. Hydrogenases

A. [FeFe]-Hydrogenases

1. Structure, Function and O_2 Sensitivity

[FeFe]-hydrogenases catalyze the activation of H_2 through the reversible reaction, $H_2 \leftrightarrows 2H^+ + 2e^-$, and function to either couple H_2 oxidation to energy-yielding processes or evolve H_2 by reducing protons as a mechanism to recycle reduced electron carriers like ferredoxin and flavodoxin that accumulate during fermentation (Vignais and Billoud 2007). Generally, hydrogenases located in the cytoplasm are associated with H_2 evolution, while those located in the periplasm or membrane are associated with H_2 uptake (Vignais et al. 2001).

[FeFe]-hydrogenases, found in higher eukaryotes, bacteria, but not archaea, share a common architecture consisting of various complements of FeS cluster domains linked to the H-cluster catalytic domain (H-domain) and can be either monomeric or part of a multimeric complex including additional FAD and NAD(P)H binding domains (Vignais and Billoud 2007). The highly conserved H-domain can be identified in primary sequences by three distinct binding motifs termed L1 (TSCCPxW), L2 (MPCxxKxxE), and L3 (ExMACxxGCxxGGGxP) (Vignais et al. 2001). Additional FeS cluster domains (F-clusters) function to mediate ET to and/or from the H-cluster active site (Pierik et al. 1992). The simplest form of [FeFe] hydrogenase, found in green algae, do not contain F-clusters and consist of only the H-cluster binding domain (Florin et al. 2001; Happe and Kaminski 2002; Forestier et al. 2003).

X-ray crystal structures of [FeFe]-hydrogenases have been determined from the anaerobic soil bacterium *Clostridium pasteurianum*, CpI (Peters et al. 1998), and the sulfate-reducing bacterium *Desulfovibrio desulfuricans*, DdH (Nicolet et al. 1999), revealing the active site H-cluster. The H-cluster consists of a [4Fe-4S] subcluster linked to a 2Fe subcluster that is coordinated by diatomic (CO and CN^-) and dithiolate ligands, and it is biologically tuned to catalyze reversible H_2 oxidation. Initial structural analysis of CO and CN^- ligands was made by FTIR spectroscopy (van der Spek et al. 1996; Happe et al. 1997). The identity of the dithiolate ligand remains undetermined by X-ray crystallography since viable ligands (dithioproprane, dithiomethylether, and dithiomethylamine) contain isotropic atoms (C, O, N) at the bridge-head position (Pandey et al. 2008). Recent EPR studies designate the dithiolate ligand to be dithiomethylamine (Silakov et al. 2009, 2012; Erdem et al. 2011).

The majority of biophysical data on the H-cluster is from CpI and DdH due to the availability of crystal structures, but more recently, hydrogenases from a larger variety of organisms are being explored, including [FeFe]-hydrogenases from *Clostridium acetobutylicum*, *C. reinhardtii* (HYDA1), and *Thermotoga maritima*. Although an X-ray crystal structure for an active algal [FeFe]-hydrogenase does not yet exist, only recently one of an immature form from *C. reinhardtii* has become available (Mulder et al. 2010).

The H-cluster undergoes structural and redox changes during H_2 catalysis, and from rigorous work, a general model for H-cluster catalytic-relevant states and H_2 activation has emerged (Lubitz et al. 2007; Vincent et al. 2007). Hydrogen is hypothesized to bind at the distal Fe atom of the 2Fe subcluster, which contains a ligand-exchangeable binding site (Nicolet et al. 2000). During oxidation and reduction, a CO ligand shifts between bridging and semi-terminal modes, while the distal Fe atom of the 2Fe subcluster switches between Fe(I) and F(II) oxidation states. At the heart of the mechanism is the formation of metal-hydride species, including Fe-H_2, Fe-H^-, and Fe-H^+, which establishes links to other activation mechanisms of small molecules by complex metallo-clusters.

Although the H-cluster primarily resides in a hydrophobic pocket, several conserved amino acid residues form hydrogen bonds to the CO/CN^- ligands and S atoms of the dithiolate ligand. Substitution of these residues in HYDA1 and CpI resulted in either the absence of an H-cluster or an H-cluster trapped in inactive states, suggesting a role for tuning H-cluster reactivity and coordination (Knorzer et al. 2012). One residue, Cys299 in CpI (TS**C**CPxW in L1 motif), resides within hydrogen bonding distance of the bridge-head atom of the dithiolate ligand. It has been suggested that this residue could donate protons for H_2 formation during catalysis (Peters et al. 1998). One attractive mechanism is for the bridgehead N atom of the dithiolate ligand to function as a catalytic base and shuttle protons between C299 and the exchangeable binding site at the 2Fe subcluster (Nicolet et al. 2001). Mutagenesis studies altering this ligand, as well as several other ligands in the proton pathway, demonstrated it to be critical toward hydrogenase activity, supporting its role in proton donation (Cornish et al. 2011).

As of now, there is no natural or engineered organism that is capable of sustaining high rates of H_2 production under completely aerobic (photosynthetic) conditions. Among several basic issues that confront the development of a completely photolytic process is the inherent sensitivity of the H_2 forming enzymes to O_2 (see Sect. IV.A). Interestingly, the algal [FeFe]-hydrogenases are known to have one of the highest sensitivities to inactivation by O_2 (Erbes et al. 1979; Ghirardi et al. 1997, 2007; Flynn et al. 2002; Cohen et al. 2005a; Goldet et al. 2009; Stripp et al. 2009). A model for the O_2-mediated inactivation of algal [FeFe]-hydrogenase has been recently proposed from freeze-quench, X-ray adsorption spectroscopy on O_2 treated samples (Haumann et al. 2005; Stripp et al. 2009; Lambertz et al. 2011). First, O_2 accesses the catalytic site through a hydrophobic channel (Montet et al. 1997; Cohen et al. 2005a, b) and binds to the distal Fe atom (Fe_D) of the H-cluster 2Fe subcluster, followed by either dissociation or conversion into reactive oxygen species (ROS) including O_2^-, OOH^-, and H_2O_2. ROS in turn can react with the 2Fe subcluster leading to oxidative loss of diatomic (CO) ligands, Fe oxidation, or oxidation of S-groups (Stiebritz and Reiher

2009, 2012). One consequence is the sequential degradation of the [4Fe-4S] subclusters into [3Fe-4S] clusters and eventually to more oxidized products. Carbon monoxide (CO) at low partial pressures is a well-known reversible inhibitor of [FeFe]-hydrogenase (Thauer et al. 1974; Adams 1990; Bennett et al. 2000; Lemon and Peters 2000; Chen et al. 2002; Baffert et al. 2011) by also binding to Fe_D. In the CO-inhibited state, [FeFe]-hydrogenase is protected from O_2 (Erbes et al. 1978; Goldet et al. 2009) due to competition for binding to Fe_D, which presumably favors CO due to faster binding kinetics and a binding mode that is resistant to enzyme oxidation. Early electron paramagnetic resonance (EPR) and biochemical studies of CpI showed oxidation by O_2 led to a loss of the H-cluster specific paramagnetic state (Adams et al. 1980), consistent with the model for O_2 binding at Fe_D of the 2Fe subcluster.

2. Maturation

Like other complex metallo-clusters (e.g. Fe-MoCo of Mo-nitrogenase), the H-cluster and unique CO, CN^-, and dithiolate ligands require complex biochemical processes for their synthesis and incorporation (Nicolet and Fontecilla-Camps 2012; Peters and Broderick 2012). During [FeFe]-hydrogenase maturation, a series of assembly, carrier, and specialized enzymes act collectively to modify simple FeS clusters to the complex H-cluster. Initially, identification of maturation proteins was achieved using screens of mutant *Chlamydomonas* strains incapable of H_2 production (Posewitz et al. 2004a). The mutations were mapped to two genes, *HYDEF* and *HYDG*, and co-expression of these genes along with the structural gene, *HYDA1*, assembled an active [FeFe]-hydrogenase (King et al. 2006). While *HYDE* and *HYDF* are fused in several green algae, in the majority of microorganisms *hydE* and *hydF* exist as separately transcribed genes (Bock et al. 2006). HydE and HydG are members of the radical S-adenosylmethionine (SAM) superfamily of enzymes, containing classical CX_3CX_2C sequence motifs. HydF, a NTPase, contains Walker A P-loop and Walker B Mg^{+2} binding motifs and a C-terminal domain with conserved Cys and His residues (King et al. 2006).

In vitro experiments demonstrated that clostridial HydA, heterologously expressed in a genetic background devoid of clostridial HydE, HydF and HydG ($HydA^{\Delta EFG}$), can be activated by addition of a cell extract containing HydE, HydF, and HydG (McGlynn et al. 2007). Further biochemical and spectroscopic characterization of $HYDA1^{\Delta EFG}$ from *C. reinhardtii* revealed that a preformed [4Fe-4S] subcluster is required for [FeFe]-hydrogenase activation (Mulder et al. 2009). These experiments suggest the role of the maturation machinery is to synthesize and insert the 2Fe subcluster after the pre-formed [4Fe-4S] cluster is synthesized and inserted by general house-keeping machinery, such as for the FeS cluster (ISC) system of *E. coli*. It was determined that HydF, expressed and purified from the genetic background $HydF^{EG}$, is capable of activating $HydA^{\Delta EFG}$ without the addition of any exogenous small molecules (McGlynn et al. 2007, 2008). These results implicate HydF as a scaffold protein, on which an H-cluster precursor could be formed by the actions of HydE and HydG and subsequently transferred to $HydA^{\Delta EFG}$. The X-ray crystal structure of $HYDA1^{\Delta EFG}$ from *C. reinhardtii* revealed that insertion of the 2Fe subcluster presumably occurs via a positively charged channel that closes following incorporation via conformational changes in two conserved loop regions (Mulder et al. 2010, 2011). This stepwise mechanism is similar to mechanisms proposed for Mo-nitrogenase (Schmid et al. 2002) and recently for [Fe]-hydrogenase maturation (Hiromoto et al. 2009), indicating overarching themes for complex metallo-cluster synthesis and evolution (Peters and Broderick 2012).

Characterization of HydF has begun to provide details on how cluster assembly may take place. Initially, reconstituted HydF from *T. maritima* was found to bind a single [4Fe-4S] cluster and to hydrolyze GTP to GDP

(Brazzolotto et al. 2006). Further, FTIR and EPR on clostridial HydF co-expressed with ($HydF^{EG}$) or without HydE and HydG ($HydF^{\Delta EG}$) indicate the presence of two clusters: a $[4Fe\text{-}4S]^{2+/1+}$ cluster, along with a $[2Fe\text{-}2S]^{1+}$ cluster in $HydF^{\Delta EG}$ that is modified to a CO and CN^- coordinated Fe species in $HydF^{EG}$ (Czech et al. 2010; Shepard et al. 2010a). HYSCORE on $HydF^{EG}$ suggests that the [4Fe-4S] cluster is ligated by three Cys thiolate ligands and one N atom supplied by a conserved His residue, which could also provide covalent attachment to the binuclear Fe center forming an H-cluster-like precursor (Czech et al. 2010). EXAFS indicates that the two clusters resemble a six Fe unit similar to the H-cluster (Czech et al. 2011) and interestingly, $HydF^{EG}$ has been demonstrated to have small amounts of H_2 evolution and uptake activity (Kuchenreuther et al. 2011). Other temperature-dependent EPR studies on $HydF^{EG}$ suggest that the [4Fe-4S] cluster and 2Fe unit are separate spin systems (Shepard et al. 2010b). While the X-ray crystal structure of an apo-form of HydF from *T. neapolitana* has recently been determined (Cendron et al. 2011), the specific nature of how [4Fe-4S] and [2Fe-2S] clusters bind to the protein are unknown.

Important insight regarding the role of HydG in H-cluster synthesis was set forward when it was determined that the sequence is similar to the radical SAM enzyme ThiH (Pilet et al. 2009), which cleaves tyrosine to yield *p*-cresol and dehydroglycine (DHG). The crystal structure of HydG remains unknown, but sequence comparisons (Pilet et al. 2009) indicate the presence of a $(\beta\alpha)_8$ TIM barrel fold along with a 90-amino acid extension on its C-terminal end that contains a motif of conserved cysteine residues ($CX_2CX_{22}C$) required for maturation (King et al. 2006). EPR studies indicate the presence a typical radical SAM [4Fe-4S] cluster, along with a second FeS cluster presumed to be a [4Fe-4S] cluster (Rubach et al. 2005; Shepard et al. 2010a). Like ThiH, HydG is able to cleave tyrosine to yield *p*-cresol and DHG (Pilet et al. 2009). From DGH, CN^-, and CO are produced (Driesener et al. 2010; Shepard et al. 2010a), and isotopic labeling, monitored by FTIR spectroscopy, demonstrates that all five CO and CN^- ligands are derived from tyrosine (Kuchenreuther et al. 2011).

Potential mechanisms for CO and CN^- production have been proposed, including a single step decarbonylation of DHG (Driesener et al. 2010; Roach 2011) or homolytic tyrosine cleavage, ultimately yielding $H_2C{=}NH$ and $^{\bullet}CO_2^-$ through decarboxylation of a protonated glycyl radical (Nicolet et al. 2010; Tron et al. 2011). Mutagenesis and SAXS on *C. acetobutylicum* HydG suggests that CO production depends on the presence of the C-terminal [4Fe-4S] cluster, whereas CN^- production occurs independently of it (Nicolet et al. 2010; Tron et al. 2011).

Given the role of HydG in CO and CN^- synthesis, one hypothesis is that HydE synthesizes the dithiolate ligand. Precedence is set by sequence comparison to radical SAM enzyme BioB (Nicolet et al. 2008) which carries out sulfur insertion reactions in the synthesis of dethiobiotin. The X-ray crystal structure of HydE from *T. maritima* has been determined and revealed a $(\alpha/\beta)_8$ TIM barrel fold with a site-differentiated [4Fe-4S] cluster bound to SAM, and an additional [2Fe-2S] cluster coordinated by three Cys residues (Nicolet et al. 2008, 2009). Site-directed mutagenesis of the Cys residues of the [2Fe-2S] cluster indicate that the cluster does not serve as the source of S atoms in the dithiolate ligand (Nicolet et al. 2008). Three anion binding sites within the $(\alpha/\beta)_8$ barrel and a substrate binding area were identified near the SAM cofactor to which thiocyanate was found to bind with high affinity (Nicolet et al. 2008). Despite this significant progress, the substrate, along with mechanism for synthesis of the dithiolate ligand, remains unknown.

The [FeFe]-hydrogenase from *Clostridium acetobutylicum* (HydA), along with its maturation machinery was recently expressed in the cyanobacterium *Synechococcus elongatus* (Ducat et al. 2011). Photosynthetic H_2 production in the presence of DCMU (which blocks PSII-catalyzed O_2-evolution) was observed with the transformant, while none was detected from the parent strain under the same conditions, suggesting a lack of native Hox hydrogenase activity. Light-driven H_2 production was further enhanced two-fold when the HydA-expressing strain was co-transformed with a ferredoxin from *C. acetobutylicum*. Collectively, these findings suggest that the foreign hydrogenase is functionally integrated in the photosynthetic machinery of the host's cells. More recently, an active [FeFe]-hydrogenase from *C. reinhardtii* (HYDA1) was also expressed in *Synechocystis,* which led to a five-fold enhancement in light-driven H_2 production (in the presence of DCMU) compared to the untransformed parent strain, which displayed native Hox hydrogenase activity (Berto et al. 2011). A C365G mutation of the H-cluster of HYDA1 abolished the improvement, hence confirming functionality of the HYDA1 in *Synechocystis*. The expression of a *C. pasteurianum* [FeFe]-hydrogenase in *Synechococcus* PCC 7942 without the co-expression of its maturation machinery (*hydEFG*) had been reported earlier (Asada et al. 2000), raising the question of how cyanobacteria can assemble a foreign [FeFe]-hydrogenase without the specific HydE, HydF and HydG maturation factors. This will perhaps be a major theme in upcoming research.

3. Regulation

The algal hydrogenases HYDA1 and HYDA2 are encoded by nuclear genes localized in separate scaffolds. They both contain transit peptides that direct the apoproteins to the chloroplast, where the inorganic catalytic cluster is proposed to be inserted (see Sect. II.A.2 above). There is limited information on the transcriptional regulation of the hydrogenase genes in algal species. HYDA1's promoter region was characterized through the expression of a reporter gene (Forestier et al. 2003; Stirnberg and Happe 2004) and shown to comprise nucleotides -128 to -21 relative to the transcription start site, but no further studies on specific transcriptional regulators had been reported until recently. The two genes are clearly regulated by anaerobiosis (Happe and Kaminski 2002; Mus et al. 2007) and by the redox state of the plastoquinone pool (Posewitz et al. 2004b). Recent studies by Pape et al. (2012) showed evidence that HYDA1's transcription is under the control of the copper response regulator 1 (CRR1). The latter contains a squamosa-binding protein domain which is known to recognize GTCA motifs on promoters of specific genes (Koprat et al. 2005). However, mutations within this motif did not completely abolish anoxic activation of HYDA1 (Pape et al. 2012), suggesting that other regulators must be available. Although O_2 sensors in *Chlamydomonas* are not well characterized, the presence of homologues of FixL genes has been reported in the *Chlamydomonas* genome. FixL proteins are responsible for O_2-sensing and transcriptional activation of nitrogenase in Rhizobia (Gillez-Gonzalez et al. 1994) and therefore their homologues could serve similar function in algae. Consistent with this hypothesis, when expressed in *E. coli*, the catalytic domains of two algal homologues were shown to bind O_2 with high affinity (Murthy et al. 2012).

B. [NiFe]-Hydrogenases

1. Structure, Function and O_2 Sensitivity

Unlike the [FeFe]-hydrogenases found in algae, cyanobacteria contain hydrogenases with [NiFe]-active sites: two-subunit uptake hydrogenases associated with nitrogenase function and/or a five-subunit bidirectional hydrogenase (Tamagnini et al. 2007). In this chapter we will focus on the bidirectional hydrogenase in cyanobacteria.

[NiFe]-hydrogenases are present throughout Archaea and Bacteria and are minimally heterodimeric, consisting of a large subunit containing the active site and a small subunit containing at least one Fe-S cluster that plays a role in ET to and from the large subunit for H_2 reduction/oxidation. In the bidirectional hydrogenase of cyanobacteria, the catalytic subunit is HoxH and the small subunit is HoxY, which contains a single [4Fe-4S] cluster. The catalytic center of HoxH contains Fe and Ni atoms coordinated by two CN^- and one CO ligands as well as sulfur atoms in conserved cysteine residues from the surrounding protein (Fontecilla-Camps et al. 2007; Germer et al. 2009; Heinekey 2009). Hydrophobic channels link the catalytic site and protein surface to allow gas diffusion (Teixeira et al. 2006; Galvan et al. 2008). Heterolytic splitting of H_2 likely occurs at the Ni atom, although this is still open to debate. Multiple redox states of the active site have been identified involving the Ni moiety and the S-donor on one of the bridging cysteines (Ogata et al. 2005; Volbeda et al. 2005; Lubitz et al. 2007; Germer et al. 2009; Ogata et al. 2009). In general, [NiFe]-hydrogenases have two oxidized states, Ni-A (unready) and Ni-B (ready). Ni-B reactivates within seconds while Ni-A reactivates much more slowly. The proton in the active site in the oxidized form is not accessible in the Ni-A state, likely due to its bridging ligand. One electron-reduction of Ni-A and Ni-B states is coupled to proton transfer in the active site and results in EPR-silent, catalytically inactive intermediates termed Ni-S_u (silent unready) and Ni-SI_r (silent ready), respectively. Reactivation in the Ni-SI_r state results in a Ni-SI_a (silent active) state, while further reduction leads to a paramagnetic Ni-C intermediate that contains a hydride bridge between the Ni and Fe atoms. Ni-C reduction results in the most reduced state, Ni-R, an EPR-silent state that maintains the bridging hydride. Ni-R exists in multiple protonated states that likely lead to the catalytic splitting of H_2 (Lubitz et al. 2007; Ogata et al. 2009; Pandelia et al. 2010). FTIR and EPR studies of the bidirectional hydrogenase of *Synechocystis* were able to detect at least four of these states but were unable to detect any paramagnetic states in the fully reduced or oxidized form (Germer et al. 2009).

Similar multisubunit bidirectional hydrogenases found in cyanobacteria have also been characterized in other bacteria such as the soil microbe *Ralstonia eutropha* and other phototrophic bacteria (Schneider et al. 1984a, b; Rakhely et al. 2004, 2007; Burgdorf et al. 2005; Long et al. 2007; Tamagnini et al. 2007; Maroti et al. 2010; Carrieri et al. 2011). The additional three subunits (HoxEFU) in these hydrogenases exhibit extensive homology to the NuoEFG components of bacterial Complex I (Appel and Schulz 1996). Similarly to the small subunit HoxY, these subunits all contain at least one FeS cluster for ET. In addition, the HoxF subunit also contains NAD and FMN binding sites that can catalyze the oxidation/reduction of NAD(P)H/NAD(P)$^+$ as a diaphorase (Massanz et al. 1998; Lauterbach et al. 2011). These so-called diaphorase subunits provide the coupled electron donor/acceptor for the hydrogenase *in vivo* (Boison et al. 1998; Massanz et al. 1998; Rakhely et al. 2004; Antal et al. 2006; Long et al. 2007) transferring electrons between the hydrogenase catalytic site and NAD(P)H/NAD(P)+. Purified *Thiocapsa roseopersicina* (Palagyi-Meszaros et al. 2009) and *Synechocystis* (Schmitz et al. 2002; Germer et al. 2009) intact HoxEFUYH complexes link to both NADH and NADPH. In *Allochromatium vinosum* (Long et al. 2007) and the cyanobacterium *Gleocapsa alpicola* CALU 743 (Serebriakova and Sheremetieva 2006) only HoxYH was purified. This HoxYH subcomplex lacked detectable hydrogenase activity with NAD(P)H as an electron donor, highlighting the role of the diaphorase subunits in linkage to NAD(P)H/NAD(P)$^+$ as an electron donor/acceptor.

The function of this bidirectional hydrogenase varies in the organism in which it is found, but it seems to act to balance reductant levels in the cell during fermentation and/or photosynthesis. The homolog in *R. eutropha* functions in uptake of H_2 under autotrophic conditions, presumably to provide reductant in the form of NAD(P)H for carbon fixation (Burgdorf et al. 2005). In the photosynthetic bacterium *T. roseopersicina,* the bidirectional hydrogenase catalyzes H_2 production under dark, fermentative conditions and in the light under nitrogenase-repressed conditions when thiosulfate is present, but functions in H_2 uptake in the light in cells actively fixing nitrogen to recycle the H_2 co-produced by nitrogenase (Rakhely et al. 2007). In cyanobacteria, the bidirectional hydrogenase has been characterized to be biased towards H_2 production (McIntosh et al. 2011). The hydrogenase is active under dark, fermentative conditions and, during the transition from dark to light, it exhibits a short burst (30 s) of production before it switches to the uptake direction and is subsequently inactivated in the presence of O_2 (Cournac et al. 2002, 2004; Schutz et al. 2004). Therefore, this hydrogenase has been hypothesized to act as an electron valve for cells under changing redox states. Mutants of the hydrogenase (*hoxH*- and *hoxEF*-) were reported to exhibit defects in growth, a 20–30 % decrease in the photochemical activity of PSII and an increased PSII/PSI ratio (Appel et al. 2000; Cournac et al. 2002, 2004; Schutz et al. 2004; Gutthann et al. 2007; Antal et al. 2006), although additional studies have been unable to reproduce these defects (Eckert et al. 2012a; Pinto et al. 2012). It has also been suggested that the bidirectional hydrogenase may play a role in respiration, since NuoE, NuoF, and NuoG are absent in cyanobacteria, and the only homologs present are HoxE, HoxF, and HoxU, respectively (Schmitz et al. 1995; Appel and Schulz 1996). Despite this homology, no evidence to date of any role of the Hox hydrogenase in respiration has been reported (Boison et al. 1998, 1999; Howitt and Vermaas 1999).

In the [NiFe]-hydrogenase from *Desulfovibrio fructosovorans*, O_2 reacts with its active site forming either a hydr(oxo) group (Ni-B) or a (hydro)-peroxo group (Ni-A) (Volbeda et al. 2005). Either oxygen radicals can be removed to restore the hydrogenase activity once the enzyme is returned to an anaerobic reducing environment. The bidirectional hydrogenase from cyanobacteria is believed to be a constitutive enzyme, with hydrogenase transcripts produced in the light (aerobic) in *Nostoc muscorum* (Axelsson and Lindblad 2002), in *Anabaena variabilis* (Boison et al. 2000) and in *Synechocystis* (Kiss et al. 2009), and the presence of the hydrogenase polypeptide is detected in *Anabaena variabilis* (Serebriakova et al. 1994). The produced enzyme presumably remains in an inactive state and as such cannot contribute to sustainable H_2 production. Dark anaerobic incubation or addition of DCMU to a light-grown culture is known to induce significantly higher bidirectional hydrogenase activity in both *Nostoc* (Sjoholm et al. 2007) and *Synechocystis* (Cournac et al. 2004; Baebprasert et al. 2010). As such, the cyanobacterial bidirectional hydrogenase is O_2 sensitive, albeit synthesized aerobically. Transient and fully reversible H_2 production is observed when a *Synechocystis* culture is switched quickly between light and dark conditions (Cournac et al. 2004). This fast transition from an inactive to an active state (within a few seconds) suggests that (a) the bidirectional hydrogenase is a holoenzyme with the maturation process fully functional aerobically; and (b) the hydrogenase must remain in the Ni-B state to afford very fast reactivation. The latter was indeed demonstrated by FTIR studies of purified bidirectional hydrogenase from *Synechocystis*, in which a "Ni-B like" state was detected even under a highly oxidized state. Both criteria (a) and (b) are therefore important considerations for developing photosynthetic H_2 production. Protein film electrochemical studies of the bidirectional hydrogenase from *Synechocystis* revealed that, despite its O_2 sensitivity, the hydrogenase functions in the proton-reduction direction at 25–50 % of the maximal rate even in the presence of 1 % O_2 (McIntosh et al. 2011). Collectively, the aerobic synthesis of a holoenzyme and its fast reactivation may allow the bidirectional hydrogenase to serve as a valuable model to gain insight as to the mechanism for fine-tuning reactivation. Recent advancements uncovered a novel mechanism of O_2 tolerance attributed to the presence of the supernumerary cysteine in the proximal cluster of the small subunit of the [NiFe]-hydrogenases in *R. eutropha* (Fritsch et al. 2011), *E. coli* (Volbeda et al. 2012), and *Aquifex aeolicus* (Pandelia et al. 2011). The unusual [4Fe-3S] cluster is poised at a very high potential with the two added cysteines functioning to deliver four extra electrons for O_2 scavenging, hence protecting the [NiFe]-active site from O_2 inactivation. This action consumes four electrons per molecule of O_2, hence at the expense of two molecules of H_2. This oxygen-resistant reaction will therefore severely impact the photosynthetic conversion efficiency if and when engineered successfully in a cyanobacterium for light-driven H_2 production.

2. Maturation

[NiFe]-hydrogenase maturation involves several biochemical steps that were first described in *E. coli* (Bock et al. 2006). The maturation chaperones required for this process are part of a set of accessory proteins encoded by the *hypABCDEF* genes (Lutz et al. 1991). Cyanobacteria contain homologous gene products which may fulfill similar functions in this organism (Maier et al. 1993; Buhrke et al. 2001; Wunschiers et al. 2003; Oliveira et al. 2004; Hoffmann et al. 2006; Leitao et al. 2006; Devine et al. 2009). In *E. coli*, the biosynthesis of [NiFe]-hydrogenase starts with the formation of an iron ligand on the HypC/HypD dimer. Simultaneously, a thiocyanate ligand is generated from carbamoyl phosphate on the HypE protein, forming a HypE-SCN complex (Paschos et al. 2001; Reissmann et al. 2003). This reaction is catalyzed by the HypF subunit and is ATP dependent. The CN ligands are subsequently transferred to the Fe-HydC/HydD dimer, which plays a role in Fe coordination of the active site. CO is also transferred to the active site, but the actual mechanism is not clear. The partially assembled metallo-cluster is then transferred from HypC to the precursor form of the catalytic subunit, pre-HycE (Magalon and Bock 2000) which is present in an open conformation that allows for metal insertion (Blokesch and Bock 2002). The next step is insertion of nickel, via the HypB-HypA-SlyD-preHycE complex, a process in which HypA acts as a scaffold and involves GTP hydrolysis by HypB, assisted by SlyD. The latter is also a Ni-storage protein (Maier et al. 1993; Chung and Zamble 2011; Kaluarachchi et al. 2011). After both metals have been coordinated to the large subunit precursor form, the C terminus is then accessible for cleavage by an endopeptidase. The cleavage reaction is a subunit-specific process and is catalyzed by HycI in *E. coli* (Vignais and Colbeau 2004). In the *Synechococcus* strain PCC 6301, the HypA, B, C, D, E, F, and HoxW proteins (homologs of the *E. coli* HypABCDE-F and HycI) are believed to be involved in the maturation of the large subunit, HoxH (Thiemermann et al. 1996). HypC possesses an N-terminal cysteine domain that could interact with the precursor of HoxH (Olson and Maier 1997), and HypD has five conserved cysteines, which could be involved in metal binding (Gubili and Borthakur 1998). Recently, the crystal structure of HypF alone or of the HypE-HypF complex revealed that HypF catalyzes the carbamoylation of the C-terminal cysteine of HypE yielding the carbamoyl-HypE molecule (Shomura and Higuchi 2012). The HypA and HypB-SlyD protein couple shows putative Ni-binding sites on HypA as well as a GTPase domain for HypB (Tibelius et al. 1993; Olson and Maier 2000; Hoffmann et al. 2006; Chung and Zamble 2011) demonstrate that HypA and HypB are required for Ni insertion into the large subunit of the *Synechocystis* sp. 6803 hydrogenase, while its two homologs, HypA2 and HypB2, probably have a role as chaperones in the maturation of a different set of metalloproteins in this organism. Transcription of *hypFCDEAB* was subjected to the same regulation as the bidirectional hydrogenase in the filamentous cyanobacterium *Lyngbya majuscula* (Ferreira et al. 2007). Finally, HoxW in *Ralstonia eutropha* (Thiemermann et al. 1996) and Slr2876 in *Synechocystis* sp. 6803 (Hoffmann et al. 2006) are identified proteases for HoxH with a role similar to that of HycI in *E. coli*.

3. Regulation

Transcriptional regulation of cyanobacterial hydrogenases was reviewed recently (Eckert et al. 2012b). The five genes *hoxEFUYH* in *Synechocystis* are clustered in an octacistronic operon that also contains three open reading frames of unknown function. Under normal laboratory growth conditions, this operon is weakly expressed as a polycistronic transcript, which initiates 168-bp upstream of the start codon of the proximal *hoxE* gene (Gutekunst et al. 2005; Oliveira and Lindblad 2005). The transcriptional regulation of the *hox* operon is complex and responds to various environmental conditions. In a circadian rhythm, *hox* transcripts increase in the light and decrease in darkness (Kucho et al. 2005). Higher levels of *hox* transcripts have been observed under microaerobic or anaerobic conditions and under high light. Additionally, lower levels of transcripts were observed when the Calvin cycle was inhibited (Kiss et al. 2009).

Regulation of *hox* transcription involves at least three proteins. The LexA-related protein (LexA; Sll1626), which appears to regulate carbon assimilation, also activates the transcription of the *hox* genes through binding to the *hox* promoter (Gutekunst et al. 2005; Oliveira and Lindblad 2005). In contrast, two AbrB-like regulators (AbrB1, Sll0359; AbrB2, Sll0822) repress *hox* transcription through binding to the same *hox* promoter (Oliveira and Lindblad 2008; Dutheil et al. 2012). Promoter-reporter fusion studies show that the *hox* promoter is weakly active, despite the presence of sequences resembling the canonical -35 (TTGctc) and -10 (TAacAa) promoter boxes (Dutheil et al. 2012). The LexA binds to a region located between nucleotides-198 and -338 bp, as well as -592 to -690 bp, with respect to the translational start point. Furthermore, a LexA-depleted mutant exhibited a decrease in Hox activity, reinforcing LexA's involvement in *hox* gene transcription (Gutekunst et al. 2005; Oliveira and Lindblad 2005) AbrB2 was found to suppress *hox* promoter activity by binding to multiple sites on the promoter (Dutheil et al. 2012). The presence of distant LexA and AbrB2 binding regions in the *hox* promoter suggests the possible involvement of a DNA looping mechanism in the regulation of *hox* transcription (Oliveira and Lindblad 2009; Dutheil et al. 2012). Deletion of AbrB2 from WT *Synechocystis* 6803 resulted in a strain exhibiting normal growth yet increased *hox* transcription and hydrogenase activity, while overexpression of AbrB2 lead to suppressed *hox* transcription and hydrogenase activity (Dutheil et al. 2012).

III. Ferredoxin Network in *Chlamydomonas reinhardtii*

Ferredoxins are small acidic proteins with a low-redox potential that harbour an FeS cluster and function as electron carriers in diverse metabolic pathways in bacteria, plants, algae, and animals. The *Chlamydomonas reinhardtii* genome contains 6 genes that encode for 6 FDXs that are categorized in 3 groups, the leaf-type (FDX1, 5), the root-type (FDX2), and the unknown-function type (FDX3, 4, and 6) (Terauchi et al. 2009). Typically, plant-type FDXs function primarily in photosynthesis where they transfer electrons from photoreduced PSI to FDX/$NADP^+$ reductase (FNR) to produce NADPH required for CO_2 assimilation, whereas the root-type FDXs can channel electrons through FNR from reduced NADPH to non-photosynthetically-active pathways such as nitrogen, sulfite, and glutamate assimilation (Hase et al. 1991; Matsubara and Saeki 1992; Onda et al. 2000; Yonekura-Sakakibara et al. 2000). Finally, the role of the third class of FDXs remains enigmatic, and they exhibit sequence similarity to cyanobacteria FDXs of unknown function.

In *Chlamydomonas*, all six FDXs have putative chloroplast transit peptides and five of them (FDX1, FDX2, FDX3, FDX5 and FDX6) have been experimentally confirmed to be localized in the chloroplast (Schmitter et al. 1988; Jacobs et al. 2009; Terauchi et al. 2009). Transcript abundance studies indicate that *FDX1* represents 98 % of the total FDX-encoded transcript pool in *Chlamydomonas* cells grown in the absence of stress and it represents the majority of the pool under stress condition, suggesting an important role of this protein in the cells (Terauchi et al. 2009). Indeed, FDX1 seems to be associated with multiple metabolic pathways, ranging from photosynthetic ET (Fischer et al. 1999; Terauchi et al. 2009) to redox-dependent enzymes involved in central metabolism such as thioredoxin (Jacquot et al. 1997), assimilatory enzymes such as nitrite reductase and glutamate synthase (Garcia-Sanchez et al. 1997; Terauchi et al. 2009), and finally, H_2 production (HYDA) (Long et al. 2008). Interestingly, all the H_2 production pathways (see previous section) diverge from FDX. However, there is no direct proof of one specific FDX interacting with pyruvate/ferredoxin oxidoreductase (PFR) under dark anaerobiosis (see previous section), and FDX1 is probably fulfilling this role. Under photosynthetic conditions, FDX1 has been demonstrated to be able to mediate ET to HYDA1 (Yacoby et al. 2011) and *in silico* docking analysis along with mutagenesis have identified probable binding complexes and several amino acid residues required for FDX/HYDA1 and FDX/HYDA2 interaction complexes (Chang et al. 2007; Long et al. 2009; Winkler et al. 2009). It appears that FDX binding to HYDA is driven mainly by electrostatic interactions. Specifically, HYDA1 Lys^{396} and FDX1 Glu^{122} make a major contribution to complex formation and ET (Winkler et al. 2009). It must be noted that HYDA2 also shows conservation of the required lysine, and all FDXs but FDX3 have the conserved glutamic acid residue.

In the case of FDX2, several studies have confirmed its role in nitrogen assimilation (Jacquot et al. 1997; Terauchi et al. 2009). FDX2 transcript and protein levels are increased in nitrate grown *Chlamydomonas* cells (Terauchi et al. 2009). FDX2 can physically interact with the *Chlamydomonas* nitrite reductase (Garcia-Sanchez et al. 1997) and efficiently transfer electrons to the enzyme to reduce nitrate at a rate 10 times higher than FDX1 (Terauchi et al. 2009). FDX2 can also receive electrons from FNR in a more efficient manner than FDX1, indicating that FDX2 is functionally a root-type FDX, as suggested by phylogenetic analysis (Terauchi et al. 2009).

FDX5 has been demonstrated to be highly induced at (both transcript and protein level) at the onset of anaerobiosis, either under dark or sulfur-deprivation conditions (Jacobs et al. 2009; Terauchi et al. 2009), and it is upregulated under copper deficiency. Indeed, the CRR1 (copper response regulator) transcription factor was shown to interact with FDX5, linking the protein to copper metabolism (Lambertz et al. 2010). However, no specific function has been described so far for FDX5, although it is known that FDX5 cannot mediate ET to the hydrogenase enzymes under anaerobiosis (Jacobs et al. 2009), and no direct interaction studies have shown the association of FDX5 with other specific proteins involved in the copper deficiency response.

Finally, despite a recent expression pattern analysis of all the FDXs, the role of FDX3, 4, and 6 remains enigmatic (Terauchi et al. 2009). Those proteins have probably evolved specialized functions, since their sequences diverge from the other FDX genes, and they do not cluster with leaf or root-type FDX (Terauchi et al. 2009). Clearly, further investigation is required to demonstrate and assign specific function for each individual FDX. Direct *in vivo* examination of mutant alleles, protein overexpression and *in vivo/in vitro* interaction analysis are essential to definitively determine the biochemical pathways and map the FDX-dependent network in *Chlamydomonas*.

IV. Barriers to H_2 Photoproduction

Although sustained H_2 photoproduction by *Chlamydomonas* can be achieved by partial inactivation of PSII (see Sect. I.A), the resulting light-conversion efficiencies are too low for commercial applications (James et al. 2008) and would require reactor areas that are too large. The barriers to commercialization are due, mostly to physiological limitations and have been listed in previous reviews (Melis and Happe 2004; Esper et al. 2006; Ghirardi et al. 2009; Rupprecht 2009; Kruse and Hankamer 2010; Lee et al. 2010; Eroglu and Melis 2011). In the next sections, we present an update on the state-of-the-art regarding the status of these barriers.

A. O_2 Sensitivity of Hydrogenases

In Sect. I.A, we described sulfur-deprivation and other methods that temporarily decrease PSII activity as a current approach to sustain H_2-photoproduction. An alternative strategy to address the O_2 sensitivity barrier while maintaining high light conversion efficiency is to heterologously express an O_2-tolerant foreign hydrogenase into an algal/cyanobacterial host. The critical criteria for success are (a) the O_2-tolerant hydrogenase should favor H_2 evolution over H_2 uptake; and (b) the foreign hydrogenase must be able to link to the host's photosynthetic ET pathway (preferentially via the low redox potential mediator, ferredoxin, for more favorable energetics). The second criterion is best met by attempting to express [FeFe]-hydrogenases in photosynthetic hosts.

Advancement has been made in the heterologous expression of O_2-tolerant [NiFe]-hydrogenases from either *Alteromonas macleodii* or *T. roseopersicina* into *Synechococcus elongatus* strains lacking their native bidirectional hydrogenase (Weyman et al. 2011). Hydrogenase activity, assayed with reduced MV was detected, but photoproduction of H_2 was not observed in either case. In both transformants, co-expression of the specific maturation machinery of the foreign [NiFe]-hydrogenase was required for activity.

In contrast to the success in developing heterologous expression of recombinant [NiFe]- and [FeFe]-hydrogenases in cyanobacteria (see Sect. II.A.2), to date there are no reports of successful heterologous expression of hydrogenases in algae. In theory, the expression of bacterial [FeFe]-hydrogenases, shown to have more O_2 tolerance than the algal enzymes, should not be constrained by maturation specificity. Both bacterial (King et al. 2006; Akhtar and Jones 2008; von Abendroth et al. 2008; Kalim Akhtar and Jones 2009; English et al. 2009; Kuchenreuther et al. 2010; Laffly et al. 2010; Yacoby et al. 2012) and algal (Agapakis et al. 2010; Ducat et al. 2011) maturation systems have been shown to activate the corresponding heterologous enzymes at relatively high efficiencies. Moreover, bacterial [FeFe]-hydrogenases were shown in early studies to effectively couple to plant-type ferredoxins (Rao et al. 1978; Fitzgerald et al. 1980; Hall et al. 1980). By far, the most challenging aspect is to develop a successful route for engineering improved O_2 tolerance in [FeFe]-hydrogenases (Flynn et al. 2002; Barstow et al. 2011; Lautier et al. 2011) which to date have not produced increases in tolerance to the levels shown for the [NiFe]-hydrogenases [(Liebgott et al. 2011) and reviewed in more detail elsewhere in this book]. Moreover, high recombinant expression levels will require optimizing the nucleotide sequences of transgenes and incorporation of transit peptides required to signal the targeting of the hydrogenase product into the chloroplast. Studies on the structure, function, and of transit peptides and linker regions (Chen and Schnell 1999; Keegstra and Cline 1999; Bruce 2000) have led to efforts in optimization (Jin et al. 2003) but this remains an area for further research.

B. Competition with Other Pathways

Electron transport from the photosynthetic chain to hydrogenases in *Chlamydomonas* is mediated by FDX, and, as discussed in Sect. III, is subject to competition with multiple other metabolic pathways. Under anaerobic and high light conditions, the major competitors are cyclic electron flow and FNR, which generates NADPH for carbon fixation.

Not surprisingly, algal cells in CEF mode exhibit low rates of light-driven H_2 evolution for two main reasons: (a) sequestration of FDX by FNR and (b) down-regulation of electron transport due to non-dissipation of the proton gradient by ATP synthesis (see Sect. IV.B). Recently, a strategy for adjusting the kinetics of electron flow to hydrogenase was reported based on the fusion of FDX to the mature form of [FeFe]-hydrogenase HYDA1 (Yacoby et al. 2011). When combined with thylakoid membranes, plastocyanin, ascorbate, and DCMU, the FDX-HYDA1 fusion catalyzed photohydrogen production, with rates being enhanced by the addition of free FDX. Addition of FNR and $NADP^+$ resulted in no change in H_2 evolution rates, whereas the rates of H_2 production by the native HYDA1 were severely depressed due to kinetic competition for reduced FDX with FNR. Whether this same effect can be observed *in vivo*, and if a synergistic effect can be observed by expression in CEF mutant backgrounds remain interesting new areas for investigation.

C. Downregulation of Electron Transport by Non-dissipation of the Proton Gradient

As mentioned in Sect. I.A, the rate of electron transport is regulated by two major mechanisms: NPQ and CEF. It has been speculated that elimination of either of these mechanisms should result in higher rates of H_2 photoproduction. This hypothesis was verified in the past by addition of the uncouplers CCCP or FCCP to anaerobically-induced algal cultures; the resulting rate of H_2 photoproduction was reported to be at least one order of magnitude higher than that in the absence of the uncoupler (Happe et al. 1994; Cournac et al. 2002; Lee and Greenbaum 2003; Kruse et al. 2005). Moreover, two *Chlamydomonas* insertional mutants, *STM6* (Kruse et al. 2005; Tolleter et al. 2011) and *PGRL1*, were recently identified as high H_2-producers. The *STM6* mutant was reported as having a complex phenotype that includes higher H_2 photoproduction rates and yields, higher starch accumulation and respiration rates, and lack of CEF (they are locked in state 1) as measured under dark anaerobic or sulfur-deprivation-induced anaerobiosis. The mutant was generated by random insertional mutagenesis and is defective in the MOC1 nuclear-encoded factor that is responsible for the assembly of the mitochondrial respiratory chain under illumination (Schonfeld et al. 2004). Unfortunately, the parental strain used to generate this mutant had low H_2-producing capabilities, and the H_2 yields by the Δ*STM6* are close to those of the cc124 WT strain (Kosourov et al. 2003).

The PGRL1 protein was first identified through quantitative proteomics as being up-regulated under iron deprivation (Merchant et al. 2007). Its involvement in CEF was initially reported by (Petroutsos et al. 2009) and confirmed by (Tolleter et al. 2011). The latter observed an aberrant fluorescence induction curve during dark-light-dark transition, and determined that the PGRL1 mutant showed higher ET rate but lower NPQ than its WT parental strain, suggesting a compromised CEF pathway (Tolleter et al. 2011). CEF was measured by two different assays: the rate of re-reduction of PSI, and the relaxation of the electrochromic shift. The results confirmed the reduced levels of CEF in the mutant. The H_2-photoproduction capability of the mutant was shown to be significantly higher than that of the WT culture, both under dark anaerobic and sulfur-deprivation-induced anaerobic conditions, mimicking the effect of the uncoupler FCCP. More recently, it has been shown that the rate of CEF correlates with the redox poise of the stroma (Johnson and Alric 2012) which can significantly impact the distribution of electrons between the Calvin cycle, the hydrogenase, and CEF. In contrast to the STM6 mutant, the parental strain for PGRL1 is the high H_2-producing WT 137C (Melis et al. 2000). Finally, work is underway in various laboratories in attempting to express a channel across the thylakoid membrane that allows proton dissipation under specific inducible conditions (Lee and Greenbaum 2003). No successful results have been reported at the time this review was submitted.

D. Other Barriers

Even if all of the above-mentioned barriers are solved, photosynthetic organisms will only photoproduce H_2 at high light conversion efficiencies under non-saturating illumination. Due to the large number of light-harvesting pigments associated with each photosystem in green algae and cyanobacteria, photosynthetic rates saturate at lower than sunlight intensities (which is about 2,000 μE m^{-2} s^{-1}). As a consequence, the rates of photosynthetic processes (such as CO_2 fixation, O_2 evolution, and H_2 photoproduction) at solar intensities become limited by the rate of ET within the photosynthetic apparatus. In mass cultures, this effect results in most of the illumination being utilized by the top few layers of cells, which shade the remainder of the culture and result in limited volumetric light conversion efficiencies (Benemann 1989; Melis et al. 1998; Ort and Melis 2011). Many laboratories are approaching this barrier by developing mutants that express truncated Chl antennae. In the case of green algae, these mutants are expected to contain a higher ratio of reaction centers/Chl (and thus higher photosynthetic rates measured on a Chl basis) and to saturate at higher light intensities (Melis et al. 1998). In cyanobacteria, truncated antennae are expected to result in lower levels of phycobilisomes and associated pigments and increased ratio of PSI/PSII (Bernat et al. 2009). This concept was demonstrated with H_2-producing *Rhodobacter* strains (Kondo et al. 2002) and with dense cultures of *Chlorella* (Nakajima and Ueda 1997; Nakajima et al. 2001). Recently, exciting results have been reported in two *C. reinhardtii* mutants, *tla1* and *tla2*, which have light-harvesting Chl antennae that are 51–66 % and 65 % smaller than those of their WT parental strains, respectively (Polle et al. 2003; Kirst et al. 2012). The H_2-production performance of the *tla1* mutant under different light intensities was investigated using sulfur-deprived, alginate-immobilized cultures and was demonstrated to be higher than that of its wild-type parental strain under high illumination, as predicted (Kosourov et al. 2011). Using *Synechocystis* 6803, three truncated antenna mutants, ΔapcE, Olive and PAL, have also been recently generated and demonstrated to have a higher PSII/PSI ratio, increased rates of LEF, lower rates of CEF in the PAL mutant, and considerably faster growth rates (Bernat et al. 2009). Although not yet tested for H_2 photoproduction, these mutants have the potential to produce H_2 at much higher rates than their respective parent strain.

Antenna truncation results in increased light utilization at high light intensities when compared to wild-type strains and, based on the results of Kosourov et al. (2011) it also leads to a change in the light intensity response curve for H_2 production. Previous reports (Laurinavichene et al. 2004) demonstrated that the maximum rates and yields of H_2 production by sulfur-deprived cultures of *C. reinhardtii* in suspension were achieved under about 30 μE m^{-2} s^{-1}, decreasing significantly at higher light intensities. However, Kosourov et al. (2011) show that the maximum rates of H_2 photoproduction in the TLA1 mutant actually occurred under 300 μE m^{-2} s^{-1}, while those of the wild-type strain peak at about 180 μE m^{-2} s^{-1}. Clearly, cell immobilization is responsible for shifting the peak rates to higher light intensities. It will be interesting to measure the effect of more significant antenna truncations, such as the *tla2* on the light response curve as well.

Acknowledgements

The authors acknowledge financial support from DOE's Office of Science's Basic Energy Sciences (MLG, PWK, DWM) and Biological Environmental Research Programs (MLG, AD), EERE's Fuel Cells Technology Office (MLG, PWK, PCM, JY), and ARPA-E (PCM, JY, CE). We are grateful for technical assistance from Dr. Damian Carrieri, Tameron Baldwin, and Lynn Westdal.

References

Adams MWW (1990) The structure and mechanism of iron-hydrogenases. Biochim Biophys Acta 1020:115–145

Adams MWW, Mortenson LE, Chen JS (1980) Hydrogenase. Biochim Biophys Acta 594:105–176

Agapakis CM, Ducat DC, Boyle PM, Wintermute EH, Way JC, Silver PA (2010) Insulation of a synthetic hydrogen metabolism circuit in bacteria. J Biol Eng 4:3

Ai X, Xu Q, Jones M, Song Q, Ding SY, Ellingson RJ, Himmel M, Rumbles G (2007) Photophysics of (CdSe) ZnS Colloidal quantum dots in an aqueous environment stabilized with amino acids and genetically-modified proteins. Photochem Photobiol Sci 6:1027–1033

Akhtar MK, Jones PR (2008) Engineering of a synthetic hydF-hydE-hydG-hydA operon for biohydrogen production. Anal Biochem 373:170–172

Akhtar MK, Jones PR (2009) Construction of a synthetic YdbK-dependent pyruvate:H_2 pathway in *Escherichia coli* BL21(DE3). Metab Eng 11:139–147

Allakhverdiev SI, Kreslavski VD, Thavasi V, Zharmukhamedov SK, Klimov VV, Nagata T, Nishihara H, Ramakrishna S (2009) Hydrogen photoproduction by use of photosynthetic organisms and biomimetic systems. Photochem Photobiol Sci 8:148–156

Allen JF (1992) Protein phosphorylation in regulation of photosynthesis. Biochim Biophys Acta 1098:275–335

Allen JF (2003) Cyclic, pseudocyclic and noncyclic photophosphorylation: new links in the chain. Trends Plant Sci 8:15–19

Alonso-Lomillo MA, Rudiger O, Maroto-Valiente A, Velez M, Rodriguez-Ramos I, Munoz FJ, Fernandez VM, De Lacey AL (2007) Hydrogenase-coated carbon nanotubes for efficient H_2 oxidation. Nano Lett 7:1603–1608

Alric J (2010) Cyclic electron flow around photosystem I in unicellular green algae. Photosynth Res 106:47–56

Amirav L, Alivisatos AP (2010) Photocatalytic hydrogen production with tunable nanorod heterostructures. J Phys Chem Lett 1:1051–1054

Antal TK, Oliveira P, Lindblad P (2006) The bidirectional hydrogenase in the cyanobacterium *Synechocystis* sp. strain PCC 6803. Int J Hydrog Energy 31:1439–1444

Appel J, Schulz R (1996) Sequence analysis of an operon of a NAD(P)-reducing nickel hydrogenase from the cyanobacterium *Synechocystis* sp. PCC 6803 gives additional evidence for direct coupling of the enzyme to NAD(P)H-dehydrogenase (complex I). Biochim Biophys Acta 1298:141–147

Appel J, Phunpruch S, Steinmuller K, Schulz R (2000) The bidirectional hydrogenase of *Synechocystis* sp. PCC 6803 works as an electron valve during photosynthesis. Arch Microbiol 173:333–338

Armstrong FA, Hirst J (2011) Reversibility and efficiency in electrocatalytic energy conversion and lessons from enzymes. Proc Natl Acad Sci U S A 108:14049–14054

Armstrong FA, Heering HA, Hirst J (1997) Reactions of complex metalloproteins studied by protein-film voltammetry. Chem Soc Rev 26:169–179

Armstrong FA, Belsey NA, Cracknell JA, Goldet G, Parkin A, Reisner E, Vincent KA, Wait AF (2009) Dynamic electrochemical investigations of hydrogen oxidation and production by enzymes and implications for future technology. Chem Soc Rev 38:36–51

Asada Y, Koike Y, Schnackenberg J, Miyake M, Uemura I, Miyake J (2000) Heterologous expression of clostridial hydrogenase in the cyanobacterium *Synechococcus* PCC7942. Biochim Biophys Acta 1490:269–278

Axelsson R, Lindblad P (2002) Transcriptional regulation of *Nostoc* hydrogenases: effects of oxygen, hydrogen, and nickel. Appl Environ Microbiol 68:444–447

Badia A, Lennox RB, Reven L (2000) A dynamic view of self-assembled monolayers. Acc Chem Res 33:475–481

Badura A, Guschin D, Esper B, Kothe T, Neugebauer S, Schuhmann W, Rögner M (2008) Photo-induced electron transfer between photosystem II via cross-linked redox hydrogels. Electroanalysis 20:1043–1047

Bae B, Kho BK, Lim TH, Oh IH, Hong SA, Ha HY (2006) Performance evaluation of passive DMFC single cells. J Power Sources 158:1256–1261

Bae S, Shim E, Yoon J, Joo H (2008) Photoanodic and cathodic role of anodized tubular titania in light-sensitized enzymatic hydrogen production. J Power Sources 185:439–444

Baebprasert W, Lindblad P, Incharoensakdi A (2010) Response of H_2 production and Hox-hydrogenase activity to external factors in the unicellular cyanobacterium *Synechocystis* sp. strain PCC 6803. Int J Hydrog Energy 35:6611–6616

Baffert C, Bertini L, Lautier T, Greco C, Sybirna K, Ezanno P, Etienne E, Soucaille P, Bertrand P, Bottin H, Meynial-Salles I, De Gioia L, Leger C (2011) CO disrupts the reduced H-Cluster of FeFe hydrogenase. A combined DFT and protein film voltammetry study. J Am Chem Soc 133:2096–2099

Baltazar CSA, Marques MC, Soares CM, DeLacey AM, Pereira IAC, Matias PM (2011) Nickel-iron-selenium hydrogenases – an overview. Eur J Inorg Chem 2011:948–962

Bandyopadhyay A, Stockel J, Min H, Sherman LA, Pakrasi HB (2010) High rates of photobiological H_2 production by a cyanobacterium under aerobic conditions. Nat Commun 1:139

Barstow B, Agapakis CM, Boyle PM, Grandl G, Silver PA, Wintermute EH (2011) A synthetic system links [FeFe]-hydrogenases to essential *E. coli* sulfur metabolism. J Biol Eng 5:7

Battchikova N, Wei L, Du L, Bersanini L, Aro EM, Ma W (2011) Identification of novel Ssl0352 protein (NdhS), essential for efficient operation of cyclic electron transport around photosystem I, in NADPH: plastoquinone oxidoreductase (NDH-1) complexes of *Synechocystis* sp. PCC 6803. J Biol Chem 286:36992–37001

Benemann J (1989) The future of microalgal biotechnology. In: Cresswell RC, Rees TAV, Shah N (eds) Algal and cyanobacterial biotechnology. Longman Scientific and Technical, Harlow, pp 317–337

Bennett B, Lemon BJ, Peters JW (2000) Reversible carbon monoxide binding and inhibition at the active site of the Fe-only hydrogenase. Biochemistry 39:7455–7460

Bernat G, Waschewski N, Rogner M (2009) Towards efficient hydrogen production: the impact of antenna size and external factors on electron transport dynamics in *Synechocystis* PCC 6803. Photosynth Res 99:205–216

Berto P, D'Adamo S, Bergantino E, Vallese F, Giacometti GM, Costantini P (2011) The cyanobacterium *Synechocystis* sp. PCC 6803 is able to express an active [FeFe] -hydrogenase without additional maturation proteins. Biochem Biophys Res Commun 405:678–683

Blackburn JL, Svedruzic D, McDonald TJ, Kim YH, King PW, Heben MJ (2008) Raman spectroscopy of charge transfer interactions between single wall carbon nanotubes and [FeFe]-hydrogenase. Dalton Trans 40:5454–5461

Blankenship RE, Tiede DM, Barber J, Brudvig GW, Fleming G, Ghirardi M, Gunner MR, Junge W, Kramer DM, Melis A, Moore TA, Moser CC, Nocera DG, Nozik AJ, Ort DR, Parson WW, Prince RC, Sayre RT (2011) Comparing photosynthetic and photovoltaic efficiencies and recognizing the potential for improvement. Science 332:805–809

Blokesch M, Bock A (2002) Maturation of [NiFe]-hydrogenases in *Escherichia coli*: the HypC cycle. J Mol Biol 324:287–296

Bock A, King PW, Blokesch M, Posewitz MC (2006) Maturation of hydrogenases. Adv Microb Physiol 51:1–71

Boison G, Schmitz O, Schmitz B, Bothe H (1998) Unusual gene arrangement of the bidirectional hydrogenase and functional analysis of its diaphorase subunit HoxU in respiration of the unicellular cyanobacterium *Anacystis nidulans*. Curr Microbiol 36:253–258

Boison G, Bothe H, Hansel A, Lindblad P (1999) Evidence against a common use of the diaphorase subunits by the bidirectional hydrogenase and by the respiratory complex I in cyanobacteria. FEMS Microbiol Lett 174:159–165

Boison G, Bothe H, Schmitz O (2000) Transcriptional analysis of hydrogenase genes in the cyanobacteria *Anacystis nidulans* and *Anabaena variabilis* monitored by RT-PCR. Curr Microbiol 40:315–321

Bolton JR, Strickler SJ, Connolly JS (1985) Limiting and realizable efficiencies of solar photolysis of water. Nature 316:495–500

Bothe H, Schmitz O, Yates MG, Newton WE (2011) Nitrogenases and hydrogenases in cyanobacteria. In: Peschek GA, Obinger C, Renger G (eds) Bioenergetic processes of cyanobacteria. Springer Science Publishing, pp 137–157

Brand JJ, Wright JN, Lien S (1989) Hydrogen-production by eukaryotic algae. Biotechnol Bioeng 33:1482–1488

Brazzolotto X, Rubach JK, Gaillard J, Gambarelli S, Atta M, Fontecave M (2006) The [Fe-Fe]-hydrogenase maturation protein HydF from *Thermotoga maritima* is a GTPase with an iron-sulfur cluster. J Biol Chem 281:769–774

Brown KA, Dayal S, Xin A, Rumbles G, King PW (2010) Controlled assembly of hydrogenase-CdTe nanocrystal hybrids for solar hydrogen production. J Am Chem Soc 132:9672–9680

Brown KA, Wilker MB, Boehm M, Dukovic G, King PW (2012) Characterization of photochemical processes for H_2 production by CdS nanorod-[FeFe]-hydrogenase complexes. J Am Chem Soc 134:5627–5636

Bruce BD (2000) Chloroplast transit peptides: structure, function and evolution. Trends Cell Biol 10:440–447

Buhrke T, Bleijlevens B, Albracht SPJ, Friedrich B (2001) Involvement of hyp gene products in maturation of the H_2-sensing [NiFe] hydrogenase of *Ralstonia eutropha*. J Bacteriol 183:7087–7093

Burgdorf T, Lenz O, Buhrke T, van der Linden E, Jones AK, Albracht SP, Friedrich B (2005) [NiFe]-hydrogenases of *Ralstonia eutropha* H16: modular enzymes for oxygen-tolerant biological hydrogen oxidation. J Mol Microbiol Biotechnol 10:181–196

Cardol P, Alric J, Girard-Bascou J, Franck F, Wollman FA, Finazzi G (2009) Impaired respiration discloses the physiological significance of state transitions in *Chlamydomonas*. Proc Natl Acad Sci U S A 106:15979–15984

Carrieri D, Wawrousek K, Eckert C, Yu J, Maness PC (2011) The role of the bidirectional hydrogenase in cyanobacteria. Bioresour Technol 102:8368–8377

Cendron L, Berto P, D'Adamo S, Vallese F, Govoni C, Posewitz MC, Giacometti GM, Costantini P, Zanotti G (2011) Crystal structure of HydF scaffold protein provides insights into [FeFe]-hydrogenase maturation. J Biol Chem 286:43944–43950

Chang CH, King PW, Ghirardi ML, Kim K (2007) Atomic resolution modeling of the ferredoxin:[FeFe] hydrogenase complex from *Chlamydomonas reinhardtii*. Biophys J 93:3034–3045

Chen XJ, Schnell DJ (1999) Protein import into chloroplasts. Trends Cell Biol 9:222–227

Chen Z, Lemon BJ, Huang S, Swartz DJ, Peters JW, Bagley KA (2002) Infrared studies of the CO-inhibited form of the Fe-only hydrogenase from *Clostridium pasteurianum* I: examination of its light sensitivity at cryogenic temperatures. Biochemistry 41:2036–2043

Chen XB, Shen SH, Guo LJ, Mao SS (2010) Semiconductor-based photocatalytic hydrogen generation. Chem Rev 110:6503–6570

Chung KCC, Zamble DB (2011) Protein interactions and localization of the *Escherichia coli* accessory protein HypA during nickel insertion to [NiFe] hydrogenase. J Biol Chem 286:43081–43090

Cohen J, Kim K, King P, Seibert M, Schulten K (2005a) Finding gas diffusion pathways in proteins: application to O_2 and H_2 transport in CpI [FeFe]-hydrogenase and the role of packing defects. Structure 13:1321–1329

Cohen J, Kim K, Posewitz M, Ghirardi ML, Schulten K, Seibert M, King P (2005b) Molecular dynamics and experimental investigation of H_2 and O_2 diffusion in [Fe]-hydrogenase. Biochem Soc Trans 33:80–82

Cornish AJ, Gartner K, Yang H, Peters JW, Hegg EL (2011) Mechanism of proton transfer in [FeFe]-hydrogenase from *Clostridium pasteurianum*. J Biol Chem 286:38341–38347

Cournac L, Mus F, Bernard L, Guedeney G, Vignais P, Peltier G (2002) Limiting steps of hydrogen production in *Chlamydomonas reinhardtii* and *Synechocystis* PCC 6803 as analysed by light-induced gas exchange transients. Int J Hydrog Energy 27:1229–1237

Cournac L, Guedeney G, Peltier G, Vignais PM (2004) Sustained photoevolution of molecular hydrogen in a mutant of *Synechocystis* sp. strain PCC 6803 deficient in the type I NADPH-dehydrogenase complex. J Bacteriol 186:1737–1746

Cracknell JA, Vincent KA, Armstrong FA (2008) Enzymes as working or inspirational electrocatalysts for fuel cells and electrolysis. Chem Rev 108:2439–2461

Cuendet P, Rao KK, Grätzel M, Hall DO (1986) Light induced H_2 evolution in a hydrogenase-TiO_2 particle system by direct electron transfer or via rhodium complexes. Biochimie 68:217–221

Czech I, Silakov A, Lubitz W, Happe T (2010) The [FeFe]-hydrogenase maturase HydF from *Clostridium acetobutylicum* contains a CO and CN- ligated iron cofactor. FEBS Lett 584:638–642

Czech I, Stripp S, Sanganas O, Leidel N, Happe T, Haumann M (2011) The [FeFe]-hydrogenase maturation protein HydF contains a H-cluster like [4Fe4S]-2Fe site. FEBS Lett 585:225–230

DalCorso G, Pesaresi P, Masiero S, Aseeva E, Schunemann D, Finazzi G, Joliot P, Barbato R, Leister D (2008) A complex containing PGRL1 and PGR5 is involved in the switch between linear and cyclic electron flow in *Arabidopsis*. Cell 132:273–285

Desplats C, Mus F, Cuine S, Billon E, Cournac L, Peltier G (2009) Characterization of Nda2, a plastoquinone-reducing type II NAD(P)H dehydrogenase in *Chlamydomonas* chloroplasts. J Biol Chem 284:4148–4157

Devine E, Holmqvist M, Stensjo K, Lindblad P (2009) Diversity and transcription of proteases involved in the maturation of hydrogenases in *Nostoc punctiforme* ATCC 29133 and *Nostoc* sp. strain PCC 7120. BMC Microbiol 9:53

Driesener RC, Challand MR, McGlynn SE, Shepard EM, Boyd ES, Broderick JB, Peters JW, Roach PL (2010) [FeFe]-hydrogenase cyanide ligands derived from S-adenosylmethionine-dependent cleavage of tyrosine. Angew Chem Int Ed Engl 49:1687–1690

Ducat DC, Sachdeva G, Silver PA (2011) Rewiring hydrogenase-dependent redox circuits in cyanobacteria. Proc Natl Acad Sci U S A 108:3941–3946

Dukovic G, Merkle MG, Nelson JH, Hughes SM, Alivisatos AP (2008) Photodeposition of Pt on colloidal CdS and CdSe/CdS semiconductor nanostructures. Adv Mater 20:4306–4311

Dutheil J, Saenkham P, Sakr S, Leplat C, Ortega-Ramos M, Bottin H, Cournac L, Cassier-Chauvat C, Chauvat F (2012) Advances in the regulation of hydrogen production in *Synechocystis* PCC6803: the AbrB2 auto-repressor (sll0822), expressed from an atypical promoter, represses the hydrogenase operon. J Bacteriol 194:5423–5433

Eberhard S, Finazzi G, Wollman FA (2008) The dynamics of photosynthesis. Annu Rev Genet 42:463–515

Eckert C, Boehm M, Carrieri D, Yu J, Dubini A, Nixon PJ, Maness PC (2012a) Genetic analysis of the Hox hydrogenase in the cyanobacterium *Synechocystis* sp. PCC 6803 reveals subunit roles in association, assembly, maturation, and function. J Biol Chem 287:43502–43515

Eckert C, Dubini A, Yu J, King P, Ghirardi ML, Seibert M, Maness PC (2012b) Hydrogenase genes and enzymes involved in solar hydrogen production. In: Levin DB, Azbar N (eds) State of the art and progress in production of biohydrogen, vol 1. Bentham Science Publisher, Sarjah, UAE, pp 8–24

Elrad D, Niyogi KK, Grossman AR (2002) A major light-harvesting polypeptide of photosystem II functions in thermal dissipation. Plant Cell 14:1801–1816

English CM, Eckert CA, Brown KA, Seibert M, King PW (2009) Recombinant and in vitro expression systems for hydrogenases: new frontiers in basic and applied studies for biological and synthetic H_2 production. Dalton Trans 45:9970–9978

Erbes DL, King D, Gibbs M (1978) Effect of carbon monoxide and oxygen on hydrogen activation by hydrogenase from *Chlamydomonas reinhardtii*. Plant Physiol 61:23

Erbes DL, King D, Gibbs M (1979) Inactivation of hydrogenase in cell-free extracts and whole cells of *Chlamydomonas reinhardtii* by oxygen. Plant Physiol 63:1138–1142

Erdem OF, Schwartz L, Stein M, Silakov A, Kaur-Ghumaan S, Huang P, Ott S, Reijerse EJ, Lubitz W (2011) A model of the [FeFe] hydrogenase active site with a biologically relevant azadithiolate bridge: a spectroscopic and theoretical investigation. Angew Chem Int Ed Engl 50:1439–1443

Eroglu E, Melis A (2011) Photobiological hydrogen production: recent advances and state of the art. Bioresour Technol 102:8403–8413

Esper B, Badura A, Rogner M (2006) Photosynthesis as a power supply for biohydrogen production. Trends Plant Sci 11:543–549

Ferreira D, Leitao E, Sjoholm J, Oliveira P, Lindblad P, Moradas-Ferreira P, Tamagnini P (2007) Transcription and regulation of the hydrogenase(s) accessory genes, hypFCDEAB, in the cyanobacterium *Lyngbya majuscula* CCAP 1446/4. Arch Microbiol 188:609–617

Finazzi G, Furia A, Barbagallo RP, Forti G (1999) State transitions, cyclic and linear electron transport and photophosphorylation in *Chlamydomonas reinhardtii*. Biochim Biophys Acta 1413:117–129

Fischer N, Setif P, Rochaix JD (1999) Site-directed mutagenesis of the PsaC subunit of photosystem I. F(b) is the cluster interacting with soluble ferredoxin. J Biol Chem 274:23333–23340

Fitzgerald MP, Rogers LJ, Rao KK, Hall DO (1980) Efficiency of ferredoxins and flavodoxins as mediators in systems for hydrogen evolution. Biochem J 192:665–672

Florin L, Tsokoglou A, Happe T (2001) A novel type of iron hydrogenase in the green alga *Scenedesmus obliquus* is linked to the photosynthetic electron transport chain. J Biol Chem 276:6125–6132

Flynn T, Ghirardi ML, Seibert M (2002) Accumulation of O_2-tolerant phenotypes in H_2-producing strains of *Chlamydomonas reinhardtii* by sequential applications of chemical mutagenesis and selection. Int J Hydrog Energy 27:1421–1430

Fontecilla-Camps JC, Volbeda A, Cavazza C, Nicolet Y (2007) Structure/function relationships of [NiFe]- and [FeFe]-hydrogenases. Chem Rev 107:4273–4303

Forestier M, King P, Zhang L, Posewitz M, Schwarzer S, Happe T, Ghirardi ML, Seibert M (2003) Expression of two [Fe]-hydrogenases in *Chlamydomonas reinhardtii* under anaerobic conditions. Eur J Biochem 270:2750–2758

Forti G, Furia A, Bombelli P, Finazzi G (2003) In vivo changes of the oxidation-reduction state of NADP and of the ATP/ADP cellular ratio linked to the photosynthetic activity in *Chlamydomonas reinhardtii*. Plant Physiol 132:1464–1474

Fouchard S, Hemschemeier A, Caruana A, Pruvost J, Legrand J, Happe T, Peltier G, Cournac L (2005) Autotrophic and mixotrophic hydrogen photoproduction in sulfur-deprived *Chlamydomonas* cells. Appl Environ Microbiol 71:6199–6205

Fritsch J, Loscher S, Sanganas O, Siebert E, Zebger I, Stein M, Ludwig M, De Lacey AL, Dau H, Friedrich B, Lenz O, Haumann M (2011) [NiFe] and [FeS] cofactors in the membrane-bound hydrogenase of *Ralstonia eutropha* investigated by X-ray absorption spectroscopy: insights into O_2-tolerant H_2 cleavage. Biochemistry 50:5858–5869

Galvan IF, Volbeda A, Fontecilla-Camps JC, Field MJ (2008) A QM/MM study of proton transport pathways in a [NiFe] hydrogenase. Proteins Struct Funct Bioinform 73:195–203

Gaponik N, Talapin DV, Rogach AL, Hoppe K, Shevchenko EV, Kornowski A, Eychmuller A, Weller H (2002) Thiol-capping of CdTe nanocrystals: an alternative to organometallic synthetic routes. J Phys Chem B 106:7177–7185

Garcia-Sanchez MI, Gotor C, Jacquot JP, Stein M, Suzuki A, Vega JM (1997) Critical residues of *Chlamydomonas reinhardtii* ferredoxin for interaction with nitrite reductase and glutamate synthase revealed by site-directed mutagenesis. Eur J Biochem 250:364–368

Germer F, Zebger I, Saggu M, Lendzian F, Schulz R, Appel J (2009) Overexpression, isolation, and spectroscopic characterization of the bidirectional [NiFe] hydrogenase from *Synechocystis* sp. PCC 6803. J Biol Chem 284:36462–36472

Gfeller RP, Gibbs M (1984) Fermentative metabolism of *Chlamydomonas reinhardtii*: I. analysis of fermentative products from starch in dark and light. Plant Physiol 75:212–218

Ghirardi ML, Togasaki R, Seibert M (1997) Oxygen sensitivity of algal H_2-production. Appl Biochem Biotechnol 63–65:141–151

Ghirardi ML, Posewitz MC, Maness PC, Dubini A, Yu J, Seibert M (2007) Hydrogenases and hydrogen photoproduction in oxygenic photosynthetic organisms. Annu Rev Plant Biol 58:71–91

Ghirardi ML, Dubini A, Yu J, Maness PC (2009) Photobiological hydrogen-producing systems. Chem Soc Rev 38:52–61

Ghysels B, Franck F (2010) Hydrogen photo-evolution upon S deprivation stepwise: an illustration of microalgal photosynthetic and metabolic flexibility and a step stone for future biotechnological methods of renewable H_2 production. Photosynth Res 106:145–154

Gillez-Gonzalez MA, Gonzalez G, Perutz MF, Kiger L, Marden MC, Poyart C (1994) Heme-based sensors, exemplified by the kinase fixL, are a new class of heme protein with distinctive ligand binding and autoxidation. Biochemistry 33:8067–8073

Godman JE, Molnar A, Baulcombe DC, Balk J (2010) RNA silencing of hydrogenase(-like) genes and investigation of their physiological roles in the green alga *Chlamydomonas reinhardtii*. Biochem J 431:345–351

Goldet G, Brandmayr C, Stripp ST, Happe T, Cavazza C, Fontecilla-Camps JC, Armstrong FA (2009) Electrochemical kinetic investigations of the reactions of [FeFe]-hydrogenases with carbon monoxide and oxygen: comparing the importance of gas tunnels and active-site electronic/redox effects. J Am Chem Soc 131:14979–14989

Greene BL, Joseph CA, Maroney MJ, Dyer RB (2012) Direct evidence of active-site reduction and photodriven catalysis in sensitized hydrogenase assemblies. J Am Chem Soc 134:11108–11111

Gubili J, Borthakur D (1998) Organization of the hupDEAB genes within the hydrogenase gene cluster of *Anabaena* sp. strain PCC7120. J Appl Phycol 10:163–167

Guldi DM, Rahman GMA, Zerbetto F, Prato M (2005) Carbon nanotubes in electron donor-acceptor nanocomposites. Acc Chem Res 38:871–878

Gust D, Moore TA, Moore AL (2000) Mimicking photosynthetic solar energy transduction. Acc Chem Res 34:40–48

Gust D, Moore TA, Moore AL (2009) Solar fuels via artificial photosynthesis. Acc Chem Res 42:1890–1898

Gutekunst K, Phunpruch S, Schwarz C, Schuchardt S, Schulz-Friedrich R, Appel J (2005) LexA regulates the bidirectional hydrogenase in the cyanobacterium *Synechocystis* sp. PCC 6803 as a transcription activator. Mol Microbiol 58:810–823

Gutierrez-Sanchez C, Olea D, Marques M, Fernandez VM, Pereira IAC, Velez M, De Lacey AL (2011) Oriented immobilization of a membrane-bound hydrogenase onto an electrode for direct electron transfer. Langmuir 27:6449–6457

Gutthann F, Egert M, Marques A, Appel J (2007) Inhibition of respiration and nitrate assimilation enhances photohydrogen evolution under low oxygen concentrations in *Synechocystis* sp. PCC 6803. Biochim Biophys Acta 1767:161–169

Hall DO, Adams MWW, Morris P, Rao KK (1980) Photolysis of water for H_2 production with the use of biological and artificial catalysts. Phil Trans R Soc A 295:473–476

Hambourger M, Liddell PA, Gust D, Moore AL, Moore TA (2007) Parameters affecting the chemical work output of a hybrid photoelectrochemical biofuel cell. Photochem Photobiol Sci 6:431–437

Hambourger M, Gervaldo M, Svedruzic D, King PW, Gust D, Ghirardi M, Moore AL, Moore TA (2008) [FeFe]-hydrogenase-catalyzed H_2 production in a photoelectrochemical biofuel cell. J Am Chem Soc 130:2015–2022

Hambourger M, Kodis G, Vaughn MD, Moore GF, Gust D, Moore AL, Moore TA (2009) Solar energy conversion in a photoelectrochemical biofuel cell. Dalton Trans 45:9979–9989

Happe T, Kaminski A (2002) Differential regulation of the Fe-hydrogenase during anaerobic adaptation in the green alga *Chlamydomonas reinhardtii*. Eur J Biochem 269:1022–1032

Happe T, Naber JD (1993) Isolation, characterization, and N-terminal amino acid sequence of hydrogenase from the green alga *Chlamydomonas reinhardtii*. Eur J Biochem 214:475–481

Happe T, Mosler B, Naber JD (1994) Induction, localization, and metal content of hydrogenase in the green alga *Chlamydomonas reinhardtii*. Eur J Biochem 222:769–774

Happe RP, Roseboom W, Pierik AJ, Albracht SP, Bagley KA (1997) Biological activation of hydrogen. Nature 385:126

Hase T, Mizutani S, Mukohata Y (1991) Expression of maize Ferredoxin cDNA in *Escherichia coli*: comparison of photosynthetic and nonphotosynthetic ferredoxin isoproteins and their chimeric molecule. Plant Physiol 97:1395–1401

Haumann M, Liebisch P, Muller C, Barra M, Grabolle M, Dau H (2005) Photosynthetic O_2 formation tracked by time-resolved X-ray experiments. Science 310:1019–1021

Heinekey DM (2009) Hydrogenase enzymes: recent structural studies and active site models. J Organomet Chem 694:2671–2680

Hemschemeier A, Happe T (2011) Alternative photosynthetic electron transport pathways during anaerobiosis in the green alga *Chlamydomonas reinhardtii*. Biochim Biophys Acta 1807:919–926

Hiromoto T, Warkentin E, Moll J, Ermler U, Shima S (2009) The crystal structure of an [Fe]-hydrogenase-substrate complex reveals the framework for H_2 activation. Angew Chem Int Ed Engl 48:6457–6460

Hoffmann D, Gutekunst K, Klissenbauer M, Schulz-Friedrich R, Appel J (2006) Mutagenesis of hydrogenase accessory genes of *Synechocystis* sp. PCC 6803. Additional homologues of hypA and hypB are not active in hydrogenase maturation. FEBS J 273:4516–4527

Howitt CA, Vermaas WFJ (1999) Subunits of the NAD(P)-reducing nickel-containing hydrogenase do not act as part of the type-1 NAD(P)H-dehydrogenase in the cyanobacterium *Synechocystis* sp. PCC 6803. In: Peschek GA, Löffelhardt W, Schmetterer G (eds) The phototrophic prokaryotes. Springer, New York, pp 595–601

Ihara M, Nakamoto H, Kamachi T, Okura I, Maeda M (2006a) Photoinduced hydrogen production by direct electron transfer from photosystem I cross-linked with cytochrome c(3) to NiFe-hydrogenase. Photochem Photobiol 82:1677–1685

Ihara M, Nishihara H, Yoon KS, Lenz O, Friedrich B, Nakamoto H, Kojima K, Honma D, Kamachi T, Okura I (2006b) Light-driven hydrogen production by a hybrid complex of a NiFe-hydrogenase and the cyanobacterial photosystem I. Photochem Photobiol 82:676–682

Iwai M, Takizawa K, Tokutsu R, Okamuro A, Takahashi Y, Minagawa J (2010) Isolation of the elusive supercomplex that drives cyclic electron flow in photosynthesis. Nature 464:1210–1213

Jacobs J, Pudollek S, Hemschemeier A, Happe T (2009) A novel, anaerobically induced ferredoxin in *Chlamydomonas reinhardtii*. FEBS Lett 583: 325–329

Jacquot JP, Stein M, Suzuki A, Liottet S, Sandoz G, Miginiac-Maslow M (1997) Residue Glu-91 of *Chlamydomonas reinhardtii* ferredoxin is essential for electron transfer to ferredoxin-thioredoxin reductase. FEBS Lett 400:293–296

Jahns P, Holzwarth AR (2012) The role of the xanthophyll cycle and of lutein in photoprotection of photosystem II. Biochim Biophys Acta 1817:182–193

James B, Baum G, Perez J, Baum K (2008) Technoeconomic boundary analysis of biological pathways to hydrogen production. subcontract Report NREL/SR-560-46674:235–239

Jin RG, Richter S, Zhong R, Lamppa GK (2003) Expression and import of an active cellulase from a thermophilic bacterium into the chloroplast both in vitro and in vivo. Plant Mol Biol 51:493–507

Johnson X, Alric J (2012) Interaction between starch breakdown, acetate assimilation, and photosynthetic cyclic electron flow in *Chlamydomonas reinhardtii*. J Biol Chem 287:26445–26452

Joo H, Bae S, Kim C, Kim S, Yoon J (2009) Hydrogen evolution in enzymatic photoelectrochemical cell using modified seawater electrolytes produced by membrane desalination process. Sol Energy Mater Solar Cells 93:1555–1561

Kaluarachchi H, Zhang JW, Zamble DB (2011) *Escherichia coli* SlyD, more than a Ni(II) reservoir. Biochemistry 50:10761–10763

Kato M, Cardona T, Rutherford AW, Reisner E (2012) Photoelectrochemical water oxidation with photosystem II integrated in a mesoporous indium-tin oxide electrode. J Am Chem Soc 134:8332–8335

Keegstra K, Cline K (1999) Protein import and routing systems of chloroplasts. Plant Cell 11:557–570

King PW, Posewitz MC, Ghirardi ML, Seibert M (2006) Functional studies of [FeFe] hydrogenase maturation in an *Escherichia coli* biosynthetic system. J Bacteriol 188:2163–2172

Kirilovsky D, Kerfeld CA (2012) The orange carotenoid protein in photoprotection of photosystem II in cyanobacteria. Biochim Biophys Acta 1817:158–166

Kirst H, Garcia-Cerdan JG, Zurbriggen A, Melis A (2012) Assembly of the light-harvesting chlorophyll antenna in the green alga *Chlamydomonas reinhardtii* requires expression of the TLA2-CpFTSY gene. Plant Physiol 158:930–945

Kiss E, Kos PB, Vass I (2009) Transcriptional regulation of the bidirectional hydrogenase in the cyanobacterium *Synechocystis* 6803. J Biotechnol 142:31–37

Knorzer P, Silakov A, Foster CE, Armstrong FA, Lubitz W, Happe T (2012) Importance of the protein framework for catalytic activity of [FeFe]-hydrogenases. J Biol Chem 287:1489–1499

Kondo T, Arakawa M, Hirai T, Wakayama T, Hara M, Miyake J (2002) Enhancement of hydrogen production by a photosynthetic bacterium mutant with reduced pigment. J Biosci Bioeng 93:145–150

Kroprat J, Tottey S, Birkenbihl RP, Depège N, Hiujser P, Merchant S (2005) A regulator of nutritional copper signaling in *Chlamydomonas* is an SBP domain protein that recognizes the GTAC core of copper response element. Proc Natl Acad Sci U S A 102:18730–18735

Kosourov S, Seibert M (2008) Hydrogen photoproduction by nutrient-deprived *Chlamydomonas reinhardtii* cells immobilized within thin alginate films under aerobic and anaerobic conditions. Biotechnol Bioeng 102:50–58

Kosourov S, Seibert M, Ghirardi ML (2003) Effects of extracellular pH on the metabolic pathways in sulfur-deprived, H_2-producing *Chlamydomonas reinhardtii* cultures. Plant Cell Physiol 44:146–155

Kosourov S, Ghirardi ML, Seibert M (2011) A truncated antenna mutant of *Chlamydomonas reinhardtii* can produce more hydrogen than the parental strain. Int J Hydrog Energy 36:2044–2048

Krassen H, Schwarze A, Friedrich B, Ataka K, Lenz O, Heberle J (2009) Photosynthetic hydrogen production by a hybrid complex of photosystem I and [NiFe] -hydrogenase. ACS Nano 3:4055–4061

Krassen H, Stripp ST, Bohm N, Berkessel A, Happe T, Ataka K, Heberle J (2011) Tailor-made modification of a gold surface for the chemical binding of a high-activity [FeFe] hydrogenase. Eur J Inorg Chem 2011:1138–1146

Krishnan S, Armstrong FA (2012) Order-of-magnitude enhancement of an enzymatic hydrogen-air fuel cell based on pyrenyl carbon nanostructures. Chem Sci 3:1015–1023

Kruse O, Hankamer B (2010) Microalgal hydrogen production. Curr Opin Biotechnol 21:238–243

Kruse O, Rupprecht J, Bader KP, Thomas-Hall S, Schenk PM, Finazzi G, Hankamer B (2005) Improved photobiological H_2 production in engineered green algal cells. J Biol Chem 280:34170–34177

Kuchenreuther JM, Grady-Smith CS, Bingham AS, George SJ, Cramer SP, Swartz JR (2010) High-yield expression of heterologous [FeFe] hydrogenases in *Escherichia coli*. PLoS One 5:e15491

Kuchenreuther JM, George SJ, Grady-Smith CS, Cramer SP, Swartz JR (2011) Cell-free H-cluster synthesis and [FeFe] hydrogenase activation: all five CO and CN(-) ligands derive from tyrosine. PLoS One 6:e20346

Kucho K, Okamoto K, Tsuchiya Y, Nomura S, Nango M, Kanehisa M, Ishiura M (2005) Global analysis of circadian expression in the cyanobacterium *Synechocystis* sp. strain PCC 6803. J Bacteriol 187:2190–2199

Laffly E, Garzoni F, Fontecilla-Camps JC, Cavazza C (2010) Maturation and processing of the recombinant FeFe hydrogenase from *Desulfovibrio vulgaris* Hildenborough (DvH) in *Escherichia coli*. Int J Hydrog Energy 35:10761–10769

Lambertz C, Hemschemeier A, Happe T (2010) Anaerobic expression of the ferredoxin-encoding FDX5 gene of *Chlamydomonas reinhardtii* is regulated by the Crr1 transcription factor. Eukaryot Cell 9:1747–1754

Lambertz C, Leidel N, Havelius KGV, Noth J, Chernev P, Winkler M, Happe T, Haumann M (2011) O_2 reactions at the six-iron active site (H-cluster) in [FeFe]-hydrogenase. J Biol Chem 286:40614–40623

Lambrev PH, Miloslavina Y, Jahns P, Holzwarth AR (2012) On the relationship between non-photochemical quenching and photoprotection of photosystem II. Biochim Biophys Acta 1817: 760–769

Laurinavichene T, Tolstygina I, Tsygankov A (2004) The effect of light intensity on hydrogen production by sulfur-deprived *Chlamydomonas reinhardtii*. J Biotechnol 114:143–151

Lauterbach L, Idris Z, Vincent KA, Lenz O (2011) Catalytic properties of the isolated diaphorase fragment of the NAD+-reducing [NiFe]-hydrogenase from *Ralstonia eutropha*. PLoS One 6:25939

Lautier T, Ezanno P, Baffert C, Fourmond V, Cournac L, Fontecilla-Camps JC, Soucaille P, Bertrand P, Meynial-Salles I, Leger C (2011) The quest for a functional substrate access tunnel in [FeFe] hydrogenase. Faraday Discuss 148:385–407

Lee JW, Greenbaum E (2003) A new oxygen sensitivity and its potential application in photosynthetic H_2 production. Appl Biochem Biotechnol 105–108:303–313

Lee HS, Vermaas WF, Rittmann BE (2010) Biological hydrogen production: prospects and challenges. Trends Biotechnol 28:262–271

Leitao E, Pereira S, Bondoso J, Ferreira D, Pinto F, Moradas-Ferreira P, Tamagnini P (2006) Genes involved in the maturation of hydrogenase(s) in the nonheterocystous cyanobacterium *Lyngbya majuscula* CCAP 1446/4. Int J Hydrog Energy 31:1469–1477

Lemeille S, Turkina MV, Vener AV, Rochaix JD (2010) Stt7-dependent phosphorylation during state transitions in the green alga *Chlamydomonas reinhardtii*. Mol Cell Proteomics 9:1281–1295

Lemon BJ, Peters JW (2000) Photochemistry at the active site of the carbon monoxide inhibited form of the iron-only hydrogenase (CpI). J Am Chem Soc 122:3793–3794

Li Z, Wakao S, Fischer BB, Niyogi KK (2009) Sensing and responding to excess light. Annu Rev Plant Biol 60:239–260

Liebgott PP, Dementin S, Leger C, Rousset M (2011) Towards engineering O_2-tolerance in Ni-Fe hydrogenases. Energy Environ Sci 4:33–41

Lojou E, Luo X, Brugna M, Candoni N, Dementin S, Giudici-Orticoni MT (2008) Biocatalysts for fuel cells: efficient hydrogenase orientation for H_2 oxidation at electrodes modified with carbon nanotubes. J Biol Inorg Chem 13:1157–1167

Long M, Liu J, Chen Z, Bleijlevens B, Roseboom W, Albracht SP (2007) Characterization of a HoxEFUYH type of [NiFe] hydrogenase from *Allochromatium vinosum* and some EPR and IR properties of the hydrogenase module. J Biol Inorg Chem 12:62–78

Long H, Chang CH, King PW, Ghirardi ML, Kim K (2008) Brownian dynamics and molecular dynamics study of the association between hydrogenase and ferredoxin from *Chlamydomonas reinhardtii*. Biophys J 95:3753–3766

Long H, King PW, Ghirardi ML, Kim K (2009) Hydrogenase/ferredoxin charge-transfer complexes: effect of hydrogenase mutations on the complex association. J Phys Chem A 113:4060–4067

Lubitz W, Reijerse E, van Gastel M (2007) [NiFe] and [FeFe] hydrogenases studied by advanced magnetic resonance techniques. Chem Rev 107:4331–4365

Lubitz W, Reijerse EJ, Messinger J (2008) Solar water-splitting into H_2 and O_2: design principles of photosystem II and hydrogenases. Energy Environ Sci 1:15–31

Lubner CE, Knorzer P, Silva PJ, Vincent KA, Happe T, Bryant DA, Golbeck JH (2010) Wiring an [FeFe]-hydrogenase with photosystem I for light-induced hydrogen production. Biochemistry 49: 10264–10266

Lubner CE, Applegate AM, Knorzer P, Ganago A, Bryant DA, Happe T, Golbeck JH (2011) Solar hydrogen-producing bionanodevice outperforms natural photosynthesis. Proc Natl Acad Sci U S A 108:20988–20991

Lutz S, Jacobi A, Schlensog V, Bohm R, Sawers G, Bock A (1991) Molecular characterization of an operon (hyp) necessary for the activity of the three hydrogenase isoenzymes in *Escherichia coli*. Mol Microbiol 5:123–135

Madden C, Vaughn MD, Diez-Perez I, Brown KA, King PW, Gust D, Moore AL, Moore TA (2012) Catalytic turnover of [FeFe] -hydrogenase based on single-molecule imaging. J Am Chem Soc 134:1577–1582

Magalon A, Bock A (2000) Dissection of the maturation reactions of the [NiFe] hydrogenase 3 from *Escherichia coli* taking place after nickel incorporation. FEBS Lett 473:254–258

Maier T, Jacobi A, Sauter M, Bock A (1993) The product of the hypB gene, which is required for nickel incorporation into hydrogenases, is a novel guanine nucleotide-binding protein. J Bacteriol 175:630–635

Markov AV, Gusakov AV, Dzedziulia EI, Ustinov BB, Antonov AA, Okunev ON, Bekkarevich AO, Sinitsyn AP (2006) Properties of hemicellulases of the enzyme complex from *Trichoderma longibrachiatum*. Prikl Biokhim Mikrobiol 42:654–664

Maroti J, Farkas A, Nagy IK, Maroti G, Kondorosi E, Rakhely G, Kovacs KL (2010) A second soluble Hox-type NiFe enzyme completes the hydrogenase set in *Thiocapsa roseopersicina* BBS. Appl Environ Microbiol 76:5113–5123

Massanz C, Schmidt S, Friedrich B (1998) Subforms and in vitro reconstitution of the NAD-reducing hydrogenase of *Alcaligenes eutrophus*. J Bacteriol 180:1023–1029

Matsubara H, Saeki K (1992) Structural and functional diversity of ferredoxins and related proteins. Adv Inorg Chem 38:223–280

Mazor Y, Toporik H, Nelson N (2012) Temperature-sensitive PSII and promiscuous PSI as a possible solution for sustainable photosynthetic hydrogen production. Biochim Biophys Acta 1817:1122–1126

McDonald TJ, Svedruzic D, Kim YH, Blackburn JL, Zhang SB, King PW, Heben MJ (2008) Wiring-up hydrogenase with single-walled carbon nanotubes. Nano Lett 7:3528–3534

McDonald AE, Ivanov AG, Bode R, Maxwell DP, Rodermel SR, Huner NP (2011) Flexibility in photosynthetic electron transport: the physiological role of plastoquinol terminal oxidase (PTOX). Biochim Biophys Acta 1807:954–967

McGlynn SE, Ruebush SS, Naumov A, Nagy LE, Dubini A, King PW, Broderick JB, Posewitz MC, Peters JW (2007) In vitro activation of [FeFe] hydrogenase: new insights into hydrogenase maturation. J Biol Inorg Chem 12:443–447

McGlynn SE, Shepard EM, Winslow MA, Naumov AV, Duschene KS, Posewitz MC, Broderick WE, Broderick JB, Peters JW (2008) HydF as a scaffold protein in [FeFe] hydrogenase H-cluster biosynthesis. FEBS Lett 582:2183–2187

McIntosh CL, Germer F, Schulz R, Appel J, Jones AK (2011) The [NiFe]-hydrogenase of the cyanobacterium *Synechocystis* sp. PCC 6803 works bidirectionally with a bias to H_2 production. J Am Chem Soc 133:11308–11319

Melis A, Happe T (2001) Hydrogen production. Green algae as a source of energy. Plant Physiol 127:740–748

Melis A, Happe T (2004) Trails of green alga hydrogen research – from Hans Gaffron to new frontiers. Photosynth Res 80:401–409

Melis A, Neidhardt J, Benemann JR (1998) *Dunaliella salina* (Chlorophyta) with small chlorophyll antenna sizes exhibit higher photosynthetic productivities and photon use efficiencies than normally pigmented cells. J Appl Phycol 10:515–525

Melis A, Zhang L, Forestier M, Ghirardi ML, Seibert M (2000) Sustained photobiological hydrogen gas production upon reversible inactivation of oxygen evolution in the green alga *Chlamydomonas reinhardtii*. Plant Physiol 122:127–136

Merchant SS, Prochnik SE, Vallon O, Harris EH, Karpowicz SJ, Witman GB, Terry A, Salamov A, Fritz-Laylin LK, Marechal-Drouard L, Marshall WF, Qu LH, Nelson DR, Sanderfoot AA, Spalding

MH, Kapitonov VV, Ren QH, Ferris P, Lindquist E, Shapiro H, Lucas SM, Grimwood J, Schmutz J, Cardol P, Cerutti H, Chanfreau G, Chen CL, Cognat V, Croft MT, Dent R, Dutcher S, Fernandez E, Fukuzawa H, Gonzalez-Ballester D, Gonzalez-Halphen D, Hallmann A, Hanikenne M, Hippler M, Inwood W, Jabbari K, Kalanon M, Kuras R, Lefebvre PA, Lemaire SD, Lobanov AV, Lohr M, Manuell A, Meir I, Mets L, Mittag M, Mittelmeier T, Moroney JV, Moseley J, Napoli C, Nedelcu AM, Niyogi K, Novoselov SV, Paulsen IT, Pazour G, Purton S, Ral JP, Riano-Pachon DM, Riekhof W, Rymarquis L, Schroda M, Stern D, Umen J, Willows R, Wilson N, Zimmer SL, Allmer J, Balk J, Bisova K, Chen CJ, Elias M, Gendler K, Hauser C, Lamb MR, Ledford H, Long JC, Minagawa J, Page MD, Pan JM, Pootakham W, Roje S, Rose A, Stahlberg E, Terauchi AM, Yang PF, Ball S, Bowler C, Dieckmann CL, Gladyshev VN, Green P, Jorgensen R, Mayfield S, Mueller-Roeber B, Rajamani S, Sayre RT, Brokstein P et al (2007) The Chlamydomonas genome reveals the evolution of key animal and plant functions. Science 318:245–251

Meuser JE, D'Adamo S, Jinkerson RE, Mus F, Yang WQ, Ghirardi ML, Seibert M, Grossman AR, Posewitz MC (2012) Genetic disruption of both *Chlamydomonas reinhardtii* [FeFe]-hydrogenases: insight into the role of HYDA2 in H_2 production. Biochem Biophys Res Commun 417:704–709

Montet Y, Amara P, Volbeda A, Vernede X, Hatchikian EC, Field MJ, Frey M, Fontecilla-Camps JC (1997) Gas access to the active site of [NiFe] hydrogenases probed by X-ray crystallography and molecular dynamics. Nat Struct Biol 4:523–526

Morra S, Valetti F, Sadeghi SJ, King PW, Meyer T, Gilardi G (2011) Direct electrochemistry of an FeFe -hydrogenase on a TiO_2 electrode. Chem Commun (Camb) 47:10566–10568

Moulis JM, Davasse V (1995) Probing the role of electrostatic forces in the interaction of *Clostridium pasteurianum* ferredoxin with its redox partners. Biochemistry 34:16781–16788

Mulder DW, Ortillo DO, Gardenghi DJ, Naumov AV, Ruebush SS, Szilagyi RK, Huynh B, Broderick JB, Peters JW (2009) Activation of HydA(DeltaEFG) requires a preformed [4Fe-4S] cluster. Biochemistry 48:6240–6248

Mulder DW, Boyd ES, Sarma R, Lange RK, Endrizzi JA, Broderick JB, Peters JW (2010) Stepwise [FeFe]-hydrogenase H-cluster assembly revealed in the structure of HydA(DeltaEFG). Nature 465: 248–251

Mulder DW, Shepard EM, Meuser JE, Joshi N, King PW, Posewitz MC, Broderick JB, Peters JW (2011) Insights into [FeFe]-hydrogenase structure, mechanism, and maturation. Structure 19:1038–1052

Murthy UM, Wecker MS, Posewitz MC, Gilles-Gonzalez MA, Ghirardi ML (2012) Novel FixL homologues in *Chlamydomonas reinhardtii* bind heme and O_2. FEBS Lett 586:4282–4288

Mus F, Dubini A, Seibert M, Posewitz MC, Grossman AR (2007) Anaerobic acclimation in *Chlamydomonas reinhardtii*: anoxic gene expression, hydrogenase induction, and metabolic pathways. J Biol Chem 282:25475–25486

Nakajima Y, Ueda R (1997) Improvement of photosynthesis in dense microalgal suspension by reduction of light harvesting pigments. J Appl Phycol 9:503–510

Nakajima Y, Tsuzuki M, Ueda R (2001) Improved productivity by reduction of the content of light-harvesting pigment in *Chlamydomonas perigranulata*. J Appl Phycol 13:95–101

Navarro RM, Alvarez-Galvan MC, Villoria de la Mano JA, Al-Zahrani SM, Fierro JLG (2010) A framework for visible-light water splitting. Energy Environ Sci 3:1865–1882

Nicolet Y, Fontecilla-Camps JC (2012) Structure-function relationships in [FeFe]-hydrogenase active site maturation. J Biol Chem 287:13532–13540

Nicolet Y, Piras C, Legrand P, Hatchikian CE, Fontecilla-Camps JC (1999) *Desulfovibrio desulfuricans* iron hydrogenase: the structure shows unusual coordination to an active site Fe binuclear center. Structure 7:13–23

Nicolet Y, Lemon BJ, Fontecilla-Camps JC, Peters JW (2000) A novel FeS cluster in Fe-only hydrogenases. Trends Biochem Sci 25:138–143

Nicolet Y, de Lacey AL, Vernede X, Fernandez VM, Hatchikian EC, Fontecilla-Camps JC (2001) Crystallographic and FTIR spectroscopic evidence of changes in Fe coordination upon reduction of the active site of the Fe-only hydrogenase from *Desulfovibrio desulfuricans*. J Am Chem Soc 123:1596–1601

Nicolet Y, Rubach JK, Posewitz MC, Amara P, Mathevon C, Atta M, Fontecave M, Fontecilla-Camps JC (2008) X-ray structure of the [FeFe]-hydrogenase maturase HydE from *Thermotoga maritima*. J Biol Chem 283:18861–18872

Nicolet Y, Amara P, Mouesca JM, Fontecilla-Camps JC (2009) Unexpected electron transfer mechanism upon AdoMet cleavage in radical SAM proteins. Proc Natl Acad Sci U S A 106:14867–14871

Nicolet Y, Martin L, Tron C, Fontecilla-Camps JC (2010) A glycyl free radical as the precursor in the synthesis of carbon monoxide and cyanide by the [FeFe]-hydrogenase maturase HydG. FEBS Lett 584:4197–4202

Ogata H, Hirota S, Nakahara A, Komori H, Shibata N, Kato T, Kano K, Higuchi Y (2005) Activation process of [NiFe] hydrogenase elucidated by high-resolution X-ray analyses: conversion of the ready to the unready state. Structure 13:1635–1642

Ogata H, Lubitz W, Higuchi Y (2009) [NiFe] hydrogenases: structural and spectroscopic studies of the reaction mechanism. Dalton Trans 37:7577–7587

Oliveira P, Lindblad P (2005) LexA, a transcription regulator binding in the promoter region of the bidirectional hydrogenase in the cyanobacterium *Synechocystis* sp. PCC 6803. FEMS Microbiol Lett 251:59–66

Oliveira P, Lindblad P (2008) An AbrB-like protein regulates the expression of the bidirectional hydrogenase in *Synechocystis* sp. strain PCC 6803. J Bacteriol 190:1011–1019

Oliveira P, Lindblad P (2009) Transcriptional regulation of the cyanobacterial bidirectional Hox-hydrogenase. Dalton Trans 45:9990–9996

Oliveira P, Leitao E, Tamagnini P, Moradas-Ferreira P, Oxelfelt F (2004) Characterization and transcriptional analysis of hupSLW in *Gloeothece* sp. ATCC 27152: an uptake hydrogenase from a unicellular cyanobacterium. Microbiology 150:3647–3655

Olson JW, Maier RJ (1997) The sequences of hypF, hypC and hypD complete the hyp gene cluster required for hydrogenase activity in *Bradyrhizobium japonicum*. Gene 199:93–99

Olson JW, Maier RJ (2000) Dual roles of *Bradyrhizobium japonicum* nickelin protein in nickel storage and GTP-dependent Ni mobilization. J Bacteriol 182:1702–1705

Onda Y, Matsumura T, Kimata-Ariga Y, Sakakibara H, Sugiyama T, Hase T (2000) Differential interaction of maize root ferredoxin:NADP(+) oxidoreductase with photosynthetic and non-photosynthetic ferredoxin isoproteins. Plant Physiol 123:1037–1045

Ort DR, Melis A (2011) Optimizing antenna size to maximize photosynthetic efficiency. Plant Physiol 155:79–85

Palagyi-Meszaros LS, Maroti J, Latinovics D, Balogh T, Klement E, Medzihradszky KF, Rakhely G, Kovacs KL (2009) Electron-transfer subunits of the [NiFe] hydrogenases in *Thiocapsa roseopersicina* BBS. FEBS J 276:164–174

Pandelia ME, Ogata H, Lubitz W (2010) Intermediates in the catalytic cycle of [NiFe] hydrogenase: functional spectroscopy of the active site. ChemPhysChem 11:1127–1140

Pandelia ME, Nitschke W, Infossi P, Giudici-Orticoni MT, Bill E, Lubitz W (2011) Characterization of a unique [FeS] cluster in the electron transfer chain of the oxygen tolerant [NiFe] hydrogenase from *Aquifex aeolicus*. Proc Natl Acad Sci U S A 108:6097–6102

Pandey AS, Harris TV, Giles LJ, Peters JW, Szilagyi RK (2008) Dithiomethylether as a ligand in the hydrogenase H-cluster. J Am Chem Soc 130:4533–4540

Papageorgiou GC, Tsimilli-Michael M, Stamatakis K (2007) The fast and slow kinetics of chlorophyll a fluorescence induction in plants, algae and cyanobacteria: a viewpoint. Photosynth Res 94:275–290

Pape M, Lambertz C, Happe T, Hemschemeier A (2012) Differential expression of the *Chlamydomonas* [FeFe]-hydrogenase-encoding HYDA1 gene Is regulated by the copper response regulator 1. Plant Physiol 159:1700–1712

Park HG, Holt JK (2010) Recent advances in nanoelectrode architecture for photochemical hydrogen production. Energy Environ Sci 3:1028–1036

Paschos A, Class RS, Bock A (2001) Carbamoyl phosphate requirement for synthesis of the active center of [NiFe]-hydrogenases. FEBS Lett 488:9–12

Pedroni P, Mura GM, Galli G, Pratesi C, Serbolisca L, Grandi G (1996) The hydrogenase from the hyperthermophilic archaeon *Pyrococcus furiosus*: from basic research to possible future applications. Int J Hydrog Energy 21:853–858

Peng XG (2010) Band gap and composition engineering on a nanocrystal (BCEN) in solution. Acc Chem Res 43:1387–1395

Peters JW (1999) Structure and mechanism of iron-only hydrogenases. Curr Opin Struct Biol 9:670–676

Peters JW, Broderick JB (2012) Emerging paradigms for complex iron-sulfur cofactor assembly and insertion. Annu Rev Biochem 81:429–450

Peters JW, Lanzilotta WN, Lemon BJ, Seefeldt LC (1998) X-ray crystal structure of the Fe-only hydrogenase (CpI) from *Clostridium pasteurianum* to 1.8 angstrom resolution. Science 282:1853–1858

Petroutsos D, Terauchi AM, Busch A, Hirschmann I, Merchant SS, Finazzi G, Hippler M (2009) PGRL1 Participates in iron-induced remodeling of the photosynthetic apparatus and in energy metabolism in *Chlamydomonas reinhardtii*. J Biol Chem 284:32770–32781

Philipps G, Happe T, Hemschemeier A (2012) Nitrogen deprivation results in photosynthetic hydrogen production in *Chlamydomonas reinhardtii*. Planta 235:729–745

Pierik AJ, Hagen WR, Redeker JS, Wolbert RB, Boersma M, Verhagen MF, Grande HJ, Veeger C, Mutsaers PH, Sands RH et al (1992) Redox properties of the iron-sulfur clusters in activated Fe-hydrogenase from *Desulfovibrio vulgaris* (Hildenborough). Eur J Biochem 209:63–72

Pilet E, Nicolet Y, Mathevon C, Douki T, Fontecilla-Camps JC, Fontecave M (2009) The role of the maturase HydG in [FeFe]-hydrogenase active site synthesis and assembly. FEBS Lett 583:506–511

Pinto F, van Elburg KA, Pacheco CC, Lopo M, Noirel J, Montagud A, Urchueguia JF, Wright PC, Tamagnini P (2012) Construction of a chassis for hydrogen production: physiological and molecular characterization of a *Synechocystis* sp. PCC 6803 mutant lacking a functional bidirectional hydrogenase. Microbiology 158:448–464

Polle JEW, Kanakagiri SD, Melis A (2003) tla1, a DNA insertional transformant of the green alga *Chlamydomonas reinhardtii* with a truncated light-harvesting chlorophyll antenna size. Planta 217:49–59

Posewitz MC, King PW, Smolinski SL, Zhang L, Seibert M, Ghirardi ML (2004a) Discovery of two novel radical S-adenosylmethionine proteins required for the assembly of an active [Fe] hydrogenase. J Biol Chem 279:25711–25720

Posewitz MC, Smolinski SL, Kanakagiri S, Melis A, Seibert M, Ghirardi ML (2004b) Hydrogen photoproduction is attenuated by disruption of an isoamylase gene in *Chlamydomonas reinhardtii*. Plant Cell 16:2151–2163

Posewitz MC, Dubini A, Meuser JE, Seibert M, Ghirardi ML (2008) Hydrogenases, hydrogen production, and anoxia. In: Stern D (ed) The *Chlamydomonas* sourcebook, vol 2, Organellar and metabolic processes. Academic, Oxford, UK, pp 217–255

Rakhely G, Kovacs AT, Maroti G, Fodor BD, Csanadi G, Latinovics D, Kovacs KL (2004) Cyanobacterial-type, heteropentameric, NAD+-reducing [NiFe] hydrogenase in the purple sulfur photosynthetic bacterium *Thiocapsa roseopersicina*. Appl Environ Microbiol 70:722–728

Rakhely G, Laurinavichene TV, Tsygankov AA, Kovacs KL (2007) The role of Hox hydrogenase in the H_2 metabolism of *Thiocapsa roseopersicina*. Biochim Biophys Acta 1767:671–676

Rao KK, Gogotov IN, Hall DO (1978) Hydrogen evolution by chloroplast-hydrogenase systems – improvements and additional observations. Biochimie 60:291–296

Reisner E (2011) Solar hydrogen evolution with hydrogenases: from natural to hybrid systems. Eur J Inorg Chem 2011:1005–1016

Reisner E, Fontecilla-Camps JC, Armstrong FA (2009a) Catalytic electrochemistry of a [NiFeSe]-hydrogenase on TiO_2 and demonstration of its suitability for visible-light driven H_2 production. Chem Commun (Camb) 5:550–552

Reisner E, Powell DJ, Cavazza C, Fontecilla-Camps JC, Armstrong FA (2009b) Visible light-driven H_2 production by hydrogenases attached to dye-sensitized TiO_2 nanoparticles. J Am Chem Soc 131: 18457–18466

Reissmann S, Hochleitner E, Wang H, Paschos A, Lottspeich F, Glass RS, Bock A (2003) Taming of a poison: biosynthesis of the [NiFe]-hydrogenase cyanide ligands. Science 299:1067–1070

Roach PL (2011) Radicals from S-adenosylmethionine and their application to biosynthesis. Curr Opin Chem Biol 15:267–275

Roessler PG, Lien S (1984) Purification of hydrogenase from *Chlamydomonas reinhardtii*. Plant Physiol 75:705–709

Rubach JK, Brazzolotto X, Gaillard J, Fontecave M (2005) Biochemical characterization of the HydE and HydG iron-only hydrogenase maturation enzymes from *Thermatoga maritima*. FEBS Lett 579:5055–5060

Rumeau D, Becuwe-Linka N, Beyly A, Louwagie M, Garin J, Peltier G (2005) New subunits NDH-M, -N, and -O, encoded by nuclear genes, are essential for plastid Ndh complex functioning in higher plants. Plant Cell 17:219–232

Rupprecht J (2009) From systems biology to fuel–Chlamydomonas reinhardtii as a model for a systems biology approach to improve biohydrogen production. J Biotechnol 142:10–20

Schmid B, Ribbe MW, Einsle O, Yoshida M, Thomas LM, Dean DR, Rees DC, Burgess BK (2002) Structure of a cofactor-deficient nitrogenase MoFe protein. Science 296:352–356

Schmitter JM, Jacquot JP, de Lamotte-Guery F, Beauvallet C, Dutka S, Gadal P, Decottignies P (1988) Purification, properties, and complete amino acid sequence of the ferredoxin from a green alga, *Chlamydomonas reinhardtii*. Eur J Biochem 172:405–412

Schmitz O, Boison G, Hilscher R, Hundeshagen B, Zimmer W, Lottspeich F, Bothe H (1995) Molecular biological analysis of a bidirectional hydrogenase from cyanobacteria. Eur J Biochem 233:266–276

Schmitz O, Boison G, Salzmann H, Bothe H, Schutz K, Wang SH, Happe T (2002) HoxE–a subunit specific for the pentameric bidirectional hydrogenase complex (HoxEFUYH) of cyanobacteria. Biochim Biophys Acta 1554:66–74

Schneider K, Cammack R, Schlegel HG (1984a) Content and localization of FMN, Fe-S clusters, and nickel in the NAD-linked hydrogenase of *Nocardia opaca* 1b. Eur J Biochem 142:75–84

Schneider K, Schlegel HG, Jochim K (1984b) Effect of nickel on activity and subunit composition of purified hydrogenase from *Nocardia opaca* 1b. Eur J Biochem 138:533–541

Schonfeld C, Wobbe L, Borgstadt R, Kienast A, Nixon PJ, Kruse O (2004) The nucleus-encoded protein MOC1 is essential for mitochondrial light acclimation in *Chlamydomonas reinhardtii*. J Biol Chem 279:50366–50374

Schutz K, Happe T, Troshina O, Lindblad P, Leitao E, Oliveira P, Tamagnini P (2004) Cyanobacterial H_2 production – a comparative analysis. Planta 218: 350–359

Selvaggi A, Tosi C, Barberini U, Franchi E, Rodriguez F, Pedroni P (1999) In vitro hydrogen photoproduction using *Pyrococcus furiosus* sulfhydrogenase and TiO_2. J Photochem Photobiol 125:107–112

Serebriakova LT, Sheremetieva ME (2006) Characterization of catalytic properties of hydrogenase isolated from the unicellular cyanobacterium *Gloeocapsa alpicola* CALU 743. Biochemistry (Mosc) 71:1370–1376

Serebriakova LT, Zorin NA, Lindblad P (1994) Reversible hydrogenase in *Anabaena variabilis* ATCC29413 – presence and localization in non-N_2-fixing cells. Arch Microbiol 161:140–144

Shepard EM, Duffus BR, George SJ, McGlynn SE, Challand MR, Swanson KD, Roach PL, Cramer SP, Peters JW, Broderick JB (2010a) [FeFe]-hydrogenase maturation: HydG-catalyzed synthesis of carbon monoxide. J Am Chem Soc 132:9247–9249

Shepard EM, McGlynn SE, Bueling AL, Grady-Smith CS, George SJ, Winslow MA, Cramer SP, Peters JW, Broderick JB (2010b) Synthesis of the 2Fe subcluster of the [FeFe]-hydrogenase H cluster on the HydF scaffold. Proc Natl Acad Sci U S A 107:10448–10453

Shikanai T (2007) Cyclic electron transport around photosystem I: genetic approaches. Annu Rev Plant Biol 58:199–217

Shomura Y, Higuchi Y (2012) Structural basis for the reaction mechanism of S-carbamoylation of HypE by HypF in the maturation of [NiFe]-hydrogenases. J Biol Chem 287:28409–28419

Shumilin IA, Nikandrov VV, Popov VO, Krasnovsky AA (1992) Photogeneration of NADH under coupled action of CdS semiconductor and hydrogenase from *Alcaligenes eutrophus* without exogenous mediators. FEBS Lett 306:125–128

Silakov A, Wenk B, Reijerse E, Lubitz W (2009) 14N HYSCORE investigation of the H-cluster of [FeFe] hydrogenase: evidence for a nitrogen in the dithiol bridge. Phys Chem Chem Phys 11:6592–6599

Silakov A, Olsen MT, Sproules S, Reijerse EJ, Rauchfuss TB, Lubitz W (2012) EPR/ENDOR, Mossbauer, and quantum-chemical investigations of diiron complexes mimicking the active oxidized state of [FeFe]-hydrogenase. Inorg Chem 51:8617–8628

Sjoholm J, Oliveira P, Lindblad P (2007) Transcription and regulation of the bidirectional hydrogenase in the cyanobacterium *Nostoc* sp. strain PCC 7120. Appl Environ Microbiol 73:5435–5446

Stiebritz MT, Reiher M (2009) Theoretical study of dioxygen induced inhibition of [FeFe]-hydrogenase. Inorg Chem 48:7127–7140

Stiebritz MT, Reiher M (2012) Hydrogenases and oxygen. Chem Sci 3:1739–1751

Stirnberg M, Happe T (2004) Identification of a cis-acting element controlling anaerobic expression of the HYDA gene from *Chlamydomonas reinhardtii*. In: Jun M, Yasuo I, Rögner M (eds) Biohydrogen III. Elsevier Science, Amsterdam, pp 117–127

Stripp ST, Goldet G, Brandmayr C, Sanganas O, Vincent KA, Haumann M, Armstrong FA, Happe T (2009) How oxygen attacks [FeFe] hydrogenases from photosynthetic organisms. Proc Natl Acad Sci U S A 106:17331–17336

Surzycki R, Cournac L, Peltier G, Rochaix JD (2007) Potential for hydrogen production with inducible chloroplast gene expression in *Chlamydomonas*. Proc Natl Acad Sci U S A 104:17548–17553

Tamagnini P, Leitao E, Oliveira P, Ferreira D, Pinto F, Harris DJ, Heidorn T, Lindblad P (2007) Cyanobacterial hydrogenases: diversity, regulation and applications. FEMS Microbiol Rev 31:692–720

Teixeira VH, Baptista AM, Soares CM (2006) Pathways of H_2 toward the active site of [NiFe]-hydrogenase. Biophys J 91:2035–2045

Terasaki N, Iwai M, Yamamoto N, Hiraga T, Yamada S, Inoue Y (2008) Photocurrent generation properties of Histag-photosystem II immobilized on nanostructured gold electrode. Thin Solid Films 516:2553–2557

Terauchi AM, Lu SF, Zaffagnini M, Tappa S, Hirasawa M, Tripathy JN, Knaff DB, Farmer PJ, Lemaire SD, Hase T, Merchant SS (2009) Pattern of expression and substrate specificity of chloroplast ferredoxins from *Chlamydomonas reinhardtii*. J Biol Chem 284:25867–25878

Thauer RK, Kaufer B, Zahringer M, Jungermann K (1974) The reaction of the iron-sulfur protein hydrogenase with carbon monoxide. Eur J Biochem 42:447–452

Thiemermann S, Dernedde J, Bernhard M, Schroeder W, Massanz C, Friedrich B (1996) Carboxyl-terminal processing of the cytoplasmic NAD-reducing hydrogenase of *Alcaligenes eutrophus* requires the hoxW gene product. J Bacteriol 178:2368–2374

Tibelius KH, Du L, Tito D, Stejskal F (1993) The *Azotobacter chroococcum* hydrogenase gene-cluster – sequences and genetic analysis of 4 accessory genes, HupA, HupB, HupY and HupC. Gene 127:53–61

Tolleter D, Ghysels B, Alric J, Petroutsos D, Tolstygina I, Krawietz D, Happe T, Auroy P, Adriano JM, Beyly A, Cuine S, Plet J, Reiter IM, Genty B, Cournac L, Hippler M, Peltier G (2011) Control of hydrogen photoproduction by the proton gradient generated by cyclic electron flow in *Chlamydomonas reinhardtii*. Plant Cell 23:2619–2630

Torzillo G, Scoma A, Faraloni C, Ena A, Johanningmeier U (2009) Increased hydrogen photoproduction by means of a sulfur-deprived *Chlamydomonas reinhardtii* D1 protein mutant. Int J Hydrog Energy 34:4529–4536

Tron C, Cherrier MV, Amara P, Martin L, Fauth F, Fraga E, Correard M, Fontecave M, Nicolet Y, Fontecilla Camps JC (2011) Further characterization of the [FeFe] hydrogenase maturase HydG. Eur J Inorg Chem 2011:1121–1127

Tsygankov AA (2007) Nitrogen-fixing cyanobacteria: a review. Appl Biochem Microbiol 43:250–259

Utschig LM, Silver SC, Mulfort KL, Tiede DM (2011) Nature-driven photochemistry for catalytic solar hydrogen production: a photosystem I transition metal catalyst hybrid. J Am Chem Soc 133:16334–16337

van der Spek TM, Arendsen AF, Happe RP, Yun S, Bagley KA, Stufkens DJ, Hagen WR, Albracht SP (1996) Similarities in the architecture of the active sites of Ni-hydrogenases and Fe-hydrogenases detected by means of infrared spectroscopy. Eur J Biochem 237:629–634

Vignais PM, Billoud B (2007) Occurrence, classification, and biological function of hydrogenases: an overview. Chem Rev 107:4206–4272

Vignais PM, Colbeau A (2004) Molecular biology of microbial hydrogenases. Curr Issues Mol Biol 6:159–188

Vignais PM, Billoud B, Meyer J (2001) Classification and phylogeny of hydrogenases. FEMS Microbiol Rev 25:455–501

Vincent KA, Cracknell JA, Clark JR, Ludwig M, Lenz O, Friedrich B, Armstrong FA (2006) Electricity from low-level H_2 in still air-an ultimate test for an oxygen tolerant hydrogenase. Chem Commun (Camb) 48:5033–5035

Vincent KA, Parkin A, Armstrong FA (2007) Investigating and exploiting the electrocatalytic properties of hydrogenases. Chem Rev 107:4366–4413

Volbeda A, Martin L, Cavazza C, Matho M, Faber BW, Roseboom W, Albracht SPJ, Garcin E, Rousset M, Fontecilla-Camps JC (2005) Structural differences between the ready and unready oxidized states of [NiFe] hydrogenases. J Biol Inorg Chem 10:239–249

Volbeda A, Amara P, Darnault C, Mouesca JM, Parkin A, Roessler MM, Armstrong FA, Fontecilla-Camps JC (2012) X-ray crystallographic and computational studies of the O_2-tolerant [NiFe]-hydrogenase 1 from *Escherichia coli*. Proc Natl Acad Sci U S A 109:5305–5310

von Abendroth G, Stripp S, Silakov A, Croux C, Soucaille P, Girbal L, Happe T (2008) Optimized over-expression of [FeFe] hydrogenases with high specific activity in *Clostridium acetobutylicum*. Int J Hydrog Energy 33:6076–6081

Wang M, Chen L, Li X, Sun L (2011) Approaches to efficient molecular catalyst systems for photochemical H_2 production using [FeFe]-hydrogenase active site mimics. Dalton Trans 40:12793–12800

Weyman PD, Vargas WA, Tong YK, Yu JP, Maness PC, Smith HO, Xu Q (2011) Heterologous expression of *Alteromonas macleodii* and *Thiocapsa roseopersicina* [NiFe} hydrogenases in *Synechococcus elongatus*. PLoS One 6:8

Winkler M, Hemschemeier A, Gotor C, Melis A, Happe T (2002) [Fe]-hydrogenases in green algae: photo-fermentation and hydrogen evolution under sulfur deprivation. Int J Hydrog Energy 27:1431–1439

Winkler M, Kuhlgert S, Hippler M, Happe T (2009) Characterization of the key step for light-driven hydrogen evolution in green algae. J Biol Chem 284:36620–36627

Woolerton TW, Sheard S, Chaudhary YS, Armstrong FA (2012) Enzymes and bio-inspired electrocatalysts in solar fuel devices. Energy Environ Sci 5:7470–7490

Wunschiers R, Batur M, Lindblad P (2003) Presence and expression of hydrogenase specific C-terminal endopeptidases in cyanobacteria. BMC Microbiol 3:8

Yacoby I, Pochekailov S, Toporik H, Ghirardi ML, King PW, Zhang S (2011) Photosynthetic electron partitioning between [FeFe]-hydrogenase and ferredoxin:NADP+-oxidoreductase (FNR) enzymes in vitro. Proc Natl Acad Sci U S A 108:9396–9401

Yacoby I, Tegler LT, Pochekailov S, Zhang S, King PW (2012) Optimized expression and purification for high-activity preparations of algal [FeFe]-hydrogenase. PLoS One 7:e35886

Yonekura-Sakakibara K, Onda Y, Ashikari T, Tanaka Y, Kusumi T, Hase T (2000) Analysis of reductant supply systems for ferredoxin-dependent sulfite reductase in photosynthetic and nonphotosynthetic organs of maize. Plant Physiol 122:887–894

Zadvornyy OA, Lucon JE, Gerlach R, Zorin NA, Douglas T, Elgren TE, Peters JW (2012) Photo-induced

H_2 production by [NiFe] -hydrogenase from *T. roseopersicina* covalently linked to a Ru(II) photosensitizer. J Inorg Biochem 106:151–155

Zhang JD, Chi Q, Kuznetsov AM, Hansen AG, Wackerbarth H, Christensen HEM, Andersen JET, Ulstrup J (2002) Electronic properties of functional biomolecules at metal/aqueous solution interfaces. J Phys Chem B 106:1131–1152

Zhang JD, Welinder AC, Chi QJ, Ulstrup J (2011) Electrochemically controlled self-assembled monolayers characterized with molecular and sub-molecular resolution. Phys Chem Chem Phys 13:5526–5545

Chapter 6

Nitrogenase-Dependent Hydrogen Production by Cyanobacteria

Hermann Bothe*
Botanical Institute, The University of Cologne, Zülpicher Str. 47b, 50923 Cologne, Germany

and

William E. Newton
Department of Biochemistry, Virginia Polytechnic Institute and State University, Blacksburg, VA 24061, USA

Summary

Cyanobacteria possess three different types of nitrogenases, two Mo- and one V-nitrogenases, all of which catalyse the reduction of the dinitrogen molecule to ammonia accompanied by the evolution of molecular hydrogen. V-nitrogenase is most effective in producing H_2 and is, therefore, suited for potential applications in solar energy conversion programs to generate molecular H_2 as a clean and renewable energy source. Intact cells of cyanobacteria often show rather little net H_2-production due to the concomitant H_2-consumption by uptake hydrogenase. The unicellular N_2-fixing Cyanothece is currently the focus of H_2-production research. Wild-type cyanobacteria are already capable of maximal H_2-production and any further enhancement of H_2-formation must be achieved by manipulating linear photosynthetic electron transport which is rate-limiting in light- and nitrogenase-dependent H_2-generation.

*Author for correspondence, e-mail: aeb75@uni-koeln.de

D. Zannoni and R. De Philippis (eds.), Microbial BioEnergy: Hydrogen Production, Advances in Photosynthesis and Respiration 38, pp. 137–153, DOI 10.1007/978-94-017-8554-9_6,

I. Introduction

Cyanobacteria have the simplest nutrient requirements among all organisms. They thrive on simple inorganic media where they can meet their nitrogen demands by (di) nitrogen fixation. Cyanobacteria have two totally different enzyme complexes that catalyze the formation of hydrogen, namely nitrogenase and hydrogenase. Whilst H_2-formation by hydrogenases is reviewed by Tamagnini et al. (Chap. 4 of this book) the current article focuses on the nitrogenase-dependent evolution of this gas. Cyanobacteria are a diverse group of organisms with extremely large deviations in genome size (Hess 2011). The smallest one, the marine *UCYN-A* organism has a single contig with 1.44 Mb, cannot be cultured as yet but performs N_2-fixation (Tripp et al. 2010). The maximum genome size of 9.05 Mb is reached in the facultative symbiotic *Nostoc punctiforme* ATCC29133 (Meeks et al. 2001). Thus components involved in H_2-formation may vary from one organism to the next. However, the structural genes of nitrogenase, *nifHDK*, are very much conserved in cyanobacteria as among all other organisms. Indeed, probes developed from *nifH* have routinely been used for a long time to screen for the occurrence of N_2 fixation in environmental samples or in as yet uncharacterized isolates (Ruvkun and Ausubel 1980).

Cyanobacteria may have evolved 2.5–2.7 billion or more years ago (Latysheva et al. 2012). The discussion continues whether the last cyanobacterial common ancestor could perform N_2-fixation and thus also H_2-formation or not. Whereas some investigators conclude that this organism was not N_2-fixing (Sanchez-Barcaldo et al. 2005; Shi and Falkowski 2008), more recent studies judge the situation as ambiguous (Larsson et al. 2011; Latysheva et al. 2012). On the other hand, the earliest nitrogenase on earth may have functioned to detoxify soils from cyanide (Kelly et al. 1967), and its capability to reduce the dinitrogen molecule may have developed afterwards (Postgate 1972). If so, nitrogen fixation and hydrogen evolution might have been traits of the earliest cyanobacterium. Since nitrogenase is inactivated by oxygen, the enzyme must have been developed in cyanobacteria at the latest before the oxygenation of the atmosphere some 2,400–2,200 million years ago (Latysheva et al. 2012). The patchy distribution of nitrogenase among the different cyanobacteria is likely due to horizontal gene transfer among the earliest cyanobacteria, followed some time later by the rate of loss of the N_2-fixation capability becoming much greater than the rate of horizontal gene transfer among closely related cyanobacteria (Latysheva et al. 2012).

H_2-production by cyanobacteria is environmentally friendly, since it is not accompanied by production of CO_2 or any other greenhouse gas. In addition, large amounts of energy can be stored in small volumes of H_2. Thus research on the conversion of solar energy to combustible H_2 has continued since the first energy crisis in 1973. Some authors even state that photobiological hydrogen production is on the way to becoming a mature technology (Eroglu and Melis 2011). The current article tries to summarize the extensive literature particularly of the last 2 years. A review with the involvement of both current authors came out in December 2010 (Bothe et al. 2010a). Other reviews on the subject are available (Madamwar et al. 2000; Dutta et al. 2005; Tamagnini et al. 2002, 2007; Tsygankov 2007; Ghirardi et al. 2007, 2009; Schwarz et al. 2010; Kim and Kim 2011; Srirangan et al. 2011; Hallenbeck 2012; Tiwari and Pandey 2012).

II. Nitrogenases in Cyanobacteria

Several types of nitrogenases exist in nature. The "classical" N_2-fixing enzyme catalyses the reaction:

$$N_2 + 8H^+ + 8e^- + 16MgATP \rightarrow 2NH_3 + H_2 + 16MgADP + 16P_i \qquad (6.1)$$

Table 6.1. Occurrence of nitrogenases in cyanobacteria and other organisms.

Organism	Mo-nitrogenase1	Mo-nitrogenase2	V-nitrogenase	Fe-nitrogenase
Cyanobacteria				
All N_2-fixing cyanobacteria besides the following:	+	–	–	–
Anabaena variabilis	+	+	+	–
Anabaena azollae	+	–	+	–
Anabaena azotica and CH1	+	–	+	–
***Azotobacter* spp.**				
A. vinelandii	+	–	+	+
A. paspali	+	–	+	+
A. chroococcum	+	–	+	–
A. salinestris	+	–	+	–
Methanosarcina spp.				
M. acetivorans	+	–	+	+
M. barkeri	+	–	+	–
Photosynthetic bacteria				
Rhodopseudomonas palustris	+	–	+	+
Rhodospirillum rubrum	+	–	–	+
Rhodobacter capsulatus	+	–	–	+
Heliobacterium gestii	+	–	–	+
Others: Non phototrophs				
Clostridium pasteurianum	+	–	–	+
Azomonas macrocytogenesis	+	–	–	+
Azospirillum brasilense Cd	+	–	–	+

This "classical" nitrogenase contains Mo in the catalytic prosthetic group and consists of two component proteins. The smaller protein component is encoded by *nifH* and is known as the Fe-protein, (di)nitrogenase reductase or component-2. It is a homodimer with a total molecular mass of 64 kDa. Its prosthetic group is a [4Fe-4S] cluster that bridges the interface of the two subunits and is ligated to both of them by two cysteinyl residues from each subunit. It serves in accepting reducing equivalents from either ferredoxin or flavodoxin, depending on the organism. The pathway by which the two low molecular-weight proteins, ferredoxin or flavodoxin, are reduced in cyanobacteria have been summarised in a previous review (Bothe et al. 2010a). Electrons from the Fe-protein of nitrogenase are transferred to the larger protein component with concurrent ATP hydrolysis. The ATP demand in nitrogen fixation is high; 2 ATP are hydrolyzed per electron transferred.

The larger nitrogenase component is termed the MoFe-protein, dinitrogenase or component-1. It is encoded by *nifDK* and is a tetrameric ($\alpha_2\beta_2$) protein of about 240 kDa. Each $\alpha\beta$-subunit pair binds one each of two prosthetic groups, the P-cluster and the FeMo-cofactor. The latter possesses homocitrate and the following atoms, 1 Mo, 7 Fe and 9 S in addition to one central light ion, recently identified as carbide (C^{4-}), (Lancaster et al. 2011; Spatzal et al. 2011; Wiig et al. 2012). The reduction of the N_2 molecule may occur at either a central 4Fe-4S face or at the Mo-homocitrate part of the FeMo-cofactor. Further details of nitrogenase catalysis can be taken from (Bothe et al. 2010a, b). This classical nitrogenase is widely distributed in cyanobacteria. However, even closely related cyanobacterial strains may or may not be N_2-fixers. A peculiarity of cyanobacteria is the occurrence of a second Mo-nitrogenase in the heterocystous *Anabaena variabilis* (Table 6.1) which was described independently by two groups in the same year (Schrautemeier et al. 1995; Thiel et al. 1995). This second Mo-nitrogenase is encoded by its own gene set (Thiel et al. 1997). *A. variabilis* expresses the classical Mo-nitrogenase in heterocysts, and the second Mo-nitrogenase in vegetative cells under reduced oxygen tension. Since N_2-fixation is light-dependent both in heterocysts and vegetative cells due to the demand for ATP and reductant (reduced ferredoxin), N_2-fixation by the second Mo-nitrogenase must proceed in parallel with photosynthetic O_2-evolution in vegetative cells. Because all nitrogenases are irreversibly damaged by O_2, vegetative cells must have developed device(s) to protect the second nitrogenase from O_2 or maybe nitrogenase resynthesis is simply faster than the destruction of the enzyme by O_2. Apart from the fact that it is encoded by a different set of *nifHDK* genes than the conventional enzyme, rather little is known about this second Mo-nitrogenase. Its distribution in diverse heterocystous cyanobaceteria has been examined (Masukawa et al. 2009). It might resemble nitrogenase from the filamentous, non-heterocystous *Plectonema* (*Leptolyngbya*) *boryanum*, which is expressed also only under reduced O_2-tension ("microaerobic conditions") (Stewart and Lex 1970).

The alternative V-nitrogenase was first discovered in the non-photosynthetic *Azotobacter vinelandii* (Joerger and Bishop 1988). In culture media without Mo but with sufficient V, this bacterium synthesizes a nitrogenase with V-Fe-S-homocitrate in the catalytic prosthetic group. This V-enzyme complex is encoded by the structural genes *vnfHDGK*. When both Mo and V are absent, this aerobic bacterium can express a third nitrogenase with Fe as the only metal in its prosthetic group, thus with an Fe-Fe-S-homocitrate cofactor in the active site (Joerger and Bishop 1988). The structural-gene cluster of this Fe-only nitrogenase is encoded by *anfHDGK*. The larger component protein of all these nitrogenases possesses the P-cluster, which is likely involved in internal electron transfer between prosthetic groups during N_2 reduction (Fisher et al. 2007).

All three nitrogenases (Mo-, V- and Fe-dependent enzymes) have similar but not identical subunit amino-acid sequences. Both alternative nitrogenases possess an additional protein encoded by the *G* gene and are either hexamers (with a $\alpha_2\beta_2\delta_2$ composition) or octamers (with a $\alpha_2\beta_2\delta_4$ composition). The G-protein might be involved during insertion of the catalytic cofactor, either V-Fe-S-homocitrate or Fe-Fe-S-homocitrate, respectively, into the apoprotein of the alternative nitrogenases. Both alternative nitrogenases support growth of *A. vinelandii* and of the other organisms where they occur (Table 6.1), however, growth rates are lower than when Mo-nitrogenase is expressed. Either alternative nitrogenase reduces N_2 or C_2H_2 with lower rates but produces more H_2 than Mo-nitrogenase (see below). According to the Thorneley-Lowe scheme of nitrogenase catalysis (Thorneley and Lowe 1984; Lukanov et al. 2012) the rate limiting step for product formation might lie in the final dissociation of the oxidized Fe-protein-MgATP from the electron transfer complex, and substitution by V or Fe might lead to the altered ratio in product concentrations.

A recent discovery is the capability of V-nitrogenase but not of the Mo-enzyme complex to reduce carbon monoxide to hydrocarbons (Hu et al. 2012).

The alternative nitrogenases have a somewhat haphazard occurrence among organisms of totally unrelated taxonomic affinities (Table 6.1). In addition to *A. vinelandii*, all three nitrogenases have been detected in *Azotobacter paspali*, the phototroph *Rhodopseudomonas palustris* and *Methanosarcina acetivorans* of the archaeons (Bothe et al. 2010a, b). Both Mo- and V-nitrogenases occur in *Azotobacter chroococcum*, *Azotobacter salinestris* and *Methanosarcina barkeri* 227. The phototrophs *Rhodospirillum rubrum*, *Rhodobacter capsulatus* and *Heliobacterium gestii* as well as the heterotrophic *Clostridium pasteurianum*, *Azospirillum brasilense* Cd and *Azomonas macrocytogenese*s posses the Mo- and Fe-nitrogenases but not the V-enzyme complex. *Klebsiella pneumoniae*, the symbiotic rhizobia and many others have only the Mo-nitrogenase.

Among cyanobacteria, only few isolates have been found to synthesize both the Mo- and the V-nitrogenase: *Anabaena variabilis* (Kentemich et al. 1988), an *Anabaena* isolate from the fern *Azolla* (Ni et al. 1990), *Anabaena* CH1 from Southern Chinese rice fields (Boison et al. 2006), *Anabaena azotica* (Boison et al. 2006), and two other *Anabaena* and one *Nostoc* strains (Masukawa et al. 2009). No Fe-nitrogenase has been annotated in any of the more than 50 cyanobacterial strains sequenced as yet (Hess 2011). However, due to the high genome diversity of cyanobacteria (Hess 2011), the discovery of Fe-nitrogenase in a newly investigated strain would not be unexpected. Hybridization of DNA from *Anabaena variabilis* with an *anfH* probe gave positive signals (Kentemich et al. 1991), but this only indicates the existence of multiple copies of *nifH* in this cyanobacterium (Bothe et al. 2010b) and not the occurrence of Fe-nitrogenase.

The molecular characterization of the V-nitrogenase from *Anabaena variabilis* (Thiel 1993) showed that *vnfDGKEN* are clustered, whereas four other *H* genes, in addition to *nifH*, are interspersed on the chromosome. The next closest *H* gene is located 27 bp from v*nfDGK*. The *nifH* or *vnfH* gene products complement either Mo- or V-nitrogenase in isolated enzymes.

Probes developed from *vnfG* and *anfG* have been used to amplify these genes from environmental samples by PCR and to screen for the occurrence of alternative nitrogenases. Both enzyme complexes were detected in diverse habitats and were found to be concentrated in the pseudomonad-azotobacteria lineage of the gammaproteobacteria (Loveless et al. 1999; Betancourt et al. 2008). Organisms may have acquired alternative nitrogenases by lateral gene transfer. This is particularly striking in *Methanosarcina barkeri* 227 where *vnfDG* have close sequence homologies to the paralog from *Anabaena variabilis* (Chien et al. 2000), whereas the *H* gene is closely related to *anfH* from *Rhodobacter capsulatus* and *Clostridium pasteurianum*. Thus, the V-nitrogenase genes may have been acquired by two different evolutionary events in *M. barkeri* 227 (Chien et al. 2000).

Molybdenum is generally non-limiting in soils or other habitats. The occurrence of alternative nitrogenases with no obvious need is an intriguing question. If really not needed, their genes should have been eliminated from the chromosome of the organisms during the long course of the evolution. Thus it cannot be ruled out that these alternative nitrogenases have other, as yet unknown, functions in cyanobacteria and all other organisms.

III. Hydrogen Production by Nitrogenases

In the absence of any other substrate, nitrogenases catalyze the reduction hydrogen ions (protons) to molecular hydrogen (Postgate 1972; Burns and Hardy 1975). This H_2-formation, as are all other substrate reductions, is ATP dependent but, unlike the others, is insensitive to carbon monoxide. When N_2 is the substrate, its reduction is accompanied by H_2-formation. Hydrogen may be generated as a prerequisite for N_2 binding or from decomposition of an enzyme-bound intermediate (Thorneley et al. 1978; Barney et al. 2005). However, the stoichiometry between NH_3 and H_2 formations may not strictly be 2:1 (as shown in Eq. 6.1). Lower ratios (more H_2) are found with the alternative nitrogenases (see Eqs. 6.2 and 6.3), which indicates their lower efficiencies for N_2 reduction. The stoichiometry between NH_3 and H_2 formation for the alternative nitrogenases is usually written as:

$$\textit{For}\ \mathrm{V}-\textit{nitrogenase}: \mathrm{N_2 + 12\,H^+ + 12\,e^- \rightarrow 2\,NH_3 + 3\,H_2} \tag{6.2}$$

$$\textit{and for Fe}-\textit{nitrogenase}: \mathrm{N_2 + 21\,H^+ + 21\,e^- \rightarrow 2\,NH_3 + 7.5\,H_2} \tag{6.3}$$

(Wall 2004; Eroglu and Melis 2011).

Although these ratios are likely to be approximations, they indicated the undoubtedly higher H_2 production by these alternative enzyme complexes which should be kept in mind in any attempts for potential commercial H_2-generations. In addition, Mo-nitrogenase produces only tiny amounts of H_2 when saturated with substrates other than N_2, like C_2H_2. Although the alternative nitrogenases also reduce C_2H_2, they produce more H_2 than does Mo-nitrogenase and, in addition, produce both C_2H_4 and C_2H_6.

Such H_2-production is observed with isolated nitrogenases, but often not with intact cells. This is especially so in N_2-fixing cyanobacteria, where H_2-production by nitrogenase means a loss of energy (4 ATP are used for every electron pair transferred to form H_2). It is, therefore, not surprising that organisms recycle the "lost" energy by means of hydrogenases catalyzing the reaction:

$$H_2 \rightarrow 2e^- + 2H^+ \qquad (6.4)$$

Cyanobacteria possess two pathways of H_2-utilization: (a) the "Knallgas" reaction where H_2 consumption is O_2-dependent and occurs via respiratory ATP formation; and (b) the light- and photosystem I-dependent H_2-uptake (Eisbrenner and Bothe 1979). Both pathways share the cytochrome *bc* complex III in cyanobacteria (Eisbrenner and Bothe 1979). Cyanobacterial hydrogenases are discussed in other chapters of this book and so are only briefly mentioned here. Most cyanobacteria possess two different hydrogenases. In intact cells, the uptake hydrogenase, encoded by *hupLS*, catalyzes only the consumption of the gas due to the fact that it couples to the respiratory (and photosynthetic) electron transport chain at complex III close to the entry of electrons from succinate dehydrogenase. The native electron acceptor for uptake hydrogenase in the thylakoid membrane has not yet been identified (possibly cytochrome *b* as in *Rhizobium*, Eisbrenner and Evans 1982). Such tight coupling to the membranes at an E_0' close to 0 mV eliminates the possibility of H_2 evolution by uptake hydrogenase. In heterocystous cyanobacteria, uptake hydrogenase is confined to these specialized cells, as recently shown for HupS with a GFP-labelled probe (Camsund et al. 2011), however, this enzyme is also present in non-heterocystous N_2-fixing species (Tamagnini et al. 2005) (Fig. 6.1).

The second hydrogenase of cyanobacteria catalyzes both the uptake and the evolution of H_2. It is encoded by *hoxEFYUH* (Schmitz et al. 1995), is NAD(P)H-dependent (Schmitz and Bothe 1996) and has recently been purified to homogeneity and characterized biochemically (Germer et al. 2009; McIntosh et al. 2011). The production of H_2 (E_0'=−420 mV for $H_2/2H^+$) by NAD(P)H (E_0'=−320 mV) for NAD(P)H/NAD(P)$^+$ represents an "uphill" reaction. Although it has been stated correctly that both H_2 and NAD(H) unlikely operate under the standard potential concentrations of 1 bar in the cells (Skizim et al. 2012), it is obvious that H_2-formation by bidirectional hydrogenase is generally marginal in whole organisms. The enzyme might function in disposing reductant generated by fermentation in the dark and under anaerobic conditions (Bothe et al. 2010a, b).

Work with mutants showed that H_2 produced by nitrogenases is recycled mainly, if not exclusively, by uptake hydrogenase with, at best, only a marginal impact by the bidirectional enzyme (Happe et al. 2000; Masukawa et al. 2002). To maximize net H_2-production by nitrogenases in intact cells, uptake hydrogenase function must be blocked by either inhibitors or molecular knock-out.

IV. Hydrogen Formation in Heterocystous Cyanobacteria

Heterocysts are irreversibly differentiated from vegetative cells which require a massive degradation of existing proteins and the synthesis of new ones. Approximately 40 % of cell proteins are degraded and others newly synthesized upon heterocyst differentiation (Thiel 1990). Heterocysts do not perform

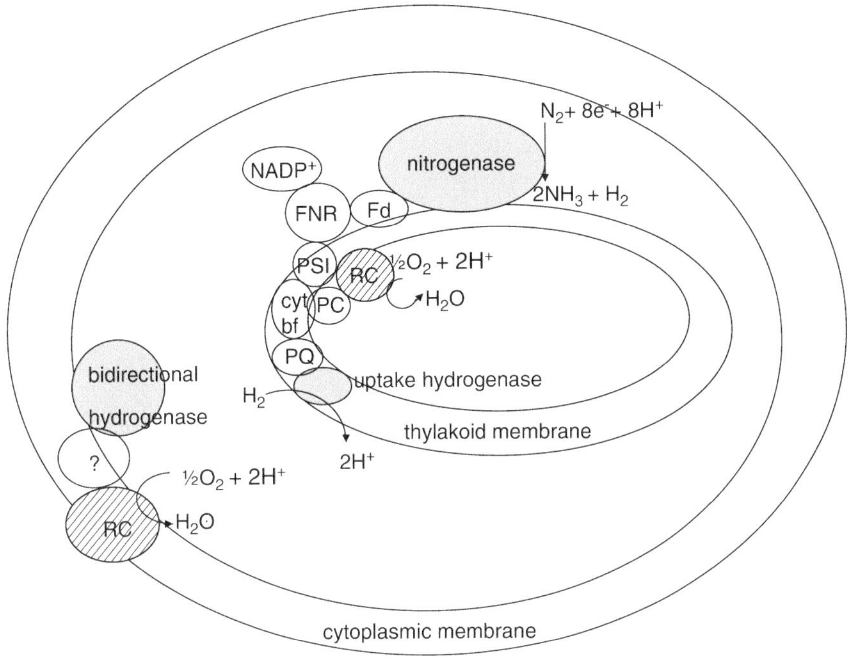

Fig. 6.1. Schematic representation of the nitrogenase–hydrogenase relationship in cyanobacteria. Uptake hydrogenase is bound to the thylakoid membrane. Due to its dependence on reduced ferredoxin or flavodoxin, an association of nitrogenase with the thylakoid membrane is also likely. Bidirectional hydrogenase may be loosely attached to the cytoplasmic membrane which is, however, not proven and its electron acceptor is unknown. For further details see Bothe et al. (2010a, b). *PQ* plastoquinone, *cyt bf* complex III in cyanobacteria, *PC* plastocyanin (or a soluble cytochrome), *PSI* photosystem I, *RC* respiratory chain (*dashed*), *FNR: NADPH* ferredoxin oxidoreductase, *Fd* ferredoxin (or flavodoxin).

CO_2-fixation and do not possess the water-splitting photosystem II. They respire at high rates and require the supply of organic carbon from vegetative cells (Fay 1992). The pathways for the transfer of organic carbon have not yet been fully elucidated. These could either employ the "microplasmodesmata" that connect vegetative cells and heterocysts (Wilcox et al. 1973; Giddings and Staehelin 1978; Giddings and Staehelin 1981) or proceed along the apoplastic route (in the periplasm) in the cell walls with an active entry of compounds across the cytoplasmic membrane of heterocysts (Mariscal et al. 2007; Mullineaux et al. 2008). Organic carbon may not be the only class of substance transported into heterocysts. These cells are packed with nitrogenase, and its synthesis might demand a massive import of Mo, Fe and S from vegetative cells which, however, has not yet been studied.

Heterocysts seem to be an ideal accommodation for nitrogenase functioning. In cell-free preparations, all nitrogenases are irreversibly damaged by O_2. This gas might diffuse into heterocysts in very small amounts due to their thick cell wall layers. This O_2 may be moped up completely by respiration. Additionally, nitrogenase-dependent H_2-evolution may serve to protect nitrogenase from damage. Years ago, R.O.D. Dixon postulated three possible functions for the uptake by hydrogenase of the H_2 produced by nitrogenase in *Rhizobium* bacteriods (Dixon 1972): (a) it may supply additional reductant and ATP for nitrogenase; (b) it may remove deleterious O_2 by the respiratory Knallgas reaction; and (c) it may prevent the built up of deleteriously high concentrations of H_2 which block nitrogenase activity. These functions in bacteroids could also well apply to heterocysts.

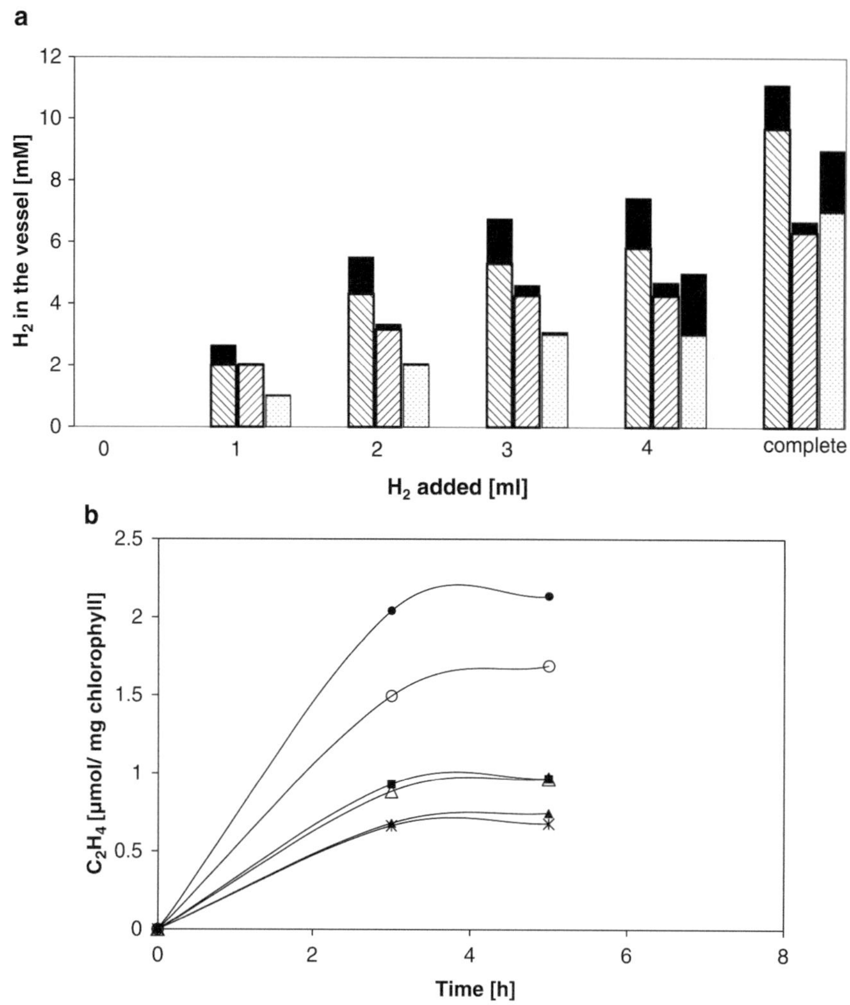

Fig. 6.2. (**a**) H_2-production by *Anabaena azotica* (V- or Mo-grown) and *A. variabilis*. The *black* part of the columns indicates the concentration of H_2 formed per (4 h × mg chlorophyll) in the 7.0-ml vessels (Fernbach flasks, gas phase argon) and the striped or dotted part of the columns the amount of H_2 added to the vessels by syringe and determined by gas chromatography at the start of the experiments. Columns with stripes rising to the *left* are V-grown *A. azotica*; columns with stripes to the *right* are Mo-grown *A. azotica*; and columns with *dots* are Mo-grown *A. variabilis*. Complete indicates a gas phase of H_2 (about 1 bar) (The data are from Bothe et al. 2008 where the exact assay conditions are described). (**b**) Inhibition of C_2H_2-reduction by increasing concentrations of H_2 added to the assays, using Mo-grown *A. azotica*. Top line (●): no H_2 injected into the 7.0-ml vessels with the cells under argon; the following lines (○, ■, △, ▲): 1, 2, 3, or 4 ml H_2 injected; the bottom line (ж): 100 % H_2 gas phase. The inhibition pattern was the same for V-grown *A. azotica* and for Mo-grown *A. variabilis* (The data are from Bothe et al. 2008 where the exact assay conditions are described).

Recently, high levels of H_2-formation were observed when *Anabaena variabilis* or *Anabaena azotica* cells were incubated with varying amounts of H_2 and C_2H_2 (Bothe et al. 2008). This additional H_2-formation upon H_2 addition was accompanied by a parallel decrease in the C_2H_4-formation rate (Fig. 6.2). This H_2-production was nitrogenase- and light-dependent, was blocked by DCMU (an electron transport inhibitor) or FCCP or CCP (energy transfer uncouplers) and was not observed with N_2 as nitrogenase substrate. It proceeded for at least 1 day and then gradually decreased within the next day. No explanation for this H_2 and C_2H_2-dependent H_2-formation is available as yet. However, the experiment shows that (almost) all electrons allocated to nitrogenase can be directed to H_2-formation without any genetic manipulation of the cells.

Other means have been described to allocate the flow of electrons within nitrogenase to H_2-production. Mutational alteration within the α-subunit of the second (vegetative cell-based) Mo-nitrogenase of *Anabaena variabilis* leads to decreased activities of both N_2-fixation and C_2H_2-reduction but has no impact on H_2-formation by this enzyme complex (Weyman et al. 2010). In cultures of the marine *Trichodesmium erythraeum*, nitrogen fixation (C_2H_2-reduction) was saturated with a certain level of light intensity, but H_2-formation and thus the H_2/C_2H_2 ratio steadily increased with either higher irradiance above this threshold value or by altering the light quality (Wilson et al. 2012), which again indicates that N_2-fixation and H_2-evoluted are not strictly coupled in nitrogenase.

Intact filamentous cyanobacteria can produce a burst of H_2 when suddenly exposed to either high light intensities or other stress conditions (Laczko 1986; Abdel-Basset and Bader 1999). These bursts of H_2-formation may come from hydrogenases or nitrogenases or both, but since they occur for only seconds or at most minutes, they are likely impractical for use in combustible energy formation in solar energy programs.

V. Hydrogen Formations by Unicellular Cyanobacteria

The discussion on H_2-production by unicellular cyanobacteria has been on-going for a long time. Many years ago, Mitsui and coworkers (Mitsui and Kumazawa 1977) reported the unusually high rate of 230 μmoles H_2 produced/(h × mg chlorophyll) by marine, non-heterocystous, nitrogen-fixing cyanobacteria (Miami strains, which presumably belong to the genus *Cyanothece*). This work was somewhat forgotten after A. Mitsui died. Most unicellular cyanobacteria separate temporally the incompatible reactions of photosynthetic O_2-production and O_2-sensitive nitrogen fixation. The extensive work of the late John Gallon and coworkers showed that *Gloeothece* (*Gloeocapsa*) performs N_2-fixation during darkness whereas light-dependent photosynthetic O_2-production proceeds during daylight (Gallon et al. 1974; Mullineaux et al. 1983; Gallon 2001). However, *Gloeothece* can also fix N_2 continuously in light, albeit with lower rates. The basis of this activity is not understood. Perhaps the extensive slime sheath surrounding *Gloeothece* may serve to limit O_2 diffusion. A temporal separation of N_2-fixation in darkness and photosynthesis during the day is known also for the marine *Crocosphaera watsonii* (Compaore and Stal 2010). In addition, besides an enhanced respiratory activity, H_2-production by nitrogenase may be utilized by uptake hydrogenase to consume deleterious O_2 in these unicellular organisms, thereby having a positive physiological function.

Special attention is currently paid to the unicellular genus *Cyanothece*. These cyanobacterial isolates are said to perform a strictly regulated temporal separation of photosynthesis during light and N_2-fixation in darkness (Mitsui et al. 1986; Skizim et al. 2012). However, *Cyanothece* also performs N_2-fixation (Toepel et al. 2008) and H_2-formation (Min and Sherman 2010a; Bandyopadhyay et al. 2011) in continuous light, but with lower rates than under light/dark cycles. Remarkably, the expression of the *psbA$_4$* gene encoding one protein of the D1 reaction centre of photosystem II is dramatically up-regulated upon transition to darkness (Toepel et al. 2008). The authors postulate that the insertion of the $PSBA_4$ protein reversibly inactivates the D1 complex and thus photosystem II.

Cyanothece sp. ATTC 51242 has been completely sequenced, a microarray is available (Toepel et al. 2008) and it is amenable to genetic transformation and mutagenesis (Min and Sherman 2010b). *Cyanothece* strains possess both the uptake and the bidirectional hydrogenase (Skizim et al. 2012). *Cyanothece* stores glycogen in granules in the light as an organic carbon reserve, which can be fermentatively degraded with concomitant H_2-release during the dark period. The extensive proteomic analysis of *Cyanothece* (Aryal et al. 2011) indicated that the expression of proteins involved in glycogen degradation and respiration parallel those for nitrogenase and are enhanced in darkness. Noteworthy, NADPH:ferredoxin oxidoreductase shows increased synthesis in the dark cycle whereas the levels of plastocyanin and proteins of the Calvin cycle are more highly expressed in the light period (Aryal et al. 2011). The highest H_2-formation rate by *Cyanothece* is observed when NAD(P)H provides the reductant via NAD(P)H dehydrogenase type 1 and photosystem I in the light when photosystem II is blocked (Skizim et al. 2012).

Recently, the world record for cyanobacterial H_2-production was reported for *Cyanothece* sp. ATTC 51242 at 375 or even 465 μmol/(h × mg chlorophyll) when the cells were supplemented with glycogen (Bandyopadhyay et al. 2010, 2011; Sherman et al. 2010). These high rates of H_2-formation were even observed under aerobic conditions (Bandyopadhyay et al. 2010) although this strain possesses an uptake hydrogenase (Min and Sherman 2010a). This H_2-production was both nitrogenase- and photosystem I-dependent, whereas hydrogenase supported significantly lower gas production rates and it was photosystem II-dependent (Min and Sherman 2010a). Notably, these latter reactions proceeded for more than 2 days in resting cells and all measurements were based on the chlorophyll determination method made by Min and Sherman (2010a). Then H_2-production based on chlorophyll concentration was 133 fold higher than when referred to dry weight as taken from the maximal values given in Table 6.1 by Min and Sherman (2010a) where 80.3 μmol H_2/(h × mg chlorophyll) corresponded to 0.6 μmol H_2/(h × mg dry weight). However, since the classical work of John Biggins on cell-free photosynthetic electron transport in cyanobacteria (Biggins 1967a, b), hundreds of cyanobacteria researchers have referred their data to the extinction coefficient (ε) of 82 $mM^{-1} \times cm^{-1}$. Whether the value of Biggins for chlorophyll A (ε) was absolutely correct or not, the common use of one extinction coefficient allowed investigators to compare the rates published for activities by different laboratories. Roughly, 1 mg chlorophyll corresponds to 25 mg protein and 40 mg dry weight in cyanobacteria (Bothe and Loos 1972). A ratio of 133 between dry weight and chlorophyll content might partially explain the exorbitantly high rates of H_2-production in *Cyanothece* (Min and Sherman 2010a; Bandyopadhyay et al. 2010, 2011). As recently pointed out (Bothe et al. 2010a), cyanobacterial cells can be recalcitrant to complete chlorophyll extraction, and low values for the chlorophyll content of cells may also give the impression of very high activities based on the (h × mg chlorophyll) unit.

Simple considerations indicate the maximal potential H_2-formation rate by nitrogenase in cyanobacteria should be around 40 μmoles or 1 ml H_2 produced/(h × mg chlorophyll) (Table 6.2). To achieve such a rate over longer periods, however, cyanobacteria have to be both fed with combined nitrogen and genetically manipulated, since combined nitrogen suppresses nitrogenase biosynthesis. In addition, the simultaneous H_2-consumption catalyzed by uptake hydrogenase has to be prevented or the trials have to be performed under argon or another O_2-free atmosphere. Moreover, 1 mg chlorophyll in living cells requires a culture volume of at least 10 ml, since cyanobacterial growth density and N_2-fixation activity are known to be dependent on quorum sensing that employs N-acylhomoserine lactone signals (Romero et al. 2011).

Fermentative H_2 formation in the dark, as catalyzed by hydrogenase, proceeds at similar rates to the light- and nitrogenase-dependent production of the gas (Table 6.2). However, here also, higher rates of fermentative H_2-production have been published,

Table 6.2. Consideration of the maximal H_2-production rate achievable in cyanobacteria.

Cyanobacterial rates	µmoles/h × mg chlorophyll
Maximal photosynthesis	100
C:N ratio in cyanobacterial cells	6:1
Maximal N_2-formation	< 20
Maximal H_2-production by nitrogenase[a]	40
Monosaccharide formation ($6\ CO_2 \rightarrow 1$ glucose)	< 20
Monosaccharide degradation via alcohol fermentation employing hydrogenase[b]	40
Maximal H_2 volume[c] both in the light and darkness	~ 1 ml

[a]Based on the fact that the production of 1 molecule of NH_3 in nitrogen fixation requires 4 e^-, whereas the production of 1 molecule H_2 from 2 H^+ requires 2 e^-
[b]Alcohol fermentation (glucose → $2\ C_2H_5OH + 2\ CO_2 + H_2$)
[c]Mol volume = 22.4 L

e.g., 7.1 mol H_2-produced/mol of glucose (Das and Veziroglu 2008), which necessitates metabolic pathways other than alcoholic fermentation. Exploitation of solar energy may be more efficient when gases other than H_2 are generated from carbohydrates. For example, approximately 80 % of the energy of glucose is retained in methane production (Thauer et al. 1977). Unfortunately, although different cyanobacteria are versatile in producing compounds (Hess 2011), they completely lack the ability to form methane. Co-cultures of cyanobacteria with methanogenic bacteria for energy production may be too complicated. Additionally, in contrast to light- and nitrogenase-dependent H_2-formation, any fermentation results in the formation of greenhouse gases such as CO_2 or CH_4.

The above calculation was based on the rate of photosynthetic CO_2-fixation being 100 µmol/(h × mg chlorophyll), and it may be argued that some cyanobacterial strains could have higher rates. However, a twofold higher activity seems to us to be outside of the range of possibility. The above described nitrogenase-dependent activity of 70 µmoles/(h × mg chlorophyll) for *Synechococcus* (*Cyanothece*) Miami BG043511 (Borodin et al. 2002), [as calculated from the data by Skizim et al. (2012)] appears to be at the utmost acceptable limit, and other much higher activities reported for *Cyanothece* apparently lack credibility.

Other unicellular cyanobacteria or species with non-heterocysts and short filaments deserve attention. Novel cyanobacterial lineages that perform N_2-fixation are still being discovered nowadays, e. g., the short filamentous forms found either in easily accessible coastal microbial mats, such as in an estuary of the Monterey Bay (Woebken et al. 2012), or in the oceans (Goebel et al. 2010). Some poorly as yet characterized unicellular cyanobacteria perform N_2-fixation and photosynthesis also in a light–dark-dependent manner (Pfreundt et al. 2012). Similarly, by employing the same mode of light–dark changes in activities, the marine *Crocosphaera watsonii* first uses Fe for photosynthesis in light but then mobilizes it for the synthesis of nitrogenase in the subsequent dark period. In addition, utilization of flavodoxin instead of ferredoxin in darkness saves iron which allows *C. watsonii* to inhabit oceanic regions with lower iron content (Saito et al. 2011). Marine cyanobacteria of the UCYN-A group have the smallest genome size among cyanobacteria. They lack major metabolic pathways, such as the tricarboxylic acid cycle and photosystem II, and are thus physiologically exciting but they still perform N_2-fixation (Bothe et al. 2010b; Zehr 2011). The unicellular *Chroococcidiopis* sp. is extremely resistant to desiccation and possibly closely related to the oldest cyanobacterium on earth and could well be an ancestor of heterocystous species (Fewer et al. 2002). Nowadays, it survives in extreme habitats as within the shards of gypsum rocks where it is exposed only to dim light and where it forms almost monocultures among the cyanobacteria (Boison et al. 2004). H_2-formation in this habitat provides an almost anaerobic environment which allows N_2-fixation to proceed. *Chroococcidiopis* has been discussed as an organism suitable for testing the survival of life on Mars in future expeditions (Billi et al. 2011; Canganella and Wiegel 2011).

Close consideration of H_2-production by many candidate cyanobacteria (Table 6.2) indicates that maximal rates are already achieved by current systems. Genetic manipulation of the acceptor side of photosystem I (Ihara et al. 2006a, b) is unlikely to enhance photosynthetic H_2-formation capacities. Heterologous expression of the Fe-hydrogenase from *Clostridium acetobutylicum* in the non-N_2-fixing, unicellular *Synechococcus elongatus* sp. 7942, which directly couples with ferredoxin that is reduced photosynthetically, will support H_2-formation with rates more than 500-times higher than that of the endogenous bidirectional hydrogenase of this cyanobacterium (Ducat et al. 2011). However, it is not clear whether this rate matches that of nitrogenase-dependent H_2-production. Further, Fe-hydrogenases are generally extremely O_2-sensitive and will need to be genetically altered in order to function in air.

The rate-limiting step of photosynthetic linear electron transport occurs within photosystem II. This presents a realistic chance to increase linear photosynthetic electron transfer by either altering or circumventing photosystem II. This could be achieved in semi-artificial systems where only photosystem I is employed as the photo-reactive element which is coupled to an O_2-insensitive H^+-reducing enzyme (hydrogenase) for H2-production (Winkler et al. 2011). Higher rates of photosynthetic electron transport can also be obtained in cyanobacterial mutants that are impaired in antenna (phycobilisome) size. In contrast, a decrease in cyclic electron-transport rates in such mutants has only minor impact on linear photosynthetic electron-transport flow (Bernat et al. 2009). Energy conversion of light and thus the rate of photosynthetic electron transport may be higher in cyanobacteria, such as the marine *Acarychloris* sp., which use chlorophyll d as a photosynthetic pigment (Pfreundt et al. 2012). Improvements of the rates of photosynthetic electron transport and H_2-formation may also be achieved by simple means, such as adjusting the cyanobacteria growth-medium composition, particularly its Ni content (Burrows et al. 2008; Carrieri et al. 2008; Marques et al. 2011), but also that of Mo or V (Attridge and Rowell 1997). The addition of monosaccharides, such as fructose (Reddy et al. 1996) or glucose (Yeager et al. 2011), to the medium may also enhance H_2-formation in some cyanobacteria. In *Anabaena siamensis*, electron flow can more be directed to H_2-production catalyzed by nitrogenase and bidirectional hydrogenase when the photosynthetic electron transport is blocked by inhibitors such as KCN, rotenone and DCMU and when the cells are supplemented with glyceraldehydes (Khetkorn et al. 2012).

The effects of sulfide on cyanobacterial electron flow have been known for a long time. *Oscillatoria limnetica* and *Aphanothece halophytica* perform photosystem I-dependent CO_2 reduction using either S^{2-} or H_2 as an electron donor (Belkin and Padan 1978). *Synechococcus* sp., strain Miami BG 043511 (*Cyanothece*) was shown to utilize sulfide for H_2-production, but only under stress conditions where light intensity and nutrient supply were growth limiting (Luo and Mitsui 1996). Sulfur deficiency in the medium causes a (partial) inactivation of photosystem II, then anaerobiosis, followed by an enhanced H_2-production. This observation was first taken with the green alga *Chlamydomonas reinhardtii* (Melis and Happe 2001) and subsequently extended to cyanobacteria (Antal and Lindblad 2005; Zhang et al. 2008).

Mutants with defects in uptake hydrogenase have been constructed several times (Mikheeva et al. 1995; Masukawa et al. 2002, 2012; Yoshino et al. 2007). These mutants uniformly show that net H_2-production by nitrogenase in intact cyanobacteria is significantly enhanced when the concomitant H_2-utilization by uptake hydogenase is prevented. Knock-out mutants of the bidirectional hydrogenase have no or rather little impact on total H_2-formation in whole cyanobacterial cells.

Cyanobacterial H_2-evolution can be significantly enhanced when the cells are immobilized on agar beads, entrapped in gels or in other matrices (Hall et al. 1995; Madamwar et al. 2000; Rashid et al. 2009).

Another possibility is to increase the number of heterocysts and thereby the concentration of nitrogenase within the filaments. This can be achieved either by supplying the filaments with 7-azatryptophan (Bothe and Eisbrenner 1977) or by site-directed mutagenesis (Buikema and Haselkorn 2001; Meeks and Elhai 2002; Liang et al. 2004), both of which increase N_2-fixation and thereby H_2-evolution activities.

VI. Conclusion

Despite intensive research over the last 30 years or so, a major break-through in the field has not been achieved yet. However, cyanobacteria still present prospects for exploitation for solar energy conversion programs and for the generation of clean energy. At the moment, it is difficult to predict when or how the required breakthroughs will appear and so when or how this research will ever result in a mature technology.

References

Abdel-Basset R, Bader KP (1999) Effects of stress conditions and calcium on the light-induced hydrogen gas exchange in *Oscillatoria chalybea*. J Plant Physiol 155:86–92

Antal TK, Lindblad P (2005) Production of H_2 by sulphur-deprived cells of the unicellular cyanobacteria *Gloeocapsa alpicola* and *Synechocystis* sp. PCC 6803 during dark incubation with methane or at various extracellular pH. J Appl Microbiol 98:114–120

Aryal UK, Stöckel J, Krovvidi RK, Gritsenko MA, Monroe ME, Moore RJ, Koppenaal DW, Smith RD, Pakrasi HB, Jacobs JM (2011) Dynamic proteomic profiling of a unicellular cyanobacterium *Cyanothece* ATCC51142 across light–dark diurnal cycles. BMC Syst Biol 5:194

Attridge EM, Rowell P (1997) Growth, heterocyst differentiation and nitrogenase activity in the cyanobacteria *Anabaena variabilis* and *Anabaena cylindrica* in response to molybdenum and vanadium. New Phytol 135:517–526

Bandyopadhyay A, Stöckel J, Min HT, Sherman LA, Pakrasi HB (2010) High rates of photobiological H_2 production by a cyanobacterium under aerobic conditions. Nat Commun 1:139

Bandyopadhyay A, Elvitigala T, Welsh E, Stöckel J, Liberton M, Min H, Sherman LA, Pakrasi HB (2011) Novel attributes of the genus *Cyanothece*, comprising a group of unicellular nitrogen-fixing cyanobacteria. mBio J 2:article e00214

Barney BM, Laryukhin M, Igarashi RY, Lee HI, Dos Santos PC, Yang TC, Hoffman BM, Dean DR, Seefeldt LC (2005) Trapping a hydrazine reduction intermediate on the nitrogenase active site. Biochemistry 44:8030–8037

Belkin S, Padan E (1978) Hydrogen metabolism in the facultative anoxygenic cyanobacteria (blue-green algae) *Oscillatoria limnetica* and *Aphanothece halophytica*. Arch Microbiol 116:109–111

Betancourt DA, Loveless TM, Brown J, Bishop PE (2008) Characterization of diazotrophs containing Mo-independent nitrogenases, isolated from diverse natural environments. Appl Environ Microbiol 74:3471–3480

Bernát G, Waschewski N, Rögner M (2009) Towards efficient hydrogen production: the impact of antenna size and external factors on electron transport dynamics in Synechocystis PCC 6803. Photosynth Res 99:205–216

Biggins J (1967a) Photosynthetic reactions by lysed protoplasts and particle preparations from the blue-green alga, *Phormidium luridum*. Plant Physiol 42:1447–1456

Biggins J (1967b) Preparation of metabolically active protoplasts from the blue-green alga, *Phormidium luridum*. Plant Physiol 42:1442–1446

Billi D, Viaggiu E, Cockel CS, Rabbow E, Horneck G, Onofri S (2011) Damage escape and repair in dried *Chroococcidiopsis* spp. from hot and cold deserts exposed to simulated space and Martian conditions. Astrobiology 11:65–73

Boison G, Mergel A, Jolkver H, Bothe H (2004) Bacterial life and dinitrogen fixation at a gypsum rock. Appl Environ Microbiol 70:7070–7077

Boison G, Steingen C, Stal LJ, Bothe H (2006) The rice field cyanobacteria *Anabaena azotica* and *Anabaena* sp.CH1 express vanadium-dependent nitrogenase. Arch Microbiol 186:367–376

Borodin VB, Rao KK, Hall DO (2002) Manifestation of behavioral and physiological functions of *Synechococcus* sp. Miami BG 043511. Mar Biol 140:455–463

Bothe H, Eisbrenner G (1977) Effect of 7-azatryptophan on nitrogen fixation and heterocyst formation in the blue-green alga *Anabaena cylindrica*. Biochem Physiol Pflanz 133:323–332

Bothe H, Loos E (1972) Effect of far red light and inhibitors on nitrogen fixation and photosynthesis in the blue-green alga *Anabaena cylindrica*. Arch Mikrobiol 86:241–254

Bothe H, Winkelmann S, Boison G (2008) Maximizing hydrogen production by cyanobacteria. Z Naturforsch 63c:226–232

Bothe H, Schmitz O, Yates MG, Newton WE (2010a) Nitrogen fixation and hydrogen metabolism in cyanobacteria. Microbiol Mol Biol Rev 74:529–551

Bothe H, Tripp HJ, Zehr JP (2010b) Unicellular cyanobacteria with a new mode of life: the lack of photosynthetic oxygen evolution allows nitrogen fixation to proceed. Arch Microbiol 192:783–790

Buikema WJ, Haselkorn R (2001) Expression of the *Anabaena hetR* gene from a copper-regulated promoter leads to heterocyst differentiation under repressing conditions. Proc Natl Acad Sci U S A 98:2729–2734

Burns RC, Hardy RWF (1975) Nitrogen fixation in bacteria and higher plants. Springer-Verlag, Berlin, Heidelberg, New York, p 189

Burrows EH, Chaplen FWR, Ely RL (2008) Optimization of media nutrient composition for increased photofermentative hydrogen production by *Synechocystis* sp PCC 6803. Int J Hydrogen Energ 33:6092–6099

Camsund D, Devine E, Holmqvist M, Johanoun P, Lindblad P, Stensjö K (2011) A HupS-GFP fusion demonstrates a heterocyst-specific location of the uptake hydrogenase in *Nostoc punctiforme*. FEMS Microbiol Lett 316:152–159

Canganella F, Wiegel J (2011) Extremophiles: from abyssal to terrestrial ecosystems and possibly beyond. Naturwissenschaften 28:253–279

Carrieri D, Ananyev G, Costas AMG, Bryant DA, Dismukes CG (2008) Renewable hydrogen production by cyanobacteria: nickel requirements for optimal hydrogenase activity. Int J Hydrog Energy 33:2014–2022

Chien Y-T, Auerbruch V, Brabban AD, Zinder SH (2000) Analysis of genes encoding an alternative nitrogenase in the archaeon *Methanosarcina barkeri* 227. J Bacteriol 182:3247–3253

Compaore J, Stal LJ (2010) Oxygen and the light–dark cycle of nitrogenase activity in two unicellular cyanobacteria. Environ Microbiol 12:54–62

Das D, Veziroglu TN (2008) Advances in biological hydrogen production processes. Int J Hydrog Energy 33:6046–6057

Dixon ROD (1972) Hydrogenase in legume root nodule bacteroids, occurrence and properties. Arch Microbiol 85:193–201

Ducat DC, Sachdeva G, Silver PA (2011) Rewiring hydrogenase-dependent redox circuits in cyanobacteria. Proc Natl Acad Sci U S A 108:3941–3946

Dutta D, De D, Chaudhuri S, Bhattacharya SK (2005) Hydrogen production by cyanobacteria. Microb Cell Fact 4:36. doi:10.1186/1475-2859-1184-1136

Eisbrenner G, Bothe H (1979) Modes of electron transfer from molecular hydrogen in *Anabaena cylindrica*. Arch Microbiol 123:37–45

Eisbrenner G, Evans HJ (1982) Spectral evidence for a component involved in hydrogen metabolism of soybean nodule bacteroids. Plant Physiol 70:1667–1672

Eroglu E, Melis A (2011) Photobiological hydrogen production: recent advances and state of the art. Bioresour Technol 102:8403–8413

Fay P (1992) Oxygen relations of nitrogen fixation in cyanobacteria. Microbiol Rev 56:340–373

Fewer D, Friedl T, Büdel B (2002) *Chroococcidiopsis* and heterocyst-differentiating cyanobacteria are each other's closest living relatives. Mol Phylogenet Evol 23:82–90

Fisher K, Lowe DJ, Tavares P, Pereira AS, Huynh BH, Edmondson D, Newton WE (2007) Conformations generated during turnover of the *Azotobacter vinelandii* nitrogenase MoFe protein and their relationship to physiological functions. J Inorg Biochem 101:1649–1656

Gallon JR (2001) N_2 fixation in phototrophs: adaptation to a specialized way of life. Plant Soil 239:39–48

Gallon JR, Larue T, Kurz W (1974) Photosynthesis and nitrogenase activity in the blue-green-alga *Gloeocapsa*. Can J Microbiol 20:1633–1637

Germer F, Zelger I, Saggu M, Lendzian F, Schulz R, Appel J (2009) Overexpression, isolation and spectroscopic characterization of the bidirectional [NiFe] hydrogenase from *Synechocystis* sp PCC 6803. J Biol Chem 284:36462–36472

Ghirardi ML, Posewitz MC, Maness PC, Dubini A, Yu J, Seibert M (2007) Hydrogenases and hydrogen photoproduction in oxygenic photosynthetic organisms. Annu Rev Plant Biol 58:71–91

Ghirardi ML, Dubini A, Yu JP, Maness PC (2009) Photobiological hydrogen-producing systems. Chem Soc Rev 38:52–61

Giddings JW, Staehelin LA (1978) Plasma membrane architecture of *Anabaena cylindrica*: occurrence of microplasmodesmata and changes associated with heterocyst development and the cell cycle. Eur J Cell Biol 16:235–249

Giddings JW, Staehelin LA (1981) Observation of microplasmodesmata in both heterocyst forming and non-heterocyst forming filamentous cyanobacteria by freeze-fracture electron -microscopy. Arch Microbiol 129:295–298

Goebel NL, Turk KA, Achilles KM, Paerl R, Hewson I, Morrison AE, Montoya JP, Edwards CA, Zehr JP (2010) Abundance and distribution of major groups of diazotrophic cyanobacteria and their potential contribution to N_2-fixation in the tropical Atlantic Ocean. Environ Microbiol 12:3272–3289

Hall DO, Markov SA, Watanabe Y, Rao KK (1995) The potential applications of cyanobacterial photosythesis for clean technologies. Photosynth Res 46:159–167

Hallenbeck P (2012) Hydrogen production by cyanobacteria. In: Hallenbeck P (ed) Microbial technologies in advanced biofuel production. Springer, Heidelberg/Berlin/New York, pp 15–28

Happe T, Schütz K, Böhme H (2000) Transcriptional and mutational analysis of the uptake hydrogenase of the filamentous cyanobacterium *Anabaena variabilis*. J Bacteriol 182:1624–1631

Hess WR (2011) Cyanobacterial genomics for ecology and biotechnology. Curr Opin Microbiol 14:608–614

Hu YL, Lee CC, Ribbe HW (2012) Vanadium nitrogenase: a two-hit wonder? Dalton Trans 41:1118–1127

Ihara M, Nishihara H, Yoon KS, Lenz O, Friedrich B, Nakamoto H, Kojima K, Honma D, Kamachi T, Okura I (2006a) Light-driven production by a hybrid complex of a [NiFe]-Hydrogenase and the cyanobacterial photosystem I. Photochem Photobiol 82:676–682

Ihara M, Nakamoto H, Kamachi T, Okura I, Maeda M (2006b) Photoinduced hydrogen production by direct electron transfer from photosystem I cross-linked with cytochrome c_3 to [NiFe]-hydrogenase. Photochem Photobiol 82:1677–1685

Joerger RD, Bishop PE (1988) Bacterial alternative nitrogen fixation systems. CRC Crit Rev Microbiol 14:1–14

Kelly M, Postgate JR, Richards RL (1967) Reduction of cyanide and isocyanide by nitrogenase of *Azotobacter chroocccum*. Biochem J 102:1–3

Kentemich T, Danneberg G, Hundeshagen B, Bothe H (1988) Evidence for the occurrence of the alternative, vanadium-containing nitrogenase in the cyanobacterium *Anabaena variabilis*. FEMS Microbiol Lett 51:19–24

Kentemich T, Haverkamp G, Bothe H (1991) The expression of a third nitrogenase in the cyanobacterium *Anabaena variabilis*. Z Naturforsch 46c:217–222

Khetkorn W, Baebprasert W, Lindblad P, Incharoensakdi A (2012) Redirection the electron flow towards the nitrogenase and bidirectional Hox-hydrogenase by using specific inhibitors results in enhanced H_2-production in the cyanobacterium *Anabaena siamensis* TISTR 8012. Bioresour Technol 118: 265–271

Kim DH, Kim MS (2011) Hydrogenases for biological hydrogen production. Bioresour Technol 102: 6423–8431

Laczko I (1986) Appearance of a reversible hydrogenase activity in *Anabaena cylindrica* grown in high light. Physiol Plant 67:634–637

Lancaster KM et al (2011) X-ray emission spectroscopy evidences a central carbon in the nitrogenase iron-molybdenum cofactor. Science 334:974–977

Larsson J, Nylander JAA, Bergman B (2011) Genome fluctuations in cyanobacteria reflect evolutionary, developmental and adaptive traits. BMC Evol Biol 11:187

Latysheva N, Junker VV, Palmer WJ, Codd GA, Barker D (2012) The evolution of nitrogen fixation in cyanobacteria. Bioinformatics 28:603–608

Liang C-M, Ekman M, Bergman B (2004) Expression of cyanobacterial genes involved in heterocyst differentiation and dinitrogen fixation along a plant symbiosis development profile. Mol Plant Microbe Interact 17:436–443

Loveless TM, Saah JR, Bishop PE (1999) Isolation of nitrogen-fixing bacteria containing molybdenum – independent nitrogenases from natural environments. Appl Environ Microbiol 65:4223–4225

Lukanov D, Yang ZY, Barney BM, Dean DR, Seefeldt LC, Hoffman BM (2012) Unification of reaction pathway and kinetic scheme for N_2 reduction catalyzed by nitrogenase. Proc Natl Acad Sci U S A 109:5583–5587

Luo YH, Mitsui A (1996) Sulfide as electron source for H_2-photoproduction in the cyanobacterium *Synechococcus* sp, strain Miami BG 043511, under stress conditions. J Photochem Photobiol B 35: 203–207

Madamwar D, Garg N, Shah V (2000) Cyanobacterial hydrogen production. World J Microbiol Biotechnol 16:757–767

Mariscal V, Herrero A, Flores E (2007) Continuous periplasm in a filamentous, heterocyst-forming cyanobacterium. Mol Microbiol 65:1139–1145

Marques AE, Barbosa AT, Jotta J, Coelho MC, Tamagnini P, Gouveia L (2011) Biohydrogen production by *Anabaena* sp PPCC 7120 wild-type and mutants under different conditions: light, nickel, propane, carbon dioxide and nitrogen. Biomass Bioenerg 35:4426–4434

Masukawa H, Mochimaru M, Sakurai H (2002) Disruption of the uptake hydrogenase gene, but not of the bidirectional hydrogenase gene, leads to enhanced photobiological hydrogen production by the nitrogen-fixing cyanobacterium *Anabaena* sp. PCC 7120. Appl Microbiol Biotechnol 58:618–624

Masukawa H, Zhang X, Yamazaki E, Iwata S, Nakamura K, Mochimaru M, Inoue K, Sakurai H (2009) Survey of the distribution of different types of nitrogenases and hydrogenases in heterocyst-forming cyanobacteria. Mar Biotechnol 11:397–409

Masukawa H, Kitashima M, Inoue K, Sakurai H, Hausinger RP (2012) Genetic engineering of cyanobacteria to enhance biohydrogen production from sunlight and water. Ambio 41(Suppl 2):169–173

McIntosh CL, Germer F, Schulz R, Appel J, Jones AK (2011) The [NiFe]-hydrogenase of the cyanobacterium *Synechocystis* sp PCC 6803 works bidirectionally with a bias to H_2-production. J Am Chem Soc 133:11308–11319

Meeks JC, Elhai J (2002) Regulation of cellular differentiation in filamentous cyanobacteria in free-living and plant-associated symbiotic growth states. Microbiol Mol Biol Rev 66:94–121

Meeks JC, Elhai J, Thiel T, Potts M, Larimer F, Lamerdin J, Predki P, Atlas R (2001) An overview of the genome of *Nostoc punctiforme*, a multicellular, symbiotic cyanobacterium. Photosynth Res 70:85–106

Melis A, Happe T (2001) Hydrogen production. Green algae as a source of energy. Plant Physiol 127:740–748

Mikheeva LE, Schmitz O, Shestakov SV, Bothe H (1995) Mutants of the cyanobacterium *Anabaena variabilis* altered in hydrogenase activities. Z Naturforsch 50c:505–510

Min HT, Sherman LA (2010a) Hydrogen production by the unicellular, diazotrophic cyanobacterium *Cyanothece* sp. strain ATCC 51142 under conditions of continuous light. Appl Environ Microbiol 76:4293–4301

Min HT, Sherman L (2010b) Genetic transformation and mutagenesis via single-stranded DNA in the unicellular, diazotrophic cyanobacteria of the genus *Cyanothece*. Appl Environ Microbiol 76: 7641–7645

Mitsui A, Kumazawa S (1977) Hydrogen production by marine photosynthetic organisms as potential energy resource. In: Mitsui A, Miyachi S, San Pietro A, Tamura S (eds) Biological solar energy conversion. Academic, London/New York/San Francisco, pp 275–289

Mitsui A, Kumazawa S, Takahashi J, Ikernoto H, Cao S, Arai T (1986) Strategy by which nitrogen-fixing unicellular cyanobacteria grow photoautotrophically. Nature 323:720–722

Mullineaux PM, Chaplin AE, Gallon JR (1983) Synthesis of nitrogenase in the cyanobacterium *Gloeothece* (*Gloeocapsa*) sp CCAP1430/3. J Gen Microbiol 129:1689–1696

Mullineaux CW, Mariscal V, Nenninger A, Khanum H, Herrero A, Flores E, Adams DG (2008) Mechanism of intercellular molecular exchange in heterocyst forming cyanobacteria. EMBO J 27:1299–1308

Ni CV, Yakuninin AF, Gogotov IN (1990) Influence of molybdenum, vanadium, and tungsten on growth and nitrogenase synthesis of the free-living cyanobacterium *Anabaena azollae*. Microbiology 59:395–398

Pfreundt U, Stal LJ, Voss B, Hess WR (2012) Nitrogen fixation in a unicellular chlorophyll d-containing cyanobacterium. ISME J 6:1367–1377

Postgate J (1972) Biological nitrogen fixation, A Merrow monograph. Watford, Herts

Rashid N, Song W, Park J, Jin HF, Lee K (2009) Characteristics of hydrogen production by immobilized cyanobacterium *Microcystis aeruginosa* through cycles of photosynthesis and anaerobic incubation. J Ind Eng Chem 15:498–503

Reddy PM, Spiller H, Albracht SL, Shanmugam KT (1996) Photodissimilation of fructose to H_2 and CO_2 by a dinitrogen-fixing cyanobacterium, *Anabaena variabilis*. Appl Environ Microbiol 62:1220–1226

Romero M, Muro-Pastor AM, Otero A (2011) Quorum sensing N-acylhomoserine lactone signals affect nitrogen fixation in the cyanobacterium *Anabaena* sp PCC7120. FEMS Microbiol Lett 315:101–108

Ruvkun GB, Ausubel FM (1980) Interspecies homology of nitrogenase genes. Proc Natl Acad Sci U S A 77:191–195

Saito MA, Bertrand EM, Dutkiewicz S, Bulygin VV, Moran DM, Monteiro FM, Follows MJ, Valois FW, Waterbury JB (2011) Iron conservation by reduction of metalloenzyme inventories in the marine diazotrophic *Crocosphaera watsonii*. Proc Natl Acad Sci U S A 108:2184–2189

Sanchez-Barcaldo P, Hayes PK, Blank CE (2005) Morphological and habitat evolution in the cyanobacteria using a compartmentalization approach. Geobiology 3:145–165

Schmitz O, Bothe H (1996) $NAD(P)^+$-dependent hydrogenase activity in extracts from the cyanobacterium *Anacystis nidulans*. FEMS Microbiol Lett 135:97–101

Schmitz O, Boison G, Hilscher R, Hundeshagen B, Zimmer W, Lottspeich F, Bothe H (1995) Molecular biological analysis of a directional hydrogenase from cyanobacteria. Eur J Biochem 233:266–276

Schrautemeier B, Neveling U, Schmitz S (1995) Distinct and differentially regulated Mo-dependent nitrogen-fixing systems evolved for heterocysts and vegetative cells of *Anabaena variabilis* ATCC 29413: characterization of the *fdX1/2* gene regions as part of the *nif1/2* gene clusters. Mol Microbiol 18:357–359

Schwarz C, Poss Z, Hoffmann D, Appel J (2010) Hydrogenases and hydrogen metabolism in photosynthetic prokaryotes. Adv Exp Med Biol 675:305–348

Sherman LA, Min H, Toepel J, Pakrasi HB (2010) Better living through *Cyanothece* – unicellular diazotrophic cyanobacteria with highly versatile metabolic systems. Adv Exp Med Biol 675:275–290

Shi T, Falkowski PG (2008) Genome evolution in cyanobacteria. The stable core and the variable shell. Proc Natl Acad Sci U S A 105:2510–2515

Skizim NJ, Anayev GM, Krishnan A, Dismukes GC (2012) Metabolic pathways for photobiological hydrogen production by nitrogenase- and hydrogenase containing unicellular cyanobacteria *Cyanothece*. J Biol Chem 287:2777–2786

Spatzal T, Aksoyoglu M, Zhang L, Andrade SL, Schleicher E, Weber S, Rees DC, Einsle O et al (2011) Evidence for interstitial carbon in nitrogenase FeMo cofactor. Science 334:940

Srirangan K, Pyne ME, Perry Chou C (2011) Biochemical and genetic engineering strategies to enhance hydrogen production in photosynthetic algae and cyanobacteria. Bioresour Technol 102: 8589–8604

Stewart WDP, Lex M (1970) Nitrogenase activity in the blue-green alga *Plectonema boryanum*. Arch Mikrobiol 73:250–260

Tamagnini P, Axelsson R, Lindberg P, Oxelfelt F, Wünschiers R, Lindblad P (2002) Hydrogenases and hydrogen metabolism of cyanobacteria. Microbiol Mol Biol Rev 66:1–20

Tamagnini P, Leitão E, Oxelfelt F (2005) Uptake hydrogenase in cyanobacteria: novel input from nonheterocystous strains. Biochem Soc Trans 33:67–68

Tamagnini P, Leitão E, Oliveira P, Ferreira D, Pinto F, Harris DJ, Heidorn T, Lindblad P et al (2007) Cyanobacterial hydrogenases: diversity, regulation and application. FEMS Microbiol Rev 31:692–720

Thauer RK, Jungermann K, Decker K (1977) Energy-conversation in chemotrophic anaerobic bacteria. Bacteriol Rev 41:100–180

Thiel T (1990) Protein-turnover and heterocyst differentiation in the cyanobacterium *Anabaena variabilis*. J Phycol 26:50–54

Thiel T (1993) Characterization of genes for an alternative nitrogenase in the cyanobacterium *Anabaena variabilis*. J Bacteriol 175:6276–6286

Thiel T, Lyons EM, Erker J, Ernst A (1995) A second nitrogenase in vegetative cells of a heterocyst-forming cyanobacterium. Proc Natl Acad Sci U S A 92:9358–9362

Thiel T, Lyons EM, Erker J (1997) Characterization of genes for a second Mo-dependent nitrogenase in the cyanobacterium *Anabaena variabilis*. J Bacteriol 179:5222–5225

Thorneley RNF, Lowe DJ (1984) The mechanism of *Klebsiella pneumoniae* nitrogenase action. Pre-steady-state kinetics of an enzyme bound intermediate in N_2 reduction and of NH_3 formation. Biochem J 224:877–886

Thorneley RNF, Eady RR, Lowe DJ (1978) Substrate hydrazine as an adduct bound to the active metal cluster of nitrogenase. Nature 272:557–558

Tiwari A, Pandey A (2012) Cyanobacterial hydrogen production – a step towards clean energy. Int J Hydrog Energy 37:139–150

Toepel J, Welsh E, Sumerfield TC, Prakasi HB, Sherman LA (2008) Differential transcriptional analysis of the cyanobacterium *Cyanothece* sp. strain ATCC51142 during light–dark and continuous-light growth. J Bacteriol 190:3904–3913

Tripp HJ, Bench SR, Turk KA, Foster RA, Desany BA, Niazi F, Affourtit JP, Zehr JP et al (2010) Metabolic streamlining in an open-ocean nitrogen-fixing cyanobacterium. Nature 464:90–94

Tsygankov A (2007) Nitrogen fixing cyanobacteria: a review. Appl Biochem Microbiol 43:250–259

Wall JD (2004) Rain or shine – a phototroph that delivers. Nat Biotechnol 22:40–41

Weyman PD, Pratte B, Thiel T (2010) Hydrogen production in nitrogenase mutants in *Anabaena variabilis*. FEMS Microbiol Lett 304:55–61

Wiig JA, Hu Y, Lee CC, Ribbe M (2012) Radical SAM-dependent carbon insertion into the nitrogenase M-cluster. Science 337:1672–1675

Wilcox M, Mitchison G, Smith RJ (1973) Pattern formation in the blue-green alga *Anabaena*. 2. Controlled proheterocyst repression. J Cell Sci 13:637–649

Wilson ST, Kolber ZS, Tozzi S, Zehr JP, Karl DM (2012) Nitrogen fixation, hydrogen cycling, and electron transport kinetics in *Trichodesmium erythraeum* (cyanobacterium) strain IMS101. J Phycol 48:595–606

Winkler M, Kawelke S, Happe T (2011) Light driven hydrogen production in protein based semi-artificial systems. Bioresour Technol 102:8493–8500

Woebken D, Burow LC, Prufert-Bebout L, Bebout BM, Hoehler TM, Pett-Ridge J, Spormann AM, Weber PK, Singer SW (2012) Identification of a novel cyanobacterial group of active diazotrophs in a coastal microbial mat using nanoSIMS analysis. ISME J 6:1427–1439

Yeager CM, Milliken CE, Bagwell CE, Staples L, Berseth PA, Sessions HT (2011) Evaluation of experimental conditions that influence hydrogen production among heterocystous cyanobacteria. Int J Hydrog Energy 36:7487–7499

Yoshino F, Ikeda H, Masukawa H, Sakurai H (2007) High photobiological hydrogen production activity of a *Nostoc* sp PCC 7422 uptake hydrogenase-deficient mutant with high nitrogenase activity. Mar Biotechnol 9:101–112

Zehr JP (2011) Nitrogen fixation by marine cyanobacteria. Trends Microbiol 19:162–173

Zhang Z, Pendse ND, Phillips KN, Cotner JB, Khodursky A (2008) Gene expression patterns of sulfur starvation in *Synechocystis* sp. PCC 6803. BMC Genomics 9:344–354

Chapter 7

Systems Biology of Photobiological Hydrogen Production by Purple Non-sulfur Bacteria

James B. McKinlay*
Department of Biology, Indiana University, 1001 E 3rd Street, Bloomington, IN 47405, USA

Summary

Photosynthetic purple non-sulfur bacteria (PNSB) can naturally convert electrons from organic compounds, protons from water, and energy from light into H_2 gas, via the enzyme nitrogenase. In 2004, the first PNSB genome sequence was reported, that of *Rhodopseudomonas palustris* strain CGA009. The CGA009 genome sequence revealed natural attributes that favored H_2 accumulation and revealed further potential for enhancing H_2 production. Since

*Author for correspondence, e-mail: jmckinla@indiana.edu

D. Zannoni and R. De Philippis (eds.), Microbial BioEnergy: Hydrogen Production, Advances in Photosynthesis and Respiration 38, pp. 155–176, DOI 10.1007/978-94-017-8554-9_7,

then, the genomes of several more *Rp. palustris* species and other PNSB have been sequenced. Comparing these genomes has led to new ideas for improving the substrate range, rate, and photosynthetic efficiency of H_2 production. Furthermore, systems biology or 'omics' approaches, including transcriptomics, proteomics, and fluxomics have been applied. Many of these systems level approaches have focused on the regulation and activity of nitrogenase – the enzyme responsible for H_2 production. Guided by these approaches, metabolic engineering has targeted metabolic pathways that compete with H_2 production for electrons, leading to strains with higher H_2 yields and potentially linking the survival of these strains to the production of H_2 biofuel. A systems level examination of PNSB is now turning to characterize largely unexplored but potentially crucial aspects involved in H_2 production including non-coding small RNAs and post-translational modifications. Systems biology approaches are also being designed to eliminate experimenter bias and highlight genes of unknown function that contribute to H_2 production, ideally providing clues to their function and their place in bacterial physiology. This chapter describes the contributions of systems biology to our understanding and application of H_2 production by PNSB, focusing on *Rhodopseudomonas palustris* and referencing examples from other PNSB.

I. Introduction

A. Background on Purple Non-sulfur Bacterial Physiology

Purple non-sulfur bacteria (PNSB) are α- and β-proteobacteria that have long fascinated researchers with their metabolic versatility. PNSB employ different metabolic modules to thrive in different environments. In the dark, PNSB grow using aerobic respiration or anaerobic respiration. The anaerobic electron acceptors that PNSB use vary between species and strains, but can include electron acceptors of denitrification (e.g, NO_3^- and N_2O), dimethylsulfoxide, and trimethylammonium oxide (McEwan et al. 1985; Ferguson et al. 1987; McEwan 1994). However, PNSB are most commonly grown and studied under photosynthetic conditions. Photosynthesis by PNSB is anoxygenic. Thus, PNSB are unlike oxygenic cyanobacteria, plants, and algae that obtain electrons from water and produce O_2 as a waste product. PNSB use a single photosystem that resembles photosystem II but it is incapable of oxidizing water and thus no O_2 is produced. Instead, PNSB use organic compounds as both a source of carbon and electrons during photosynthetic growth – a photoheterotrophic lifestyle (Fig. 7.1a). Alternatively, PNSB can use inorganic electron donors other than water (e.g., H_2, thiosulfate, or Fe^{2+}) as an electron source and CO_2 as a carbon source – a photoautotrophic lifestyle (Fig. 7.1b). Electrons pulled from the electron donor are energized by the photosystem using light and channeled through a H^+-pumping electron transfer chain. The resulting proton motive force can be used to make ATP via ATP synthase or to power other energy-requiring processes (e.g., solute uptake via H^+-symport). The electrons can be donated to $NADP^+$ to generate NADPH for biosynthesis through reverse electron transfer (a process that utilizes the proton motive force). Alternatively, the electrons can be repeatedly energized and cycled through the electron transfer chain. This cycling allows for the continuous maintenance of the proton motive force and ATP pools in a process called cyclic photophosphorylation. Cyclic photophosphorylation is particularly advantageous under starvation conditions as cycling a few electrons can generate usable energy for cell maintenance and repair.

Abbreviations: αKG – α-ketoglutarate or 2-oxoglutarate; GOGAT – Glutamine 2-oxoglutarate aminotransferase; PNSB – Purple nonsulfur bacterium/bacteria; Rubisco – Ribulose 1,5 bisphosphate carboxylase

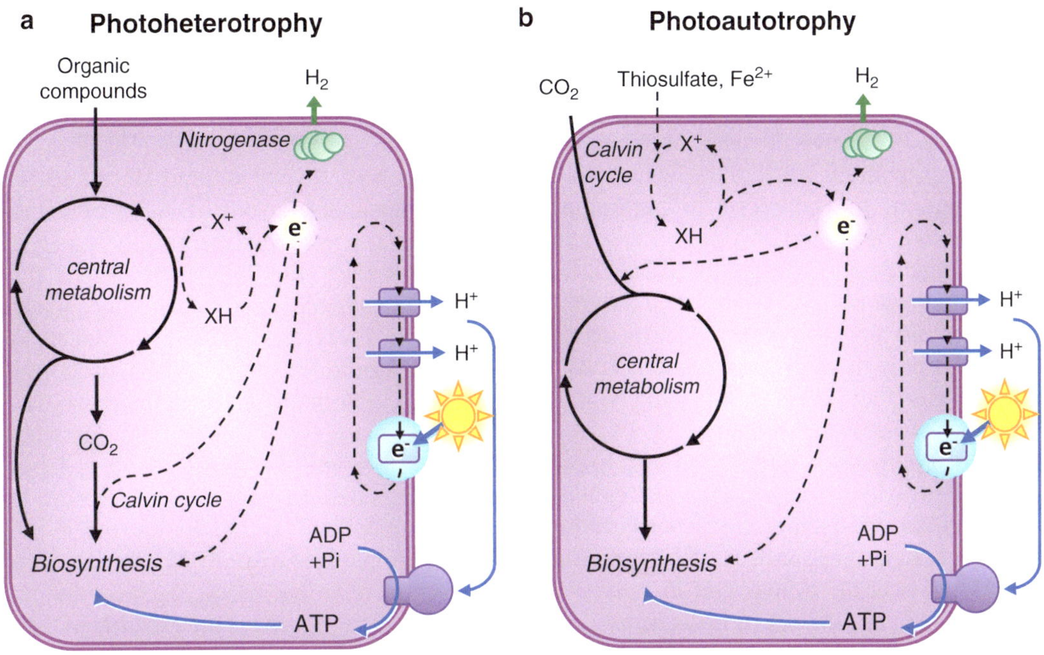

Fig. 7.1. PNSB like *Rp. palustris* can grow using light for energy and either organic or inorganic carbon and electron sources. (**a**) During photoheterotrophic growth, organic compounds serve as carbon and energy sources. Excess reductant can be oxidized through CO_2 fixation via the Calvin cycle or through H_2 production via nitrogenase. This oxidation of excess reductant is an essential process required for growth. (**b**) During photoautotrophic growth, CO_2 serves as the carbon source and is reduced to organic biosynthetic intermediates by the Calvin cycle. Inorganic compounds other than water, such as thiosulfate, serve as the electron source. Under certain conditions, some electrons can be channeled to H_2 production, resulting in simultaneous fixation of CO_2 greenhouse gas and production of H_2 biofuel (Figure modified from McKinlay and Harwood 2010a).

During photoheterotrophic growth, PNSB can use a wide variety of organic acids and alcohols but the ability to utilize sugar is less common trait observed in PNSB, though some PNSB like *Rhodobacter sphaeroides* are routinely grown with sugars (Fuhrer et al. 2005; Kontur et al. 2011). *Rp. palustris* is relatively unique among PNSB for its ability to degrade aromatic compounds (e.g., p-coumarate), released during the degradation of lignin by certain fungi (Harwood 2009). Additionally, *Rp. palustris* can degrade some chlorinated aromatic compounds. As a result, much of the earlier research on *Rp. palustris* was devoted to understanding the biochemistry of its aromatic compound degrading pathways which are potentially useful in bioremediating sites contaminated with chlorinated and aromatic pollutants (Harwood 2009). During photoautotrophic growth, PNSB obtain electrons from inorganic electron donors such as H_2 (most if not all PNSB), thiosulfate (i.e., *Rp. palustris* (van Niel 1944; Rolls and Lindstrom 1967)), and in some cases Fe^{2+} (i.e., *Rp. palustris* TIE-1 (Jiao et al. 2005) and *Rhodobacter* sp. SW2 (Croal et al. 2007)). When using inorganic electron donors, PNSB employ the Calvin cycle to utilize CO_2 as a carbon source. As will be described in detail later on, CO_2 fixation can actually be essential during photosynthetic growth on organic compounds, as it allows the cell to deal with the excess electrons that are invariably generated during growth on organic compounds. PNSB can also obtain nitrogen from atmospheric N_2 using the enzyme, nitrogenase. As described below, it is nitrogenase that is most often exploited by researchers when using PNSB to produce H_2 gas.

B. Hydrogen Gas Production by PNSB

As PNSB primarily consume compounds other than sugars, most research on H_2 production by PNSB has focused on using fermented agricultural waste as a feedstock. In this situation, the cellulose in agricultural material has been broken down into sugars and fermented mainly into organic acids. It is these organic acids (e.g., acetate and butyrate) that serve as the carbon and electron source for PNSB growth and H_2 production. Thus, it is envisioned that PNSB could couple H_2 production to waste-water remediation.

The two classes of enzymes known to produce H_2 are hydrogenase and nitrogenase. Though there are examples of PNSB producing H_2 via hydrogenase (e.g. *Rhodospirillum rubrum* (Fox et al. 1996a) and *Rp. palustris* BisB18 (Oda et al. 2008)), H_2 is more commonly produced via nitrogenase. Nitrogenase is better known for producing NH_4^+ from atmospheric N_2. However, H_2 is an obligate product of the nitrogenase reaction (Eq. 7.1). Even in the presence of 50 ATM of N_2, nitrogenase will continue to produce H_2 at a 1:2 ratio with NH_4^+ (Simpson and Burris 1984). In an atmosphere devoid of N_2, nitrogenase behaves like a hydrogenase, producing H_2 as the sole product (Eq. 7.2). An argument that is sometimes made against using nitrogenase to produce H_2 is that its turnover rate is an order of magnitude lower than Ni-hydrogenase and three orders of magnitude lower than Fe-hydrogenase (McKinlay and Harwood 2010b). However, this comparison is based on in vitro rates. In vivo rates of specific H_2 production show a much narrower gap (McKinlay and Harwood 2010b), perhaps because bacteria that use nitrogenase tend to produce large amounts of the enzyme to compensate for its slow rate.

$$\text{Mo-nitrogenase:}\ N_2 + 8H^+ + 8e^- + 16\,ATP \rightarrow 2\,NH_3 + H_2 + 16\,ADP \quad (7.1)$$

$$\text{Mo-nitrogenase in absence of}\ N_2:\ 8H^+ + 8e^- + 16\,ATP \rightarrow 4H_2 + 16\,ADP \quad (7.2)$$

$$\text{Hydrogenase:}\ 2H^+ + 2e^- \longleftrightarrow H_2 \quad (7.3)$$

Unlike hydrogenase, which simply requires an electron donor (Eq. 7.3), nitrogenase also requires ATP (Eqs. 7.1 and 7.2). This ATP requirement allows nitrogenase to generate very high levels of H_2 without the reaction slowing and eventually running in reverse, as is the case with hydrogenase. The high ATP requirement is not a barrier to H_2 production for PNSB, since they can produce ample ATP from recycled electrons via cyclic photophosphorylation, provided that they are illuminated. However, a significant hurdle in producing H_2 via nitrogenase is the repressive effects of NH_4^+. The repression of nitrogenase in response to NH_4^+ and strategies to bypass this regulation are described in Sect. III.A.

II. Purple Non-sulfur Bacteria in the Light of Genomics and Systems Biology

A. Rhodopseudomonas palustris *CGA009 – The First Purple Genome Sequence*

Rp. palustris CGA009 was the first PNSB genome to be published in 2004 (Larimer et al. 2004). Prior to that time most work on *Rp. palustris* had focused on the biochemistry of anaerobic pathways for the degradation of aromatic compounds and the biophysics of its photosynthetic apparatus. The genome sequence of *Rp. palustris* CGA009 revealed several features that made it naturally suited for H_2 production such as (i) an inactive uptake hydrogenase, (ii) multiple nitrogenase isozymes, (iii) multiple pathways for consumption of aromatic compounds, and (iv) pathways for the utilization of inorganic electron donors. This section will describe these key features and the insights made into *Rp. palustris* physiology and H_2 production from comparative genomics, functional genomics, and targeted biochemical and mutational analyses.

1. An Inactive Uptake Hydrogenase

N_2 fixation by nitrogenase is electron-intensive – each enzymatic cycle requiring six electrons to make two NH_4^+ and another two electrons for the obligate production of H_2. From the perspective of maximizing cell growth, the production of H_2 is 'wasteful' as it would be beneficial to instead use the electrons in H_2 to fix more N_2. Indeed, N_2-fixing prokaryotes tend to encode a Ni-containing uptake hydrogenase to recapture H_2 electrons. Eliminating uptake hydrogenase is often the first step in increasing H_2 yields in PNSB but this was unnecessary in *Rp. palustris* CGA009. The CGA009 genome sequence revealed that *Rp. palustris* had an uptake hydrogenase but curiously CGA009 was incapable of growing photoautotrophically with H_2 as an electron donor (Rey et al. 2006). Upon closer examination, a 4-nucleotide deletion was noticed in *hupV*, which encodes a subunit of a hydrogen sensor protein needed to activate transcription of the hydrogenase gene cluster. Without uptake hydrogenase activity, H_2 produced via nitrogenase could escape the cell and accumulate in the sealed growth container (Rey et al. 2006). When the mutated gene was replaced with a 'repaired' sequence, *Rp. palustris* was able to grow photoautotrophically on H_2 and accumulated less H_2 when grown under N_2-fixing conditions (Rey et al. 2006). This repaired strain, CGA010, is sometimes referred to as the wild-type strain though it is derived from CGA009.

A microarray analysis was used to compare CGA009 with CGA010 grown under N_2-fixing conditions where the uptake hydrogenase is expected to capture H_2 from nitrogenase (Rey et al. 2006). The comparison confirmed that *hupV* is required for the expression of the hydrogenase gene cluster. Curiously, five other genes were differentially expressed between the two strains. Two genes encoding a putative dicarboxylic acid transporter, a predicted formate transporter, and a glutamine synthetase were all upregulated 2–8-fold in CGA010 relative to the *hupV*-defective CGA009, suggesting that HupV is involved in activating transcription of these genes under N_2-fixing conditions. It was speculated that the dicarboxylic acid transporter and glutamine synthetase could allow *Rp. palustris* to better assimilate oxidized organic acids and N_2 gas in the presence of H_2. In support of this hypothesis, CGA010 had slightly higher growth rates than CGA009 (Rey et al. 2006). The fifth gene encoded a hypothetical protein and showed 47-fold lower expression in CGA010 relative to CGA009, suggesting that HupV is involved in strong repression of this gene. As of yet, no phenotype has been associated with the differential expression of these genes and it is worth noting that CGA010 is indistinguishable from CGA009 if Ni^{2+} is not added to the growth medium. In other words, CGA010 cannot consume H_2 if Ni^{2+} levels are insufficient to support synthesis of Ni-containing uptake hydrogenase.

2. An Arsenal of H_2 Producing Nitrogenases

Perhaps the biggest surprise from the CGA009 genome sequence was the presence of genes encoding all three nitrogenase isozymes. Most N_2 fixing prokaryotes encode Mo-nitrogenase, which has a Fe-Mo cluster in the active site. However, there are two 'alternative' nitrogenases, V-nitrogenase and Fe-nitrogenase, named for the metals used in place of Mo in the active site. Many bacteria encode one alternative nitrogenase in addition to Mo-nitrogenase but prior to *Rp. palustris* CGA009, the only organisms known to harbor all three were non-photosynthetic *Azotobacter vinelandii* (Joerger et al. 1989) and *Methanosarcina acetivorans* (Galagan et al. 2002). *Rp. palustris* CGA009 remains the only example of a photosynthetic microbe that encodes all three nitrogenase isozymes.

Alternative nitrogenases are advantageous for H_2 production because they are better producers of H_2 than NH_4^+ (Eqs. 7.4, 7.5, and 7.6). Furthermore, in vivo rates of H_2 production are comparable between strains expressing individual forms of each nitrogenase (Table 7.1). The alternative nitrogenases were shown to be expressed in *Rp. palustris* in response to genetic mutations rendering Mo-nitrogenase non-functional (Oda et al. 2005), similar to what was found for the expression of Fe-nitrogenase in *Rs. rubrum* (Lehman and Roberts 1991). This observation suggests that the alternative nitrogenases are expressed in response to severe nitrogen starvation (e.g., when Mo availability limits Mo-nitrogenase function). Microarray analysis of *Rp. palustris* cells expressing either of the two alternative nitrogenases show upregulation of genes involved in acquiring diverse forms of nitrogen, supporting the hypothesis that the alternative nitrogenases are part of a general response to nitrogen starvation. In other bacteria like *Rhodobacter capsulatus* the presence of Mo represses alternative nitrogenase gene expression (Masepohl et al. 2002) but this does not hold true for *Rp. palustris* (Oda et al. 2005) or *Rs. rubrum* (Lehman and Roberts 1991).

Table 7.1. Strains that use alternative nitrogenases grow more slowly but have higher specific productivities.

Nitrogenase expressed	Growth rate (h^{-1})[a]	H_2 production (μmol (mg protein)$^{-1}$)[a]	Specific H_2 productivity (μmol (mg protein)$^{-1}$ h^{-1})[b]
Mo-only	0.048	30	1.44
V-only	0.036	51	1.84
Fe-only	0.028	140	3.92

[a]Data taken from Oda et al. (2005)
[b]Specific H_2 productivities were calculated according to the Monod model where the specific rate of product formation is equal to the amount of product produced per unit biomass multiplied by the growth rate. This equation assumes a constant ratio between H_2 and biomass during the period in which the growth rate was measured

$$\text{Mo-nitrogenase: } N_2 + 8H^+ + 8e^- + 16\,ATP \rightarrow 2NH_3 + H_2 + 16\,ADP \quad (7.4)$$

$$\text{V-nitrogenase: } N_2 + 12H^+ + 12e^- + 24\,ATP \rightarrow 2NH_3 + 3H_2 + 24\,ADP \quad (7.5)$$

$$\text{Fe-nitrogenase: } N_2 + 24H^+ + 24e^- + 48\,ATP \rightarrow 2NH_3 + 9H_2 + 48\,ADP \quad (7.6)$$

As mentioned earlier, Mo-nitrogenase can produce H_2 as the sole product in the absence of N_2. In such a situation, all nitrogenase isozymes are equally matched in terms of electrons devoted to H_2 production. Thus, alternative nitrogenases are most advantageous under conditions where N_2 is plentiful, thereby allowing the electron balance to be shifted towards H_2 production while still assimilating enough nitrogen for growth. There is still much to be learned in regards to the regulation of the alternative nitrogenases. Through understanding how the enzymes are regulated it may be possible to express all three isozymes at once, increasing the copy number of these relatively slow enzymes and therefore the rate of H_2 production.

3. Finding the Route from Lignin Monomers to H_2

Rp. palustris is distinguished from most other PNSB by its ability to grow phototrophically on aromatic compounds. Many of these aromatic compounds are lignin monomers released during lignin degradation by fungi. The CGA009 genome sequence suggested two distinct routes for the degradation of the lignin monomer *p*-coumarate: a β-oxidation route and a non-β-oxidation route. Microarray and quantitative

^{15}N-proteomic analyses were used to identify which routes were used by comparing transcript and protein levels during growth on succinate versus *p*-coumarate (Pan et al. 2008). The agreement between the transcriptional and proteomic data sets pointed to the non-β-oxidation route for *p*-coumarate degradation and putative genes were identified for every step in the pathway (Pan et al. 2008). This approach greatly narrows the targets for identifying and characterizing lignin monomer-degrading enzymes through genetic and biochemical approaches, more so than basing predictions on a genome sequence alone. For example, the CGA009 genome was predicted to have a single β-oxidation pathway for degrading fatty acids encoded by *pimFABCDE*. Elimination of this gene cluster resulted in slower growth on several straight chain fatty acids and on benzoate – a common intermediate for many aromatic compound degradation pathways that itself is degraded by β-oxidation after ring cleavage (Harrison and Harwood 2005). However, even without this gene cluster, growth on these compounds was not eliminated and growth was unimpaired on some fatty acids like 8-carbon caprylate (Harrison and Harwood 2005). Thus, other pathways must exist to degrade long chain fatty acids and transcriptomic and proteomic approaches could help identify them.

4. Removing Greenhouse Gases While Producing Biofuels Through the Use of Inorganic Feedstocks

The genome sequence also revealed genes for utilizing some inorganic electron donors. A carbon monoxide dehydrogenase was found that could be used to convert CO (e.g., from syngas) into H_2 (Larimer et al. 2004). Thus far, the functionality of the *Rp. palustris* CO dehydrogenase has not been tested but *Rs. rubrum* has a CO dehydrogenase that has been intensively characterized (Bonam et al. 1989; Kerby et al. 1995; Shelver et al. 1995; Spangler et al. 1998; Munk et al. 2011). Although not mentioned in the original annotation, *Rp. palustris* CGA009 also encodes genes for the utilization of ferrous iron (RPA0746-4). Attempts to grow CGA009 on Fe^{2+} were unsuccessful, but closely related *Rp. palustris* TIE-1 grows photoautotrophically on Fe^{2+} using the homologous *pio* operon (Jiao et al. 2005; Jiao and Newman 2007). The CGA009 genome annotation also pointed to a *sox* operon, encoding a thiosulfate oxidizing complex (Larimer et al. 2004). In 1944, Van Neil demonstrated that *Rp. palustris* could use CO_2 as the sole carbon source and inorganic thiosulfate as the electron donor, setting it apart from purple non-sulfur bacteria like *Rb. capsulatus, Rb. sphaeroides*, *Rp. gelatinosa* (van Niel 1944), and *Rs. rubrum* (Rolls and Lindstrom 1967). Recently, it was shown that *Rp. palustris* can grow autotrophically on CO_2 and thiosulfate while fixing N_2/producing H_2 (Huang et al. 2010). Thus, when grown on inorganic electron donors like thiosulfate, PNSB can produce H_2 biofuel while removing CO_2 greenhouse gas – a claim usually reserved for processes using cyanobacteria and algae.

B. Comparative Rp. palustris *Genomics*

As of 2012, the genome sequences for seven *Rp. palustris* strains were available – more than any other PNSB species (genomes sequences were available for five *Rb. sphaeroides* strains, two *Rs. rubrum* strains, and single strains of several other PNSB species). Furthermore there are genome sequences available for 14 *Bradyrhizobium* species, which are more closely related to *Rp. palustris* than most PNSB are. In BLAST alignments performed in 2008, less than 1 % of the genes in a given *Rp. palustris* genome had top hits to *Rb. sphaeroides* or *Rs. rubrum* but about 80 % of the genes had a top hit to a *Bradyrhizobium* or *Nitrobacter* (excluding comparisons with other *Rp. palustris* genomes) (Oda et al. 2008). A specialized visual tool for comparing the first six *Rp. palustris* genomes to be sequenced is publically available (http://public.tableausoftware.com/views/rhodo_palustris/uniquepfam2strain) (Simmons et al. 2011) as well as more general web-based resources for comparative genomics such as Integrated Microbial Genomes (http://img.jgi.doe.gov/). Comparative genomic analysis between *Rp. palustris* CGA009, closely related strain TIE-1, and four strains isolated from the Netherlands: BisA53, BisB5, BisB18, HA2 (with the two BisB strains isolated from the same half-gram of sediment) have been described (Oda et al. 2008; Simmons et al. 2011). About half of the genes were shared by all five strains while 10–18 % of the genes in each genome were strain specific. Surprisingly, the two strains isolated from the same half-gram of sediment shared fewer orthologs than any other comparison.

Overall, the genome of each strain suggested a specialization to a specific environment. Some of these specializations present potentially beneficial features for H_2 production. CGA009 was the only strain found to encode three nitrogenase isozymes (more recently sequenced strains, TIE-1 and DX-1, are missing the V-nitrogenase). CGA009 is thus specialized for H_2 production in diverse environments, with the potential to activate all three enzymes through genetic engineering. The V-nitrogenase in CGA009 was likely acquired by horizontal gene transfer. One comparative genomic study highlighted the presence of an aquaporin in CGA009 and BisB5 that could impart an advantage under freezing conditions, whereas the other strains lacking the aquaporin could have an advantage in an environment with high sugar concentrations (Simmons et al. 2011). Strain BisB18 showed some capacity for fermentative metabolism, encoding pyruvate formate-lyase and formate hydrogen-lyase. *Rs. rubrum* also encodes a fermentative hydrogenase that allows it to produce H_2 from CO in the dark (Fox et al. 1996b). This fermentative capacity raises the possibility of producing hydrogen both via nitrogenase and hydrogenase, which has been shown to result in a twofold higher H_2 yield when the *Rs. rubrum* hydrogenase was expressed in *Rb. sphaeroides* (Kim et al. 2008). BisB18 also has the ability to grow on methanol. Strain BisB5 encoded a larger repertoire of enzymes for degrading aromatic compounds under anaerobic conditions, perhaps making it ideally suited for using lignin monomers as a renewable feedstock for H_2 production. Strain BisA53 was able to absorb light at additional wavelengths not absorbed by the other strains, giving it the potential to have a higher efficiency in converting light energy into chemical H_2 energy by harnessing more of the light spectrum. As mentioned above, TIE-1 can potentially use Fe^{2+} as an electron source for H_2 production while fixing CO_2. CGA009 was unable to use Fe^{2+} as an electron donor despite encoding the necessary *pio* operon. BisB18 and BisA53 also encode the *pio* operon but growth on Fe^{2+} has not been tested. Finally, strain DX-1 is reported to interact with electrodes in a microbial fuel cell allowing for electricity generation (Xing et al. 2008) which can also be converted to H_2 via electrolysis (Cheng and Logan 2007).

The genome sequences of various PNSB genomes have revealed an impressive inventory of metabolic and physiological attributes that allow for the production of H_2 under a wide range of conditions. However, this metabolic versatility also introduces a challenge to identify and harness attributes that would enhance H_2 production while distinguishing them from those attributes that would work against H_2 production.

III. Deciphering and Engineering the Metabolic and Regulatory Mechanisms Involved in H_2 Production

A. Regulation of Nitrogenase in Response to NH_4^+

Although PNSB can photosynthetically generate ample ATP to run nitrogenase, the enzyme is subject to negative feedback by NH_4^+ (and other nitrogen compounds) at multiple levels. Nitrogenase is a complicated enzyme. It requires over 20 accessory genes for its proper assembly and the individual subunits of the active enzyme must associate and disassociate eight times in one catalytic cycle to convert one N_2 into 2 NH_4^+, expending 16 ATP in the process. Thus, any microbe has good reason not to synthesize the enzyme if it can obtain NH_4^+ from the environment.

The inhibition of nitrogenase in response to NH_4^+ in PNSB has been most intensively studied in *Rs. rubrum* (Munk et al. 2011) (the PNSB in which nitrogenase was first discovered by Howard Gest (Gest 1999)) and *Rb. capsulatus* (Masepohl et al. 2002). Nitrogenase regulation has also been examined in other PNSB with seemingly subtle but sometimes important differences. Nitrogenase regulation is closely tied to the intracellular levels of α-ketoglutarate (αKG) and glutamine, which respectively signal nitrogen starvation and abundant NH_4^+. Both of these signal metabolites serve as substrates for the enzyme that sets the stage for most of the aminotransferase reactions in the cell: glutamine 2-oxoglutarate aminotransferase or GOGAT (note: 2-oxoglutarate is another name for αKG). As depicted in Fig. 7.2 the enzyme transfers an amino group from glutamine to αKG, producing two molecules of glutamate. Glutamate then serves as the amino donor for the synthesis of nearly all amino acids. If NH_4^+ is abundant, glutamine synthetase provides ample glutamine to move the GOGAT reaction forward (Fig. 7.2). If NH_4^+ is low, then the GOGAT reaction stalls, waiting for glutamine substrate. As a result, αKG accumulates and triggers a nitrogen starvation response including the synthesis of nitrogenase. The ratio of αKG to glutamine is first sensed by the uridylyltransferase, GlnD. GlnD then transmits the nitrogen status through the uridylylation state of small trimeric signal transduction proteins called PII proteins. PII proteins are uridylylated by GlnD when αKG is abundant and de-uridylylated by GlnD when glutamine is abundant. The PII uridylylation state determines how they will interact with downstream regulatory proteins involved in nitrogen metabolism.

In general for PNSB, the PII proteins are involved in nitrogenase regulation at three levels (Fig. 7.3): (i) transcriptional regulation involving the two-component regulatory system NtrBC, (ii) transcriptional regulation involving the σ^{54}-enhancer-binding protein NifA, and (iii) post-translational covalent modification of nitrogenase by DraT and DraG. When NH_4^+ is scarce and αKG accumulates, the uridylylated PII proteins cannot interact with NtrB. NtrB is then free to phosphorylate NtrC. Phosphorylated NtrC then activates the transcription of a regulon

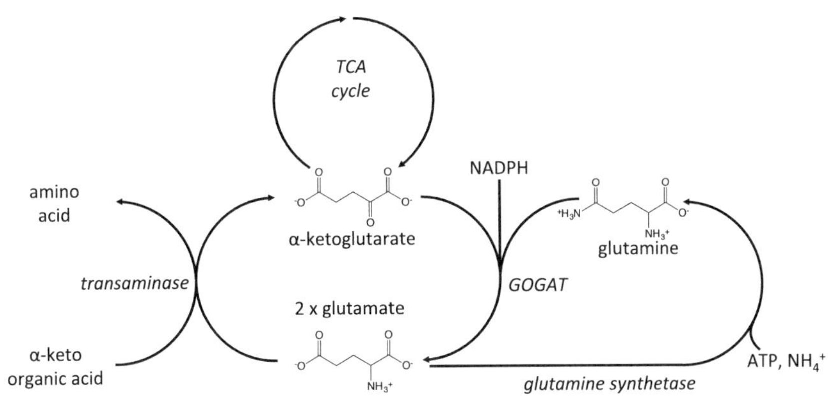

Fig. 7.2. The ammonium assimilation cycle. The nitrogen status of the cell (abundant ammonium or nitrogen starvation) is signaled through the levels of the two substrates for the glutamine 2-oxoglutarate aminotransferase (*GOGAT*) reaction: αKG and glutamine. The reaction produces two glutamate. Glutamate serves as an amino donor for the synthesis of nearly all amino acids via transaminase reactions. If NH_4^+ is scarce, glutamine cannot be synthesized via glutamine synthetase and αKG accumulates, signaling nitrogen starvation and nitrogenase is expressed.

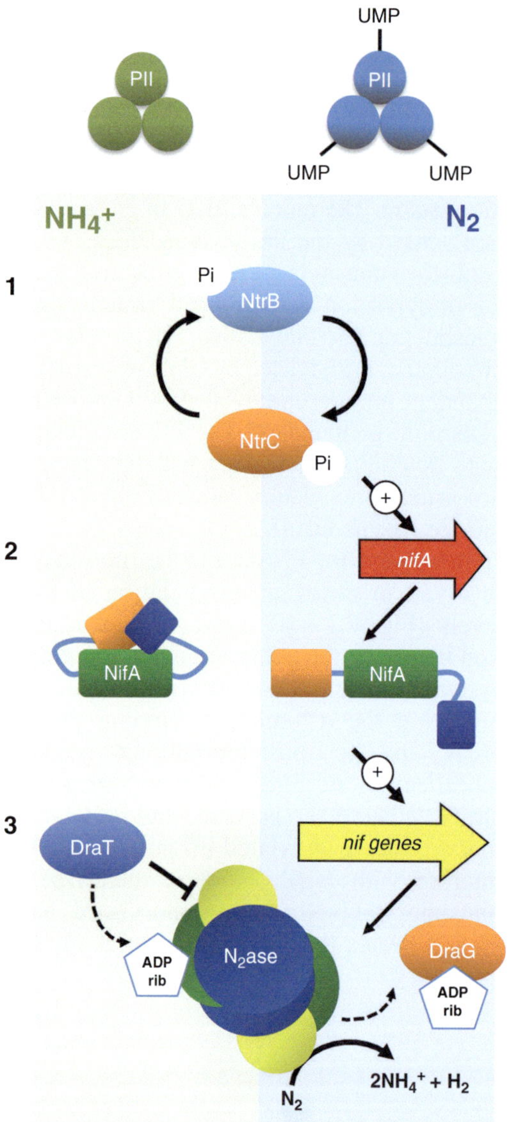

Fig. 7.3. Nitrogenase is regulated at three levels. (*1*) In the presence of NH_4^+ PII proteins respond to high glutamine levels and prevent phosphorylation of NtrC by NtrB. During nitrogen starvation, high α-ketoglutarate levels lead to the uridylylation of PII proteins and allow NtrB to phosphorylate NtrC. NtrC then promotes the transcription of genes involved in nitrogen fixation, including *nifA*. (*2*) PII proteins respond to the nitrogen status of the cell and either allow or prevent NifA from activating the transcription of nitrogenase-encoding genes. (*3*) Nitrogenase enzyme activity can be switched off if glutamine levels rise. PII proteins interact with DraT, which halts nitrogenase activity by adding ADP-ribosyl groups to nitrogenase. If α-ketoglutarate levels rise, PII proteins are uridylylated and DraG removes the ADP-ribosyl groups to allow nitrogenase activity to continue (Reprinted with permission from the American Society for Microbiology (Microbe, January 2006, p. 20–24)).

involved in nitrogen starvation, including *nifA*, which encodes the master transcriptional activator of nitrogenase. PII proteins are also generally thought to interact with NifA and during nitrogen-starvation cause NifA to bind to the enhancer of the nitrogenase operon, leading to the expression of nitrogenase and its many accessory proteins. Once nitrogenase is expressed and active there is a post-translational mechanism to switch off its activity in response to ammonium. If NH_4^+ becomes available, deuridylylated PII proteins interact with DraT, which adds an ADP-ribosyl group to nitrogenase, preventing its activity. If NH_4^+ becomes scarce before the inactivated nitrogenase is degraded, DraG can reactivate nitrogenase by removing the ADP-ribosyl groups.

The above regulatory network presented a major hurdle for photobiological production of H_2. In the lab, environmental conditions can be easily modified to induce nitrogenase expression. For example, providing glutamate as the nitrogen source and omitting N_2 gas (e.g., by growing cultures under argon) is a common method used to induce nitrogenase activity and maximize its hydrogenase activity. This technique carries over from the serendipitous discovery of nitrogenase in PNSB where Howard Gest observed H_2 production because he used a growth medium with glutamate as the sole nitrogen source (Gest 1999). However, most PNSB-based strategies for H_2 production envision using agricultural or industrial waste as a feedstock. These wastes invariably contain nitrogen compounds at concentrations that can repress nitrogenase and therefore H_2 production (Adessi et al. 2012). Overcoming this complicated and multilayered nitrogenase regulatory network appeared to be a monumental task. However, in *Rp. palustris* all that was necessary was a single nucleotide change.

B. Bypassing the Repression of Nitrogenase in Response to NH_4^+

The repression of nitrogenase in response to NH_4^+ is entirely due to regulatory mechanisms. The repression is not a direct chemical or thermodynamic effect of NH_4^+. To bypass this regulatory network in *Rp. palustris* the Harwood lab applied a strong selective pressure for spontaneous mutations that would require *Rp. palustris* to produce H_2 to grow (Rey et al. 2007). Since the 1930s it has been known that PNSB require an electron acceptor to grow photosynthetically on organic compounds that contain more electrons per carbon than the average carbon in cellular biomass (Muller 1933). The cell must dispose of these excess electrons in order to maintain a pool of oxidized electron carrier molecules (e.g., NAD^+) required by crucial metabolic reactions. CO_2 is the traditional electron acceptor used in most experiments but the production of H_2 also suffices for eliminating excess electrons (McKinlay and Harwood 2011). The Harwood lab used this knowledge to select for *Rp. palustris* strains that constitutively produce H_2 by incubating cells in growth medium with NH_4^+ and an electron rich carbon source but without CO_2. After several months under constant illumination, some cultures suddenly grew and produced H_2 in the presence of NH_4^+ (Rey et al. 2007). Sequencing genes involved in nitrogenase regulation revealed that each mutant had a single nucleotide change in *nifA* – the gene encoding the master transcriptional activator of nitrogenase. A single nucleotide was confirmed to be all that was necessary for constitutive H_2 production by introducing the mutated gene into a wild-type genetic background. These mutants that produce H_2 constitutively are called NifA* strains. Since then a NifA* strain containing a 48-nucleotide deletion in *nifA* was constructed and has a more stable phenotype than the original spontaneous NifA* strains (McKinlay and Harwood 2010a).

It is remarkable that a single nucleotide change could bypass the entire nitrogenase regulatory network. As it turns out, this is a feature that may be unique to *Rp. palustris*. In *Rs. rubrum* a similar *nifA* mutation is required but the DraT activity must also be disrupted to prevent post-translational repression (Zou et al. 2008). Microarray and genetic approaches have been used to determine why nitrogenase is not switched off in *Rp. palustris* NifA* strains. In *Rp. palustris*, NtrBC activates the expression of the PII protein, GlnK2 (one of three PII proteins encoded in the CGA009 genome) which in turn controls DraT2 (one of two DraT proteins encoded in the genome) (Heiniger et al. 2012). When *Rp. palustris* is growing by N_2 fixation, NtrC is phosphorylated and GlnK2 is expressed. Under these conditions, the introduction of NH_4^+ causes GlnD to remove the uridylyl groups from GlnK2, and GlnK2 can activate DraT2 to switch off nitrogenase. However, when NifA* cells are grown with NH_4^+ (i.e., prolonged exposure to NH_4^+), NtrC is not phosphorylated and GlnK2 levels are low. Thus, there is insufficient GlnK2 to activate the switch off mechanism in NifA* strains grown with NH_4^+. When NifA* cells are grown with N_2 and then exposed to NH_4^+, a switch off response occurs but it is not nearly strong enough to prevent H_2 production. One reason for this low switch off activity appears to be insufficient DraT2 to completely switch off nitrogenase in NifA* strains. Comparisons of NifA* strains grown with NH_4^+ to wild-type cells grown with N_2 show that nitrogenase activity is about threefold higher in NifA* strains while DraT2 levels are similar. Indeed, expressing both *glnK2* and *draT2* from a plasmid in the NifA* strain resulted in H_2 production levels at 22 % that of the NifA* strain with an empty vector (Heiniger et al. 2012). Knocking out *draT2* resulted in a 1.3-fold increase in H_2 production indicating that NifA* strains were still subject to a low level of switch-off activity when grown with NH_4^+ (Heiniger et al. 2012).

Microarray analysis of NifA* strains has also been useful in defining the NifA regulon. When *Rp. palustris* is switched from growth on NH_4^+ to N_2, over 200 genes are differentially expressed – about 4 % of the genome (Oda et al. 2005). However, microarray comparisons between the NifA* strain and the wild type, both grown with NH_4^+, show that only 18 genes outside of the nitrogenase gene cluster increase their expression levels (Rey et al. 2007). Thus, it appears that most of the genes involved in N_2 fixation are not essential for the functioning of nitrogenase but are more likely part of a broad response to nitrogen starvation. In contrast, the small regulon revealed by the NifA* strains potentially points to genes that are involved in nitrogenase function, and therefore H_2 production. These genes encode proteins that could form novel electron transfer chains delivering electrons to nitrogenase or in iron scavenging and storage to meet the high iron demands of a large pool of functional nitrogenase (Rey et al. 2007). Proteomic analysis of *Rp. palustris* also pointed to the importance of iron acquisition for nitrogenase activity with the detection of 14 different TonB-dependent iron transporters (VerBerkmoes et al. 2006). Other genes upregulated in the NifA* strains included those encoding light harvesting complex II proteins, perhaps to meet the energetic demands of nitrogenase, and hypothetical proteins of unknown function (Rey et al. 2007).

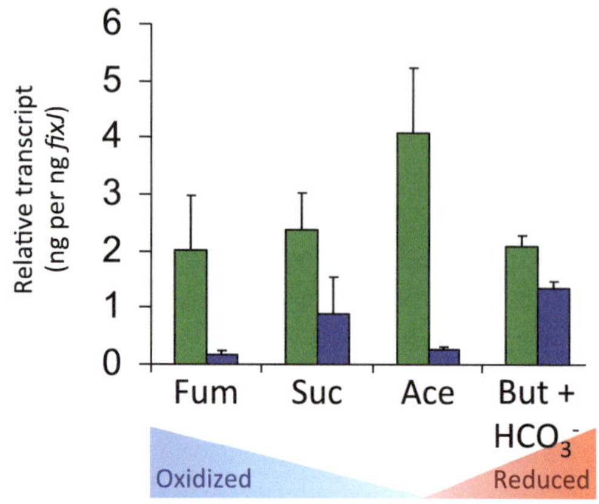

Fig. 7.4. Calvin cycle gene expression levels are lower during H_2 production. Examples for Type I Rubisco transcript levels (*cbbL*) determined by RT-qPCR analysis are shown for wild-type *Rp. palustris* (*green*) and a H_2-producing NifA* strain (*blue*) grown in the presence of NH_4^+ on substrates having different electron contents (McKinlay and Harwood 2011). Similar trends were observed for other Calvin cycle genes in other NifA* strains and in wild-type during N_2 fixation by microarray analysis (McKinlay and Harwood 2010a) (Figure reproduced and amended with permission from the American Society for Microbiology under a Creative Commons Attribution Non-commercial Share Alike license (McKinlay and Harwood 2011)).

C. Identifying and Eliminating Pathways That Compete with H_2 Production

In addition to the increased expression levels of 18 genes in NifA* strains, microarray comparisons also indicated that several genes had lower transcript levels in the NifA* strains compared to wild-type (Rey et al. 2007). Among these genes were those encoding the CO_2-fixing Calvin cycle (Fig. 7.4).

Similar decreases in Calvin cycle gene expression were observed during N_2 fixation/ H_2 production in microarray analyses of *Rb. sphaeroides* (Kontur et al. 2011). The Calvin cycle is best known for its role in allowing autotrophic organisms like plants, algae, and some bacteria to grow on CO_2 as the sole carbon source at the expense of ATP and reductant. In PNSB, the Calvin cycle also functions to maintain electron balance during photosynthetic growth on organic compounds (photoheterotrophic growth; Fig. 7.1a). When PNSB grow photoheterotrophically, the organic substrates are oxidized, resulting in reduction of electron carriers such as $NAD(P)^+$ to NAD(P)H. In respiring organisms, this reductant would be oxidized by H^+-pumping electron transfer chains, intimately associated with the formation of ATP. In photoheterotrophic PNSB, ATP is formed by cyclic photophosphorylation, without the need for a terminal electron acceptor. Even so, the reduced electron carriers must be oxidized to maintain metabolic flow and avoid cell death. By fixing CO_2 via ribulose 1,5 bisphosphate carboxylase (Rubisco), the Calvin cycle eventually forms glyceraldehyde-3-phosphate that can accept electrons from NAD(P)H. The CO_2 'electron acceptor' is ultimately incorporated into biomass. Elimination of Calvin cycle genes encoding Rubisco and phosphoribuokinase can disable PNSB from growing on organic carbons sources. However, growth of such Calvin cycle mutants can be rescued by the addition of electron acceptors or by allowing the cells to rid themselves of excess electrons through H_2 production (Hallenbeck et al. 1990a, b; Falcone and Tabita 1991; McKinlay and Harwood 2010a). In fact, uncharacterized constitutive H_2-producing strains of *Rb. sphaeroides* were obtained through the long-term incubation of Rubisco mutants (Joshi and Tabita 1996) – a similar strategy to what the Harwood lab used to obtain NifA* strains of *Rp. palustris* (Rey et al. 2007). An electron balancing activity such as the Calvin cycle or H_2 production is required for photoheterotrophic growth even on organic compounds that have less electrons per carbon than the average carbon in cellular biomass. In the absence of added CO_2, *Rp. palustris* relies on a Rubisco type I enzyme to scavenge CO_2 released by oxidative metabolic pathways rather than the Rubisco type II enzyme it encodes, as indicated by proteomic analysis and followed up with biochemical and mutational approaches (VerBerkmoes et al. 2006; Joshi et al. 2009). These results are consistent with Rubisco type I having a higher affinity for CO_2 than the type II enzyme (Tabita 1988).

Recent observations with *Rs. rubrum* have led to the argument that preventing Rubisco activity results in an accumulation of ribulose-1,5-bisphosphate and it is the toxic effect of this compound rather than an inability to maintain electron balance that disrupts growth (Wang et al. 2011). As observed in other PNSB, a Rubisco mutant of *Rs. rubrum* had severe growth defects (Wang et al. 2010, 2011). However, knocking out phosphoribulokinase, the enzyme that produces the ribulose 1,5-bisphosphate substrate for Rubisco, restored normal growth (Wang et al. 2011). Though toxic accumulation of ribulose-1,5-bisphosphate could disrupt growth it does not rule out the fact that electrons must be balanced to obey conservation of mass. We have since confirmed the observations made with *Rs. rubrum* and suggest that it has alternative mechanisms to maintain electron balance since phosphoribulokinase mutants of other PNSB including *Rp. palustris* (G.C. Gordon and J.B. McKinlay, unpublished), *Rb. sphaeroides* (Hallenbeck et al. 1990a), and *Rb. capsulatus* (Öztürk et al. 2012) do not grow or show severe growth defects under photoheterotrophic conditions with NH_4^+.

Importantly, the Calvin cycle and H_2 production are both vital mechanisms by which PNSB deal with excess electrons during photoheterotrophic growth. Thus, because of their common roles they can potentially compete for reductant. Even though the Calvin cycle is down-regulated when nitrogenase is active (Fig. 7.4), the Calvin cycle could still consume electrons that could otherwise be used to produce H_2. Rarely do genomic transcript levels correlate with metabolic activity in a quantitative manner. To determine the effect of the Calvin cycle on H_2 production ^{13}C-metabolic flux analysis or 'fluxomics' was performed. This approach provides a quantitative view of the in vivo flow of carbon (and associated electrons and ATP) through a metabolic network. Metabolic flux distributions with and without H_2 production (i.e., NifA* vs wild-type *Rp. palustris*) were compared on four different carbon sources having different oxidation states – fumarate, succinate, acetate, and butyrate (McKinlay and Harwood 2011).

In the absence of H_2 production, and in the absence of added CO_2 or bicarbonate, the Calvin cycle fixes a significant amount of the CO_2 released by other metabolic reactions as the organic carbon source is oxidized (i.e., ranging from 20 % on fumarate to 70 % on acetate). When H_2 is produced, the Calvin cycle flux always decreased (Fig. 7.5a), supporting the microarray observations (Fig. 7.4). However, the Calvin cycle flux magnitude depended on the carbon source used. For example, during growth on acetate, H_2 production resulted in a Calvin cycle flux that was ~20 % of that in the absence of H_2 production (Fig. 7.5a). However, during growth on succinate, H_2 production only resulted in a decrease of Calvin cycle flux to 60 % of the level observed during the absence of H_2 production. Thus, depending on the growth conditions the Calvin cycle can divert a considerable portion of available electrons away from H_2 production. Calvin cycle flux was prevented by deleting the genes encoding Rubisco enzymes resulting in increased H_2 yields that were proportional in magnitude to the Calvin cycle fluxes observed in the parental NifA* strain (Fig. 7.5b). On some carbon sources such as succinate and

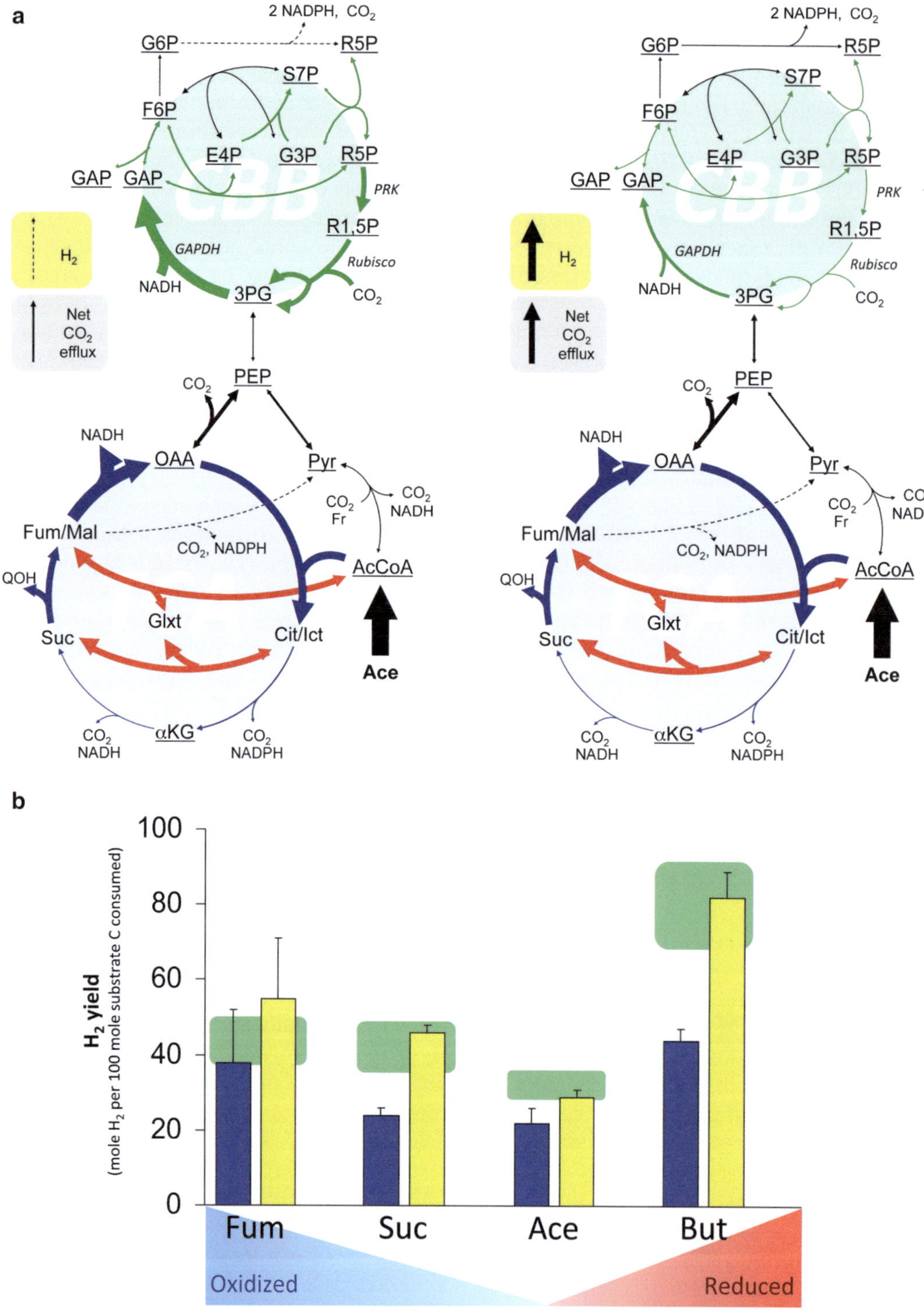

Fig. 7.5. **The Calvin cycle competes with H_2 production for electrons.** (**a**) A comparison of metabolic fluxes between wild-type *Rp. palustris* (*left*) and an H_2-producing NifA* strain (*right*) from ^{13}C-metabolic flux analyses (^{13}C-labeling experiments) show that there is less metabolic flow through the Calvin cycle (*green arrows*) when H_2 is being produced. *Arrow thickness* is proportional to the flux through corresponding pathway or reaction. *Dotted arrows* signify fluxes that are less than 5 molar percent of the acetate uptake rate. (**b**) When Calvin cycle flux is prevented by deleting the genes encoding Rubisco the H_2 yields from the Calvin cycle NifA* mutant (*yellow*) are higher than the NifA* parent strain (*blue*). The *green boxes* show the 90 % confidence intervals for H_2 yields that were predicted to result from eliminating Calvin cycle flux based on flux maps obtained for the four different carbon sources (Figures reproduced and amended with permission from the American Society for Microbiology under a Creative Commons Attribution Non-commercial Share Alike license (McKinlay and Harwood 2011)).

butyrate, preventing Calvin cycle activity resulted in a twofold increase in the H_2-yield (Fig. 7.5b) (McKinlay and Harwood 2011). This study illustrated how functional genomics and systems biology approaches like fluxomics can be used to identify targets for improving product formation. Several *in silico* flux balance models of PNSB have also pointed to the importance of the Calvin cycle (and other reductive pathways including sulfide production and an alternative CO_2-fixing ethylmalonyl-CoA pathway found in some PNSB) in maintaining electron balance (Klamt et al. 2002; Hädicke et al. 2011; Imam et al. 2011; Rizk et al. 2011). Although these models require more assumptions than those that incorporate experimental data from ^{13}C-labeling experiments, they can identify potentially good targets for metabolic engineering. Like other 'omics' approaches, these models are most effective when resulting hypotheses are tested through follow up biochemical or mutational experiments and the resulting information used to refine the models.

An important benefit of producing H_2 in the absence of the Calvin cycle is that formation of the desired product (i.e., H_2) is required for cell viability by maintaining redox balance. This is a rare scenario for engineered biofuel-producing microbes. Typically, the production of the desired product is at odds with cell growth. Thus, H_2-producing PNSB strains that lack Calvin cycle activity may be unique examples of biofuel-producing microbes with a stable phenotype. However, one must always consider that if a more efficient way to maintain electron balance exists, it will likely evolve. For example, a *Rb. sphaeroides* Calvin cycle mutant was reported to have evolved the ability to reduce sulfate to sulfide as an electron balancing mechanism instead of producing H_2 (Rizk et al. 2011). It is not clear whether this is indeed a more efficient means of dealing with excess electrons than producing H_2. Accumulation of internal polyhydroxybutyrate is also another potential alternative electron balancing mechanism (De Philippis et al. 1992; Imam et al. 2011). Knocking out this pathway in some PNSB has resulted in higher H_2 yields (Yilmaz et al. 2010).

IV. Future Directions for a System-Level Understanding of Photobiological H_2 Production

Systems biology approaches such as microarrays and ^{13}C-metabolic flux analysis have provided insights into how PNSB produce H_2 and have guided the successful engineering of PNSB to improve H_2 production. In the process, these approaches have also provided fundamental knowledge about the physiology of PNSB and the regulatory mechanisms that govern their diverse metabolic modules. With advances in systems biology technologies we can expect further complementary insights into the fundamental and applied aspects of PNSB. As systems level experimental and computational approaches grow in popularity and garner more attention we must not lose sight of the importance of rigorous biochemical and genetic approaches, particularly in verifying the trends and testing the predictions that emerge from systems level approaches.

A. RNAseq Analysis of Coding and Non-coding RNAs

New sequencing technologies have made it cost-effective to sequence and count cDNA copies of transcripts in a process called RNAseq. A major challenge in RNAseq is to focus the sequencing power on non-ribosomal RNA molecules. Recently, a method called not-so-random (NSR) priming was developed, which uses a random selection of hexamer primers to convert RNA into cDNA, but excludes those sequences predicted to bind rRNA (Armour et al. 2009). This approach makes use of only 1 μg of RNA. Combining NSR random hexamer primers with short 5′ barcode tags (e.g., so that multiple samples can be read per lane in an Illumina sequencer) also allows for directional sequencing and the identification of antisense RNA molecules. This new method is being applied to *Rp. palustris* with results now starting to reach publication (Hirakawa et al. 2011).

Thus far, RNAseq has revealed an antisense transcript involved in the quorum sensing response of *Rp. palustris* CGA009. *Rp. palustris* has an unusual quorum sensing system involving production of a homoserine lactone with an aromatic side chain instead of a typical acyl side chain (Schaefer et al. 2008). Interestingly this aromatic p-coumaryl homoserine lactone signal is only made when aromatic compounds like p-coumarate are provided (*Rp. palustris* cannot synthesize p-coumarate itself). The RNAseq analysis revealed a small transcript that is anti-sense to *rpaR* which encodes the quorum sensing receptor/regulatory protein (Hirakawa et al. 2011). A follow up study confirmed the antisense activity of this transcript as it repressed the expression of *rpaR* and therefore production of the quorum sensing signal (Hirakawa et al. 2012).

Aside from the antisense transcript, this single use of RNAseq on *Rp. palustris* revealed several intergenic regions that are differentially expressed and could encode small RNA molecules (sRNA) (Hirakawa et al. 2011). A computational approach was also used to predict sRNA coding regions, four of which were confirmed to exist using RNA gel blots (Madhugiri et al. 2012). Noncoding RNAs are receiving increasing attention for their regulatory roles in diverse prokaryotes (Gottesman 2005). sRNAs add an additional layer of complexity that must be understood to successfully engineer a microbial system. sRNA and other functional RNAs including aptamers and riboswitches also offer the potential for new tools to redirect metabolic flux towards desired products like H_2 (Win et al. 2009).

B. Proteomic Analysis of Post-translational Regulatory Mechanisms

Though traditional transcriptomics and proteomics have made invaluable contributions to our understanding and application of microbial systems, these functional genomic approaches often do not explain phenotypic trends, such as metabolic flux. For example, a study that compared multiple omics with various *E. coli* mutants found that it was an exception, rather than the rule, for transcriptomics, proteomics, and metabolomics data sets to correlate with metabolic fluxes (Ishii et al. 2007). Metabolic fluxes are often determined by thermodynamics, kinetic parameters, and through post-translational modifications of the enzymes.

Mass spectrometry has been used to examine post-translational modifications of specific proteins in *Rp. palustris.* For example, focusing on the regulation of nitrogenase (and therefore H_2 production) the uridylylation state of the three *Rp. palustris* PII proteins, GlnK1, GlnK2, and GlnB were examined in the presence and absence of ammonium (Connelly et al. 2006). In the future, mass spectrometry-based proteomic analyses will likely target post-translational modifications on a systems scale. For example, Crosby et al. recently used such a proteomic approach to identify N-lysine acetylated proteins in *Rp. palustris* (Crosby et al. 2012). N-lysine acylation (i.e., modification of lysine residues with acetyl from acetyl-CoA, or in some cases, propionyl, butyryl, succinyl, or malonyl groups) can be used to activate or deactivate activity of a wide variety of proteins in both prokaryotes and eukaryotes with deacylation having the opposite effect (Albaugh et al. 2011; Kim and Yang 2011). In *Rp. palustris*, biochemical and mutational analyses had previously revealed that N-lysine acetylation inhibits activity of three enzymes involved in degradation of lignin monomers such as benzoate and cyclohexanecarboxylate, and deacetylation reactivates these enzymes (Crosby et al. 2010). The proteomics analysis originally identified hundreds of potential targets for acylation. However, after stringent parameters were applied to the screen, requiring that acetylated proteins be identified by two different proteomics software packages and that proteins show elevated levels of acylation in a mutant strain lacking two known deacetylases, 14 candidate acetylation targets were identified. Biochemical assays then verified that nine of these proteins were actual targets for acetylation by two known *Rp. palustris* acetyltransferases (Crosby et al. 2012). One of the candidate proteins (that was not a target for the known acetyltransferases) was glyceraldehyde-3-phosphate dehydrogenase – a key enzyme of the Calvin cycle and other metabolic pathways. The importance of the acetylation of this enzyme and others identified in the study for *Rp. palustris* physiology and H_2 production have not yet been reported.

C. Global Identification and Characterization of Ligand-Binding Proteins

Part of the metabolic versatility of PNSB involves their ability to use a wide range of organic and inorganic substrates. *Rp. palustris* was the focus of a recent study that attempted to characterize all of the solute-binding proteins associated with transporters in the cell (Giuliani et al. 2011). 107 genes encoding candidate *Rp. palustris* solute binding proteins were expressed in *E. coli*, 75 of the resulting proteins were screened and ligands were ultimately identified for 45 proteins. The proteins were assayed for their ability to bind compounds in a fluorescence-based thermal shift assay using a library of small molecules including metals, aromatic compounds, amino acids, fatty acids, and other compounds. Related to H_2 production, this approach identified a vanadate transporter that is likely important for supplying vanadium required by V-nitrogenase. This gene product was previously incorrectly annotated as a phosphate transporter. Two solute binding proteins were found to bind a wide range of fatty acids, and not coincidentally, were encoded near the *pimFABCDE* gene cluster encoding enzymes for the β-oxidation of a broad range of fatty acids (Harrison and Harwood 2005; Giuliani et al. 2011). The screen also identified six gene products capable of binding aromatic compounds. These proteins are therefore potentially important for H_2 production from lignin monomers. Four of these proteins were further studied to determine their thermodynamic and structural properties using isothermal calorimetry and small/wide angle X-ray scattering (Pietri et al. 2012). The data was used to guide the development of structural models that may prove useful for identifying aromatic compound-binding proteins based on sequences in other genomes (Pietri et al. 2012).

A surprising result to emerge from the screen was that only 11 gene products bound ligands that were predicted from their original annotation (Giuliani et al. 2011). This study is therefore invaluable for refining our knowledge of the function of poorly annotated gene products – knowledge that can ideally be extended to other genomes.

D. The Physiology of Non-growing Cells – Approaching the Maximum Theoretical H_2 Yield

Although advances have been made in understanding and engineering *Rp. palustris* to produce H_2, most work has focused on producing H_2 using growing cells – in other words, cells that are dedicating electrons to biosynthesis that could otherwise be used to produce H_2. Even the NifA* Calvin cycle mutants only use 20–36 % of the consumed electrons for H_2 production and the rest are almost exclusively used for biosynthesis (McKinlay and Harwood 2011). Certainly biosynthesis must occur to have biocatalysts to produce H_2, and the cells themselves may even have value as fertilizer or animal feed (Honda et al. 2006). However, once cell numbers accumulate, there are great advantages to using them to produce H_2 in a non-growing state. First, when growth is prevented by nitrogen starvation, wild-type CGA009 will divert about 50 % of the electrons from acetate to H_2 and up to 75 % to H_2 from an inorganic substrate like thiosulfate (Huang et al. 2010). Second, *Rp. palustris* is ideally suited for long-term use in a non-growing state. As long as *Rp. palustris* is illuminated, it can repeatedly energize and recycle electrons through an electron transfer chain, maintaining a proton motive force and ATP pools to repair itself. This ability to maintain itself may explain why no classical stress proteins were detected in a proteomics analysis of stationary phase (carbon-starved) *Rp. palustris* cells (VerBerkmoes et al. 2006). *Rp. palustris* cells have been immobilized in artificial latex biofilms and maintained in a non-growing state for months at a time. These biological solar panels remained metabolically active when transferred between batches of fresh medium losing little activity after the first 6 days maintaining H_2 productivity at 2.08 ± 0.08 mmol H_2 m^{-2} h^{-1} for over 5 months (Gosse et al. 2010). It is worth noting that the mode of starvation can greatly influence H_2 yields. Unlike the favorable effects of nitrogen starvation on H_2 production, sulfur starvation did not result in high H_2 yields but rather favored polyhydroxybutyrate synthesis (Melnicki et al. 2008).

Techniques such as RNAseq and recent advances in ^{13}C-metabolic flux analysis procedures (Rühl et al. 2012) have facilitated our ability to examine non-growing cells. Such approaches are expected to reveal novel targets for improving H_2 yields by non-growing cells and to lead to a better general understanding of the physiology of bacteria under stressful starvation conditions.

E. Non-biased Interpretation and Utilization of Systems Biology Data

Though the immense volumes of data from systems biology is undeniably useful, is also introduces a challenge in effectively interpreting the data and applying it to a given problem. For example, when comparing transcriptomic datasets in which gene expression levels vary for hundreds of genes it is impractical to follow up on each and every variation. Thus, experimenters tend to focus on well-annotated and characterized gene products than genes of unknown function. As a result, some potentially important aspects of microbial physiology are overlooked and metabolic engineering efforts are heavily weighted on hypotheses that can be biased by a researcher's background knowledge. However, there are approaches to identify targets on a systems scale without introducing experimenter bias. For example, screening transposon mutant libraries can link a gene of unknown function to a phenotype. Such approaches have turned up useful targets for metabolic engineering *E. coli* for lycopene production (Alper et al. 2005). Integrating systems biology approaches such as genomic variation with transcriptomic data can also highlight key drivers of complex functions through predictions of causal relationships (Schadt et al. 2005). This approach led to the identification of three genes previously not known to be involved in obesity in mice, which were validated through analyses of mice with mutations in these genes (Schadt et al. 2005). It may be possible to use similar approaches to identify key, and potentially unexpected, drivers of H_2 production in PNSB. Identifying targets that an experimenter would not otherwise predict is not only useful for improving product formation but can lead to a functional characterization of gene products with previously unknown function and define their role in the greater context of microbial physiology.

Acknowledgements

I am grateful to Caroline Harwood, University of Washington, for comments on this manuscript and for being an outstanding mentor. I am also grateful for financial support through the US Department of Energy Early Career Research Program (DE-SC0008131), an Oak Ridge Associated Universities Ralph E. Powe Junior Faculty Enhancement Award, and the College of Arts and Sciences at Indiana University.

References

Adessi A, McKinlay JB, Harwood CS, De Philippis R (2012) A *Rhodopseudomonas palustris nifA** mutant produces H_2 from NH_4^+-containing vegetable wastes. Int J Hydrog Energy 37:5893–15900

Albaugh BN, Arnold KM, Denu JM (2011) KAT(ching) metabolism by the tail: insight into the links between lysine acetyltransferases and metabolism. Chembiochem 12:90–298

Alper H, Miyaoku K, Stephanopoulos G (2005) Construction of lycopene-overproducing *E. coli* strains by combining systematic and combinatorial gene knockout targets. Nat Biotechnol 23:612–616

Armour CD, Castle JC, Chen R, Babak T, Loerch P, Jackson S, Shah JK, Dey J, Rohl CA, Johnson JM, Raymond CK (2009) Digital transcriptome profiling using selective hexamer priming for cDNA synthesis. Nat Methods 6:647–649

Bonam D, Lehman L, Roberts GP, Ludden PW (1989) Regulation of carbon monoxide dehydrogenase and hydrogenase in *Rhodospirillum rubrum*: effects of CO and oxygen on synthesis and activity. J Bacteriol 171:3102–3107

Cheng S, Logan BE (2007) Sustainable and efficient biohydrogen production via electrohydrogenesis. Proc Natl Acad Sci U S A 104:18871–18873

Connelly HM, Pelletier DA, Lu TY, Lankford PK, Hettich RL (2006) Characterization of pII family (GlnK1, GlnK2, and GlnB) protein uridylylation in response to nitrogen availability for *Rhodopseudomonas palustris*. Anal Biochem 357:93–104

Croal LR, Jiao Y, Newman DK (2007) The *fox* operon from *Rhodobacter* strain SW2 promotes phototrophic Fe(II) oxidation in *Rhodobacter capsulatus* SB1003. J Bacteriol 189:1774–1782

Crosby HA, Heiniger EK, Harwood CS, Escalante-Semerena JC (2010) Reversible *N*ε-lysine acetylation regulates the activity of acyl-CoA synthetases involved in anaerobic benzoate catabolism in *Rhodopseudomonas palustris*. Mol Microbiol 76:874–888

Crosby HA, Pelletier DA, Hurst GB, Escalante-Semerena JC (2012) System-wide studies of *N*-Lysine acetylation in *Rhodopseudomonas palustris* reveals substrate specificity of protein acetyltransferases. J Biol Chem 287:15590–15601

De Philippis R, Ena A, Guastini M, Sili C, Vincenzini M (1992) Factors affecting poly-ß-hydroxybutyrate accumulation in cyanobacteria and in purple non-sulfur bacteria. FEMS Microbiol Rev 103:187–194

Falcone DL, Tabita FR (1991) Expression of endogenous and foreign ribulose 1,5-bisphosphate carboxylase-oxygenase (RubisCO) genes in a RubisCO deletion mutant of *Rhodobacter sphaeroides*. J Bacteriol 173:2099–2108

Ferguson SJ, Jackson JB, McEwan AG (1987) Anaerobic respiration in the Rhodospirillaceae: characterisation of pathways and evaluation of roles in redox balancing during photosynthesis. FEMS Microbiol Rev 46:117–143

Fox JD, He Y, Shelver D, Roberts GP, Ludden PW (1996a) Characterization of the region encoding the CO-induced hydrogenase of *Rhodospirillum rubrum*. J Bacteriol 178:6200–6208

Fox JD, Kerby RL, Roberts GP, Ludden PW (1996b) Characterization of the CO-induced, CO-tolerant hydrogenase from *Rhodospirillum rubrum* and the gene encoding the large subunit of the enzyme. J Bacteriol 178:1515–1524

Fuhrer T, Fischer E, Sauer U (2005) Experimental identification and quantification of glucose metabolism in seven bacterial species. J Bacteriol 187:1581–1590

Galagan JE, Nusbaum C, Roy A, Endrizzi MG, Macdonald P, FitzHugh W, Calvo S, Engels R, Smirnov S, Atnoor D, Brown A, Allen N, Naylor J, Stange-Thomann N, DeArellano K, Johnson R, Linton L, McEwan P, McKernan K, Talamas J, Tirrell A, Ye W, Zimmer A, Barber RD, Cann I, Graham DE, Grahame DA, Guss AM, Hedderich R, Ingram-Smith C, Kuettner HC, Krzycki JA, Leigh JA, Li W, Liu J, Mukhopadhyay B, Reeve JN, Smith K, Springer TA, Umayam LA, White O, White RH, Conway de Macario E, Ferry JG, Jarrell KF, Jing H, Macario AJ, Paulsen I, Pritchett M, Sowers KR, Swanson RV, Zinder SH, Lander E, Metcalf WW, Birren B (2002) The genome of *M. acetivorans* reveals extensive metabolic and physiological diversity. Genome Res 12:532–542

Gest H (1999) Memoir of a 1949 railway journey with photosynthetic bacteria. Photosynth Res 61:91–96

Giuliani SE, Frank AM, Corgliano DM, Seifert C, Hauser L, Collart FR (2011) Environment sensing and response mediated by ABC transporters. BMC Genomics 12(Suppl 1):S8

Gosse JL, Engel BJ, Hui JC, Harwood CS, Flickinger MC (2010) Progress toward a biomimetic leaf: 4,000 h of hydrogen production by coating-stabilized nongrowing photosynthetic *Rhodopseudomonas palustris*. Biotechnol Prog 26:907–918

Gottesman S (2005) Micros for microbes: non-coding regulatory RNAs in bacteria. Trends Genet 21:399–404

Hädicke O, Grammel H, Klamt S (2011) Metabolic network modeling of redox balancing and biohydrogen production in purple nonsulfur bacteria. BMC Syst Biol 5:150. doi:10.1186/1752-0509-5-150

Hallenbeck PC, Lerchen R, Hessler P, Kaplan S (1990a) Phosphoribulokinase activity and regulation of CO_2 fixation critical for photosynthetic growth of *Rhodobacter sphaeroides*. J Bacteriol 172:1749–1761

Hallenbeck PC, Lerchen R, Hessler P, Kaplan S (1990b) Roles of CfxA, CfxB, and external electron acceptors in regulation of ribulose 1,5-bisphosphate carboxylase/oxygenase expression in *Rhodobacter sphaeroides*. J Bacteriol 172:1736–1748

Harrison FH, Harwood CS (2005) The *pimFABCDE* operon from *Rhodopseudomonas palustris* mediates dicarboxylic acid degradation and participates in anaerobic benzoate degradation. Microbiology 151:727–736

Harwood CS (2009) Degradation of aromatic compounds by purple nonsulfur bacteria. In: Hunter CN, Daldal F, Thurnauer MC, Beatty JT (eds) The purple phototrophic bacteria, vol 28, Advances in photosynthesis and respiration. Springer, Dordrecht, pp 577–594

Heiniger EK, Oda Y, Samanta SK, Harwood CS (2012) How posttranslational modification of nitrogenase is circumvented in *Rhodopseudomonas palustris* strains that produce hydrogen gas constitutively. Appl Environ Microbiol 78:1023–1032

Hirakawa H, Oda Y, Phattarasukol S, Armour CD, Castle JC, Raymond CK, Lappala CR, Schaefer AL, Harwood CS, Greenberg EP (2011) Activity of the *Rhodopseudomonas palustris p*-coumaroyl-homoserine lactone-responsive transcription factor RpaR. J Bacteriol 193:2598–2607

Hirakawa H, Harwood CS, Pechter KB, Schaefer AL, Greenberg EP (2012) Antisense RNA that affects *Rhodopseudomonas palustris* quorum-sensing signal receptor expression. Proc Natl Acad Sci U S A 109:12141–12146

Honda R, Fukushi K, Yamamoto K (2006) Optimization of wastewater feeding for single-cell protein production in an anaerobic wastewater treatment process utilizing purple non-sulfur bacteria in mixed culture condition. J Biotechnol 125:565–573

Huang JJ, Heiniger EK, McKinlay JB, Harwood CS (2010) Production of hydrogen gas from light and the inorganic electron donor thiosulfate by *Rhodopseudomonas palustris*. Appl Environ Microbiol 76:7717–7722

Imam S, Yilmaz S, Sohmen U, Gorzalski AS, Reed JL, Noguera DR, Donohue TJ (2011) iRsp1095: a genome-scale reconstruction of the *Rhodobacter sphaeroides* metabolic network. BMC Syst Biol 5:116. doi:10.1186/1752-0509-5-116

Ishii N, Nakahigashi K, Baba T, Robert M, Soga T, Kanai A, Hirasawa T, Naba M, Hirai K, Hoque A, Ho PY, Kakazu Y, Sugawara K, Igarashi S, Harada S, Masuda T, Sugiyama N, Togashi T, Hasegawa M, Takai Y, Yugi K, Arakawa K, Iwata N, Toya Y, Nakayama Y, Nishioka T, Shimizu K, Mori H, Tomita M (2007) Multiple high-throughput analyses monitor the response of *E. coli* to perturbations. Science 316:593–597

Jiao Y, Newman DK (2007) The *pio* operon is essential for phototrophic Fe(II) oxidation in *Rhodopseudomonas palustris* TIE-1. J Bacteriol 189:1765–1773

Jiao Y, Kappler A, Croal LR, Newman DK (2005) Isolation and characterization of a genetically tractable photoautotrophic Fe(II)-oxidizing bacterium, *Rhodopseudomonas palustris* strain TIE-1. Appl Environ Microbiol 71:4487–4496

Joerger RD, Jacobson MR, Premakumar R, Wolfinger ED, Bishop PE (1989) Nucleotide sequence and mutational analysis of the structural genes (*anfH-DGK*) for the second alternative nitrogenase from *Azotobacter vinelandii*. J Bacteriol 171:1075–1086

Joshi HM, Tabita FR (1996) A global two component signal transduction system that integrates the control of photosynthesis, carbon dioxide assimilation, and nitrogen fixation. Proc Natl Acad Sci U S A 93:14515–14520

Joshi GS, Romagnoli S, Verberkmoes NC, Hettich RL, Pelletier D, Tabita FR (2009) Differential accumulation of form I RubisCO in *Rhodopseudomonas palustris* CGA010 under photoheterotrophic growth conditions with reduced carbon sources. J Bacteriol 191:4243–4250

Kerby RL, Ludden PW, Roberts GP (1995) Carbon monoxide-dependent growth of *Rhodospirillum rubrum*. J Bacteriol 177:2241–2244

Kim GW, Yang XJ (2011) Comprehensive lysine acetylomes emerging from bacteria to humans. Trends Biochem Sci 36:211–220

Kim EJ, Kim MS, Lee JK (2008) Hydrogen evolution under photoheterotrophic and dark fermentative conditions by recombinant *Rhodobacter sphaeroides* containing the genes for fermentative pyruvate metabolism of *Rhodospirillum rubrum*. Int J Hydrog Energy 33:5131–5136

Klamt S, Schuster S, Gilles ED (2002) Calculability analysis in underdetermined metabolic networks illustrated by a model of the central metabolism in purple nonsulfur bacteria. Biotechnol Bioeng 77:734–751

Kontur WS, Ziegelhoffer EC, Spero MA, Imam S, Noguera DR, Donohue TJ (2011) Pathways involved in reductant distribution during photobiological H_2 production by *Rhodobacter sphaeroides*. Appl Environ Microbiol 77:7425–7429

Larimer FW, Chain P, Hauser L, Lamerdin J, Malfatti S, Do L, Land ML, Pelletier DA, Beatty JT, Lang AS, Tabita FR, Gibson JL, Hanson TE, Bobst C, Torres JL, Peres C, Harrison FH, Gibson J, Harwood CS (2004) Complete genome sequence of the metabolically versatile photosynthetic bacterium *Rhodopseudomonas palustris*. Nat Biotechnol 22:55–61

Lehman LJ, Roberts GP (1991) Identification of an alternative nitrogenase system in *Rhodospirillum rubrum*. J Bacteriol 173:5705–5711

Madhugiri R, Pessi G, Voss B, Hahn J, Sharma CM, Reinhardt R, Vogel J, Hess WR, Fischer HM, Evguenieva-Hackenberg E (2012) Small RNAs of the *Bradyrhizobium/Rhodopseudomonas* lineage and their analysis. RNA Biol 9:47–58

Masepohl B, Drepper T, Paschen A, Gross S, Pawlowski A, Raabe K, Riedel KU, Klipp W (2002) Regulation of nitrogen fixation in the phototrophic purple bacterium *Rhodobacter capsulatus*. J Mol Microbiol Biotechnol 4:243–248

McEwan AG (1994) Photosynthetic electron transport and anaerobic metabolism in purple non-sulfur phototrophic bacteria. Antonie Van Leeuwenhoek 66:151–164

McEwan AG, Greenfield AJ, Wetzstein HG, Jackson JB, Ferguson SJ (1985) Nitrous oxide reduction by members of the family Rhodospirillaceae and the nitrous oxide reductase of *Rhodopseudomonas capsulata*. J Bacteriol 164:823–830

McKinlay JB, Harwood CS (2010a) Carbon dioxide fixation as a central redox cofactor recycling mechanism in bacteria. Proc Natl Acad Sci U S A 107:11669–11675

McKinlay JB, Harwood CS (2010b) Photobiological production of hydrogen gas as a biofuel. Curr Opin Biotechnol 21:244–251

McKinlay JB, Harwood CS (2011) Calvin cycle flux, pathway constraints, and substrate oxidation state together determine the H_2 biofuel yield in photoheterotrophic bacteria. mBio 2(2):e00323–10. doi:10.1128/mBio.00323-10

Melnicki MR, Bianchi L, De Philippis R, Melis A (2008) Hydrogen production during stationary phase in purple photosynthetic bacteria. Int J Hydrog Energy 33:6525–6534

Muller FM (1933) On the metabolism of purple sulphur bacteria in organic media. Arch Microbiol 4:131–166

Munk AC, Copeland A, Lucas S, Lapidus A, Del Rio TG, Barry K, Detter JC, Hammon N, Israni S, Pitluck S, Brettin T, Bruce D, Han C, Tapia R, Gilna P, Schmutz J, Larimer F, Land M, Kyrpides NC, Mavromatis K, Richardson P, Rohde M, Goker M, Klenk HP, Zhang Y, Roberts GP, Reslewic S, Schwartz DC (2011) Complete genome sequence of *Rhodospirillum rubrum* type strain (S1). Stand Genomic Sci 4:293–302

Oda Y, Samanta SK, Rey FE, Wu L, Liu X, Yan T, Zhou J, Harwood CS (2005) Functional genomic analysis of three nitrogenase isozymes in the photosynthetic bacterium *Rhodopseudomonas palustris*. J Bacteriol 187:7784–7794

Oda Y, Larimer FW, Chain PS, Malfatti S, Shin MV, Vergez LM, Hauser L, Land ML, Braatsch S, Beatty JT, Pelletier DA, Schaefer AL, Harwood CS (2008) Multiple genome sequences reveal adaptations of a phototrophic bacterium to sediment microenvironments. Proc Natl Acad Sci U S A 105:18543–18548

Öztürk Y, Gökçe A, Peksel B, Gürgan M, Özgür E, Gündüz U, Eroğlu I, Yücel M (2012) Hydrogen production properties of *Rhodobacter capsulatus* with genetically modified redox balancing pathways. Int J Hydrog Energy 37:2014–2020

Pan C, Oda Y, Lankford PK, Zhang B, Samatova NF, Pelletier DA, Harwood CS, Hettich RL (2008) Characterization of anaerobic catabolism of p-coumarate in *Rhodopseudomonas palustris* by integrating transcriptomics and quantitative proteomics. Mol Cell Proteomics 7:938–948

Pietri R, Zerbs S, Corgliano DM, Allaire M, Collart FR, Miller LM (2012) Biophysical and structural characterization of a sequence-diverse set of solute-binding proteins for aromatic compounds. J Biol Chem 287:23748–23756

Rey FE, Oda Y, Harwood CS (2006) Regulation of uptake hydrogenase and effects of hydrogen utilization on gene expression in *Rhodopseudomonas palustris*. J Bacteriol 188:6143–6152

Rey FE, Heiniger EK, Harwood CS (2007) Redirection of metabolism for biological hydrogen production. Appl Environ Microbiol 73:1665–1671

Rizk ML, Laguna R, Smith KM, Tabita FR, Liao JC (2011) Redox homeostasis phenotypes in RubisCO-deficient *Rhodobacter sphaeroides* via ensemble modeling. Biotechnol Prog 27:15–22

Rolls JP, Lindstrom ES (1967) Effect of thiosulfate on the photosynthetic growth of *Rhodopseudomonas palustris*. J Bacteriol 94:860–869

Rühl M, Le Coq D, Aymerich S, Sauer U (2012) ^{13}C-flux analysis reveals NADPH-balancing transhydrogenation cycles in stationary phase of nitrogen-starving *Bacillus subtilis*. J Biol Chem 287:27959–27970

Schadt EE, Lamb J, Yang X, Zhu J, Edwards S, Guhathakurta D, Sieberts SK, Monks S, Reitman M, Zhang C, Lum PY, Leonardson A, Thieringer R, Metzger JM, Yang L, Castle J, Zhu H, Kash SF, Drake TA, Sachs A, Lusis AJ (2005) An integrative genomics approach to infer causal associations between gene expression and disease. Nat Genet 37:710–717

Schaefer AL, Greenberg EP, Oliver CM, Oda Y, Huang JJ, Bittan-Banin G, Peres CM, Schmidt S, Juhaszova K, Sufrin JR, Harwood CS (2008) A new class of homoserine lactone quorum-sensing signals. Nature 454:595–599

Shelver D, Kerby RL, He Y, Roberts GP (1995) Carbon monoxide-induced activation of gene expression in *Rhodospirillum rubrum* requires the product of *cooA*, a member of the cyclic AMP receptor protein family of transcriptional regulators. J Bacteriol 177:2157–2163

Simmons SS, Isokpehi RD, Brown SD, McAllister DL, Hall CC, McDuffy WM, Medley TL, Udensi

UK, Rajnarayanan RV, Ayensu WK, Cohly HH (2011) Functional annotation analytics of *Rhodopseudomonas palustris* genomes. Bioinform Biol Insight 5:115–129

Simpson FB, Burris RH (1984) A nitrogen pressure of 50 atmospheres does not prevent evolution of hydrogen by nitrogenase. Science 224:1095–1097

Spangler NJ, Meyers MR, Gierke KL, Kerby RL, Roberts GP, Ludden PW (1998) Substitution of valine for histidine 265 in carbon monoxide dehydrogenase from *Rhodospirillum rubrum* affects activity and spectroscopic states. J Biol Chem 273:4059–4064

Tabita FR (1988) Molecular and cellular regulation of autotrophic carbon dioxide fixation in microorganisms. Microbiol Rev 52:155–189

van Niel CB (1944) The culture, general physiology, morphology, and classification of the non-sulfur purple and brown bacteria. Bacteriol Rev 8:1–118

VerBerkmoes NC, Shah MB, Lankford PK, Pelletier DA, Strader MB, Tabb DL, McDonald WH, Barton JW, Hurst GB, Hauser L, Davison BH, Beatty JT, Harwood CS, Tabita FR, Hettich RL, Larimer FW (2006) Determination and comparison of the baseline proteomes of the versatile microbe *Rhodopseudomonas palustris* under its major metabolic states. J Proteome Res 5:287–298

Wang D, Zhang Y, Welch E, Li J, Roberts GP (2010) Elimination of Rubisco alters the regulation of nitrogenase activity and increases hydrogen production in *Rhodospirillum rubrum*. Int J Hydrog Energy 35:7377–7385

Wang D, Zhang Y, Pohlmann EL, Li J, Roberts GP (2011) The poor growth of *Rhodospirillum rubrum* mutants lacking Rubisco is due to the accumulation of ribulose-1,5-bisphosphate. J Bacteriol 193:3293–3303

Win MN, Liang JC, Smolke CD (2009) Frameworks for programming biological function through RNA parts and devices. Chem Biol 16:298–310

Xing D, Zuo Y, Cheng S, Regan JM, Logan BE (2008) Electricity generation by *Rhodopseudomonas palustris* DX-1. Environ Sci Technol 42:4146–4151

Yilmaz LS, Kontur WS, Sanders AP, Sohmen U, Donohue TJ, Noguera DR (2010) Electron partitioning during light and nutrient-powered hydrogen production by *Rhodobacter sphaeroides*. Bioenerg Res 3:55–66

Zou X, Zhu Y, Pohlmann EL, Li J, Zhang Y, Roberts GP (2008) Identification and functional characterization of NifA variants that are independent of GlnB activation in the photosynthetic bacterium *Rhodospirillum rubrum*. Microbiology 154:2689–2699

Chapter 8

The Extremely Thermophilic Genus *Caldicellulosiruptor*: Physiological and Genomic Characteristics for Complex Carbohydrate Conversion to Molecular Hydrogen

Jeffrey V. Zurawski, Sara E. Blumer-Schuette, Jonathan M. Conway, and Robert M. Kelly*
Department of Chemical and Biomolecular Engineering, North Carolina State University, Raleigh, NC 27695-7905, USA

Summary

Extremely thermophilic, carbohydrate-utilizing bacteria from the genus *Caldicellulosiruptor* should be considered for biohydrogen production to take advantage of their broad growth substrate range and high substrate conversion efficiency. In fact, *Caldicellulosiruptor* species produce molecular hydrogen at yields approaching the Thauer limit of 4 mol H_2/mol

*Author for correspondence, e-mail: rmkelly@ncsu.edu

D. Zannoni and R. De Philippis (eds.), Microbial BioEnergy: Hydrogen Production, Advances in Photosynthesis and Respiration 38, pp. 177–195, DOI 10.1007/978-94-017-8554-9_8,

glucose equivalent. *Caldicellulosiruptor* species can utilize pentoses, hexoses, di/oligosaccharides, as well as complex polysaccharides, including crystalline cellulose. The broad appetite of these organisms relates to the natural environment of *Caldicellulosiruptor*, where they thrive at high temperatures (65–78 °C), utilizing the variable saccharide composition of lignocellulosic biomass as growth substrate. The ability to degrade recalcitrant plant biomass and utilize a wide variety of polysaccharides in their fermentation pathways sets *Caldicellulosiruptor* species apart from many other candidate biofuel-producing microorganisms. The conversion of lignocellulose to fuels in *Caldicellulosiruptor* is driven by an array of novel multi-domain glycoside hydrolases that work synergistically to degrade plant polysaccharides into oligo/monosaccharides that enter the cytoplasm via an array of carbohydrate specific ABC sugar transporters. These carbohydrates are then processed through a series of catabolic pathways, after which they enter the EMP pathway to produce reducing equivalents in the form of NADH and Fd^{red}. The reducing equivalents are ultimately utilized by both cytoplasmic and membrane-bound hydrogenases to form molecular hydrogen. Recently completed genome sequences for a number of *Caldicellulosiruptor* species have revealed important details concerning how plant biomass is deconstructed enzymatically and shown significant diversity within the genus with respect to lignocellulose conversion strategies.

I. Introduction

The genus *Caldicellulosiruptor* is comprised of extremely thermophilic, gram-positive bacteria with optimal growth temperatures between 65 and 78 °C (Blumer-Schuette et al. 2010; Hamilton-Brehm et al. 2010). Members of the genus are associated with plant debris in high temperature terrestrial hot springs and mud flats worldwide (Fig. 8.1). Currently, eight *Caldicellulosiruptor* species have sequenced genomes, providing important insights into the metabolic and physiological traits of these extreme thermophiles (van de Werken et al. 2008; Kataeva et al. 2009; Elkins et al. 2010; Blumer-Schuette et al. 2011). Common to all species is the capability to convert complex polysaccharides into simple sugars, which are then fermented to molecular hydrogen, acetate, lactate and small amounts of alcohol (Rainey et al. 1994; Ahring 1995; Huang et al. 1998; Bredholt et al. 1999; Miroshnichenko et al. 2008; Hamilton-Brehm et al. 2010; Yang et al. 2010). *Caldicellulosiruptor* species have potential importance for biofuels production, since they produce H_2 near the Thauer limit of 4 mol H_2 per mol glucose (Ivanova et al. 2009; de Vrije et al. 2009; Zeidan and van Niel 2010; Willquist and van Niel 2012).

The long list of complex polysaccharides serving as growth substrates for members of the genus *Caldicellulosiruptor* includes α- and β-glucans, mannans, xylans, pectin and, for some species, crystalline cellulose (Rainey et al. 1994; Ahring 1995; Huang et al. 1998; Bredholt et al. 1999; Miroshnichenko et al. 2008; Hamilton-Brehm et al. 2010; Yang et al. 2010; Blumer-Schuette et al. 2012). The genus collectively contains 106 glycoside hydrolases (GH), representing 43 GH families, and an array of ATP-binding

Abbreviations: ABC – ATP binding cassette; ADH – Alcohol dehydrogenase; CAZy – Carbohydrate-active enzyme; CBM – Carbohydrate binding module; CCR – Carbon catabolite repression; CE – Carbohydrate esterase; CUT – Carbohydrate uptake; DPP – Di-peptide; EMP – Embden-Meyerhoff-Parnas; FD^{red} – Reduced ferredoxin; GH – Glycoside hydrolase; LDH – Lactate dehydrogenase; OPP – Oligo-peptide; PL – Polysaccharide lyase; PPP – Pentose phosphate pathway; PTS – Phosphoenolpyruvate-dependent phosphotransferase; SLH – S-layer homology; TCA – Tricarboxylic acid

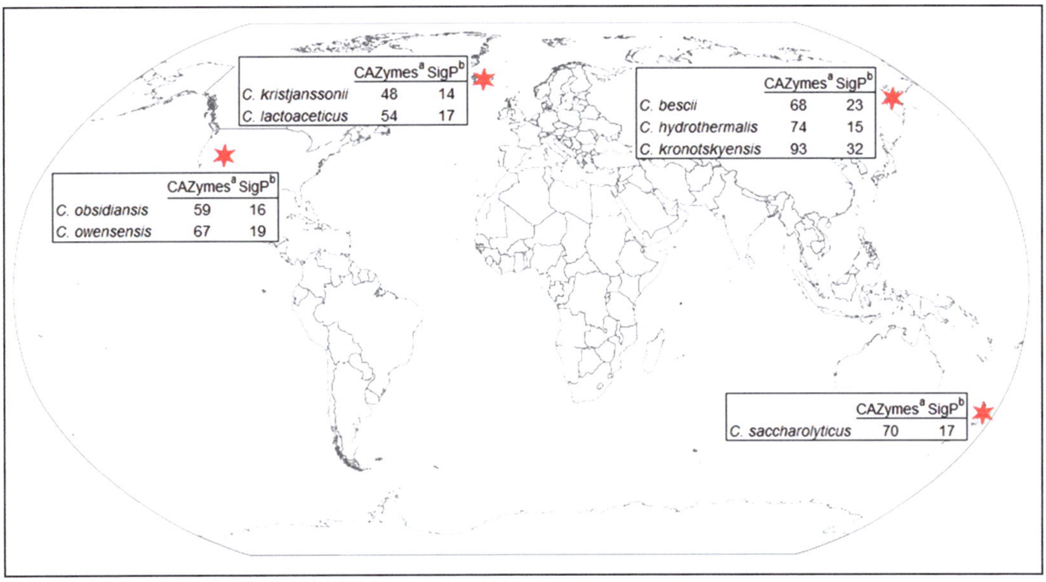

Fig. 8.1. **Geographic distribution of *Caldicellulosiruptor* species and Carbohydrate Active Enzymes** ([a]=Number of ORFs encoding either a CBM, CE, GH or PL; [b]=signal peptide is encoded in the ORF).

cassette (ABC) transporters belonging to the Carbohydrate Uptake 2 (CUT 2), Carbohydrate Uptake 1 (CUT 1), and Di/Oligopeptide (Dpp/Opp) families (Vanfossen et al. 2009). These GHs and transporters are deployed to synergistically process complex polysaccharides prior to entering into fermentation pathways (Blumer-Schuette et al. 2012). While many microorganisms preferentially utilize hexose over pentose sugars and often exhibit carbon catabolite repression (CCR) (Gancedo 1998; Brückner and Titgemeyer 2002), this is not the case for *Caldicellulosiruptor* species (Vanfossen et al. 2009). The lack of CCR makes *Caldicellulosiruptor* species especially promising in decomposing characteristically heterogeneous plant biomass to molecular hydrogen.

Although the discovery and initial isolation of *Caldicellulosiruptor* species (*C. saccharolyticus* formerly *Caldocellum saccharolyticum*) occurred more than 20 years ago (Donnison et al. 1986; Rainey et al. 1994), it was only within the past 5 years, concomitant with the increased interest in biofuels, that these bacteria have received intense interest. An overview of current progress in studying *Caldicellulosiruptor* is provided here, with an eye towards how these bacteria produce molecular hydrogen from complex carbohydrates, especially lignocellulosic biomass.

II. Extracellular Deconstruction of Lignocellulosic Biomass

The production of molecular hydrogen from plant biomass begins with extraction and deconstruction of the carbohydrate content of lignocellulose into fermentable sugars (Fig. 8.2).

A. Lignocellulose Composition and Recalcitrance

Lignocellulose is primarily composed of cellulose, hemicellulose and lignin; the physical and chemical properties of these polymers varies between plant species, stages of growth, and environmental conditions (Reddy and Yang 2005). Cellulose, the major structural component, is a long chain of glucose molecules linked by β–1,4 glycosidic bonds (van Wyk 2001). Hydrogen bonds between the polysaccharide chains form

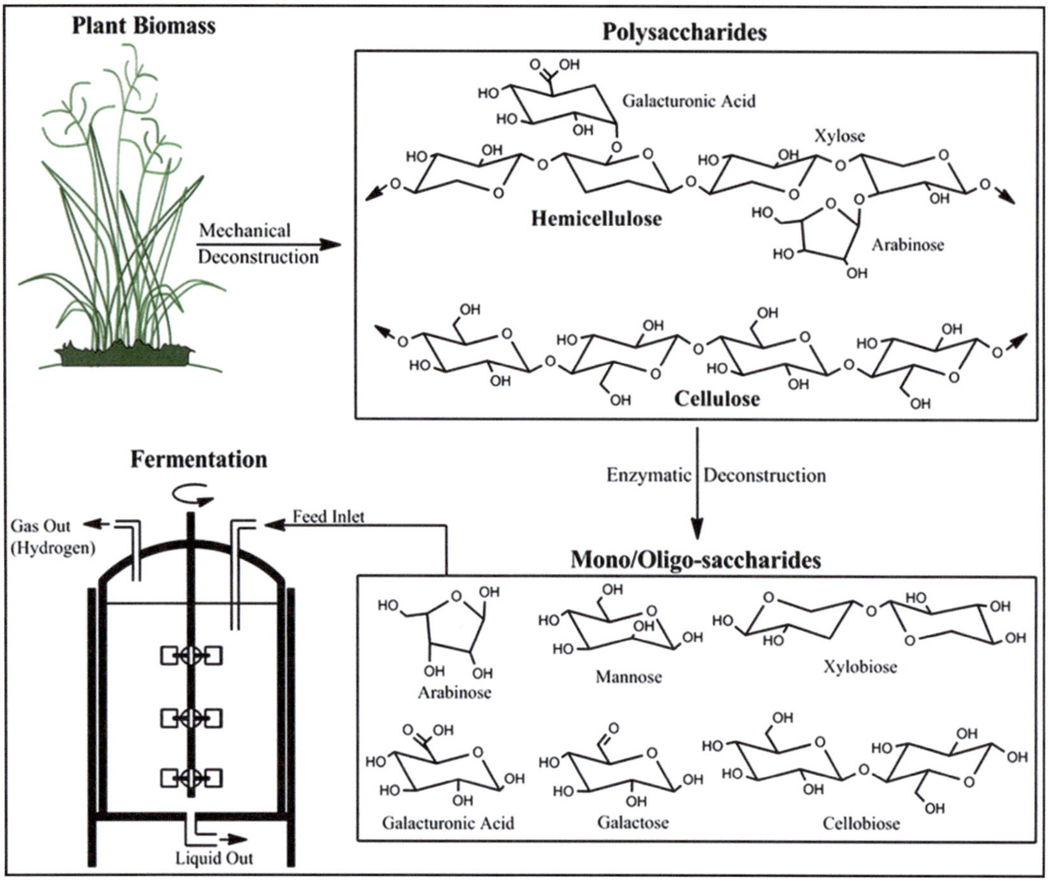

Fig. 8.2. **Conversion process of plant biomass to fuels**. Plant biomass is first mechanically degraded to yield long chained polysaccharides, which are then enzymatically deconstructed to shorter chained mono/oligo saccharides. Shorter chained saccharides are fermented to produce biofuels.

crystalline cellulose, conferring an increased resistance to degradation (Rubin 2008). Hemicellulose is a heteropolymer, consisting of xylose, mannose, galactose, glucose, arabinose and glucuronic and galacturonic acids. These sugars are linked primarily by β-1,4 and β-1,3 glycosidic bonds. Lignin is an amorphous, water-insoluble heteropolymer, consisting of phenylpropane units joined by different types of linkages. Lignin acts as molecular "glue", conferring structural support, impermeability and resistance to microbial attack (Fig. 8.3) (Pérez et al. 2002; Rubin 2008). The antimicrobial characteristics of lignin and crystallinity of cellulose are the two major challenges in the lignocellulosic deconstruction process. Microorganisms that can overcome the recalcitrance of cellulose in the context of potentially toxic lignin moieties are especially interesting for biofuels production.

B. Enzymatic Lignocellulose Deconstruction

The deconstruction of lignocellulose by *Caldicellulosiruptor* initially involves extracellular enzymatic attack of the plant biomass substrate. Members of the genus utilize an array of extracellular glycoside hydrolases (GHs), polysaccharide lyases (PLs), and carbohydrate esterases (CEs) that break the glycosidic linkages of long-chained polysaccharides to eventually yield oligosaccharides and simple sugars (Blumer-Schuette et al. 2012). These are then transported into the cell for utilization in metabolic pathways.

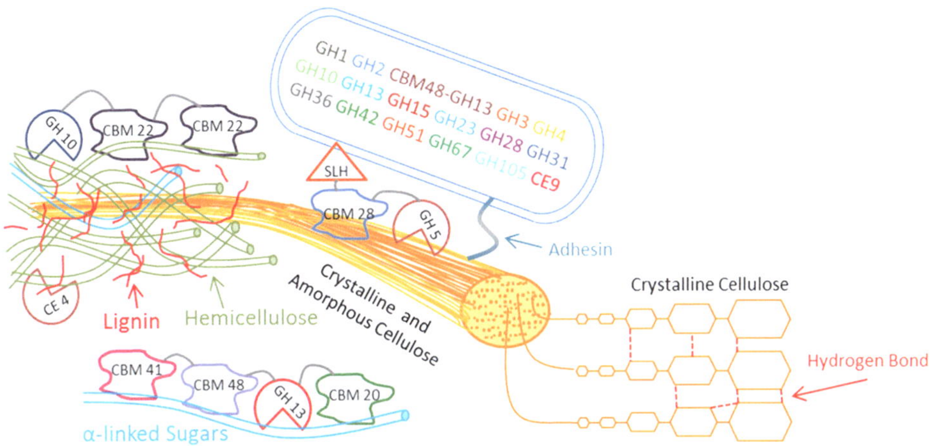

Fig. 8.3. **Lignocellulose microfibril with *Caldicellulosiruptor* core extracellular enzymes**. Hemicellulose and lignin form a protective sheath around cellulose. Core enzymes have activity against α-linked sugars, xylan and amorphous cellulose.

The degradation of crystalline cellulose and other recalcitrant plant polysaccharides requires the synergistic action of multiple catalytic domains, often within the same enzyme. The efficacy of these enzymes can be enhanced through the conjugation of the catalytic subunit(s) with one or more carbohydrate binding modules (CBM). The CBMs act to increase the catalytic efficiency by targeting the catalytic GH unit toward accessible polysaccharide, disrupting the polysaccharide structure, and maintaining the substrate in prolonged intimate contact with the catalytic GH (Shoseyov et al. 2006). The end goal of extracellular polysaccharide degradation is the production of carbohydrates in a transportable form, typically with six or fewer saccharide units.

The strategy used to generate small, transportable saccharides differs across the microbial world. For example, the cellulolytic fungus, *Trichoderma reesei*, utilizes extracellular enzymes, not associated with the cell, that contain a single catalytic domain and, in many cases, a single CBM (Martinez et al. 2008). The cellulosome, initially discovered in *Clostridium thermocellum*, is a multi-protein complex constructed around an enzymatically inactive scaffoldin. It contains cohesin domains for the attachment of enzyme subunits and a CBM to mediate attachment to the substrate. Enzyme subunits, which contain dockerin domains, attach to the scaffoldin via cohesin-dockerin interactions. Similarly, interactions between a dockerin domain on the scaffoldin and a cell-associated cohesin domain anchor the cellulosome complex to the cell (Bayer et al. 1983, 1998; Fontes and Gilbert 2010). Members of the genus *Caldicellulosiruptor* are non-cellulosomal, but do employ several multi-domain enzymes that mediate cellular attachment to plant biomass through S-layer homology (SLH) domains (Ozdemir et al. 2012). The S-layer containing enzymes in *Caldicellulosiruptor* are much smaller than the cellulosome, and have one or two catalytic domains coupled with one or more CBM (Blumer-Schuette et al. 2010; Dam et al. 2011; VanFossen et al. 2011). The *Caldicellulosiruptor* SLH-domain containing proteins with additional GH and/or CBM domains, contribute to biomass degradation by localizing the substrate and holding the cell in close proximity (Fig. 8.3). The majority of extracellular GHs encoded in *Caldicellulosiruptor* genomes lack SLH domains, such that they freely diffuse in the biomass-containing milieu. The presence of multiple catalytic domains within a single

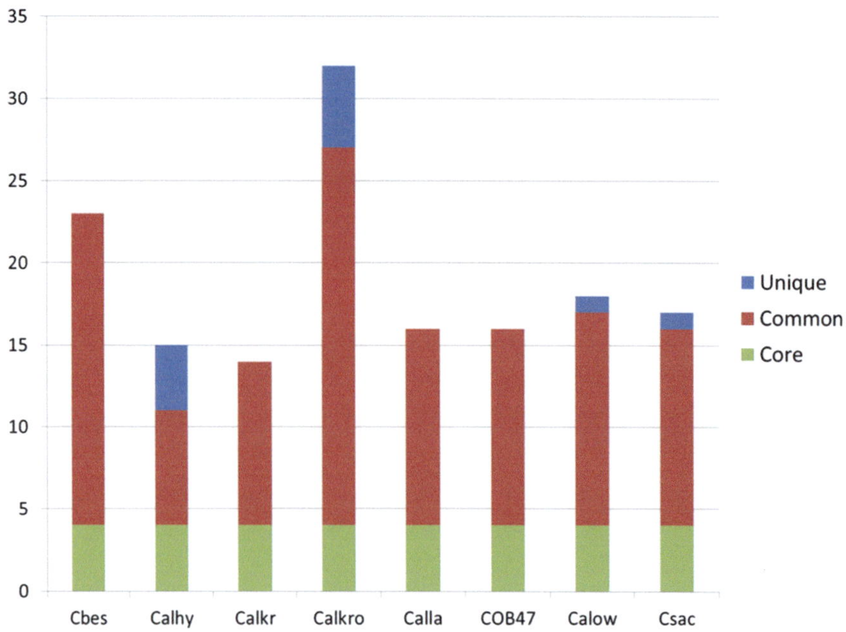

Fig. 8.4. **Extracellular glycoside hydrolases of *Caldicellulosiruptor* species**. Core GHs are common to all species. Common GHs are possessed by one or more species, while unique GHs are only in present in a particular species. Abbreviations follow the assigned locus tags and are as follows: Cbes *C. bescii,* Calhy C. *hydrothermalis,* Calkr *C. kristjanssonii,* Calkro *C. kronotskyensis,* Calla *C. lactoaceticus*, COB47 *C. obsidiansis,* Calow *C. owensensis,* Csac *C. saccharolyticus*.

extracellular enzyme imparts the capacity to degrade complex heterogeneous polysaccharides synergistically.

The pan-genome of *Caldicellulosiruptor* encodes 134 carbohydrate-active enzymes (CAZy) (GHs, CEs, PLs and CBMs), of which 106 are GHs, representing 43 GH families. However, only 26 GHs from 17 families are included in the core genome (Cantarel et al. 2009; Blumer-Schuette et al. 2012). Presumably, the core set of enzymes contains the basic catalytic capacity required for growth on plant biomass by members of the genus. It may be necessary, but not sufficient, for plant biomass deconstruction, since all *Caldicellulosiruptor* species contain additional GHs in the core genome. The core set of GHs include four out of the five known GH families that hydrolyze the β-1,4 xyloside linkages characteristic of xylan, three out of the four GH families that hydrolyze the β-1,4 mannoside linkages of mannan, and four out of the five xyloglucanase families that hydrolyze β-1,4 glucan linkages (Blumer-Schuette et al. 2010).

1. Core Caldicellulosiruptor *Hydrolytic Enzymes*

The core carbohydrate active enzyme component of the *Caldicellulosiruptor* genome includes four extracellular enzymes (Figs. 8.3 and 8.4), identified by the presence of a signal peptide at the N-terminus, directing the protein to be secreted into the extracellular environment (Navarre and Schneewind 1999). The extent to which the core set extracellular enzymes can degrade lignocellulosic substrates is based on biochemical characteristics, homology and phenotypic characteristics of the genus. Csac_0678, a bi-functional GH5 conjugated to a CBM28 and S-layer homology (SLH) domains (Fig. 8.3), has orthologs in all *Caldicellulosiruptor* genomes. As mentioned above, the S-layer homology domains of this enzyme act to anchor the enzyme to the cell surface, while the CBM facilitates attachment of the multi-domain enzyme to the substrate (Sára and Sleytr 2000). Biochemical characterization of Csac_0678 showed that the GH5 domain exhibited both

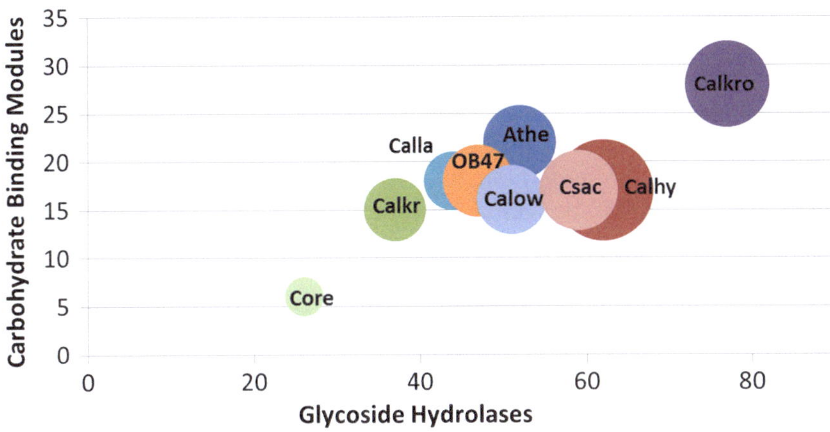

Fig. 8.5. **Number of ORFs containing glycoside hydrolases, carbohydrate binding modules and ABC transporters**. *Bubble size* correlates to the number of ABC transporters in each species.

endoglucanase and xylanase activity, while the CBM28 was required for activity and binding to crystalline cellulose (Ozdemir et al. 2012). Two other extracellular core GHs are a putative xylanase, containing a GH10 domain conjugated to two CBM22 domains, and a putative amylase with a GH13 domain conjugated to a CBM41, a CBM48 and a CBM20 (Janecek 1997; Andrews et al. 2000). The remaining extracellular core enzyme is a CE family 4 enzyme with putative xylanase activity (Caufrier et al. 2003; Cantarel et al. 2009). This core set of extracellular enzymes theoretically provides the genus with the ability to hydrolyze α- and β-glucan linkages of starch and cellulose, respectively, in addition to β-xyloside linkages of xylan. It should be noted, even though the core extracellular enzyme set of *Caldicellulosiruptor* contains biocatalysts active against the β-glucan linkages of cellulose, this does not necessarily mean crystalline cellulose deconstruction is possible, as not all species are able to efficiently hydrolyze this substrate.

2. Cellulolytic Caldicellulosiruptor *Enzymes*

Beyond the core genome, the presence and absence of specific types of extracellular GHs in *Caldicellulosiruptor* species correlates to the capacity to utilize crystalline cellulose (Blumer-Schuette et al. 2010). In particular, growth on Avicel and filter paper differentiates the cellulolytic members of the genus. For example, the strongly cellulolytic species: *C. bescii*, *C. kronotskyensis*, *C. saccharolyticus* and *C. obsidiansis* grow well on Avicel and filter paper, while *C. lactoaceticus* grows to a lesser extent on these substrates. The weakly cellulolytic species, *C. hydrothermalis*, *C. kristjanssonii* and *C. owensensis*, grow to a limited extent on filter paper, with no visible deconstruction of the solid substrate. Within the sequenced *Caldicellulosiruptor* genomes, *C. kronotskyensis* contains the most carbohydrate-active encoded enzymes, indicating the ability to degrade a wide range of polysaccharides (Figs. 8.4 and 8.5) (Blumer-Schuette et al. 2012). The genomes of the four strongly cellulolytic species contain a shared set of seven GHs, three of which are extracellular. These extracellular multi-domain enzymes each contain different GH domains (GH9 and GH48, GH74 and GH48, or GH9 and GH5) linked by CBM3 modules. The activity of one or more of these extracellular GHs presumably confers the ability to degrade crystalline cellulose. In order to determine which of these enzymes confers the degradation of crystalline cellulose, the weakly cellulolytic species were inspected for the presence of these four GH families. All *Caldicellulosiruptor* genomes sequenced to date harbor GH5-containing enzymes.

However, while the *C. kristjanssonii* genome encodes a putative extracellular enzyme containing GH9 and GH74 domains linked to CBM3 domains, this bacterium is weakly cellulolytic. As such, the presence of GH5, GH9 and GH74 enzyme families is not necessarily indicative of crystalline cellulose hydrolytic capacity in the genus *Caldicellulosiruptor*. On the other hand, GH48 family enzymes cannot be identified in the genomes of any of the weakly cellulolytic species, suggesting the presence of a GH48 domain is an essential determinant for the ability to hydrolyze crystalline cellulose by *Caldicellulosiruptor* species (Blumer-Schuette et al. 2010, 2012). Furthermore, the coupling of GH48 with CBM3 domains is indicative of strong cellulolytic capacity. Along these lines, CelA, the GH9-, GH48- and CBM3-containing enzyme present in the cellulolytic species, has been characterized biochemically. CelA, isolated from *C. bescii* culture supernatants, as well as specific GH domains produced recombinantly in *E. coli,* had activity against crystalline cellulose and other β-linked glucans (Te'o et al. 1995; Zverlov et al. 1998), demonstrating the importance of CelA to the cellulolytic phenotype in *Caldicellulosiruptor*. As genetic tools for this genus become available, it will be interesting to see if the insertion of a GH48-domain containing enzyme can impart a strong celluloytic capacity on the weakly celluloytic species in this genus or if the absence of CelA results in loss of capacity to degrade crystalline cellulose.

III. Carbohydrate Transport

Upon degradation of long-chained polysaccharides to di/oligosaccharides by extracellular enzymes of *Caldicellulosiruptor* species, the simpler sugars are transported into the cell via transmembrane carbohydrate transport systems for use in anabolism or catabolism (VanFossen et al. 2011). Given the wide-ranging inventory of GHs found in the various *Caldicellulosiruptor* species, it is not surprising that there is also significant variability in the number and specificity of substrate transporters across the genus. ABC and phosphoenolpyruvate-dependent phosphotransferase (PTS) carbohydrate transport systems can be identified in *Caldicellulosiruptor* genomes, although the presence of PTS transporters in the genus is sparse and variable. ABC carbohydrate transporters typically belong to one of two groups, the carbohydrate uptake transporter (CUT) family and the Di/Oligopeptide transporter family (Dpp/Opp) (Schneider 2001). The CUT-family transporters are further divided into two sub-families, differentiated in architecture and substrate specificity. CUT sub-family 1 (CUT1) systems, in *Caldicellulosiruptor,* transport both di/oligosaccharides and monosaccharides (Vanfossen et al. 2009). CUT1 transporters consist of an extracellular substrate binding protein, two membrane proteins forming the translocation path, and a single ATP binding subunit likely in the form of a homodimer. The CUT2 sub-family is solely involved in monosaccharide transport, containing a single membrane protein, presumably a homodimer, and two fused ATPase domains. The Dpp/Opp transport family has been implicated in the transport of di- and oligopeptides, nickel, heme, as well as sugars. Its architecture is a combination of CUT1 and CUT2 sub-family features, with an extracellular binding protein, two membrane domains and two ATPase domains that form a heterodimer (Koning et al. 2002). The genus *Caldicellulosiruptor* collectively contains 45 ABC transporters, with the core genome consisting solely of 6 CUT1 transporters (Fig. 8.5) (Blumer-Schuette et al. 2012). The weakly cellulolytic *C. hydrothermalis* contains the greatest number of ABC transporters, indicating carbohydrate transporter inventory is not necessarily correlated to a strongly cellulolytic phenotype (Fig. 8.5). Across the genus, CUT1 transporters appear to be responsible for the majority of carbohydrate transport into the cell, making up 37 of the 45 identifiable transporter systems in *Caldicellulosiruptor* genomes. Dpp/Opp and CUT 2 systems account for 3 and 5 of the ABC transporters present in the genus, respectively.

Currently, none of the *Caldicellulosiruptor* ABC transporters have been biochemically characterized. Even with the lack of specific biochemical knowledge, bioinformatics analysis can be used to map transport substrates and transport mechanisms through homology with other characterized transporters. VanFossen et al. (2009) analyzed the transcriptomes of *C. saccharolyticus* grown on glucose, fructose, mannose, xylose, arabinose, galactose and a mixture of all these sugars, in addition to xylan, xylose, xyloglucan and xylogluco-oligosaccharides. These data-sets, using metrics developed with previous work on *Thermatoga maritima*, a heterotrophic hyperthermophile (Conners et al. 2005), could be used to predict carbohydrate preference of the majority of transporters in *C. saccharolyticus*. It was concluded that the genome of *C. saccharolyticus* contained transporters for all the substrates tested. The carbohydrate specificities of the ABC transporters had either limited specificity for only one substrate, as is often observed with oligosaccharide transporters, or broad specificity for a variety of substrates, as is often the case with monosaccharide transporters. Ultimately, *C. saccharolyticus* is able to transport and utilize the wide variety of carbohydrates, simple or complex, that result from lignocellulosic biomass hydrolysis.

Phenotypic and genotypic differences can provide insight into the role of specific ABC transporters in carbohydrate transport. For example, *C. lactoaceticus* is incapable of growth on glucose, even though it hydrolyzes cellulose, raising the prospect that glucose catabolism could be transport-limited. *C. lactoaceticus* also has the fewest number of carbohydrate ABC transporters within the genus (Fig. 8.5) (Blumer-Schuette et al. 2012). Closely related *C. kristjanssonii* is capable of growth on glucose and only contains three ABC transporters not present in the *C. lactoaceticus* genome, suggesting that one of these three transporters imparts the capacity for glucose transport. Two of these transporters are members of the CUT1 and CUT2 transporter families with orthologs in all other *Caldicellulosiruptor* species. VanFossen et al. (2009) predicted that these two transporters are involved in glucose, fructose and xylose transport. In fact, these are the only transporters identified to transport glucose into *C. saccharolyticus*. Taken together, these transporters seem to enable growth on glucose by *C. kristjanssonii* and most likely other *Caldicellulosiruptor* species.

The pan-genome of *Caldicellulosiruptor* contains one identified PTS (Blumer-Schuette et al. 2012). In many organisms, the PTS is bi-functional, playing roles in carbohydrate transport and as a starting point in regulating carbon catabolism (Stulke and Hillen 2000; Kotrba et al. 2001; Brückner and Titgemeyer 2002). The PTS consists of two cytosolic energy coupling proteins (Enzyme I and histidine-containing protein (HPr)) and carbohydrate specific, Enzyme II, which catalyzes concomitant carbohydrate translocation and phosphorylation at the expense of PEP (Kotrba et al. 2001). In *Caldicellulosiruptor*, the PTS is currently the only identified mannose transporter and has been implicated in fructose transport (Vanfossen et al. 2009). The possible role of the PTS in carbohydrate catabolite regulation in *Caldicellulosiruptor* has not been established, although in a mixture of saccharides including galactose, glucose, mannose, xylose, arabinose and fructose, *C. saccharolyticus* utilized fructose to the greatest extent, followed by arabinose and xylose (Vanfossen et al. 2009). Whether the fructose specific PTS plays a role in regulation of substrate utilization is not known. In addition, genomes of *Caldicellulosiruptor* species encode the genes required for the carbon control protein A (CcpA)-dependent CCR present in *B. subtilis* and other gram-positive bacteria. The components of the CcpA-dependent CCR signaling cascade present in *Caldicellulosiruptor* include the fructose specific PTS transporter, HPr(Ser) kinase, catabolic repression HPr protein (CrH) and the CcpA (Warner et al. 2003; van de Werken et al. 2008). Though *Caldicellulosiruptor* does not exhibit traditional CCR, the combination of the fructose specific PTS, the genes encoding CcpA-dependent CCR, and proclivity for fructose utilization implies that this system plays a role in the carbohydrate preferences of these bacteria.

IV. Intermediary Metabolism

The genus *Caldicellulosiruptor* can utilize a range of carbohydrates for growth, as such, an array of metabolic pathways are implicated in bioenergetics. Genome sequence data and ^{13}C-NMR analysis revealed *C. saccharolyticus* contains a complete Embden-Meyerhoff-Parnas (EMP) pathway (de Vrije et al. 2007; van de Werken et al. 2008). The EMP pathway, which serves as the primary generator of ATP and reducing equivalents, is conserved within the genome sequenced members of the genus. The central role of the EMP pathway in *Caldicellulosiruptor* metabolism requires all carbohydrate growth substrates be directly or indirectly fed into the pathway for energy generation. Glucose, liberated from cellulose and starch, can be oxidized directly by the EMP pathway. The hydrolysis products of hemicellulose, such as xylose, pectin and galactose, must first be processed through alternative pathways. The products of these sub-pathways are then funneled into the EMP pathway at different levels (Fig. 8.6). These sub-pathways are often incomplete and have varying levels of conservation across the genus. The ability to metabolize xylose is conserved within the genus. Xylose, the major constituent of hemicellulose, is readily available during growth on lignocellulosic biomass. Xylose enters the non-oxidative branch of the pentose phosphate pathway (PPP) via conversion by a xylose isomerase and xylulokinase. Arabinose, often associated with xylan, is also funneled into the non-oxidative branch of the pentose phosphate pathway. Unlike xylose, it is converted into PPP intermediates by means of a bifunctional L-fucose/D-arabinose isomerase and a L-ribulokinase that are not conserved in the North American or Icelandic *Caldicellulosiruptor* species (Figs. 8.1 and 8.6) (van de Werken et al. 2008). The lack of these enzymes in the Icelandic species correlates with their inability to grow on arabinose (Ahring 1995; Bredholt et al. 1999). In contrast, the North American species are capable of growth on arabinose, indicating the presence of alternative enzymes for arabinose metabolism (Huang et al. 1998; Hamilton-Brehm et al. 2010). Metabolism of xylose and arabinose through the non-oxidative PPP yields β-D-fructose-6P or glyceraldehyde-3P, early metabolites in glycolysis (Fig. 8.6). Though members of the genus *Caldicellulosiruptor* rely on the non-oxidative PPP for the metabolism of many carbohydrates, the oxidative branch of the PPP is not present, akin to other anaerobic biomass degraders in the class Clostridia (Hemme et al. 2011). The oxidative branch of the PPP pathway in many organisms is the sole generator of NADPH, the primary source of reducing equivalents for cellular biosynthetic pathways (Kruger and von Schaewen 2003). There appear to be other enzymes in *Caldicellulosiruptor* with the capability for generating NADPH, but the exact physiological roles of these enzymes is unclear (van de Werken et al. 2008). This raises questions as to the mode and extent of NADPH generation within the cell.

Uronic acids, the building blocks of pectin, are primarily composed of galacturonic acid (Ridley and O'Neill 2001). All *Caldicellulosiruptor* species have been described to support growth on pectin (Rainey et al. 1994; Ahring 1995; Huang et al. 1998; Bredholt et al. 1999; Miroshnichenko et al. 2008; Hamilton-Brehm et al. 2010; Yang et al. 2010). Galacturonate, the anion of galacturonic acid, enters metabolism through isomerization to tagaturonate. Upon conversion to tagaturonate, the pertinent metabolic pathway becomes unclear, as tagaturonate reductase and altronate hydrolase, have not been identified in the genus. This implies the use of a novel pathway or unidentified enzymes for the conversion of galacturonate.

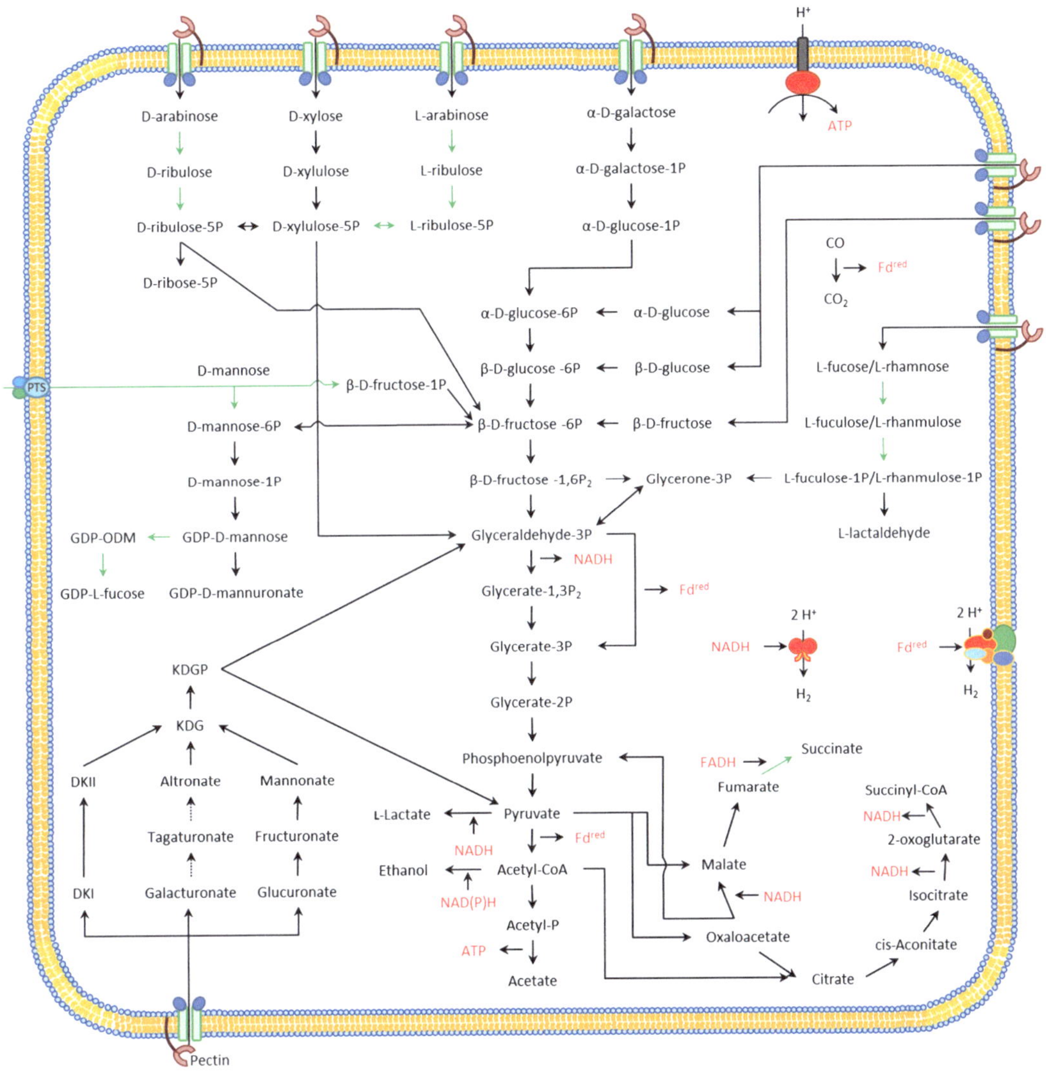

Fig. 8.6. **Metabolic features of *Caldicellulosiruptor* species**. *Green arrows* indicate reactions not conserved in all species. Abbreviations: *DKI* 5-keto-4-deoxyuronate, *DKII* 2,5-Diketo-3-deoxy-D-gluconate, Fd^{red} reduced ferrodoxin, *KDG* -2-Dehydro-3-deoxy-D-gluconate, *KDGP* KDG phosphate, *NADH* reduced nicotinamide adenine dinucleotide, *P* phosphate.

Similar to the conversion of xylose and arabinose, the metabolism of the deoxysugars, such as fucose and rhamnose, is variable within the genus. Fucose, is found as a subunit of xyloglucans (Hisamatsu et al. 1991) and rhamnose is a common component of pectin (Komalavilas and Mort 1989; Ridley and O'Neill 2001). Icelandic *Caldicellulosiruptor* species are incapable of growth on rhamnose (Ahring 1995; Bredholt et al. 1999), so it was not surprising to correlate the lack of rhamnose isomerase and rhamnulokinase from their genome sequences to this physiological trait. Limited information is available for growth of *Caldicellulosiruptor* species on fucose, which presumably requires fucose isomerase for this phenotype. North American and Icelandic species both lack fucose isomerase and consistent with this observation, *C. obsidiansis* is unable to utilize fucose as a growth substrate (Hamilton-Brehm et al. 2010). Mannose and galactose are also found as constituents of hemicellulose, but in smaller amounts than xyloglucan. Galactose is metabolized to glucose-6P through the Leloir pathway (Holden et al. 2003). The Leloir pathway is conserved in all sequenced *Caldicellulosiruptor* species. Mannose is typically carried across the cell membrane via a PTS transporter where it is phosphorylated to mannose-6P. Several *Caldicellulosiruptor* species (*C. kristjanssonii*, *C. lactoaceticus* and *C. obsidiansis*) lack homologs to a PTS, yet have the ability of growth on mannose. An alternative system for mannose phosphorylation has not yet been reported in the genus. Fructose can also be transported via the same PTS to yield fructose-1P, which is then shuttled directly into the EMP pathway.

While the tricarboxylic acid (TCA) cycle is not involved directly in substrate utilization, it is important because essential precursors to biosynthetic pathways are produced. The TCA cycle in *Caldicellulosiruptor* species is incomplete, however all species have an oxidative branch to succinyl-CoA and a reductive branch to Fumarate (Fig. 8.6). The incomplete TCA cycle present in *Caldicellulosiruptor* likely functions to generate amino acid biosynthesis precursors, such as 2-oxoglutarate (alpha-ketoglutaric acid) and oxaloacetate, rather than reducing equivalents. The production of excess reducing equivalents in the TCA cycle could overwhelm the fermentative *Caldicellulosiruptor* without the presence of an aerobic electron transport chain.

V. Metabolism of Fuel Production

The degradation of recalcitrant plant biomass and subsequent utilization of polysaccharides in *Caldicellulosiruptor* fermentation pathways produces several metabolic products including ethanol and molecular hydrogen.

A. Ethanol

The genus *Caldicellulosiruptor* has the ability to produce small amounts of ethanol, indicating pathways to this fermentation product exist or ethanol the result of promiscuous enzymes. Instead of ethanol production, most carbon is directed toward acetate, and as a consequence, large quantities of molecular hydrogen are produced as a fermentation product (Fig. 8.6). The primary role of hydrogen and ethanol production in anaerobic metabolism is to re-oxidize reducing equivalents generated during the fermentation of sugars. The production of these compounds is dependent on environmental conditions and growth state. Ethanol production occurs via the reduction of acetyl-CoA by alcohol dehydrogenase (ADH). Ethanol production serves as an efficient means to recycle reducing equivalents in many other organisms, but ethanol has only been detected in very low to trace levels in *Caldicellulosiruptor* (Rainey et al. 1994; Ahring 1995; Huang et al. 1998; Bredholt et al. 1999; Hamilton-Brehm et al. 2010; Yang et al. 2010) and thus, has not been studied in detail.

Caldicellulosiruptor species contain several putative ADHs, but the specific enzyme responsible for the conversion of acetyl-CoA to ethanol is unknown. In *Thermoanaerobacter psuedethanolicus* (formerly *Thermoanaerobacter ethanolicus* 39E) ethanol production is NADPH-dependent, through the activity of a bi-functional alcohol dehydrogenase/acetyl-CoA thioesterase (Burdette and Zeikus 1994). A putative ADH in *C. saccharolyticus* (Csac_0395) contains a NADPH-binding domain and sequence similarity to the bi-functional enzyme from *T. pseudethanolicus* (van de Werken et al. 2008). This suggests that ethanol production in *Caldicellulosiruptor* is NADPH-dependent, and targeted to oxidizing NADPH, rather than the NADH generated during glycolysis. However, due to the lack of an oxidative branch of the PPP, the mode of generation and levels of NADPH are unknown. It is likely that the amount of NADPH produced is limited, leaving NADPH regeneration to $NADP^+$ to biosynthetic pathways, resulting in minimal NADPH levels available for ethanol production. In another example, ADH activity of an ethanol adapted mutant strain of *Clostridium thermocellum* shifted from NADH to NADPH dependence, suggesting similarities for ethanol tolerance mechanisms and redox homeostasis (Brown et al. 2011). The specific role of ethanol production in *Caldicellulosiruptor* has not been explored; it is not likely a means of controlling the cellular redox balance, since homeostasis is maintained through hydrogen and lactate production.

B. Hydrogen

Caldicellulosiruptor species produce significant amounts of molecular hydrogen as a fermentation product, such that the maximum yield of H_2 is among the highest for hydrogen-producing microorganisms. *Caldicellulosiruptor* species employ the EMP pathway to achieve a maximum theoretical yield (Thauer limit) of 4 moles H_2 per mol glucose (Thauer et al. 1977). *Caldicellulosiruptor* can utilize both the less energetic NADH and preferential reduced ferredoxin (Fd^{red}) for the reduction of protons to produce molecular hydrogen. Both of these reducing equivalents are generated during the oxidation of sugars in the EMP pathway. Fd^{red} is generated from the oxidation of pyruvate to acetyl-CoA by pyruvate: ferredoxin oxidoreductase (PFOR) and the oxidation of glyceraldehyde-3P to glycerate-3P by aldehyde ferredoxin oxidoreductase. Alternatively, glyceraldehyde-3P can be oxidized to glycerate-3P via glyceraldehyde phosphate dehydrogenase (GAPDH) and phosphoglycerate kinase to generate NADH and ATP (Fig. 8.6).

1. Hydrogen Production and Carbohydrate Transport

Hydrogen production in *Caldicellulosiruptor* may be linked to the primary use of ABC transporters for carbohydrate translocation. The translocation of substrate by ABC transporters, and subsequent phosphorylation, requires two molecules of ATP. Alternatively, import of monosaccharides by PTS requires phosphoenolpyruvate as a phosphate donor to achieve transport and phosphorylation in one step. Currently, as mentioned above, there is one orthologous PTS identified in some species of the genus. *Caldicellulosiruptor* relies primarily on ABC transporters for carbohydrate transport, making the generation of a supplementary source of ATP molecules for carbohydrate transport advantageous to the cell. The oxidation of sugars to acetate generates an extra ATP (2 moles/mole hexose), offsetting the consumption by ABC transporters, while at the same time generating Fd^{red}. The production of H_2 is then used to re-oxidize Fd^{red} generated as a byproduct of ATP generation for carbohydrate transport.

2. Hydrogenases in Caldicellulosiruptor

In *Caldicellulosiruptor* species, the reduction of protons to molecular hydrogen occurs via two distinct hydrogenases, a cytoplasmic Fe-only hydrogenase (HydA to HydD), and a membrane-bound Ni-Fe hydrogenase (EchA to EchF). Though neither of these hydrogenases have been biochemically characterized, homologs in *Caldanaerobacter subterraneus* subsp. *tengcongensis* (formerly *Thermoanaerobacter tengcongensis*) were found to be NADH- and Fd^{red}-dependent, respectively (Fardeau et al. 2004; Soboh et al. 2004). A third, putative hydrogenase cluster, containing an NADH-binding protein, also exists, but the function of this cluster is unknown and is theorized to be redundant (van de Werken et al. 2008); however, this remains to be confirmed experimentally. The production of H_2 from Fd^{red} is energetically favorable; making H_2 production by the membrane bound Ni-Fe hydrogenase preferable. In contrast, the utilization of the NADH-specific, Fe-only hydrogenase is less favorable; only under a very limited set of conditions is the production of hydrogen from NADH thermodynamically favorable (Verhaart et al. 2010). It is interesting that this Fe-only hydrogenase has approximately 50 % amino acid sequence identity to a bifuricating hydrogenase in *T. maritima*. This bifuricating hydrogenase uses the exergonic oxidation of ferredoxin to drive the unfavorable oxidation of NADH to produce H_2 (Schut and Adams 2009). If the Fe-only hydrogenase of *Caldicellulosiruptor* is, indeed, bifuricating, NADH would serve as an energetically favorable substrate for the reduction of protons to H_2.

C. Growth Conditions and Hydrogen Production

During exponential growth, *Caldicellulosiruptor* produces H_2, CO_2 and acetate, almost exclusively as fermentation products (Van Niel et al. 2002; Zeidan and van Niel 2009). However, there are additional fermentation end products that are produced under specific physiological conditions. For example, increased H_2 concentrations and the transition to stationary phase, modulates NAD^+ regeneration and metabolic flux of pyruvate toward lactate formation via lactate dehydrogenase (LDH) (Willquist and van Niel 2010). Lactate formation consumes NADH and bypasses the production of Fd^{red} and ATP (Fig. 8.6). The regulation of flux at the pyruvate node is a function of LDH and hydrogenase activity.

1. Regulation of Lactate Dehydrogenase

The activity of LDH plays a key role in cellular ATP levels and redox potential, making its regulation important and complex. LDH is regulated by metabolic energy carriers: inorganic phosphate (PPi), ATP and NAD^+. The utilization of the energy carrier PPi is an alternative strategy used in *Caldicellulosiruptor* and other bacteria, to conserve energy (Mertens 1991; Bielen et al. 2010). The primary source of PPi is anabolic reactions, such as poly-nucleic acid biosynthesis and the activation of fatty acids and amino acids for lipid and protein synthesis (Heinonen 2001). Regulation of LDH occurs by both activation and inhibition; competitive inhibition occurs by PPi and NAD^+ and allosteric activation by fructose 1,6-bisphosphate, ATP and ADP (Willquist and van Niel 2010). The multitude of pathways generating and consuming these molecules results in variable activity of LDH. LDH activity has been shown to follow PPi levels and growth phase. For example, during exponential growth, high anabolic flux leads to increased generation of PPi, thereby inactivating LDH, and maximizing flux to acetate and hydrogen. As growth factors trigger stationary phase, PPi levels decrease and ATP levels increase (Bielen et al. 2010), enhancing the affinity of LDH to NADH redirecting carbon flux to lactate.

2. Hydrogen Concentration Affects Hydrogen Production

If the removal of metabolic H_2 from the growth environment is insufficient, levels of dissolved hydrogen in liquid and partial pressure in the gas phase will begin to increase. Increasing levels of H_2 severely inhibit hydrogen production through product inhibition (Ljunggren et al. 2011; van Niel et al. 2003). The decrease in hydrogen production results in accumulation of reducing equivalents, requiring changes in metabolic flux to balance the reactive species. The critical threshold value of hydrogen partial pressure varies with growth phase and study to study (Ljunggren et al. 2011; Willquist et al. 2011), but is typically 10–20 kPa, as determined in batch cultures of *C. saccharolyticus* (van Niel et al. 2003). H_2 inhibition is more directly related to dissolved H_2 concentrations. Ljunggren et al. (2011) found a critical dissolved H_2 concentration of 2.2 mmol/L results in complete inhibition of hydrogen production. Gas sparging can be used to alleviate rising H_2 concentrations (Chou et al. 2008), and specifically, N_2 sparging can increase hydrogen yields (Zeidan and van Niel 2010; Ljunggren et al. 2011; Willquist and van Niel 2012). However, at a process level, inert gas sparging is expensive and economically unfavorable. Alternatively, CO_2 is readily available from many industrial processes and can be relatively easily separated in downstream processing of the gas stream (Hallenbeck and Benemann 2002). However, sparging with CO_2 negatively affects growth and H_2 production in *C. saccharolyticus*. Dissolved CO_2, in the form of bicarbonate and protons, inhibits growth

Table 8.1. Reported hydrogen yields of *Caldicellulosiruptor* species.

Culture Type	Species	Substrate	Yield[a] (mol H_2/mol C_6)	Reference
Continuous	*saccharolyticus*	Glucose	3.8	Willquist et al. (2011)
Continuous	*saccharolyticus*	Glucose[b]	3.5	Willquist and van Niel (2012)
Trickle bed reactor	*saccharolyticus* (non-sterile)	Sucrose	2.8	van Groenestijn et al. (2009)
Batch	*saccharolyticus*	Miscanthus hydrolysate	3.4	de Vrije et al. (2009)
Batch	*saccharolyticus*	Paper sludge hydrolysate	2–3.8	Kádár et al. (2004)
Batch	*saccharolyticus*	Wheat straw	3.8	Ivanova et al. (2009)
Batch	*saccharolyticus*	Pretreated maize leaves	3.7	Ivanova et al. (2009)
Continuous	*kristjanssonii*	Glucose	3.5	Zeidan et al. (2010)
Batch	*kristjanssonii*	Glucose + Xylose	3.0	Zeidan and van Niel (2009)
Batch	*owensensis*	Glucose[b]	4.0	Zeidan and van Niel (2009)
Batch	*owensensis*	Xylose[b]	3.5	Zeidan and van Niel (2009)
Batch	*owensensis*	Glucose + Xylose	2.7	Zeidan and van Niel (2009)
Continuous	*saccharolyticus* + *kristjanssonii*	Glucose	3.7	Zeidan et al. (2010)
Continuous	*saccharolyticus* + *kristjanssonii*	Glucose + Xylose	3.6	Zeidan et al. (2010)
Batch	*saccharolyticus* + *kristjanssonii*	Glucose + Xylose	3.8	Zeidan and van Niel (2009)
Batch	*saccharolyticus* + *owensensis*	Glucose + Xylose	3.3	Zeidan and van Niel (2009)

[a]Maximum hydrogen yield reported at varying culture conditions (dilution rate, gas sparging etc.)
[b]Defined growth medium

through a decrease in pH and an increase in osmotic pressure, rendering CO_2 sparging infeasible (Willquist et al. 2009). Ljunggren et al. (2011) found an osmolarity between 0.27 and 0.29 mol/L to be inhibitory to the growth of *C. saccharolyticus*. Engineering *Caldicellulosiruptor* strains to be insensitive to increased osmotic pressures and pH changes and/or hydrogenases with a greater hydrogen tolerance will likely be a requirement of a *Caldicellulosiruptor*-based H_2 production process.

3. Hydrogen Yields

Experimental studies of hydrogen production in *Caldicellulosiruptor* have looked at both batch (Ivanova et al. 2009; Zeidan and van Niel 2009, 2010; Willquist and van Niel 2012) and continuous (de Vrije et al. 2007; Willquist et al. 2009; Zeidan et al. 2010) cultures. Hydrogen yields vary with species, substrate and growth conditions. Yields obtained in these experiments generally range from 80 to 95 % of the 4 mol H_2/mol C_6 theoretical maximum. Note that a batch culture of *C. owensensis* in defined medium, with glucose as substrate, achieved the Thauer limit of 4 mol H_2/mol C_6 sugar using continuous N_2 sparging (Table 8.1) (Zeidan and van Niel 2010). The maximum stoichiometric yield of H_2 from glucose is 12 mol H_2 per mol of glucose (Thauer et al. 1977), even so yields in vivo have not exceeded the Thauer limit. In vitro studies, using enzymes of the pentose phosphate pathway and a $NADP^+$ dependent hydrogenase from *P. furiosus*, achieved 11.6 mol H_2 per mol glucose-6-phosphate demonstrating the ability to produce near maximum H_2 yields in biological systems (Woodward and Mattingly 1996).

In continuous culture, H_2 production varies with dilution rate (i.e. growth rate), such that lower dilution rates result in lower growth rates and an increase in H_2 yield, albeit with a decrease in productivity. The inverse is true at higher dilution rates (de Vrije et al. 2007). At low growth rates, the majority of substrate is directed toward cell maintenance, during which many biosynthetic pathways remain dormant directing many of the reducing equivalents generated during glycolysis toward H_2 production. Thus, maximizing hydrogen production is a balance between the high productivities of fast growth rates and the high yields of slow growth rates. A proposed solution to increase both yield and productivity is to inoculate slow growing cultures at high cell densities (Chou et al. 2008). *Caldicellulosiruptor* species have also been found to persist in H_2-producing co-cultures. These co-cultures have shown synergy, such that the co-culture had higher hydrogen yields than the monoculture (Table 8.1) (Zeidan and van Niel 2009; Zeidan et al. 2010). For example, continuous co-culture of *C. saccharolyticus* and *C. kristjanssonii* found that both species persisted for 70 days with a hydrogen yield 6 % greater than either species alone. More importantly, cell-free growth supernatants of *C. saccharolyticus* were found to enhance the growth of *C. kristjanssonii* by decreasing its lag phase and increasing the maximum cell concentration by 18 % (Zeidan et al. 2010). Hydrogen yields from various *Caldicellulosiruptor* species have reached the Thauer limit (Zeidan and van Niel 2010). Increasing H_2 productivity in these bacteria, while maintaining high yields, will be a significant challenge in the development of a *Caldicellulosiruptor* bio-hydrogen production process, and may be possible through strategic metabolic engineering of these bacteria.

Acknowledgements

This work was funded in part by the BioEnergy Science Center, a U.S. Department of Energy Bioenergy Research Center supported by the Office of Biological and Environmental Research in the DOE Office of Science.

References

Ahring BK (1995) Isolation and characterization of *Caldicellulosiruptor lactoaceticus* sp. nov., an extremely thermophilic, cellulolytic, anaerobic bacterium. Arch Microbiol 163:223–230

Andrews SR, Charnock SJ, Lakey JH, Davies GJ, Claeyssens M, Nerinckx W, Underwood M, Sinnott ML, Warren AJ, Gilbert HJ (2000) Substrate specificity in glycoside hydrolase family 10. J Biol Chem 275:23027–230333

Bayer EA, Kenig R, Lamed R (1983) Adherence of *Clostridium thermocellum* to cellulose. J Bacteriol 156:818–827

Bayer EA, Shimon LJW, Shoham Y, Lamed R (1998) Cellulosomes – structure and ultrastructure. J Struct Biol 124:221–234

Bielen AAM, Willquist K, Engman J, van der Oost J, van Niel EWJ, Kengen SWM (2010) Pyrophosphate as a central energy carrier in the hydrogen-producing extremely thermophilic *Caldicellulosiruptor saccharolyticus*. FEMS Microbiol Lett 307:48–54

Blumer-Schuette SE, Lewis DL, Kelly RM (2010) Phylogenetic, microbiological, and glycoside hydrolase diversities within the extremely thermophilic, plant biomass-degrading genus *Caldicellulosiruptor*. Appl Environ Microbiol 76:8084–8092

Blumer-Schuette SE, Ozdemir I, Mistry D, Lucas S, Lapidus A, Cheng JF, Goodwin LA, Pitluck S, Land ML, Hauser LJ, Woyke T, Mikhailova N, Pati A, Kyrpides NC, Ivanova N, Detter JC, Walston-Davenport K, Han S, Adams MWW, Kelly RM (2011) Complete genome sequences for the anaerobic, extremely thermophilic plant biomass-degrading bacteria *Caldicellulosiruptor hydrothermalis*, *Caldicellulosiruptor kristjanssonii*, *Caldicellulosiruptor kronotskyensis*, *Caldicellulosiruptor owensensis* and *Caldicellulosiruptor lactoaceticus*. J Bacteriol 193:1483–1484

Blumer-Schuette SE, Giannone RJ, Zurawski JV, Ozdemir I, Ma Q, Yin Y, Xu Y, Kataeva I, Poole FL, Adams MWW, Hamilton-Brehm SD, Elkins JG, Larimer FW, Land ML, Hauser L, Cottingham RW, Hettich RL, Kelly RM (2012) *Caldicellulosiruptor* core and pan genomes reveal determinants for noncellulosomal thermophilic deconstruction of plant biomass. J Bacteriol 194:4015–4028

Bredholt S, Sonne-Hansen J, Nielsen P, Mathrani M, Ahring BK (1999) *Caldicellulosiruptor kristjanssonii* sp. nov., a cellulolytic, extremely thermophilic, anaerobic bacterium. Int J Syst Bacteriol 49:991–996

Brown SD, Guss AM, Karpinets TV, Parks JM, Smolin N, Yang S, Land ML, Klingeman DM, Bhandiwad A, Rodriguez M, Raman B, Shao X, Mielenz JR, Smith JC, Keller M, Lynd LR (2011) Mutant alcohol dehydrogenase leads to improved ethanol tolerance in *Clostridium thermocellum*. Proc Natl Acad Sci U S A 108:13752–13757

Brückner R, Titgemeyer F (2002) Carbon catabolite repression in bacteria: choice of the carbon source and autoregulatory limitation of sugar utilization. FEMS Microbiol Lett 209:141–148

Burdette D, Zeikus JG (1994) Purification of acetaldehyde dehydrogenase and alcohol dehydrogenases from *Thermoanaerobacter ethanolicus* 39E and characterization of the secondary-alcohol dehydrogenase (2 degrees Adh) as a bifunctional alcohol dehydrogenase–acetyl-CoA reductive thioestesterase. Biochem J 302:163–170

Cantarel BL, Coutinho PM, Rancurel C, Bernard T, Lombard V, Henrissat B (2009) The Carbohydrate-Active EnZymes database (CAZy): an expert resource for Glycogenomics. Nucleic Acids Res 37:D233–D238

Caufrier F, Martinou A, Dupont C, Bouriotis V (2003) Carbohydrate esterase family 4 enzymes: substrate specificity. Carbohydr Res 338:687–692

Chou CJ, Jenney FE, Adams MWW, Kelly RM (2008) Hydrogenesis in hyperthermophilic microorganisms: implications for biofuels. Metab Eng 10:394–404

Conners SB, Montero CI, Donald A, Shockley KR, Johnson MR, Chhabra R, Kelly RM (2005) An expression-driven approach to the prediction of carbohydrate transport and utilization regulons in the hyperthermophilic bacterium *Thermotoga maritima*. J Bacteriol 187:7267–7282

Dam P, Kataeva I, Yang SJ, Zhou F, Yin Y, Chou W, Poole FL, Westpheling J, Hettich R, Giannone R, Lewis D, Kelly R, Gilbert HJ, Henrissat B, Xu Y, Adams MWW (2011) Insights into plant biomass conversion from the genome of the anaerobic thermophilic bacterium *Caldicellulosiruptor bescii* DSM 6725. Nucleic Acids Res 39:3240–3254

de Vrije T, Mars AE, Budde MAW, Lai MH, Dijkema C, de Waard P, Claassen PAM (2007) Glycolytic pathway and hydrogen yield studies of the extreme thermophile *Caldicellulosiruptor saccharolyticus*. Appl Microbiol Biotechnol 74:1358–1367

de Vrije T, Bakker RR, Budde MAW, Lai MH, Mars AE, Claassen PA (2009) Efficient hydrogen production from the lignocellulosic energy crop Miscanthus by the extreme thermophilic bacteria *Caldicellulosiruptor saccharolyticus* and *Thermotoga neapolitana*. Biotechnol Biofuels 2:12–27

Donnison AM, Brockelsby CM, Morgan HW (1986) Abstract, pp 14–15 (page 203). Microbe 86, XIV international congress of microbiology, Manchester, England

Elkins JG, Lochner A, Hamilton-Brehm SD, Davenport KW, Podar M, Brown SD, Land ML, Hauser L, Klingeman DM, Raman B, Goodwin LA, Tapia R, Meincke LJ, Detter JC, Bruce DC, Han CS, Palumbo AV, Cottingham RW, Keller M, Graham DE (2010) Complete genome sequence of the cellulolytic thermophile *Caldicellulosiruptor obsidiansis* OB47T. J Bacteriol 192:6099–6100

Fardeau ML, Salinas MB, L'Haridon S, Jeanthon C, Verhé F, Cayol JL, Patel BKC, Garcia JL, Ollivier B (2004) Isolation from oil reservoirs of novel thermophilic anaerobes phylogenetically related to Thermoanaerobacter subterraneus: reassignment of T. subterraneus, Thermoanearobacter yonseiensis, Thermoanaerobacter tengcongensis and Carboxydibrachium pacificum to Caldanaerobacter subterraneus gen. nov., sp. nov., comb. nov as four novel species. Int J Syst Evol Microbiol 54:467–474

Fontes CMGA, Gilbert HJ (2010) Cellulosomes: highly efficient nanomachines designed to deconstruct plant cell wall complex carbohydrates. Annu Rev Biochem 79:655–681

Gancedo JM (1998) Yeast carbon catabolite repression. Microbiol Mol Biol R62:334–361

Hallenbeck PC, Benemann JR (2002) Biological hydrogen production; fundamentals and limiting processes. Int J Hydrogen Energ 27:1185–1193

Hamilton-Brehm SD, Mosher JJ, Vishnivetskaya T, Podar M, Carroll S, Allman S, Phelps TJ, Keller M, Elkins J (2010) *Caldicellulosiruptor obsidiansis* sp. nov., an anaerobic, extremely thermophilic, cellulolytic bacterium isolated from Obsidian Pool, Yellowstone National Park. Appl Environ Microbiol 76:1014–1020

Heinonen JK (2001) Biological role of inorganic pyrophosphate. Kluwer, Norwell

Hemme CL, Fields MW, He Q, Deng Y, Lin L, Tu Q, Mouttaki H, Zhou A, Feng X, Zuo Z, Ramsay BD, He Z, Wu L, Van Nostrand J, Xu J, Tang YJ, Wiegel J, Phelps T, Zhou J (2011) Correlation of genomic and physiological traits of *Thermoanaerobacter* species with biofuel yields. Appl Environ Microbiol 77:7998–8008

Hisamatsu M, Impallomeni G, York WS, Albersheim P, Darvill AG (1991) A new undecasaccharide subunit of xyloglucans with two alpha-L-fucosyl residues. Carbohydr Res 211:117–129

Holden HM, Rayment I, Thoden JB (2003) Structure and function of enzymes of the Leloir pathway for galactose metabolism. J Biol Chem 278: 43885–43888

Huang C, Patel BK, Mah RA, Baresi L (1998) *Caldicellulosiruptor owensensis* sp. nov., an anaerobic, extremely thermophilic, xylanolytic bacterium. Int J Syst Bacteriol 48:91–97

Ivanova G, Rákhely G, Kovács KL (2009) Thermophilic biohydrogen production from energy plants by *Caldicellulosiruptor saccharolyticus* and comparison with related studies. Int J Hydrog Energy 34:3659–3670

Janecek S (1997) Alpha-amylase family: molecular biology and evolution. Prog Biophys Mol Biol 67:67–97

Kádár Z, de Vrije T, van Noorden GE, Budde MAW, Szengyel Z, Réczey K, Claassen PAM (2004) Yields from glucose, xylose, and paper sludge hydrolysate during hydrogen production by the extreme thermophile *Caldicellulosiruptor saccharolyticus*. Appl Biochem Biotechnol 113–116:497–508

Kataeva IA, Yang SJ, Dam P, Poole FL, Yin Y, Zhou F, Chou W, Xu Y, Goodwin L, Sims DR, Detter JC, Hauser LJ, Westpheling J, Adams MWW (2009) Genome sequence of the anaerobic, thermophilic, and cellulolytic bacterium *Anaerocellum thermophilum* DSM 6725. J Bacteriol 191:3760–3761

Komalavilas P, Mort AJ (1989) The acetylation at O-3 of galacturonic acid in the rhamnose-rich portion of pectins. Carbohydr Res 189:261–272

Koning SM, Albers SV, Konings WN, Driessen AJM (2002) Sugar transport in (hyper)thermophilic archaea. Res Microbiol 153:61–67

Kotrba P, Inui M, Yukawa H (2001) Bacterial phosphotransferase system (PTS) in carbohydrate uptake and control of carbon metabolism. J Biosci Bioeng 92:502–517

Kruger NJ, von Schaewen A (2003) The oxidative pentose phosphate pathway: structure and organisation. Curr Opin Plant Biol 6:236–246

Ljunggren M, Willquist K, Zacchi G, van Niel EW (2011) A kinetic model for quantitative evaluation of the effect of hydrogen and osmolarity on hydrogen production by *Caldicellulosiruptor saccharolyticus*. Biotechnol Biofuels 4:31–46

Martinez D, Berka RM, Henrissat B, Saloheimo M, Arvas M, Baker SE, Chapman J, Chertkov O, Coutinho PM, Cullen D, Danchin EDJ, Grigoriev IV, Harris P, Jackson M, Kubicek CP, Han CS, Ho I, Larrondo LF, de Leon AL, Magnuson JK, Merino S, Misra M, Nelson B, Putnam N, Robbertse B, Salamov AA, Schmoll M, Terry A, Thayer N, Westerholm-Parvinen A, Schoch CL, Yao J, Barabote R, Barbote R, Nelson MA, Detter C, Bruce D, Kuske CR, Xie G, Richardson P, Rokhsar DS, Lucas SM, Rubin EM, Dunn-Coleman N, Ward M, Brettin TS (2008) Genome sequencing and analysis of the biomass-degrading fungus *Trichoderma reesei* (syn. *Hypocrea jecorina*). Nat Biotechnol 26:553–560

Mertens E (1991) Pyrophosphate-dependent phosphofructokinase, an anaerobic glycolytic enzyme? FEBS Lett 285:1–5

Miroshnichenko ML, Kublanov IV, Kostrikina NA, Tourova TP, Kolganova TV, Birkeland NK, Bonch-Osmolovskaya EA (2008) *Caldicellulosiruptor kronotskyensis* sp. nov. and *Caldicellulosiruptor hydrothermalis* sp. nov., two extremely thermophilic, cellulolytic, anaerobic bacteria from Kamchatka thermal springs. Int J Syst Evol Microbiol 58:1492–1496

Navarre WW, Schneewind O (1999) Surface proteins of gram-positive bacteria and mechanisms of their targeting to the cell wall envelope. Microbiol Mol Biol Rev 63:174–229

Ozdemir I, Blumer-Schuette SE, Kelly RM (2012) S-layer homology domain proteins Csac_0678 and Csac_2722 are implicated in plant polysaccharide deconstruction by the extremely thermophilic bacterium *Caldicellulosiruptor saccharolyticus*. Appl Environ Microbiol 78:768–777

Pérez J, Muñoz-Dorado J, de la Rubia T, Martínez J (2002) Biodegradation and biological treatments of cellulose, hemicellulose and lignin: an overview. Int Microbiol 5:53–63

Rainey FA, Donnison AM, Janssen PH, Saul D, Rodrigo A, Bergquist PL, Daniel RM, Stackenbrandt E, Morgan HW (1994) Description of *Caldicellulosiruptor saccharolyticus* gen. nov., sp. nov: an obligately anaerobic, extremely thermophilic, cellulolytic bacterium. FEMS Microbiol Lett 120:263–266

Reddy N, Yang Y (2005) Biofibers from agricultural byproducts for industrial applications. Trends Biotechnol 23:22–27

Ridley B, O'Neill M (2001) Pectins: structure, biosynthesis, and oligogalacturonide-related signaling. Phytochemistry 57:929–967

Rubin EM (2008) Genomics of cellulosic biofuels. Nature 454:841–845

Sára M, Sleytr UB (2000) S-layer proteins. J Bacteriol 182:859–868

Schneider E (2001) ABC transporters catalyzing carbohydrate uptake. Res Microbiol 152: 303–310

Schut GJ, Adams MWW (2009) The iron-hydrogenase of *Thermotoga maritima* utilizes ferredoxin and NADH synergistically: a new perspective on anaerobic hydrogen production. J Bacteriol 191:4451–4457

Shoseyov O, Shani Z, Levy I (2006) Carbohydrate binding modules: biochemical properties and novel applications. Microbiol Mol Biol Rev 70:283–295

Soboh B, Linder D, Hedderich R (2004) A multisubunit membrane-bound [NiFe] hydrogenase and an NADH-dependent Fe-only hydrogenase in the fermenting bacterium *Thermoanaerobacter tengcongensis*. Microbiology 150:2451–2463

Stulke J, Hillen W (2000) Regulation of carbon catabolism in *Bacillus* species. Annu Rev Microbiol 54:849–880

Te'o VS, Saul DJ, Bergquist P (1995) celA, another gene encoding for a multidomain cellulase from the extreme thermophile *Caldocellum saccharolyticum*. Appl Microbiol Biotechnol 43:291–296

Thauer RK, Jungermann K, Decker K, Decker K (1977) Energy conservation in chemotrophic anaerobic bacteria. Bacteriol Rev 41:100–180

van de Werken HJG, Verhaart MRA, VanFossen AL, Willquist K, Lewis DL, Nichols JD, Goorissen HP, Mongodin EF, Nelson KE, van Niel EWJ, Stams AJM, Ward DE, de Vos WM, van der Oost J, Kelly RM, Kengen SWM (2008) Hydrogenomics of the extremely thermophilic bacterium *Caldicellulosiruptor saccharolyticus*. Appl Environ Microbiol 74:6720–6729

van Groenestijn JW, Geelhoed JS, Goorissen HP, Meesters KPM, Stams AJM, Claassen PAM (2009) Performance and population analysis of a non-sterile trickle bed reactor inoculated with *Caldicellulosiruptor saccharolyticus*, a thermophilic hydrogen producer. Biotechnol Bioeng 102:1361–1367

van Niel EWJ, Budde MAW, de Haas GG, van der Wal FJ, Claassen PAM, Stams AJM (2002) Distinctive properties of high hydrogen producing extreme thermophiles, *Caldicellulosiruptor saccharolyticus* and *Thermotoga elfii*. Int J Hydrog Energy 27:1391–1398

van Niel EWJ, Claassen PAM, Stams AJM (2003) Substrate and product inhibition of hydrogen production by the extreme thermophile, *Caldicellulosiruptor saccharolyticus*. Biotechnol Bioeng 81:255–262

van Wyk JPH (2001) Biotechnology and the utilization of biowaste as a resource for bioproduct development. Trends Biotechnol 19:172–177

Vanfossen AL, Verhaart MRA, Kengen SMW, Kelly RM (2009) Carbohydrate utilization patterns for the extremely thermophilic bacterium *Caldicellulosiruptor saccharolyticus* reveal broad growth substrate preferences. Appl Environ Microbiol 75:7718–7724

VanFossen AL, Ozdemir I, Zelin SL, Kelly RM (2011) Glycoside hydrolase inventory drives plant polysaccharide deconstruction by the extremely thermophilic bacterium *Caldicellulosiruptor saccharolyticus*. Biotechnol Bioeng 108:1559–1569

Verhaart MRA, Bielen AAM, van der Oost J, Stams AJM, Kengen SWM (2010) Hydrogen production by hyperthermophilic and extremely thermophilic bacteria and archaea: mechanisms for reductant disposal. Environ Technol 31:993–1003

Warner JB, Lolkema JS, Warner JB, Lolkema JS (2003) CcpA-dependent carbon catabolite repression in bacteria. Microbiol Mol Biol R67:475–490

Willquist K, van Niel EWJ (2010) Lactate formation in *Caldicellulosiruptor saccharolyticus* is regulated by the energy carriers pyrophosphate and ATP. Metab Eng 12:282–290

Willquist K, van Niel EWJ (2012) Growth and hydrogen production characteristics of *Caldicellulosiruptor saccharolyticus* on chemically defined minimal media. Int J Hydrog Energy 37:4925–4929

Willquist K, Claassen PAM, van Niel EWJ (2009) Evaluation of the influence of CO_2 on hydrogen production by *Caldicellulosiruptor saccharolyticus*. Int J Hydrog Energy 34:4718–4726

Willquist K, Pawar SS, Niel EWJ (2011) Reassessment of hydrogen tolerance in *Caldicellulosiruptor saccharolyticus*. Microb Cell Fact 10:111–122

Woodward J, Mattingly S (1996) In vitro hydrogen production by glucose dehydrogenase and hydrogenase. Nat Biotechnol 14:872–874

Yang SJ, Kataeva I, Wiegel J, Yin Y, Dam P, Xu Y, Westpheling J, Adams MWW (2010) Classification of "*Anaerocellum thermophilum*" strain DSM 6725 as *Caldicellulosiruptor bescii* sp. nov. Int J Syst Evol Microbiol 60:2011–2015

Zeidan AA, van Niel EWJ (2009) Developing a thermophilic hydrogen-producing co-culture for efficient utilization of mixed sugars. Int J Hydrog Energy 34:4524–4528

Zeidan AA, van Niel EWJ (2010) A quantitative analysis of hydrogen production efficiency of the extreme thermophile *Caldicellulosiruptor owensensis* OLT. Int J Hydrog Energy 35:1128–1137

Zeidan AA, Rådström P, van Niel EWJ (2010) Stable coexistence of two *Caldicellulosiruptor* species in a de novo constructed hydrogen-producing co-culture. Microb Cell Fact 9:102–115

Zverlov V, Mahr IS, Riedel K, Bronnenmeier K (1998) Properties and gene structure of a bifunctional cellulolytic enzyme (CelA) from the extreme with separate glycosyl hydrolase family 9 and 48 catalytic domains. Microbiology 144:457–465

Chapter 9

Members of the Order Thermotogales: From Microbiology to Hydrogen Production

Martina Cappelletti
Laboratory of General and Applied Microbiology, Department of Pharmacy and BioTechnology (FaBiT), University of Bologna, 40126 Bologna, Italy

Davide Zannoni*
Department of Pharmacy and BioTechnology (FaBiT), University of Bologna, 40126 Bologna, Italy

and

Anne Postec, Bernard Ollivier
Aix-Marseille Université, Université du Sud Toulon-Var, CNRS/INSU, IRD, MIO, UM 110, 13288 Marseille, Cedex 09, France

*Author for correspondence, e-mail: davide.zannoni@unibo.it

D. Zannoni and R. De Philippis (eds.), Microbial BioEnergy: Hydrogen Production, Advances in Photosynthesis and Respiration 38, pp. 197–224, DOI 10.1007/978-94-017-8554-9_9,

Summary

Members of the deep-branching order *Thermotogales* are widespread in various terrestrial, submarine and subterrestrial extreme environments. This bacterial order included both thermophilic and hyperthermophilic anaerobic microorganisms so far pertaining to ten genera. It is only recently (2011) that cultivation of a mesophilic member of this order belonging to a novel genus, *Mesotoga*, has been successful. All members, with the exception of Mesotoga spp., are recognized as high hydrogen producers having possible applications in biotechnology with a peculiar emphasis for members of the genus *Thermotoga* (e.g. *T. maritima* and *T. neapolitana*). The ecology, phylogeny and metabolism linked to hydrogen production of these bacteria, are reviewed.

I. Introduction

Members of the order *Thermotogales* (Fig. 9.1) were represented as a deep-branching lineage within the phylogenetic tree (Huber et al. 1986; Reysenbach et al. 2001; Huber and Hannig 2006) thus suggesting that representatives of this order might have arisen early during the first steps of bacterial evolution. However the phylogenetic position of *Thermotogales*, as the nature (hyperthermophile or mesophile) of the ancestor of *Bacteria*, still remains a matter of debate (Brochier and Philippe 2002; Zhaxybayeva et al. 2009). Most of these Gram-negative anaerobic bacteria possess an outer sheath like structure called a "toga" ballooning over the ends of the cell (e.g. *Thermotoga* and *Thermosipho* spp.) (Huber et al. 1986; Antoine et al. 1997). Terminal protuberances on one end of the cells and single sphere containing several cells have also been observed in particular in *Fervidobacterium* spp. (Patel et al. 1985). *Thermotogales* range from thermophiles to hyperthermophiles having optimum temperature for growth above 80 °C with possible growth up to 90 °C (e.g. *Thermotoga maritima*, *T. neapolitana*, and *T. hypogea*). They are recognized as non-sporing rods occurring singly, in pairs or in chains with the absence of meso-diaminopimelic acid in the peptidoglycan (Reysenbach et al. 2001; Huber and Hannig 2006). This order comprises ten genera: *Thermotoga*, *Thermosipho*, *Fervidobacterium*, *Geotoga*, *Petrotoga*, *Marinitoga*, *Thermococcoides* and the recently described genera *Kosmotoga*, *Oceanotoga* and *Defluviitoga* (Di Pippo et al. 2009; Jayasinghearachchi and Lal 2011; Ben Hania et al. 2012). Because of the almost identical 16S rRNA gene sequences of *K. olearia* and *Thermoccoides shengliensis*, and their many shared phenotypic features, *T. shengliensis* has been proposed to be reclassified within the genus *Kosmotoga* and named *K. shengliensis* (Feng et al. 2010; Nunoura et al. 2010). Recently, a mesophilic lineage (*Mesotoga*) within the *Thermotogales* has been evidenced by the detection of *Thermotogales* 16S rRNA gene sequences in many mesothermic environments (Nesbø et al. 2006, 2010). The mesophilic nature of such microorganisms has been established by their isolation and cultivation in 2011 ("*Mesotoga sulfurireducens*" strain PhosAc3, Ben Hania et al. 2011), in 2012 (*Mesotoga prima* strain MesG1.Ag.4.2, Nesbø et al. 2012) and in 2013 (*Mesotoga infera* strain VNs100T, Ben Hania et al. 2013) with *Mesotoga prima* being the first described representative and type species of genus *Mesotoga* (Nesbø et al. 2012). All isolated *Mesotoga* species were confirmed to grow optimally at mesothermic conditions (40 °C for "*Mesotoga sulfurireducens*", 45 °C for *M. infera* and 37 °C for *M. prima*) (Ben Hania et al. 2011, 2013; Nesbø et al. 2012) making them microorganisms of notable interest to understand bacterial evolution from mesophily to thermophily or *vice versa* (Nesbø et al. 2006; Ben Hania et al. 2011).

Abbreviations: CMC – Carboxy methyl cellulose; GAP deh – Glyceraldehyde-3-phosphate dehydrogenase; GghA – 1,4-β-D-glucan glucohydrolase; HEPES – 4-(2-hydroxyethyl)-1-piperazineethanesulfonic acid; MBS – Metabisulfite; NRO – NADH oxidoreductase; ORF – Open reading frames; RET – Reversed electron transport; ROS – Reactive oxygen species

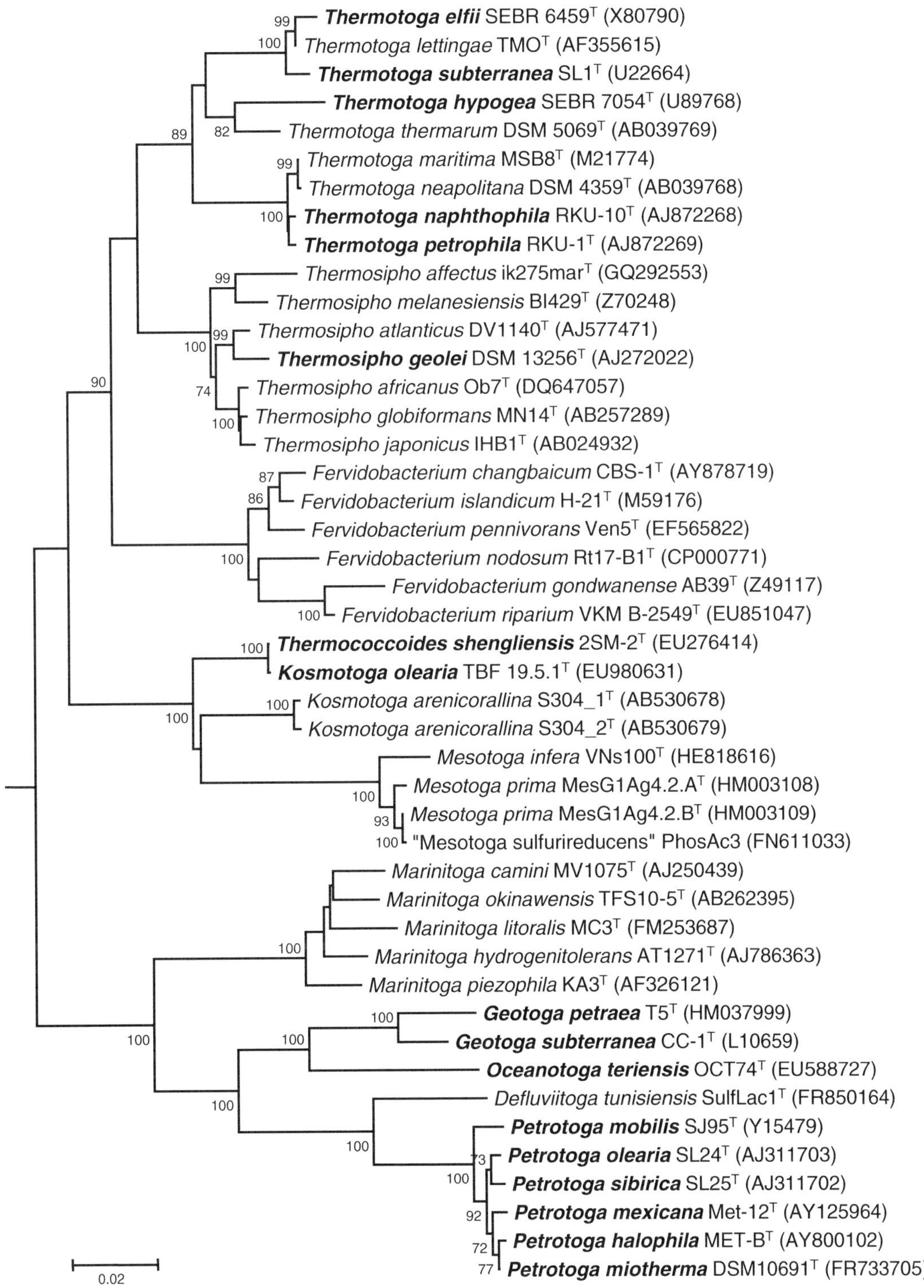

Fig. 9.1. Phylogenetic tree based on 16 rRNA gene sequences representing the position of oilfield microorganisms (*bold characters*) within the order *Thermotogales*. Neighbor joining method was used, calculation from 1,190 aligned pb, bootstrap from 1,000 replicates. *Aquifex aeolicus* (AJ309733) and *Bacillus subtilis* (K00637) were used as outgroup (not shown). *Bar scale*, 0.02 substitution per nucleotide.

II. Habitat

Most of the thermophilic *Thermotogales* have been detected by molecular approaches and/or cultural approaches from geothermally heated environments (Table 9.1) (Huber and Hannig 2006). Most *Fervidobacterium* spp. have been isolated from low saline terrestrial hot springs (Patel et al. 1985; Huber et al. 1990; Andrews and Patel 1996; Friedrich and Antranikian 1996), which is also the case of *Thermotoga thermarum* (Windberger et al. 1989). Numerous slightly halophilic bacteria pertaining to the genera *Marinitoga* (Wery et al. 2001; Alain et al. 2002; Postec et al. 2005, 2010; Nunoura et al. 2007) and *Thermosipho* originated from marine ecosystems (e.g. deep-sea hydrothermal vents) (Huber et al. 1989; Antoine et al. 1997; Takai and Horikoshi 2000; Urios et al. 2004). Many thermophilic *Thermotogales* have been recovered from oilfield waters and facilities (Ollivier and Cayol 2005). They include *Thermotoga* spp. (*T. elfii*, *T. petrophila*, *T. hypogea*) (Magot et al. 2000; Ollivier and Cayol 2005), and *Kosmotoga olearia* (Di Pippo et al. 2009); species of both these genera were also isolated from submarine thermal vents (e.g. *T. neapolitana* and *K. arenicorallina*) (Windberger et al. 1989; Nunoura et al. 2010). Interestingly *Petrotoga*, *Geotoga*, and *Oceanotoga* spp. have representatives which originated only from oil reservoirs thus suggesting that such bacteria might be indigenous to these extreme environments and might possibly be relevant for enhancing oil recovery in the petroleum industry (Magot et al. 2000; Ollivier and Cayol 2005; Jayasinghearachchi and Lal 2011). However the indigenous character of *Thermotogales* to the oil field ecosystems should be considered with caution as their presence may result from anthropogenic activities after drilling operations or water injections during oil exploration (Magot et al. 2000). Besides geothermally heated sediments, only a few thermophilic *Thermotogales* have been isolated from anaerobic digesters. These include *Thermotoga lettingae* (Balk et al. 2002) and the recently described *Defluviitoga tunisiensis* (Ben Hania et al. 2012). *Thermotoga lettingae* was isolated from a thermophilic sulfate-reducing bioreactor operating at 65 °C with methanol as the sole substrate. *D. tunisiensis* was isolated from a mesothermic bioreactor (37 °C) treating lactoserum and phosphogypsum (Balk et al. 2002; Ben Hania et al. 2012). As mentioned above, there is now evidence for the presence of *Thermotogales* (*Mesotoga*) in mesothermic environments such as enrichment cultures degrading chlorinated compounds, temperate hydrocarbon–impacted sites, oil reservoirs, oil gas storage, anaerobic bioreactors treating solvent-containing pharmaceutical wastewater, and a gas-fed bioreactor treating sulfate- and zinc-rich wastewater (Ben Hania et al. 2011, 2013; Nesbø et al. 2006, 2012). In this respect, we may expect *Mesotoga* spp. to be of ecological relevance in the bioremediation of polluted sites (Ben Hania et al. 2011).

III. Metabolic Features

A. Electron Donors

Members of the *Thermotogales* are considered as heterotrophic fermentative microorganisms able to ferment sugars (Table 9.1) (Reysenbach et al. 2001; Huber and Hannig 2006). All thermophilic *Thermotogales* have been demonstrated to grow on complex organic substrates such as peptone and yeast extract, the latter being required to ferment sugars (Huber and Hannig 2006). Besides monosaccharides (e.g. glucose, fructose, xylose), di- and trisaccharides (e.g. sucrose, lactose, cellobiose, raffinose), *Thermotogales* can also ferment polysaccharides (Huber and Hannig 2006). Their use of cellulose has been reported, in particular for *Thermotoga maritima*, *T. neapolitana*, *Fervidobacterium islandicum*, and *Marinitoga camini* (Huber et al. 1986, 1990; Windberger et al. 1989; Wery et al. 2001; Huber and Hannig 2006). *Thermotoga maritima*, *T. neapolitana*, and *M. camini* were also reported to ferment glycogen (Huber et al. 1986; Windberger et al. 1989; Wery et al. 2001; Huber and Hannig 2006). The use of starch has been reported many times within the *Thermotogales* (e.g. *Thermotoga*, *Geotoga*,

Table 9.1. Characteristics of ten genera of Thermotogales order.

Characteristic	"*Mesotoga*"	*Defluviitoga*	*Oceanotoga*	*Kosmotoga*	*Marinitoga*	*Geotoga*	*Petrotoga*	*Thermosipho*	*Thermotoga*	*Fervido-bacterium*
Isolation source (s)	Sediments, Mesothermic, Bioreactor	Mesothermic, Bioreactor	Oil reservoir	Oil reservoir, Shallow hydrothermal vent	Hydrothermal vents	Oil reservoirs	Oil reservoirs	Hydrothermal vents and oil reservoir	Hydrothermal vents, oil reservoirs and bioreactor	Terrestrial hot springs
Temperature optimum (°C)	37–40 °C	55	55–58	65	55–65	45–50	55–60	65–75	65–80	65–70
Oxygen tolerance (v/v)	up to 2 %	up to 0.5 %	nd	Growth at 15 % O_2	No growth at 4 %	Strict anaerobes	No growth with 0.2–1 %	No growth at 0.2–8 %	Do not grow aerobically	Strict anaerobes
Electron acceptors										
Elemental sulfur	+/–	+	+	–	+	–	+	+/–	+/–	+
Thiosulfate	+/–	+	+	+	+/–	nd	+/–	+/–	+	+/–
Nitrate	–	nd	–	–	–	nd	–	–	nd	nd
Nitrite	–	nd	nd	–	–	nd	–	–	nd	nd
Sulfite	+/–	–	nd	–	–	nd	+/–	–	nd	nd
Cystine	–	nd	nd	–	+	nd	–	+/–	+/–	nd
Sulfate	–	–	–	+[a]	–	nd	–	–	–	–

(continued)

Table 9.1. (continued)

Characteristic	"*Mesotoga*"	*Defluviitoga*	*Oceanotoga*	*Kosmotoga*	*Marinitoga*	*Geotoga*	*Petrotoga*	*Thermosipho*	*Thermotoga*	*Fervido-bacterium*
Substrate utilization										
Complex organic compounds	+	+	+	+	+/–	+	+	+	+	+
Carbohydrates	+	+	+	+	+	+	+	+	+	+
Alcohols	nd	nd	nd	nd	nd	nd	nd	nd	+	+
Metabolic products	**Acetate**, butyrate, isobutyrate, isovalerate, 2-methyl-butyrate[b]	**Acetate, H_2, CO_2**	**Acetate, H_2, CO_2**, ethanol[b]	**Acetate, H_2, CO_2**, ethanol, isobutyrate, isovalerate[b]	**Acetate, H_2, CO_2**, lactate, iso-butyrate, isovalerate[b]	**Acetate, H_2, CO_2**, ethanol[b]	**Acetate, H_2, CO_2**, ethanol[b]	**Acetate, H_2, CO_2**, L-alanine[b]	**Acetate, H_2, CO_2**, lactate, L-alanine[b]	**Acetate, H_2, CO_2**, L-alanine[b]
DNA G+C content (mol%)	45.3	33.6	26.8	42.5	28–29	29.5–29.9	31–39.8	29–33	39.2–50	33.7–40

Data were taken from Ben Hania et al. (2011, 2013); Nesbø et al. (2012) ("*Mesotoga*"); Ben Hania et al. (2012) (*Defluviitoga*); Jayasinghearachchi and Lal (2011) (*Oceanotoga*); Alain et al. (2002); Nunoura et al. (2007); Postec et al. (2005) and Wery et al. (2001) (*Marinitoga*); Davey et al. (1993) (*Geotoga*); L'Haridon et al. (2002); Lien et al. (1998) and Miranda-Tello et al. (2004, 2007) (*Petrotoga*); Antoine et al. (1997); Huber et al. (1989); L'Haridon et al. (2001); Takai and Horikoshi (2000) and Urios et al. (2004) (*Thermosipho*); Balk et al. (2002); Fardeau et al. (1997); Huber et al. (1986); Jannasch et al. (1988); Jeanthon et al. (1995); Ravot et al. (1995); Takahata et al. (2001) and Windberger et al. (1989) (*Thermotoga*) and Andrews and Patel (1996); Friedrich and Antranikian (1996); Huber et al. (1990) and Patel et al. (1985) (*Fervidobacterium*)

nd No data available; – does not enhance growth; + enhanced growth; ± enhanced growth for some, but not all, species

[a] Slight growth enhancement, but no sulfide produced

[b] Metabolic products which are written in bold characters are produced by all species. For those which are not written in bold characters, they are produced depending on species

Petrotoga, *Thermosipho*, *Fervidobacterium*, *Marinitoga*, *Kosmotoga* spp.) (Huber and Hannig 2006). Xylanolytic activity has been detected in *T. maritima*, *T. hypogea*, and *Petrotoga* spp. (e.g. *P. mobilis* and *P. olearia*) (Huber et al. 1986; Davey et al. 1993; Fardeau et al. 1997; Lien et al. 1998). *Marinitoga camini* was shown to degrade chitin and *F. pennovorans* to degrade keratin (Friedrich and Antranikian 1996; Wery et al. 2001).

The alcohols possibly fermented by *Thermotogales* include mannitol (e.g. *T. naphtophila*), and glycerol (e.g. *T. lettingae, T. neapolitana* and *F. nodosum*) (Patel et al. 1985; Takahata et al. 2001; Van Ooteghem et al. 2004; Huber and Hannig 2006). Methanol was poorly fermented by *T. lettingae*, however, in the presence of thiosulfate or elemental sulfur as terminal electron acceptors, or methanoarchaea as hydrogenotrophic partners, methanol was more quickly used (12 days for the oxidative process instead of 30 days for the fermentative process) (Balk et al. 2002). Methanol utilization has also been demonstrated for *T. subterranea*, *T. elfii*, *T. maritima* and *T. neapolitana* by the same authors (Balk et al. 2002).

Regarding organic acids utilization, pyruvate served as energy source for some *Thermotoga* and *Petrotoga* spp. (Huber and Hannig 2006), and there is one report on formate utilization by *Thermotoga lettingae* (Balk et al. 2002). Besides formate, which was used only in the presence of thiosulfate as electron acceptor, *T. lettingae* was the first *Thermotogales* reported to ferment lactate (Balk et al. 2002). Thereafter, the use of lactate has been also evidenced in "*Mesotoga sulfurireducens*" and in *M. infera* where the presence of elemental sulfur as terminal electron acceptor was required thus suggesting that lactate was oxidized, but not fermented by this bacterium (Ben Hania et al. 2011, 2013). *T. lettingae* was also shown to oxidize acetate in the presence of thiosulfate or hydrogenotrophic methanoarchaea as terminal electron acceptors in agreement with results for previously described thermophilic or mesophilic acetate-degrading bacteria coupled to a methanogenic partner or an electron acceptor (Balk et al. 2002).

B. Electron Acceptors

Thermotogales do not only use a wide range of organic substrates as energy sources but have also the ability to reduce sulfur-containing compounds (e.g. elemental sulfur, and thiosulfate, but not sulfate) into sulfide (Table 9.1) (Huber and Hannig 2006). Amongst *Thermotogales* genera, *Geotoga* is the only genus in which no thiosulfate-reducing species was reported (Davey et al. 1993). The ability of *Thermotogales* to reduce elemental sulfur in the presence of sugars was first reported for *T. maritima* and was found stimulatory for its growth (Huber et al. 1986). However, this reductive process was suggested to be a detoxifying process preventing H_2 accumulation rather than an energy-yielding electron sink reaction (Huber et al. 1986; Huber and Hannig 2006). A similar conclusion could have been drawn regarding thiosulfate reduction by *Thermotogales*. However growth experiments with *T. maritima* and *T. neapolitana* in the presence of thiosulfate suggested that thiosulfate reduction could be regarded as an energy-yielding reaction through an oxidative phosphorylation process from sugars (Ravot et al. 1995). Indeed, for both bacteria, despite important improvements of growth observed in the presence of thiosulfate, there was no significant change in the end-products of sugar metabolism (Ravot et al. 1996). Besides thiosulfate and elemental sulfur, Fe(III) was also used as electron acceptor by *T. maritima* and in its presence, hydrogen oxidation was demonstrated (Vargas et al. 1998). While one study indicated that *T. maritima* may gain energy by iron respiration (Vargas et al. 1998), other reports suggested Fe(III) as an additional electron sink together with thiosulfate and elemental sulfur to prevent the inhibitory effect of hydrogen on growth (Huber et al. 1986; Schröder et al. 1994; Huber and Stetter 2001). The ability of *T. maritima* to reduce Fe(III) raises questions on its possible involvement in reducing heavy metals and should merit further attention. In the presence of thiosulfate as terminal electron acceptor, *T. lettingae* was shown to oxidize hydrogen (Balk et al. 2002). While the possible use of cystine as electron acceptor has been established for *Marinitoga*, *Thermotoga,* and *Thermosipho* species (Huber and Hannig 2006), only *T. lettingae* was reported to use anthraquinone −2,6-disulfonate as electron acceptor (Balk et al. 2002). Results obtained with *Thermotogales* regarding the possible use of sulfur compounds as electron acceptors, indicate they may play a crucial ecological role in mineralizing organic matter in hot ecosystems. This is particularly true for *Thermotogales* originating from shallow or deep sea hydrothermal vents (e.g. *Thermosipho* and *Marinitoga* spp.) where different oxidized forms of sulfur compounds, including elemental sulfur and thiosulfate, are not limiting.

C. End-Products of Metabolism

The major end-products of sugar metabolism by thermophilic *Thermotogales* are acetate, hydrogen, and CO_2 (Table 9.1) (Reysenbach et al. 2001; Huber and Hannig 2006). Surprisingly, while the recent isolated mesophiles "*Mesotoga sulfurireducens*", *M. prima* and *M. infera* were shown to produce mainly/only acetate from sugar utilization, there is no report on H_2 production by these microorganisms (Ben Hania et al. 2011, 2013; Nesbø et al. 2012). Moreover, in contrast to *M. prima* which was described as a fermentative bacterium (Nesbø et al. 2012), *M. infera* was shown to rather oxidize its substrates in the presence of elemental sulfur as terminal electron acceptor (Ben Hania et al. 2013). Further researches on the physiological traits of *Mesotoga* spp. will be therefore of interest to understand the overall metabolic capabilities within the *Thermotogales*. It was found, for example, that acetate was also produced during methanol fermentation by *Thermotoga lettingae* (Balk et al. 2002). Lactate production was detected in particular by *T. maritima* and was dependent on culture conditions (e.g. H_2 partial pressure) (Janssen and Morgan 1992; Schröder et al. 1994). It was also produced by *Marinitoga camini* when growing on sugars (Wery et al. 2001). Ethanol has been measured at many occasions (e.g. *Geotoga*, *Petrotoga*, *Kosmotoga*, and *Oceanotoga* spp.) together with isovalerate, isobutyrate, and/or propionate (e.g. *M. camini*, *K. olearia*), but also alpha-aminobutyrate, hydroxyphenyl-acetate or phenylacetate (Huber and Hannig 2006) as end-products of sugar metabolism. Studies with *T. maritima* indicated that this bacterium used glucose *via* the Embden-Meyerhof glycolytic pathway and, to a lesser extent, *via* the Entner-Doudouroff pathway (Schröder et al. 1994; Selig et al. 1997; Huber and Hannig 2006). However the importance of the use of both pathways in other *Thermotogales* when fermenting sugars is still poorly documented and deserves further investigation. Notably, it was demonstrated in *T. neapolitana* that glucose was taken up *via* an active transport system that was energized by an ion gradient generated by ATP derived from substrate-level phosphorylation (Galperin et al. 1996; Huber and Hannig 2006). It is established, that several *Thermotogales* have a high ratio of acetate produced *versus* sugar consumed, thus indicating that they are good candidates for H_2 production from the biomass approaching the theoretical maximum yield of 4 mol H_2 per mol glucose consumed (Schröder et al. 1994; Van Ooteghem et al. 2004; Eriksen et al. 2011). This is emphasized by the capacities of many species to use various carbohydrates including cellulose, hemicelluloses and starch together with proteinaceous compounds. Besides acetate, lactate and hydrogen, L-alanine was also found as a significant end-product of sugar fermentation by *Thermotoga elfii*, *Fervidobacterium islandicum, F. nodosum*, *F. gondwanense*, and *Thermosipho africanus* with up to 0.52 mol L-alanine produced per mol glucose consumed (*T. africanus*). In contrast, *T. maritima* and *T. neapolitana* were found to be poor L-alanine producers (Ravot et al. 1996). In the presence of thiosulfate, a decrease of the L-alanine/acetate ratio was observed for *F. islandicum*, *T. africanus*, *T. elfii*, *Thermotoga* SEBR 7054, and

Thermotoga lettingae (Ravot et al. 1996; Balk et al. 2002). For all these bacteria, the presence of thiosulfate caused a shift of metabolism with more acetate and less L-alanine being produced from sugars thus enabling them to obtain more ATP from substrate level phosphorylation *via* the formation of acetyl-CoA. It was hypothesized that similarly to the hyperthermophilic archaeon, *Pyrococcus furiosus*, L-alanine production from glucose fermentation resulted from alanine transferase activity coupled with glutamate dehydrogenase activity (Ravot et al. 1996). Therefore, because of a similar type of sugar metabolism by members of the *Thermococcales* (e.g. *Pyrococcus furiosus* and *Thermococcus profundus*), domain *Archaea*, placed as a deep-branching lineage within the phylogenetic tree, L-alanine production by *Thermotogales* has been interpreted as a remnant of an ancestral metabolism (Ravot et al. 1996). Interestingly, L-alanine was also produced when *T. lettingae* grew on methanol as energy source in the presence of thiosulfate or elemental sulfur and this was the first report of L-alanine formation from a C1 substrate. In contrast, in the presence of a methanogenic partner (e.g. *Methanothermobacter thermoautotrophicus*) methanol was completely oxidized to CO_2 (Balk et al. 2002). A complete oxidation of acetate was also observed when *T. lettingae* was cocultured with an hydrogenotrophic methanoarchaea, while L-alanine was produced from acetate in the presence of thiosulfate (Balk et al. 2002).

D. Oxygen Tolerance

Although *Thermotogales* are recognized as strict anaerobes, there are evidences that they may cope with limited amount of oxygen during their growth (Table 9.1). *T. maritima* may grow in the presence of 0.5 % of oxygen (Le Fourn et al. 2008), *T. neapolitana* was shown to grow under oxygen concentrations ranging from 1 to 6 % (Van Ooteghem et al. 2004) (see Sect. IV.A). *Kosmotoga olearia*, an oilfield isolate, was reported to grow in the presence of 15 % oxygen (Di Pippo et al. 2009). This indicates that *Thermotogales* may use biochemical strategy(ies) to face oxidative stress as already reported for other strict anaerobes (e.g. sulfate-reducing bacteria) (Krekeler et al. 1998; Teske et al. 1998; Cypionka 2000; Dolla et al. 2006). *Thermotoga maritima* has been studied in recent years with regard to the strategy(ies) possibly used to deal with O_2 and reactive oxygen species (ROS) such as peroxides. Yang and Ma (2005) have purified and characterized a heterodimeric NADH oxidase catalyzing oxygen to hydrogen peroxide. They proposed this enzyme together with other hydrogen peroxide scavenging enzymes as acting in an oxygen-removing system. Le Fourn and collaborators (2008) using differential proteomics analyses identified a flavoprotein, homologous to the rubredoxin oxygen reductase (FprA) of *Desulfovibrio* sp. that was overproduced when *T. maritima* was cultivated under oxic conditions. They provided evidence that by reducing oxygen to water, this enzyme had a crucial role in protecting the bacterium against oxygen and suggested that NADH oxidase and rubredoxin oxygen reductase were involved in this process (Le Fourn et al. 2008). However, they outlined that the direct reduction of oxygen to water by FprA might be a preferred system as compared to the production of hydrogen peroxide, known to be highly toxic to cells. Later on, Le Fourn et al. (2011) showed that *T. maritima* cells could consume oxygen at a rate of 41.5 nmol min^{-1} per mg of total protein and demonstrated that this bacterium reduced oxygen *via* a three-partner chain involving an NADH oxidoreductase (NRO), a rubredoxin (Rd) together with the rubredoxin oxidoreductase mentioned above (FprA), known as a flavor-diiron protein. They concluded that the genes coding for the three components O_2 reduction system were acquired by the *Thermococcales*, domain *Archaea*, through a single horizontal gene transfer event (Le Fourn et al. 2011). Because of the position of *Thermotogales* and *Thermococcales* within the phylogenetic tree, it has been suggested that such mechanism has been important for the first anaerobes to adjust to the presence of traces of oxygen on the "primordial" Earth. Oxygen uptake has also been observed when *T. maritima* was grown in a 2.3-L bioreactor under controlled oxygen exposure (Lakhal et al. 2010). Transcriptomic analysis, indicated that when exposed to oxygen for a short time, *T. maritima* had to deal with oxygen but also with the peroxides produced. The oxygen reductase FprA appeared as primary consumer of oxygen, followed by alkyl hydroperoxide reductase and peroxiredoxin-encoding genes as main ROS-scavenging systems when higher concentrations of O_2 were reached (Lakhal et al. 2011). It is noteworthy that the expression of the gene *hyd* that encodes the single hydrogenase of *T. maritima* was drastically affected by the presence of oxygen (Lakhal et al. 2011). These data are in accordance with the known sensitivity of hydrogenases towards oxygen (Vincent et al. 2005). Regarding the anaerobic metabolism of *T. maritima*, batch cultures indicated that it significantly decreased the redox potential (E_h) of the culture medium to about −480 mV (Lakhal et al. 2010) similarly to what was observed for the strict anaerobic methanoarchaea (Fetzer and Conrad 1993). Finally, under oxidative conditions, Lakhal et al. (2010) observed that glucose consumption rate by *T. maritima* decreased with a concomitant shift in glucose metabolism towards lactate production.

From these results, we may conclude that despite the high sensitivity of *T. maritima* to oxygen, this bacterium adapted an adequate strategy to face exposures to this gas (see Sect. IV.B for more details). This may explain why this hyperthermophilic bacterium can survive and even grow in shallow hydrothermal vents where partially oxygenated conditions cannot be precluded.

E. Hydrogen Sensitivity

Hydrogen which accumulates during the fermentation processes of carbohydrates by *Thermotogales*, with the exception of *Mesotoga* spp. (see also § III.D) is known to inhibit the growth of most of them (Huber and Hannig 2006). No growth occurred when cultures of *T. maritima* were pressurized with hydrogen-containing gas (H_2/CO_2=80:20; 300 kPa) (Huber et al. 1986). Similar observations were done with *Thermotoga thermarum* and *T. neapolitana* (Windberger et al. 1989), *Petrotoga miotherma*, *Geotoga petraea*, and *Thermosipho melaniensis* (Davey et al. 1993; Antoine et al. 1997). However for all these microorganisms, growth inhibition could be overcome by gassing the headspace with N_2 or N_2/CO_2 or by the addition of S° or thiosulfate (e.g. *Petrotoga mobilis*) which served as an electron acceptor in the culture medium; concomitantly H_2S was formed. Depending on microorganisms (see comments above), the addition of elemental sulfur and/or thiosulfate may or may not have a stimulatory effect on growth. H_2 removal in the gas phase may also result from co-cultures of *Thermotogales* with thermophilic to hyperthermophilic methanoarchaea (e.g. *Methanococcus*, *Methanopyrus* spp.) or other hydrogen oxidizing archaeons (e.g. *Archaeoglobus* or *Ferroglobus* spp.) (Huber and Hannig 2006). Some genera are hydrogen tolerant. *Geotoga subterranea* was not inhibited by the hydrogen concentration mentioned above (Davey et al. 1993). *Marinitoga camini* and *M. piezophila* reached maximum cell concentrations with 0 % H_2 while no growth occurred with 80 % H_2 (Wery et al. 2001; Alain et al. 2002), *M. hydrogenotolerans* can grow in the presence of 100 % H_2 in the gas phase (Postec et al. 2005) and this makes it the most hydrogen tolerant member of the *Thermotogales*. This confirms that *Thermotogales*, depending on species, have different H_2 sensitivities as reported earlier by Ravot et al. (1996).

IV. Hydrogen Production by *Thermotogales* spp.

A. Thermodynamic Features

The organisms described in this chapter are mainly thermophilic or hyperthermophilic anaerobes with temperature optima above 65 °C. As stated in the preceding paragraphs, many of these, such as *Thermotoga maritima* and *T. neapolitana* are capable of performing fermentative hydrogen production from a wide variety of substrates. Indeed, thermophilic hydrogen production benefits from some general advantages of performing such a process at elevated temperatures thank to a lower viscosity, better mixing, less risk of contamination, higher reaction rates and no need for reactor cooling (Wiegel et al. 1985). As already stated in Chap. 1, to make hydrogen production economically sustainable, organisms are needed that can generate hydrogen or directly or indirectly from biomass. As cellulose and hemicellulose are the most abundant polysaccharides in nature, xylose and glucose are the predominant monomeric sugars available (Kapdan and Kargi 2006). Further, starch and sucrose can be abundantly present in plants as storage material. Interestingly, the bacterial species belonging to *Thermotoga* have the capacity to hydrolyze most of the substrates derived from biomass. For example, as also reported in Chap. 8, both *Caldicellulosiruptor* and *Thermotoga* spp. contain a variety of glycoside hydrolases and transferases stating their metabolic capacity (Vanfossen et al. 2008).

A few thermodynamic considerations clearly indicate that under standard conditions (reactants concentration equal to 1 M, at 25 °C and pH 7.0), glucose oxidation to CO_2 and H_2 has a positive Gibbs energy change (reaction 9.1). This means that H_2 production requires an input of extra energy.

$$\text{Glucose} + 12H_2O \rightarrow 6HCO_3^- + 6H^+ + 12H_2 \qquad (9.1)$$
$$\Delta G^{0'} = +3.2\,\text{KJ/mol}$$

As shown in Table 9.2, most of the available literature reports that under optimal growth conditions, the oxidation of one hexose molecule will result in the formation of a variable number of hydrogen molecules ($>2 \leq 4$) in addition to acetate and CO_2. The maximum amount of ATP is obtained through production of acetate but this can only occur if all the reducing equivalents are disposed in the form of hydrogen molecules. Apparently, the *a priori* requirement to get a significant hydrogen production yield is to keep a low hydrogen partial pressure through the use of a hydrogen-consuming system (Schink and Stams 2006). Under this latter condition, as shown in reaction 9.2, glucose oxidation to acetate, CO_2 and H_2 has a significantly negative Gibbs energy change.

$$\text{Glucose} + 4H_2O \rightarrow \text{Acetate} + 2HCO_3^- + 4H^+ + 4H_2 \qquad (9.2)$$
$$\Delta G^{0'} = -206.1\,\text{KJ/mol}$$

In most fermentative hydrogen producers, catabolism *via* the Embden-Meyerhof pathway generates reducing equivalents in the form of NADH at the level of the glyceraldehyde-3-phosphate dehydrogenase (GAP deh) and reduced ferredoxin by the pryruvate:ferredoxin oxidoreductase reaction. Under standard conditions, the mid-point potentials of NAD^+/NADH and ferredoxin^{ox}/ferredoxin^{red} are −320 mV and −380 mV, respectively (Thauer et al. 1977). Recycling of these redox couples can be accomplished by various reactions such as, for example, the production of lactate though reduction of pyruvate by NADH; however, the feasibility of these reactions is *a priori* determined by the standard Gibbs free energy change ($\Delta G^{0'}$) of each specific conversion step. This latter consideration predicts that the production of H_2 by reduction of H^+ with NADH is a thermodynamically unfavorable process as expected by the low mid-point potential ($E^{0'}=-414$ mV) of the H^+/H_2 redox couple. Although the situation is more favorable in the case of ferredoxin ($E^{0'}=-380$ mV) the formation of other products such as ethanol and lactate is thermodynamically more feasible. Thus, although the microbial reduction of protons to lead H_2 generation is a metabolic unexpected process, for certain thermophiles (see Table 9.1) the amount of hydrogen produced is close to

Table 9.2. H_2 production rates and H_2 yields from various sugars conversion by *Thermotoga* spp.

Strain	Carbon source	T°	Conc.	Initial pH	Type of reactor	H_2 production rate ($mmol_{H2}/L/h$)	H_2 yield ($mol_{H2}/mol_{substrate}$)	By-products	References
T. maritima	Glucose	80	10–15 mM	6.5	Batch	10.0	4.0	Acetate	Schröder et al. (1994)
T. elfii	Glucose	65	10 g/L	7.5	Controlled batch	2.7	3.3	Acetate	Van Niel et al. (2002)
T. elfii	Glucose	65	7 g/L	7.5	Batch	–	3.3	Acetate	De Vrije et al. (2002)
T. petrophila	Glucose	80	0.1 % w/v	7.0	Batch	–	3.8	Acetate, lactate	Takahata et al. (2001)
T. naphtophila	Glucose	80	0.1 % w/v	7.0	Batch	–	4	Acetate, lactate	Takahata et al. (2001)
T. maritima	Glucose	80	7.5 g/L	6.5	Batch	8.2	1.7	–	Nguyen et al. (2008a)
T. neapolitana	Glucose	75	7.5 g/L	7.0	Batch	8.7	1.8	–	Nguyen et al. (2008a)
T. neapolitana	Glucose	77	7.0 g/L	7.5	Batch	–	3.2	Acetate butyrate	Nguyen et al. (2010a)
T. neapolitana	Glucose	75	5 g/L	7.5	Controlled batch	4.25	3.2	Acetate, lactate	Ngo et al. (2011b)
T. neapolitana	Glucose	85	2.5 g/L	7.5	Batch	0.9	3.8	Acetate, lactate	Munro et al. (2009)
T. neapolitana	Glucose	80	5 g/L	8.0	Batch	–	2.4	Acetate, lactate	Eriksen et al. (2008)
T. maritima	Arabinose	80	5 g/L	8.0	Batch	0.6	3.2	Acetate, lactate	Eriksen et al. (2011)
T. neapolitana	Arabinose	80	5 g/L	8.0	Batch	1.0	3.8	Acetate, lactate	Eriksen et al. (2011)
T. maritima	Xylose	80	5 g/L	8.0	Batch	0.4	2.7	Acetate, lactate	Eriksen et al. (2011)
T. neapolitana	Xylose	80	5 g/L	8.0	Batch	1.5	3.4	Acetate, lactate	Eriksen et al. (2011)
T. neapolitana	Xylose	75	5 g/L	7.5	Controlled batch	3.9	2.8	Acetate, lactate	Ngo et al. (2012)
T. neapolitana	Xylose	75	5 g/L	7.5	Controlled batch	3.3	2.2	Acetate, lactate	Ngo et al. (2011b)
T. neapolitana	Sucrose	75	5 g/L	7.5	Controlled batch	1.9	5.0	Acetate, lactate	Ngo et al. (2011b)
T. neapolitana	Glycerol	75	5 g/L	7.5	Batch	–	1.0	Acetate, lactate	Ngo and Sim (2011)
T. neapolitana	Cellulose	80	0.5 % w/v	7.5	Batch	–	2.2	–	Nguyen et al. (2008b)

four molecules per molecule of glucose oxidized, suggesting that in these microorganisms the thermodynamic constraints are somehow overcome as the growth conditions are quite far from the standard physiological values. Indeed, the actual Gibbs energy change as a function of both the substrates and products concentrations (see Eq. 9.3) predicts that even the reduction of H^+ by NADH becomes exergonic (−4.7 kJ/mol) if the partial H_2 concentration [$P(H_2)$] is kept as low as 10^{-2} kPa.

$$\Delta G = \Delta G + RT\ln([C][D]/[A][B]) \quad (9.3)$$

Further, the thermodynamic of the process is affected by temperature at which the reaction takes place according to Eq. (9.4)

$$\Delta G^0 = \Delta H - T\Delta S^0 \quad (9.4)$$

which says that at temperatures higher than standard conditions (25 °C), the Gibbs free energy change for the overall reaction from glucose to acetate (Eq. 9.1) is more favorable. A second consideration that might explain why thermophiles show such unexpected H_2 yields, is based on the fact that as previously shown by Thauer et al. (1977) and Amend and Plyasunov (2001), the hydrogen partial pressure needed to make reaction (9.2) feasible varies from 0.022 kPa at 25 °C to 2.2 kPa at 100 °C. Thus, at room temperature, hydrogen must be rapidly removed to avoid the inhibition of reaction (9.2) while the presence of 10^2 higher hydrogen concentration is tolerated at 100 °C.

Another possible explanation for the hydrogen formation from redox couples having a mid-point potential higher than −414 mV ($E^{0'}$ of H^+/H_2) might be the presence of a reversed electron transport (RET) mechanism linked to membrane bound NAD-dependent hydrogenases. Although this mechanism has never been described in thermophiles, its presence in the genus *Clostridium* (*Cl. tetanomorphum*), where a sodium gradient is used to drive the reduction first of ferredoxin and then for hydrogen production (Boiangiu et al. 2005), does not exclude *a priori* that other fermenting bacteria support the uphill reduction of H^+ by NADH through the use of RET. Alternatively, it has clearly been shown by Schut and Adams (2009) that in cells of *Thermotoga maritima* (*T. maritima*) ferredoxin is a more suitable reducing agent for hydrogen production than NADH.

B. The Hydrogenases of Thermotoga *spp.*

As overviewed in Chaps. 2, 3, 4, and 8, the enzymes responsible for hydrogen production (H_2) combining hydrogen protons and reducing equivalents ($2H^+ + 2e^-$) are the hydrogenases (EC 1.12.99.6 and EC 1.12.7.2) also catalyzing the reversible oxidation of molecular hydrogen. In anaerobic thermophiles, two main types of hydrogenases, based on their metal content, are found: [Fe-Fe] and [Ni-Fe] hydrogenases. Further, hydrogenases can use different types of electron carriers, e.g. NAD, NADP, FAD and ferredoxin (Fd), which are reduced in the glycolytic pathway and in particular during the conversions of both glyceraldehyde-3-P to 3-P-glycerate and pyruvate to acetyl-CoA. In most fermentative hydrogen producers, reduced electron carriers generated in these steps (NADH and Fd_{red}) need to be re-oxidized to keep the glycolytic pathway functioning and this disposal mechanism can be different among the different thermophilic hydrogen producers (Jenney and Adams 2008).

As many other bacterial species, *Thermotoga maritima* uses the Embden-Meyerhof pathways for glycolysis resulting in acetate, lactate, ethanol, CO_2 and H_2. However, recycling of reducing equivalents is performed by a trimeric [Fe-Fe] hydrogenase which uses both NADH and Fd_{red} in a 1:1 ratio to generate hydrogen (see Eq. 9.5).

$$NADH + 2Fd_{red} + 3H^+ \rightarrow NAD^+ + 2Fd_{ox} + 2H_2 \quad (9.5)$$

This so-called "flavin-based bifurcating enzyme" is coupling the exergonic oxidation of Fd_{red} ($E^{0'}$ of $Fd_{ox}/Fe_{red}=-453$ mV) to drive the unfavorable oxidation of NADH ($E^{0'}$ of $NAD^+/NADH+H^+=-320$ mV) to produce H_2 ($E^{0'}=-420$ mV) (Schut and Adams 2009). As this mechanism is favored by low hydrogen pressures, in the case of higher H_2 pressures, a switch from acetate to lactate production is seen (Huber et al. 1986). This however does not seem to affect the bifurcating mechanism of the hydrogenase, which presumably remains the same.

The anaerobically purified "bifurcating" hydrogenase of *T. maritima*, the enzyme being inactivated in the presence of even trace amounts of oxygen, is composed of three subunits – HydA (73 kDa), HydB (68 kDa) and HydC (18 kDa) – in a 1:1:1 ratio stoichiometry. The holoenzyme showed an apparent molecular mass of 500 kDa at pH 7.0 and one of 150 kDa at pH 10.0. The enzyme contained loosely bound FMN along with more than 30 Fe per heterotrimer in line with sequence analysis prediction (Buckel and Thauer 2012). Based on this latter approach, HydA subunit should harbor the active site interacting with hydrogen (H_2). This prediction was confirmed by showing that HydA alone, after dissociation of the trimeric-complex with urea, can catalyze the reduction of viologen dyes with H_2. Besides the hydrogen interacting site, HydA contains 3x[4Fe-4S] and 2x[2Fe-2S] iron-sulfur clusters with a 43 % sequence similarity to the monomeric [Fe-Fe] hydrogenase of *Clostridium pasteurianum*. HydA of *T. maritima* differs however in having a C-terminal extension with a [2Fe-2S] binding site which is lacking in the monomeric enzyme from *C. pasteurianum*.

The HydB subunit shows a 70 % similarity to the gene product HndC of the $NADP^+$-reducing [Fe-Fe] hydrogenase from *Desulfovibrio fructosovorans* and 60 % similarity to NuoF of the NADH:ubiquinone oxidoreductase from *E. coli*. Within the sequence, there are two highly conserved stretches, one featuring NAD^+ binding sites and the other recalling orthodox FMN binding sites. The C-terminal part contains Cys motifs that could bind three [4Fe-4S] clusters. At its N-terminus HydB contains four Cys residues that are suggested to be involved in binding a [2Fe-2S] cluster (Verhagen et al. 1999).

The smaller subunit of this trimeric bifurcating hydrogenase, HydC, contains fours Cys residues arranged in a motif which is highly similar (58 %) to motifs in *E. coli* Nuo and *D. fructosovorans* HndA, which are supposed to bind a [2Fe-2S] cluster.

The above reported information, gives a picture of the HydABC complex that is tentatively shown in Fig. 9.2 where a series of assumptions were made, namely: (a) the binding site for reduced ferredoxin (Fd_{red}) at HydC; (b) the presence of a second FMN loosely bound to HydB; (c) the [Fe-Fe] center plus [4Fe-4S] cluster as part of the active site of the hydrogenase in subunit HydA.

It has been proposed that the three genes encoding HydABC in *T. maritima* are arranged in a cluster *hydCBA* that is most likely a transcription unit (Verhagen et al. 1999). Clustered genes for proteins with sequence similarity to HydABC product are also found in many other anaerobic bacteria such as for example *Clostridium ljungdahlii* (Kopke et al. 2010), *Acetobacterium woodii* (Poehlein et al. 2012), and *Moorella thermoacetica* (Pierce et al. 2008), although these gene products have not been characterized. Interestingly, the enzyme complex from the anaerobe *Thermoanaerobacter tengcongensis* is composed of four subunits rather than three. Most likely, this is due to the fact that in this latter species the HydA homolog lacks the C-terminal extension with the [2Fe-2S] cluster so that a fourth subunit, homologous to the C-terminal extension, is required (Soboh et al. 2004).

Another interesting observation recently done by Thauer et al. (2010) is that electron-bifurcating [Fe-Fe] hydrogenases are not found in the Archaea domain which appear to only contain [Ni-Fe]- and/or [Fe]-hydrogenases (Thauer et al. 2010).

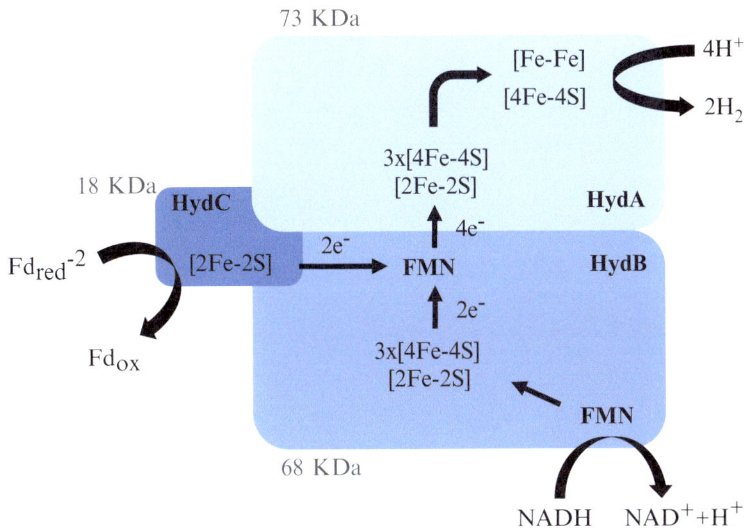

Fig. 9.2. Tentative representation of the structure and function of the HydABC complex from *Thermotoga maritimae* (See text for details).

As mentioned in Sect. III.D, several strains of *Thermotogales*, such as *Petrotoga miotherma*, *Thermosipho africanus*, *Thermotoga elfii*, *Fervidobacterium pennavorans* and *Thermotoga neapolitana*, are able to tolerate microaerophilic growth conditions and efficiently produce H_2 as a by-product of their metabolism (Van Ooteghem et al. 2002). In particular, *T. neapolitana* showed the highest H_2 production (25–30 % v/v) in these conditions (Van Ooteghem et al. 2004). Through the use of a bioinformatics approach it has been shown that the operon structure of *T. neapolitana* is the same as *T. maritima* in both the ordering and spacing of the ORFs (open reading frames) (Tosatto et al. 2008). In details, a high sequence conservation is preserved from a minimum of 75 % to a maximum of 91 % for all gene products with a sequence identity attributed to the [Fe-Fe] hydrogenase subunits of 85–91 %. Notably, the HydA subunits of both species share a 91 % sequence identity, are of the same length, and can be aligned without gaps. At the DNA level, the two sequences share 82 % identity, for a total of 375 mutated nucleotides, comprising three fully mutated codons corresponding to mutations R363E (GAA → AGG), E475S (GAG → TCC) and T539L (ACA → GTG). Taking into account the sequences of *T. petrophila* and *T. maritima* showing that R363 is not conserved between the two species (GAA → AAA), it has been concluded that only two residues, E475S and T539L of the *T. neapolitana* HydA subunit appear to be subjected to strong selection (Tosatto et al. 2008). Apparently, the functional differences between *T. neapolitana* and *T. maritima* reside in these latter subtle changes possibly involved in the conformation of the active site (H-cluster) near the surface of the protein. As suggested by Cohen et al. (2005) in the *C. pasteurianum* [FeFe] hydrogenase crystal structure, there might be two alternative gas diffusion pathways defining a hydrophobic channel toward the H-cluster active site. As suggested in Fig. 9.3, the possibility of a selective effect on gas accessibility based on the size of side-chains and charge distribution on the protein surface at the entrance of the hypothetical gas channels, B (Threonine 539) and A (Serine 475) has been discussed in detail by Tosatto et al. (2008).

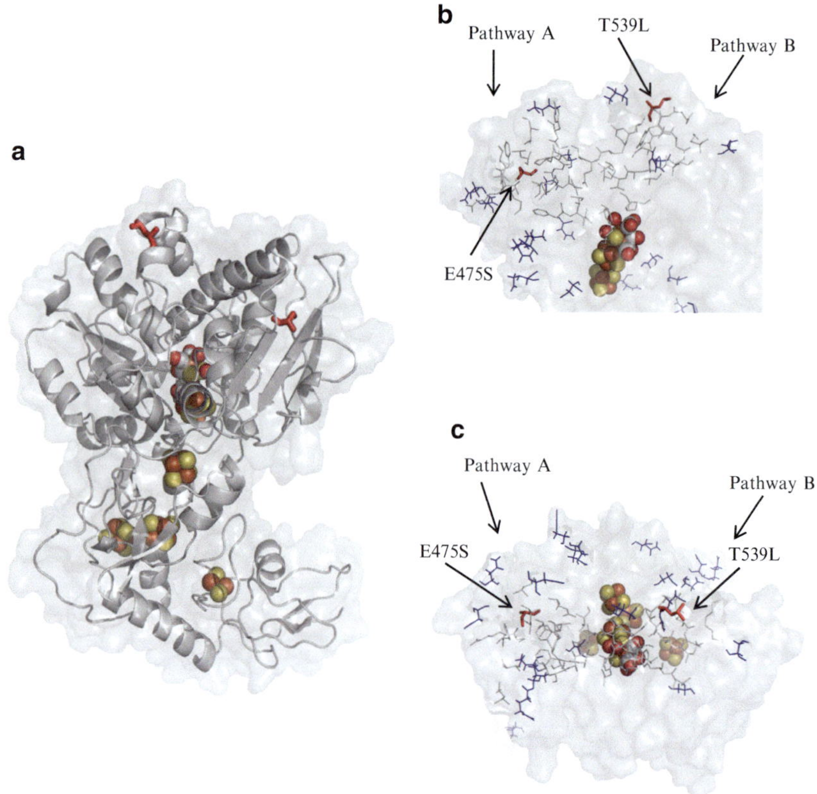

Fig. 9.3. Structural model of *Thermotoga neapolitana* HydA protein. *Panel A*. Model is shown as cartoon beneath a semi-transparent surface. Iron–sulfur clusters are shown as *spheres*. *Panel B*. Close-up of upper half of model. Residues forming part of hydrophobic channel pathways *A* and *B* are shown as sticks. Residues mutating between *T. neapolitana* and *Thermotoga maritima* HydA proteins are shown as *blue lines*. Two mutated residues forming part of hydrophobic channel pathways A (E475S) and B (T539L), are in *red* and indicated by *arrows*. *Panel C*. Same model as in *panel B*, rotated by 90° around the x-axis to show a *top view* of the molecule, where hydrophobic channel entrances are located (Tosatto et al. 2008).

C. Hydrogen Production as a Function of Variable Substrate

As reported above, *Thermotoga* spp. are able to grow on a wide array of simple sugars and polysaccharides, including starch, β-1,4 glucan (cellulose), and hemicellulose (xylan, laminarin and mannan) (Conners et al. 2006). This capacity is consistent with the production of a diverse set of proteins and enzymes that are devoted to the uptake and processing of carbohydrates (Vanfossen et al. 2008). The use of functional genomics-based approaches has provided important insights into the various mechanisms employed by these microorganisms to assimilate and metabolize carbohydrates, and has helped to identify the specific genes and operons involved (Nguyen et al. 2001, 2004). *T. maritima* genome encodes for the largest number of glycoside hydroxylases of any sequenced thermophile (Chhabra et al. 2003). These enzymes specifically hydrolyze the glycosidic bond between two or more carbohydrates or between a carbohydrate and a non-carbohydrate moiety and function during the mobilization of complex carbohydrates for subsequent metabolism. T. *neapolitana* produces enzymes such as α-galactosidase, a laminarinase, and two cellulases (endo-1,4-β-glucanases) that have orthologs in *T. maritima* (Zverlov et al. 1997; Bok et al. 1998; King et al. 1998), but this microorganism also produces several unique glycoside hydrolases. One example is a 1,4-β-D-glucan glucohydrolase (GghA), which hydrolyzes cellotetraose, cellotriose, cellobiose, and lactose (McCarthy et al. 2004). *T. petrophila* slightly differs from *T. maritima* and *T. neapolitana* with respect to the utilization of certain sugars, growing weakly on cellulose and, in the presence of a mixture of monosaccharides, utilizing glucose to a significantly lesser extent than the other two species (Takahata et al. 2001; Frock et al. 2012). Unlike *T. maritima* and *T. neapolitana*, *T. elfii* failed to grow on sucrose and carboxy methyl cellulose (CMC) (Van Niel et al. 2002). *T. maritima* was shown to have a preference for complex carbohydrates, as growth in the presence of monosaccharides was slower than growth in the presence of oligo/polysaccharides (Chhabra et al. 2003).

1. Use of Simple Sugars and Polysaccharides

Thermotoga spp. produce hydrogen from a wide range of organic materials including complex carbohydrates and wastes/biomass rich in sugars. Simple sugars such as glucose and xylose are readily biodegradable and thus preferred as reference substrates for studying biochemical and physiological aspects of the hydrogen production by these bacteria. A wide variety exists among *Thermotogales* with respect to the utilization of sugars for growth and H_2 production that is consistent with the genetic diversity between the strains and the degree of optimization of the process for H_2 production (Frock et al. 2012). Table 9.2 summarizes the results of hydrogen production by *Thermotoga* spp. obtained with different types of simple and complex carbohydrates and different conditions of growth. Under optimal conditions, the oxidation of glucose and xylose will at best result in the formation of 4 mol of H_2 per mole of hexose and 3.33 mol^{-1} mol of H_2 per mole of pentose, respectively. Maximum hydrogen yields, both from hexoses or pentoses, are obtained with acetate as fermentation product. Lower yields are associated to the formation of more reduced end products compared to acetate, such as butyrate, propionate and alcohols (ethanol, butanol) and lactic acid. It would be therefore important to establish the actual bacterial metabolism resulting in acetate as end product (Kaushik and Debabrata 2004).

T. neapolitana converted effectively sucrose to H_2 with H_2 yield of 4.95 $mol_{H2}/mol_{sucrose}$ (Table 9.2) despite the poor fructose based metabolism reported for this strain (de Vrije et al. 2009, 2010). Woodward et al. (2002) reported the list of sugars that were used by *T. maritima* for H_2 production. This strain was shown to preferentially use glucose, fructose, and galactose, whereas mannose and lactose metabolisms mainly produced carbon dioxide (Woodward et al. 2002). *T. neapolitana* and *T. maritima* were able to use the glucose-based complex carbohydrates, starch and cellulose, for hydrogen fermentation, although H_2 yields obtained with raw cellulose (30 $mL_{H2}/g_{cellulose}$) were significantly lower than with starch (180 mL_{H2}/g_{starch}) (Nguyen et al. 2008a). The pretreatment of cellulose with chemical agents (e.g. ionic liquids) that could disrupt the hydrogen-bonding network of the polysaccharide significantly increased its degradability (Nguyen et al. 2008b).

2. Use of Carbon Sources from Varies Waste-Residues

Most studies on hydrogen production in *Thermotoga* have used glucose as carbon source although hydrogen production at a large scale should be based on cheap and renewable substrates, such as industrial/agricultural wastes and residues. These materials have often high contents of hexose and pentoses stored in carbohydrate polymers that are ideal conversion substrates for H_2 generation (Ntaikou et al. 2010). The major criteria that have to be met for the selection of substrates suitable for fermentative biohydrogen production are availability, cost, carbohydrate content, biodegradability and concentration of inhibitory compounds (Hawkes et al. 2002). As shown in Table 9.3, *T. neapolitana* was widely used in technical processes with variable feedstock sources owing to its many advantageous properties including tolerance of moderate oxygen amount (6–12 %) and resistance to high H_2 partial pressure (Van Ooteghem et al. 2002, 2004). This strain was shown to successfully produce H_2 from lignocellulosic materials such as rice straw, wheat straw and *Mischantus* (de Vrije et al. 2002, 2009; Nguyen et al. 2010b; Eriksen et al. 2011), algal biomasses (Nguyen et al. 2010c; Dipasquale et al. 2012) and food waste materials, such carrot pulp, cheese whey and molasses (Table 9.3) (de Vrije et al. 2010; Cappelletti et al. 2012). Many physical and chemical along with structural and compositional factors in complex feedstock sources often hinder the biological digestibility (Chang et al. 2001). Because of that reason, pre-treatment methods including wet oxidation under alkaline conditions, mechanical pre-treatment, mild and concentrated acid hydrolysis and solvent extractions were required for an efficient utilization of these feedstock sources promoting the accessibility of polysaccharides in the substrates for enzymatic hydrolysis (Ntaikou et al. 2010). Molasses and cheese whey were effectively converted into H_2 by both attached- and suspended-cells of *T. neapolitana* without any pre-treatment needed (Cappelletti et al. 2012). Besides high yields in terms of H_2 production, the *Thermotoga* ability of utilizing different sugars in complex mixtures is essential in making H_2 production from biomass successful. Simultaneous utilization of glucose and xylose has also been observed in *T. neapolitana* (de Vrije et al. 2009), while in this microorganism, catabolite repression of lactose has been demonstrated in the presence of glucose (Vargas and Noll 1996). A preference for glucose was also shown by *T. neapolitana* when a mixture of glucose and fructose was present in the medium although both sugars were consumed at the same time (de Vrije et al. 2010). No mechanism of carbon catabolite repression has yet been defined for *T. maritima*.

In addition to carbohydrate-rich residues, *T. neapolitana* was also shown to produce H_2 by fermenting waste glycerol that is the main byproduct of the large-scale productions of bio-diesel (Ngo et al. 2011a; Ngo and Sim 2011). As compared with mesophilic *Enterobacter aerogenes* fermentation of glycerol, higher H_2 yield was obtained with *T. neapolitana* (2.7 mol_{H2} $mol^{-1}{}_{glycerol}$ instead of 0.9 mol_{H2} $mol^{-1}{}_{glycerol}$) (Ito et al. 2005) (Table 9.3).

Non-sugar substrates, such as yeast extract and trypticase that are components of the typical *Thermotoga* growth media, were shown to contribute to 9–12 % of the total H_2 production (d'Ippolito et al. 2010; Cappelletti et al. 2012). These media components represent undefined sources of nitrogen and carbon for bacteria that were shown to increase cell biomass and H_2 production in *T. maritima* and *T. neapolitana* cultures growing on glucose, glycerol, cheese whey and molasses (Nguyen et al. 2008a; d'Ippolito et al. 2010; Ngo and Sim 2011; Cappelletti et al. 2012). van Niel et al. (2002) reported that growth of *T. elfii* was completely dependent on yeast extract, while in the absence of tryptone, lower H_2 yields were obtained. Alternative nitrogen sources such as soybean meal or canola meal alone supported growth but H_2 production rates were reduced (Drapcho et al. 2008).

Because of the complexity and richness of some industrial/agricultural wastes, the utilization of these complex feedstock sources for H_2 production by *Thermotoga* could allow the reduction of the process-associated cost by simplifying the culture medium. A growth medium composed only by NH_4Cl, K_2HPO_4, NaCl, buffer and cysteine-HCl (see Sect. IV.D.b) led to an efficient H_2 production from molasses with *T. neapolitana* (Cappelletti et al. 2012). The omission of vitamin and trace elements solutions, some inorganic elements and nitrogen sources reduced the fermentation cost without a significant lost in H_2 production. Further cost

Table 9.3. H_2 production rates and H_2 yields from various industrial/agricultural wastes conversion by *Thermotoga* spp. described in the literature.

Strain	T°	Feedstock	Main carbon source	Additional treatment	H_2 production rate (mmol/L/h)	H_2 yield	References
T. elfii	65	*Miscanthus*	Glucose, xylose	Mechanical and NaOH	–	1.1 mol_{H2}/mol_{sugar}	de Vrije et al. (2002)
T. neapolitana	80	*Miscanthus*	Glucose, xylose, arabinose	NaOH, $Ca(OH)_2$ and enzymatic hydrolysis	13.1	3.2 mol_{H2}/mol_{hexose}	de Vrije et al. (2009)
T. neapolitana	75	Rice straw	Glucose, xylose	Combined treatment	4.7	2.7 $mmol_{H2}/g_{straw}$	Nguyen et al. (2010b)
T. neapolitana	75	Algal biomass	Starch, dextrins, oligosaccharides	Enzymatic hydrolysis	227.3	2.5 $mol_{H2}/mol_{glucose}$	Nguyen et al. (2010c)
T. neapolitana	80	Carrot pulp	Glucose, fructose, sucrose	Enzymatic hydrolysis	12.5	2.7 mol_{H2}/mol_{hexose}	de Vrije et al. 2010
T. neapolitana	75	Biodiesel manufacturing waste	Glycerol	Removal of methanol/ ethanol and solids	–	2.73 $mol_{H2}/mol_{glycerol}$	Ngo et al. (2011a) and Ngo and Sim (2011)
T. neapolitana	77	Molasses	Sucrose, glucose, fructose	–	1.2	2.95 mol_{H2}/mol_{hexose}	Cappelletti et al. (2012)
T. neapolitana	77	Cheese whey	Lactose, glucose, galactose	–	0.9	2.5 mol_{H2}/mol_{hexose}	Cappelletti et al. (2012)

reductions were achieved by replacing the lab-grade NaCl with non-refined sea salt and the cysteine-HCl with metabisulfite (see Sect. IV.D.b) (Cappelletti et al. 2012).

D. H_2 Production Process and Culture Parameters

Environmental parameters such as pH, hydrogen partial pressure, media components and temperature, are key factors as they influence the metabolism and therefore the fermentation end products. Thus, optimization of these processes and culture parameters is required to give enhanced H_2 yields.

1. Product Inhibition

Strategies for growth under H_2 inhibition conditions have been developed in hypethermophiles including *T. maritima* and *T. neapolitana*. H_2-producing archea and bacteria can use sulfur compounds such as elemental sulfur, polysulfides, and cystine as alternative electron acceptors (Adams 1990; Drapcho et al. 2008). The use of Fe(III) as electron acceptor was also observed in *T. maritima* when H_2 levels became inhibitory (Vargas et al. 1998). Nevertheless, the metabolic pathways of hydrogen formation are sensitive to H_2 concentrations and are subject to end-product inhibition. Therefore, the H_2 partial pressure is an extremely important factor for hydrogen synthesis. H_2 production is a means by which bacteria re-oxidise reduced ferredoxin and hydrogen-carrying coenzymes, and these reactions are less favourable as the H_2 concentration in the liquid rises (Hawkes et al. 2002). Consequently, H_2 production decreases and the metabolic pathways shift to production of more reduced substrates such as lactate, ethanol, acetone, butanol, or alanine (Levin et al. 2004). Several strategies have been developed to avoid the negative effect of H_2 accumulation. These include vigorous mixing to avoid super-saturation (Lay 2000), utilization of H_2 permeable membrane to remove dissolved H_2 from mixed liquor (Liang et al. 2002) and sparging with inert nitrogen. The application of the latter technique to *T. neapolitana* growing on either glucose or waste glycerol increased the H_2 production by 50–80 % (Nguyen et al. 2010a; Ngo et al. 2011b). However, d'Ippolito et al. (2010) showed that sparging with nitrogen had a little influence on H_2 yield at low ratio between the volumes of culture and headspace. Gas sparging became more important with the increase of the liquid fraction. They observed the highest H_2 yield with a culture/headspace volume ratio of 1:3. An increased ratio between gas and liquid phase volumes was also associated to a decreased synthesis of lactic acid in cultures of *Thermotoga maritima* (Schröder et al. 1994).

In addition to culture/headspace volume ratio, the stirring regime and the substrate concentration were shown to influence H_2 production most probably because of their correlation to the dissolved H_2 concentration (Hawkes et al. 2002). A moderate agitation of the cultures (at 75 rpm) was shown to almost double the H_2 production by *T. neapolitana* cultures in 20 h of growth on glucose (Van Ooteghem et al. 2002). Using the same type of culture and carbon source, H_2 production rate was improved by increasing the stirring speed from 300 to 400 rpm. However, speeds of 500 and 600 rpm did negatively affect this fermentative process (Ngo et al. 2011b).

Considering the influence of the substrate concentration on the H_2 production process, Nguyen et al. (2008a) reported that concentrations of glucose over 15 g/L had inhibitory effects on cell growth and H_2 production in *T. neapolitana* and *T. maritima* strains batch cultures. The growth and H_2 content showed the highest values at the initial glucose concentration of 7.5 g/L for both strains. In *T. neapolitana* cultures growing on xylose, the maximal values of biomass and cumulative H_2 production were obtained at an initial substrate concentration of 5.0 g/L, while higher concentrations of xylose were not favorable for *T. neapolitana* growth and H_2 formation (Ngo et al. 2012). However, at substrate concentration of 5.0 g/L, the converted H_2 yield from xylose was lower than at the initial xylose concentration of 2 g/L suggesting that the change in xylose concentration remarkably affected not only H_2 production itself but also the substrate utilisation (Ngo et al. 2012). The best performing initial concentration of waste glycerol (3.0 g/L) was less than half than the pure glycerol (7.0 g/L) despite they resulted in comparable H_2 productions (Ngo and Sim 2011). This indicates different potentials of the two types of glycerol sources to release a specific amount of H_2.

2. pH Buffering System

A rapid decrease in pH is observed in cultures of *Thermotoga* spp. fermenting sugars to H_2 causing, in some cases, the process to stop before all the substrate is consumed (Eriksen et al. 2008). Experiments with pH adjustment showed that when cultures were neutralized with either increased initial buffer or injection of $NaHCO_3$ or NaOH, glucose was completely consumed and H_2/acetic acid productions increased proportionally. Both the type of buffer and the initial pH have significant effect on H_2 production process. Organic and inorganic buffer systems have been tested in literature resulting in different H_2 productions depending on the growth conditions and buffers concentration in the culture medium. HEPES resulted to be the best performing buffer when compared to HPO_4:H_2PO_4, Tris-HCl, Mops, Pipes buffers in experiments using *T. neapolitana* batch cultures growing on glucose (Cappelletti et al. 2012). Effective H_2 productions were also obtained with HEPES-buffered medium containing raw material feedstocks such as cheese whey, molasses and glycerol waste as carbon sources for *T. neapolitana* (Ngo et al. 2011a; Cappelletti et al. 2012). The good buffering properties of HEPES might be due to its pK (7.55) that is near the optimum for the growth of *T. neapolitana*. Itaconic acid was also successfully used to overcome pH-induced limitations of H_2-producing *T. neapolitana* cultures growing glucose. The buffering capacity of this carbohydrate was tested after that it was found to be poorly metabolized by this strain (Van Ooteghem et al. 2004). Its applicability for H_2 productions from industrial residues was demonstrated by Ngo and Sim (2011). The addition of itaconic acid into the culture medium of *T. neapolitana* growing on waste glycerol increased the process performance by almost 40 % (Ngo and Sim 2011).

In addition to the buffer, the initial pH value was shown to have a significant effect on growth and H_2 production of *Thermotoga* spp. Anna et al. (1991) indicated that the pH control is crucial to the H_2 formation pathway because of the effects of pH on the hydrogenase activity. The optimum initial pH for *T. neapolitana* suspended-cells ranged from 6.5 to 7.5 depending on both the substrate and the conditions of growth (Ngo et al. 2011b). Interestingly, an higher pH range (7.7–8.5) yielded the highest hydrogen production in experiments with glucose-grown *T. neapolitana* attached-cells on ceramic supports (Cappelletti et al. 2012). An initial pH of 7.0 provided the most promising results in terms of H_2 and acetic acid productions in *T. neapolitana* growing on xylose (Ngo et al. 2012). The optimum initial pH values for growth and hydrogen production from glucose were 6.5–7.0 for *T. maritima* and 6.5–7.5 for *T. neapolitana*, respectively. The best performing initial pH for H_2 production from glycerol by *T. neapolitana* was 7.0–7.5 (Ngo and Sim 2011). The application of initial pH above 8 resulted in a decrease of cumulative H_2 production as well as cell concentration suggesting the influence of pH over the metabolism pathway of the bacteria (Ngo and Sim 2011).

3. Oxygen Exposure

Hydrogen production by *Thermotoga* strains is a hyperthermophilic anaerobic process, thus, high temperatures (70–90 °C) and strictly anaerobic conditions must be initiated and maintained in the reactor vessel during production.

Some researchers have reported that low concentrations of oxygen are tolerated by both *T. neapolitana* (Tosatto et al. 2008) and *T. maritima* (Le Fourn et al. 2008) and an O_2

insensitive hydrogenase have been described in *T. neapolitana* (Käslin et al. 1998). Van Ooteghem et al. (2002, 2004) reported that microaerobic metabolism increased the H_2 yield from *T. neapolitana* up to values higher than the theoretical 4 $mol_{H2}/mol_{glucose}$ possible from fermentative metabolism (Table 9.2); however, this result was not confirmed by other researchers who observed H_2 production after the injection of O_2, but with rate and extent that were lower than those found in cultures without O_2 (Eriksen et al. 2008; Munro et al. 2009).

Prevention of O_2 exposure could be difficult on an industrial scale and requires expensive reducing agents (cysteine-HCl). Anaerobic conditions can be initiated, maintained, and monitored in the reactor vessels by (1) flushing media with nitrogen gas, (2) heating or boiling of the media to remove dissolved oxygen, (3) adding chemical agents such as sodium sulfite or cysteine-HCl to consume residual O_2 in the liquid, (4) adding resazurin to act as visible redox indicator, and (5) maintaining positive pressure in headspace to prevent air contamination (Drapcho et al. 2008). Cysteine-HCl at concentration of 0.5–1 g/L was commonly added to provide reducing conditions in media and consume residual oxygen (de Vrije et al. 2002; Van Ooteghem et al. 2002, 2004). The addition of a reducing agent to H_2 production resulted to be fundamental even when complex feedstock sources were supplemented to the culture medium. However, to reduce the cost of the process, the utilization of cheaper reducing agents was attempted. For example, the replacement of cysteine-HCl with metabisulfite (MBS) gave promising results in terms of both medium cost (90 % reduction) and H_2 yield in *T. neapolitana* cultures growing on molasses (Cappelletti et al. 2012). Conversely, in the cheese whey tests, the attempt to replace cysteine with MBS led to poor performances (Cappelletti et al. 2012).

4. Growth-Temperature and H_2 Production

Optimum temperature of growth differs among *Thermotoga* species and has to be considered in the bioreactor operating process to maximize H_2 production rates (Munro et al. 2009; Nguyen et al. 2008a). Optimum temperature of growth and H_2 production is 77 °C for *T. maritima*, 88 °C for *T. petrophila* and *T. naphthophila* and only 66 °C for *T. elfii* (Jannasch et al. 1988; Huber and Hannig 2006). Cultures of T. neapolitana grown at 77 and 85 °C exhibited the greatest rate and extent of H_2 production (Munro et al. 2009).

Advantages to using high temperatures for fermentation process include (1) to reduce the likelihood of contamination by H_2-consuming organisms that lessens the need for sterilization of media and equipment, (2) to favour the catalytic activity of hydrogenase of evolving H_2 (Adams 1990) (3) to directly utilize industrial organic wastewaters that are often discharged at elevated temperatures; (4) to avoid cooling down processes usually required by mesophilic fermentations that in large scale generate excess heat (Drapcho et al. 2008).

V. Future Perspectives

This Chapter summarizes our present knowledge on the microbiology, physiology, and biochemistry of the *Thermotogales* order with the aim to better define problems and challenges linked to H_2 production. In this respect, it is worth noting that important improvements have been recently made through optimization of both the bioprocess parameters and *Thermotoga* spp. to be used. Although the identification of suitable feed-stocks for fermentative hydrogen production was done, more research work to improve hydrogen production rates and yields, is required. The use, for example, of *Thermotoga* strains metabolically engineered and the development of a "two stage process" is likely to improve H_2 production. Indeed, this latter approach involves the fermentation of the selected substrate to both hydrogen and organic acids by *Thermotoga* spp. in the first stage and, in a second stage, either an additional energy extraction or the generation of highly-valuable products by exploiting the effluent of the first stage reactor. An alternative approach to improve H_2 production might also be achieved through a specific bioreactor configuration ameliorating both biomass concentration and substrate conversion efficiency by employing biomass retention systems such as granules, flocs or biofilm-formation supports.

Acknowledgements

The research leading to results obtained by D.Z. and M.C. on H_2-production by *Thermotoga* spp. has received funding from the Italian Ministry of Agriculture, Food and Forestry (MIPAAF) under grant 'Combined production of hydrogen and methane from agricultural and zootechnical wastes through biological processes (BIO-HYDRO)'.

References

Adams MWW (1990) The metabolism of hydrogen by extremely thermophilic, sulfur-dependent bacteria. FEMS Microbiol Rev 75:219–237

Alain K, Marteinsson VT, Miroshnichenko ML, Bonch-Osmolovskaya EA, Prieur D, Birrien JL (2002) *Marinitoga piezophila* sp. nov., a rod-shaped, thermo-piezophilic bacterium isolated under high hydrostatic pressure from a deep-sea hydrothermal vent. Int J Syst Evol Microbiol 52:1331–1339

Amend JP, Plyasunov AV (2001) Carbohydrates in thermophile metabolism: calculations of the standard molal thermodynamic properties of aqueous pentoses and hexoses at elevated temperatures and pressures. Geochim Cosmochim Acta 65:3901–3917

Andrews KT, Patel BKC (1996) *Fervidobacterium gondwanense* sp. nov., a new thermophilic anaerobic bacterium isolated from non-volcanically heated geothermal waters of the Great Artesian Basin of Australia. Int J Syst Bacteriol 46:265–269

Anna J, Shigetoshi A, Adams MWW (1991) The extremely thermophilic eubacterium, *Thermotoga maritima*, contains a novel iron-hydrogenase whose cellular activity is dependent upon tungsten. J Biol Chem 266:13834–13841

Antoine E, Cilia V, Meunier J, Guezennec J, Lesongeur F, Barbier G (1997) *Thermosipho melanesiensis* sp. nov., a new thermophilic anaerobic bacterium belonging to the order Thermotogales, isolated from deep-sea hydrothermal vents in the southwestern Pacific Ocean. Int J Syst Bacteriol 47:1118–1123

Balk M, Weijma J, Stams AJ (2002) *Thermotoga lettingae* sp. nov., a novel thermophilic, methanol-degrading bacterium isolated from a thermophilic anaerobic reactor. Int J Syst Evol Microbiol 52:1361–1368

Ben Hania W, Ghodbane R, Postec A, Brochier-Armanet C, Hamdi M, Fardeau M-L, Ollivier B (2011) Cultivation of the first mesophilic representative ("mesotoga") within the order Thermotogales. Syst Appl Microbiol 34:581–585

Ben Hania W, Ghodbane R, Postec A, Hamdi M, Ollivier B, Fardeau M-L (2012) *Defluviitoga tunisiensis* gen. nov., sp. nov., a thermophilic bacterium isolated from a mesothermic and anaerobic whey digester. Int J Syst Evol Microbiol 62:1377–1382

Ben Hania W, Postec A, Aullo T, Ranchou-Peyruse A, Erauso G, Brochier-Armanet C, Hamdi M, Ollivier B, Saint-Laurent S, Margot M, Fardeau ML (2013) Mesotoga infera sp. nov., a novel mesophilic member of the order Thermotogales, isolated from an underground gas storage in France. Int J Syst Evol Microbiol 63:3003–3008. doi:1099/ijs.0.047993-0

Boiangiu CD, Jayamani E, Brugel D, Hermann G, Kim J, Forzi L, Hedderich R, Vgenopoulou I, Pierik AJ, Steuber J, Buckel W (2005) Sodium ion pumps and hydrogen production in glutamate fermenting anaerobic bacteria. J Mol Microbiol Biotechnol 10:105–119

Bok JD, Yernool DA, Eveleigh DE (1998) Purification, characterization, and molecular analysis of thermostable cellulases CelA and CelB from *Thermotoga neapolitana*. Appl Environ Microbiol 64:4774–4781

Brochier C, Philippe H (2002) Phylogeny: a non-hyperthermophilic ancestor for Bacteria. Nature 417:244

Buckel W, Thauer RK (2012) Energy conservation via electron bifurcating ferredoxin reduction and proton/Na+ translocating ferredoxin oxidation. Biochim Biophys Acta 1827:94–113. doi:10.1016/j.bbabio.2012.07.002

Cappelletti M, Bucchi G, Mendes JDS, Alberini A, Fedi S, Bertin L, Frascari D (2012) Biohydrogen production from glucose, molasses and cheese whey by suspended and attached cells of four hyperthermophilic *Thermotoga* strains. J Chem Technol Biotechnol 87:1291–1301

Chang VS, Nagwani M, Kim CH, Holtzapple MT (2001) Oxidative lime pretreatment of high-lignin biomass. Appl Biochem Biotechnol 94:1–28

Chhabra SR, Shockley KR, Conners SB, Scott KL, Wolfinger RD, Kelly RM (2003) Carbohydrate-induced differential gene expression patterns in the hyperthermophilic bacterium *Thermotoga maritima*. J Biol Chem 278:7540–7552

Cohen J, Kim K, King P, Seibert M, Schulten K (2005) Finding gas diffusion pathways in proteins: application to O_2 and H_2 transport in CpI [FeFe]-

hydrogenase and the role of packing defects. Structure 13:1321–1329
Conners SB, Mongodin EF, Johnson MR, Montero CI, Nelson KE, Kelly RM (2006) Microbial biochemistry, physiology, and biotechnology of hyperthermophilic *Thermotoga* species. FEMS Microbiol Rev 30:872–905
Cypionka H (2000) Oxygen respiration by *Desulfovibrio* species. Annu Rev Microbiol 54:827–848
d'Ippolito G, Dipasquala L, Vella FM, Romano I, Gambacorta A, Fontana A (2010) Hydrogen metabolism in the extreme thermophile *Thermotoga neapolitana*. Int J Hydrog Energy 35:2290–2295
Davey ME, Wood WA, Key R, Nakamura K, Stahl D (1993) Isolation of three species of Geotoga and Petrotoga: two new genera, representing a new lineage in the bacterial line of descent distantly related to the 'Thermotogales'. Syst Appl Microbiol 16:191–200
de Vrije T, De Haas GG, Tan GB, Keijsers ERP, Claassen PAM (2002) Pretreatment of *Miscanthus* for hydrogen production by *Thermotoga elfii*. Int J Hydrog Energy 27:1381–1390
de Vrije T, Bakker RR, Budde MAW, Lai MH, Mars AE, Claassen PAM (2009) Efficient hydrogen production from the lignocellulosic energy crop *Miscanthus* by the extreme thermophilic bacteria *Caldicellulosiruptor saccharolyticus* and *Thermotoga neapolitana*. Biotechnol Biofuels 2:12
de Vrije T, Budde MAW, Lips SJ, Bakker RR, Mars AE, Claassen PAM (2010) Hydrogen production from carrot pulp by the extreme thermophiles *Caldicellulosiruptor saccharolyticus* and *Thermotoga neapolitana*. Int J Hydrog Energy 35:13206–13213
Di Pippo JL, Nesbo CL, Dahle H, Doolittle WF, Birkland N-K, Noll KM (2009) *Kosmotoga olearia* gen. nov., sp. nov., a thermophilic, anaerobic heterotroph isolated from an oil production fluid. Int J Syst Evol Microbiol 59:2991–3000
Dipasquale L, d'Ippolito G, Gallo C, Vella FM, Gambacorta A, Picariello G, Fontana A (2012) Hydrogen production by the thermophilic eubacterium *Thermotoga neapolitana* from storage polysaccharides of the CO_2-fixing diatom *Thalassiosira weissflogii*. Int J Hydrog Energy 37:12250–12257
Dolla A, Fournier M, Dermoun Z (2006) Oxygen defense in sulfate-reducing bacteria. J Biotechnol 126:87–100
Drapcho CM, Nhuan NP, Walker TH (eds) (2008) Hydrogen production by fermentation. In: Biofuels engineering process technology. McGraw-Hill, New York, pp 269–299
Eriksen NT, Nielsen TM, Iversen N (2008) Hydrogen production in anaerobic and microaerobic *Thermotoga neapolitana*. Biotechnol Lett 30:103–109
Eriksen NT, Leegaard Riis M, Kindby Holm N, Iversen N (2011) H2 synthesis from pentoses and biomass in *Thermotoga* spp. Biotechnol Lett 33:293–300
Fardeau M-L, Ollivier B, Patel BKC, Magot M, Thomas P, Rimbault A, Rocchiccioli F, Garcia J-L (1997) *Thermotoga hypogea* sp. nov., a xylanolytic, thermophilic bacterium from an oil-producing well. Int J Syst Bacteriol 47:1013–1019
Feng Y, Cheng L, Zhang X, Li X, Deng Y, Zhang H (2010) Thermococcoides shengliensis gen. nov., sp. nov., a new member of the order Thermotogales isolated from oil-production fluid. Int J Syst Evol Microbiol 60:932–937
Fetzer S, Conrad R (1993) Effect of redox potential on methanogenesis by *Methanosarcina barkeri*. Arch Microbiol 160:108–113
Friedrich AB, Antranikian G (1996) Keratin degradation by *Fervidobacterium pennavorans*, a novel thermophilic anaerobic species of the order Thermotogales. Appl Environ Microbiol 62:2875–2882
Frock AD, Gray SR, Kelly RM (2012) Hyperthermophilic Thermotoga species differ with respect to specific carbohydrate transporters and glycoside hydrolases. Appl Environ Microbiol 78:1978–1986
Galperin MY, Noll KM, Romano AH (1996) The glucose transport system of the hyperthermophilic anaerobic bacterium *Thermotoga neapolitana*. Appl Environ Microbiol 62:2915–2918
Hawkes F, Dinsdale R, Hawkes D, Hussy I (2002) Sustainable fermentative hydrogen production: challenges for process optimisation. Int J Hydrog Energy 27:1339–1347
Huber R, Hannig M (2006) Thermotogales. In: Dworkin M, Falkow S, Rosenberg E, Schleifer KH, Stackebrandt E (eds) The prokaryotes. Springer, New York, pp 899–922
Huber R, Stetter KO (2001) Discovery of hyperthermophilic microorganisms. In: Adams MWW, Kelly RM (eds) Methods in enzymology. Academic, San Diego, pp 11–24
Huber R, Langworthy TA, Konig H, Thomm M, Woese CR, Sleytr UB, Stetter KO (1986) *Thermotoga maritima* sp. nov. represents a new genus of unique extremely thermophilic eubacteria growing up to 90°C. Arch Microbiol 144:324–333
Huber R, Woese CR, Langworthy TA, Fricke H, Stetter KO (1989) *Thermosipho africanus* gen. nov., represents a new genus of thermophilic Eubacteria within the Thermotogales. Syst Appl Microbiol 12:32–37
Huber R, Woese CR, Langworthy TA, Kristjansson JK, Stetter KO (1990) *Fervidobacterium islandicum* sp. nov., a new extremely thermophilic eubacterium belonging to the Thermotogales. Arch Microbiol 154:105–111

Ito T, Nakashimada Y, Senba K, Matsui T, Nishio N (2005) Hydrogen and ethanol production from glycerol-containing wastes discharged after biodiesel manufacturing process. J Biosci Bioeng 100:260–265

Jannasch HW, Huber R, Belkin S, Stetter KO (1988) *Thermotoga neapolitana* sp. nov. of the extremely thermophilic, eubacterial genus *Thermotoga*. Arch Microbiol 150:103–104

Janssen PH, Morgan HW (1992) Heterotrophic sulfur reduction by *Thermotoga* sp. strain FjSS3.B1. FEMS Microbiol Lett 96:213–217

Jayasinghearachchi HS, Lal B (2011) *Oceanotoga teriensis* gen. nov., sp. nov., a thermophilic bacterium isolated from offshore oil-producing wells. Int J Syst Evol Microbiol 61:554–560

Jeanthon C, Reysenbach AL, L'Haridon S, Gambacorta A, Pace NR, Glenat P, Prieur D (1995) *Thermotoga subterranea* sp. nov., a new thermophilic bacterium isolated from a continental oil reservoir. Arch Microbiol 164:91–97

Jenney FE, Adams MWW (2008) Hydrogenases of the model hyperthermophiles. Incredible anaerobes: from physiology to genomics to fuels. Ann N Y Acad Sci 1125:252–266

Kapdan IK, Kargi F (2006) Bio-hydrogen production from waste materials. Enzyme Microb Technol 38:569–582

Käslin SA, Childers SE, Noll KM (1998) Membrane-associated redox activities in *Thermotoga neapolitana*. Arch Microbiol 170:297–303

Kaushik N, Debabrata D (2004) Improvement of fermentative hydrogen production: various approaches. Appl Microbiol Biotechnol 65:520–529

King MR, Yernool DA, Eveleigh DE, Chassy BM (1998) Thermostable alpha-galactosidase from *Thermotoga neapolitana*: cloning, sequencing and expression. FEMS Microbiol Lett 163:37–42

Kopke M, Held C, Hujer S, Liesegang H, Wiezer A, Wollherr A, Ehrenreich A, Liebl W, Gottschalk G, Durre P (2010) *Clostridium ljiungdahlii* represents a microbial production platform based on syngas. Proc Natl Acad Sci U S A 107:13087–13092

Krekeler D, Teske A, Cypionka H (1998) Strategies of sulfate-reducing bacteria to escape oxygen stress in a cyanobacterial mat. FEMS Microbiol Ecol 25:89–96

L'Haridon S, Miroshnichenko M, Hippe H, Fardeau M-L, Bonch-Osmolovskaya EA, Stackebrandt E, Jeanthon C (2001) *Thermosipho geolei* sp. nov., a thermophilic bacterium isolated from a continental petroleum reservoir in Western Siberia. Int J Syst Evol Microbiol 51:1327–1334

L'Haridon S, Miroshnichenko ML, Hippe H, Fardeau ML, Bonch-Osmolovskaya EA, Stackebrandt E, Jeanthon C (2002) *Petrotoga olearia* sp. nov. and *Petrotoga sibirica* sp. nov., two thermophilic bacteria isolated from a continental petroleum reservoir in Western Siberia. Int J Syst Evol Microbiol 52:1715–1722

Lakhal R, Auria R, Davidson S, Ollivier B, Dolla A, Hamdi M, Combet-Blanc Y (2010) Effect of oxygen and redox potential on glucose fermentation in thermotoga maritima under controlled physico-chemical conditions. Int J Microbiol 2010:896510. doi:10.1155/2010/896510

Lakhal R, Auria R, Davidson S, Ollivier B, Durand M-C, Dolla A, Hamdi M, Combet-Blanc Y (2011) Oxygen uptake rates in the hyperthermophilic anaerobe *Thermotoga maritima* grown in a bioreactor under controlled oxygen exposure: clues to its defence strategy against oxidative stress. Arch Microbiol 193:429–438

Lay JJ (2000) Modeling and optimization of anaerobic digested sludge converting starch to hydrogen. Biotechnol Bioeng 68:269–278

Le Fourn C, Fardeau M-L, Ollivier B, Lojou E, Dolla A (2008) The hyperthermophilic anaerobe *Thermotoga maritima* is able to cope with limited amount of oxygen: insights into its defence strategies. Environ Microbiol 10:1877–1887

Le Fourn C, Brasseur G, Brochier-Armanet C, Pieulle L, Brioukhanov A, Ollivier B, Dolla A (2011) An oxygen reduction chain in the hyperthermophilic anaerobe *Thermotoga maritima* highlights horizontal gene transfer between Thermococcales and Thermotogales. Environ Microbiol 13:2132–2145

Levin DB, Pitt L, Love M (2004) Biohydrogen production: prospects and limitations to practical application. Int J Hydrog Energy 29:173–185

Liang TM, Cheng SS, Wu KL (2002) Behavioral study on hydrogen fermentation reactor installed with silicone rubber membrane. Int J Hydrog Energy 27:1157–1165

Lien T, Madsen M, Rainey FA, Birkeland N-K (1998) *Petrotoga mobilis* sp. nov., from a North Sea oil-production well. Int J Syst Bacteriol 48:1007–1013

Magot M, Ollivier B, Patel BK (2000) Microbiology of petroleum reservoirs. Antonie Van Leeuwenhoek 77:103–116

McCarthy JK, Uzelac A, Davis DF, Eveleigh DE (2004) Improved catalytic efficiency and active site modification of 1,4-beta-D-glucan glucohydrolase A from *Thermotoga neapolitana* by directed evolution. J Biol Chem 279:11495–11502

Miranda-Tello E, Fardeau M-L, Thomas P, Ramirez F, Casalot L, Cayol J-L, Garcia J-L, Ollivier B (2004) *Petrotoga mexicana* sp. nov., a novel thermophilic, anaerobic and xylanolytic bacterium isolated from an oil-producing well in the Gulf of Mexico. Int J Syst Evol Microbiol 54:169–174

Miranda-Tello E, Fardeau M-L, Joulian C, Magot M, Thomas P, Tholozan JL, Ollivier B (2007) *Petrotoga halophila* sp. nov., a thermophilic, moderately halophilic, fermentative bacterium isolated from an offshore oil well in Congo. Int J Syst Evol Microbiol 57:40–44

Munro SA, Zinder SH, Walker LP (2009) The fermentation stoichiometry of *Thermotoga neapolitana* and influence of temperature, oxygen, and pH on hydrogen production. Biotechnol Prog 25:1035–1042

Nesbø CL, Dlutek M, Zhaxybayeva O, Doolittle WF (2006) Evidence for existence of "Mesotogas," members of the order Thermotogales adapted to low-temperature environments. Appl Environ Microbiol 72:5061–5068

Nesbø CL, Kumaraswamy R, Dlutek M, Doolittle WF, Foght J (2010) Searching for mesophilic Thermotogales bacteria: "Mesotogas" in the wild. Appl Environ Microbiol 76:4896–4900

Nesbø C, Bradnan D, Adebusuyi A, Dlutek M, Petrus A, Foght J, Doolittle W, Noll K (2012) *Mesotoga prima* gen. nov., sp. nov., the first described mesophilic species of the Thermotogales. Extremophiles 16:387–393

Ngo TA, Sim SJ (2011) Dark fermentation of hydrogen from waste glycerol using hyperthermophilic eubacterium *Thermotoga neapolitana*. Environ Prog Sustain Energy 31:466–473

Ngo TA, Kim MS, Sim SJ (2011a) High-yield biohydrogen production from biodiesel manufacturing waste by *Thermotoga neapolitana*. Int J Hydrog Energy 36:5836–5842

Ngo TA, Kim MS, Sim SJ (2011b) Thermophilic hydrogen fermentation using *Thermotoga neapolitana* DSM 4359 by fed-batch culture. Int J Hydrog Energy 36:14014–14023

Ngo TA, Nguyen TH, Bui HTV (2012) Thermophilic fermentative hydrogen production from xylose by *Thermotoga neapolitana* DSM 4359. Renew Energy 37:174–179

Nguyen TN, Borges KM, Romano AH, Noll KM (2001) Differential gene expression in *Thermotoga neapolitana* in response to growth substrate. FEMS Microbiol Lett 195:79–83

Nguyen TN, Ejaz AD, Brancieri MA, Mikula AM, Nelson KE, Gill SR, Noll KM (2004) Whole-genome expression profiling of *Thermotoga maritima* in response to growth on sugars in a chemostat. J Bacteriol 186:4824–4828

Nguyen TAD, Kim JP, Kim MS, Oh YK, Sima SJ (2008a) Optimization of hydrogen production by hyperthermophilic eubacteria, *Thermotoga maritima* and *Thermotoga neapolitana* in batch fermentation. Int J Hydrog Energy 33:1483–1488

Nguyen TAD, Kim JP, Kim MS, Oh YK, Sim SJ (2008b) Hydrogen production by the hyperthermophilic eubacterium, *Thermotoga neapolitana*, using cellulose pretreated by ionic liquid. Int J Hydrog Energy 33:5161–5168

Nguyen TAD, Han SJ, Kim JP, Kim MS, Sim SJ (2010a) Hydrogen production of the hyperthermophilic eubacterium, *Thermotoga neapolitana* under N2 sparging condition. Bioresour Technol 101:S38–S41

Nguyen TAD, Kim KR, Kim MS, Sim SJ (2010b) Thermophilic hydrogen fermentation from Korean rice straw by *Thermotoga neapolitana*. Int J Hydrog Energy 35:13392–13398

Nguyen TAD, Kim KR, Nguyen MT, Kim MS, Kim D, Sim SJ (2010c) Enhancement of fermentative hydrogen production from green algal biomass of *Thermotoga neapolitana* by various pretreatment methods. Int J Hydrog Energy 35:13035–13040

Ntaikou I, Antonopoulou G, Lyberatos G (2010) Biohydrogen production from biomass and wastes via dark fermentation: a review. Waste Biomass Valor 1:21–39

Nunoura T, Oida H, Miyazaki M, Suzuki Y, Takai K, Horikoshi K (2007) *Marinitoga okinawensis* sp. nov., a novel thermophilic and anaerobic heterotroph isolated from a deep-sea hydrothermal field, Southern Okinawa Trough. Int J Syst Evol Microbiol 57:467–471

Nunoura T, Hirai M, Imachi H, Miyazaki M, Makita H, Hirayama H, Furushima Y, Yamamoto H, Takai K (2010) Kosmotoga arenicorallina sp. nov. a thermophilic and obligately anaerobic heterotroph isolated from a shallow hydrothermal system occurring within a coral reef, southern part of the Yaeyama Archipelago, Japan, reclassification of *Thermococcoides shengliensis* as *Kosmotoga shengliensis* comb. nov., and emended description of the genus Kosmotoga. Arch Microbiol 192:811–819

Ollivier B, Cayol J-L (2005) The fermentative, iron-reducing, and nitrate-reducing microorganisms. In: Ollivier B, Magot M (eds) Petroleum microbiology. ASM Press, Washington, DC, pp 71–88

Patel BKC, Morgan HW, Daniel RM (1985) *Fervidobacterium nodosum*; gen. nov. and spec. nov., a new chemoorganotrophic, caldoactive, anaerobic bacterium. Arch Microbiol 141:63–69

Pierce E, Xie G, Barabote RD, Saunders E, Han CS, Detter JC, Richardson P, Brettin TS, Das A, Ljungdahl LG, Ragsdale SW (2008) The complete genome sequence of *Moorella thermoacetica* (f. *Clostridium thermoaceticum*). Environ Microbiol 10:2550–2573

Poehlein A, Schmidt S, Kaster AK, Goenrich M, Vollmers J, Thurmer A, Bertsch J, Schuchmann K, Voigt B, Hecker M, Daniel R, Thauer RK, Gottschalk G, Muller V (2012) An ancient pathway combining carbon dioxide fixation with the generation and utilization of a

sodium ion gradient for ATP synthesis. PLoS One 7:e33439. doi:10.1371/journal.pone.0033439
Postec A, Le Breton C, Fardeau M-L, Lesongeur F, Pignet P, Querellou J, Ollivier B, Godfroy A (2005) *Marinitoga hydrogenitolerans* sp. nov., a novel member of the order Thermotogales isolated from a black smoker chimney on the Mid-Atlantic Ridge. Int J Syst Evol Microbiol 55:1217–1221
Postec A, Ciobanu MC, Birrien J-L, Bienvenu N, Prieur D, Le Romancer M (2010) *Marinitoga litoralis* sp. nov., a thermophilic, heterotrophic bacterium isolated from a coastal thermal spring on Ile Saint-Paul, Southern Indian Ocean. Int J Syst Evol Microbiol 60:1778–1782
Ravot G, Ollivier B, Magot M, Patel B, Crolet J, Fardeau M-L, Garcia J (1995) Thiosulfate reduction, an important physiological feature shared by members of the order Thermotogales. Appl Environ Microbiol 61:2053–2055
Ravot G, Ollivier B, Fardeau M-L, Patel BKC, Andrews KT, Magot M, Garcia JL (1996) L-alanine production from glucose fermentation by hyperthermophilic members of the domains Bacteria and Archaea: a remnant of an ancestral metabolism? Appl Environ Microbiol 62:2657–2659
Reysenbach A-L, Boone DR, Castenholz RW, Garrity GM (2001) Phylum BII. Thermotogae phy. nov. In: Boone DR, Castenholz RW (eds) Bergey's manual of systematic bacteriology. Springer, New York, pp 369–387
Schink B, Stams A (2006) Syntrophism among prokaryotes. In: Schleifer K-H, Stackebrandt E (eds) The prokaryotes. Springer, New York, pp 322–337
Schröder C, Selig M, Schönheit P (1994) Glucose fermentation to acetate, CO_2 and H_2 in the anaerobic hyperthermophilic eubacterium *Thermotoga maritima*: involvement of the Embden-Meyerhof pathway. Arch Microbiol 161:460–470
Schut GJ, Adams MW (2009) The iron-hydrogenase of *Thermotoga maritima* utilizes ferredoxin and NADH synergistically: a new perspective on anaerobic hydrogen production. J Bacteriol 191:4451–4457
Selig M, Xavier KB, Santos H, Schonheit P (1997) Comparative analysis of Embden-Meyerhof and Entner-Doudoroff glycolytic pathways in hyperthermophilic archaea and the bacterium *Thermotoga*. Arch Microbiol 167:217–232
Soboh B, Linder D, Hedderich R (2004) A multisubunit membrane-bound [NiFe] hydrogenase and an NADH-dependent Fe-only hydrogenase in the fermenting bacterium *Thermoanaerobacter tengcongensis*. Microbiology 150:2451–2463
Takahata Y, Nishijima M, Hoaki T, Maruyama T (2001) *Thermotoga petrophila* sp. nov. and *Thermotoga naphthophila* sp. nov., two hyperthermophilic bacteria from the Kubiki oil reservoir in Niigata, Japan. Int J Syst Evol Microbiol 51:1901–1909
Takai K, Horikoshi K (2000) *Thermosipho japonicus* sp. nov., an extremely thermophilic bacterium isolated from a deep-sea hydrothermal vent in Japan. Extremophiles 4:9–17
Teske A, Ramsing NB, Habicht K, Fukui M, Kuver J, Jorgensen BB, Cohen Y (1998) Sulfate-reducing bacteria and their activities in cyanobacterial mats of solar Lake (Sinai, Egypt). Appl Environ Microbiol 64:2943–2951
Thauer RK, Jungermann K, Decker K (1977) Energy conservation in chemotrophic anaerobic bacteria. Microbiol Mol Biol Rev 41:100–180
Thauer RK, Kaster AK, Goenrich M, Schick M, Hiromoto T, Shima S (2010) Hydrogenases from methanogenic archaea, nickel, a novel cofactor, and H2 storage. Annu Rev Biochem 79:507–536
Tosatto SCE, Toppo S, Carbonera D, Giacometti GM, Costantini P (2008) Comparative analysis of the [FeFe] hydrogenase from Thermotogales indicates the molecular bases of resistance to oxygen inactivation. Int J Hydrog Energy 33:570–578
Urios L, Cueff-Gauchard V, Pignet P, Postec A, Fardeau M-L, Ollivier B, Barbier G (2004) *Thermosipho atlanticus* sp. nov., a novel member of the Thermotogales isolated from a Mid-Atlantic Ridge hydrothermal vent. Int J Syst Evol Microbiol 54:1953–1957
Van Niel EWJ, Budde MAW, de Haas GG, van der Wal FJ, Claassen PAM, Stams AJM (2002) Distinctive properties of high hydrogen producing extreme thermophiles, *Caldicellulosiruptor saccharolyticus* and *Thermotoga elfii*. Int J Hydrog Energy 27:1391–1398
Van Ooteghem SA, Beer SK, Yue PC (2002) Hydrogen production by the thermophilic bacterium *Thermotoga neapolitana*. Appl Biochem Biotechnol 98–100:177–189
Van Ooteghem SA, Jones A, van der Lelie D, Dong B, Mahajan D (2004) H2 production and carbon utilization by *Thermotoga neapolitana* under anaerobic and microaerobic growth conditions. Biotechnol Lett 26:1223–1232
Vanfossen AL, Lewis DL, Nichols JD, Kelly RM (2008) Polysaccharide degradation and synthesis by extremely thermophilic anaerobes. Ann N Y Acad Sci 1125:322–337
Vargas M, Noll KM (1996) Catabolite repression in the hyperthermophilic bacterium *Thermotoga neapolitana* is independent of cAMP. Microbiology 142:139–144
Vargas M, Kashefi K, Blunt-Harris EL, Lovley DR (1998) Microbiological evidence for Fe(III) reduction on early Earth. Nature 395:65–67

Verhagen MF, O'Rourke T, Adams MW (1999) The hyperthermophilic bacterium *Thermotoga maritima* contains an unusually complex iron-hydrogenase: amino acid sequence analyses versus biochemical characterization. Biochim Biophys Acta 1412:212–229

Vincent KA, Parkin A, Lenz O, Albracht SPJ, Fontecilla-Camps JC, Cammack R, Friedrich B, Armstrong FA (2005) Electrochemical definitions of O_2 sensitivity and oxidative inactivation in hydrogenases. J Am Chem Soc 127:18179–18189

Wery N, Lesongeur F, Pignet P, Derennes V, Cambon-Bonavita M, Godfroy A, Barbier G (2001) *Marinitoga camini* gen. nov., sp. nov., a rod-shaped bacterium belonging to the order Thermotogales, isolated from a deep-sea hydrothermal vent. Int J Syst Evol Microbiol 51:495–504

Wiegel J, Ljungdahl LG, Demain AL (1985) The importance of thermophilic bacteria in biotechnology. Crit Rev Biotechnol 3:39–108

Windberger E, Huber R, Trincone A, Fricke H, Stetter KO (1989) *Thermotoga thermarum* sp. nov. and *Thermotoga neapolitana* occurring in African continental solfataric springs. Arch Microbiol 151:506–512

Woodward J, Heyer NI, Getty JP, O'Neill HM, Pinkhassik E, Evans BR (2002) Efficient hydrogen production using enzymes of the pentose phosphate pathway. NREL/CP-610–32405. US Department of Energy, Washington, DC

Yang X, Ma K (2005) Purification and characterization of an NADH oxidase from extremely thermophilic anaerobic bacterium *Thermotoga hypogea*. Arch Microbiol 183:331–337

Zhaxybayeva O, Swithers KS, Lapierre P, Fournier GP, Bickhart DM, DeBoy RT, Nelson KE, NesbØ CL, Doolittle WF, Gogarten JP, Noll KM (2009) On the chimeric nature, thermophilic origin, and phylogenetic placement of the Thermotogales. Proc Natl Acad Sci U S A 106:5865–5870

Zverlov VV, Volkov IY, Velikodvorskaya TV, Schwarz WH (1997) Highly thermostable endo-1,3-β-glucanase (chrysolaminaranase) LamA from *Thermotoga neapolitana*: nucleotide sequence of the gene and characterization of the recombinant gene product. Microbiology 143:1701–1708

Chapter 10

Bioelectrochemical Systems for Indirect Biohydrogen Production

John M. Regan*
Department of Civil and Environmental Engineering, The Pennsylvania State University, University Park, PA 16802, USA

and

Hengjing Yan
Institute for Collaborative Biotechnologies, Department of Chemistry and Biochemistry, University of California Santa Barbara, Santa Barbara, CA 93106, USA

Summary

Bioelectrochemical systems involve the use of exoelectrogenic (i.e., anode-reducing) microbes to produce current in conjunction with the oxidation of reduced compounds. This current can be used directly for power in a microbial fuel cell, but there are alternate uses of this current. One such alternative is the production of hydrogen in a microbial electrolysis cell (MEC), which accomplishes cathodic proton reduction with a slight applied potential by exploiting the low redox potential produced by exoelectrogens at the anode. As an indirect approach to biohydrogen production, these systems are not subject to the hydrogen yield constraints of fermentative processes and have been proven to work with virtually any biodegradable organic substrate. With continued advancements in reactor design to reduce the system internal resistance, increase the specific surface area for anode biofilm development, and decrease the material costs, MECs may emerge as a viable alternative technology for biohydrogen production. Moreover, these systems can also incorporate other value-added functionalities for applications in waste treatment, desalination, and bioremediation.

*Author for correspondence, e-mail: jregan@engr.psu.edu

D. Zannoni and R. De Philippis (eds.), Microbial BioEnergy: Hydrogen Production, Advances in Photosynthesis and Respiration 38, pp. 225–233, DOI 10.1007/978-94-017-8554-9_10,

I. Principles of Microbial Electrolysis Cells

Bioelectrochemical systems (BESs) can be configured to present an approach to biological hydrogen production that is distinctly different from fermentative and photoheterotrophic routes. These systems involve microbially catalyzed electrode reactions that produce current, and microbial electrolysis cell (MEC) designs use that current to reduce protons. This chapter introduces the basic microbial and electrochemical features, performance, and potential applications of this emerging technology.

A. Microbially Catalyzed Anode Reduction with Applied Potential

The essential feature of BESs is the microbially catalyzed reduction of an anode and/or oxidation of a cathode to sustain an electrical current. Over the past decade, this extracellular redox capability of some microbes has been exploited in the development of numerous BES designs having a myriad of applications and end products. Potential applications include waste treatment (Logan 2008), remote data collection (Lowy et al. 2006; Tender et al. 2008), water desalination (Cao et al. 2009), and bioremediation (Lovley 2011). Depending on the application, systems can be configured to directly recover power (microbial fuel cells (MFCs)); conserve the electrical current in the production of reduced energy carriers or chemicals such as hydrogen (Liu et al. 2005b; Logan et al. 2008b; Rozendal et al. 2006), methane (Cheng et al. 2009), acetate (Nevin et al. 2011), and hydrogen peroxide (Rozendal et al. 2009); use current to support desired cathodic reactions such as nitrate reduction (Clauwaert et al. 2007), chromate reduction (Hsu et al. 2012), or reductive dechlorination (Aulenta et al. 2009); and/or couple charge separation with the recovery of products such as caustic soda (Pikaar et al. 2011; Rabaey et al. 2010) or desalinated water (Kim and Logan 2011; Luo et al. 2011).

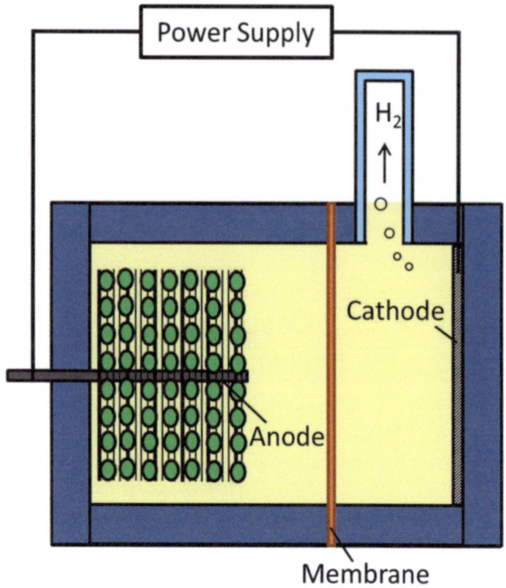

Fig. 10.1. Schematic of a two-chamber microbial electrolysis cell.

Hydrogen can be produced in a BES design commonly referred to as an MEC (Fig. 10.1). In these systems, exoelectrogenic microbes are used to oxidize an electron donor and reduce an anode. These reducing conditions result in a low anode potential (E_{an}), only slightly higher than the reduction potential for protons at pH 7 (i.e., −0.414 V). Taking advantage of this low E_{an}, the addition of a small applied voltage from a power source (E_{ps}) and maintenance of anaerobic conditions in the cathode will reduce the cathode potential below this value and result in hydrogen evolution from the cathode. E_{ps} can be as small as approximately 0.25 V, much smaller than the roughly 1.6 V that

---

Abbreviations: BES – BioElectrochemical system; COD – Chemical oxygen demand; E – Cell voltage; E_{an} – Anode potential; E_{ps} – Applied voltage from the power source; MEC – Microbial electrolysis cell; MFC – Microbial fuel cell; R_{ex} – External resistance; SHE – Standard hydrogen electrode; SS – Stainless steel; VSS – Volatile suspended solids must be added in practice to electrolyze water, though the current density and hydrogen production rate increase with E_{ps} (Liu et al. 2005b). By capturing the exoelectrogen-produced current and boosting the potential slightly, MECs offer an indirect (i.e., not strictly hydrogenase dependent) means of biological hydrogen production.

B. Substrate Versatility

One of the main advantages of MECs relative to other biohydrogen production approaches is the substrate versatility they offer. Given that the hydrogen in these systems is derived from microbially produced current, and is not a direct product of microbial metabolism, MECs do not have the substrate limitations associated with fermentative or photofermentative hydrogen production routes. Exoelectrogenesis has been demonstrated in BESs using virtually any biodegradable organic electron donor. These include simple substrates such as acetate (Liu et al. 2005b), butyrate (Liu et al. 2005a), glucose (Selembo et al. 2009b), and ethanol (Kim et al. 2007). Current production has also been demonstrated using polymeric materials such as cellulose (Ren et al. 2007) and proteins (Heilmann and Logan 2006) as well as complex mixtures such as dairy manure (Kiely et al. 2011), swine wastewater (Wagner et al. 2009), domestic wastewater (Hays et al. 2011), brewery wastewater (Feng et al. 2008), and sediments (Bond et al. 2002).

A related benefit of MECs relative to fermentative hydrogen production routes is that the process does not encounter the fundamental fermentative barrier in hydrogen yield associated with the production of other reduced products. For example, in a fermentation that yields acetate and butyrate, each mole of acetate and butyrate retains 8 and 20 e^- eq that will not contribute to hydrogen yield (representing a loss of 4 mol H_2 mol acetate^{-1} and 10 mol H_2 mol butyrate^{-1}) unless followed by a photoheterotrophic route. In an MEC, even if fermentations are occurring in the system, the products are compatible substrates for exoelectrogenesis and can contribute to current production and indirect hydrogen formation at the cathode. MECs are capable of converting a high percentage of the substrate into current and hydrogen, with recoveries depending on the substrate complexity and system operating conditions.

C. Hydrogen Production Rates and Yield and Energy Efficiency

The hydrogen generation performance in MECs is usually evaluated in terms of production rate, hydrogen yield, and energy recovery. The production rate of hydrogen in MECs, Q_{H2} (m^3 m^{-3} d^{-1}), is calculated based on the volume of hydrogen produced, the volume of catholyte, and the duration of the process. Reported Q_{H2} values have generally ranged from 0.01 to 6.3 m^3 m^{-3} d^{-1}, and with optimized volumetric current density achieved by reducing electrode spacing to 2 cm, the maximum hydrogen production rate of 17.8 m^3 m^{-3} d^{-1} was reported (Cheng and Logan 2011) at an E_{ps} of 1 V. Although this is still lower than the production rate of dark fermentation (up to 64.5 m^3 m^{-3} d^{-1} (Li and Fang 2007)), factors such as reactor architecture (Call and Logan 2008; Hu et al. 2008; Lee et al. 2009; Tartakovsky et al. 2009), electrode materials (Cheng and Logan 2007; Lee and Rittmann 2010), solution chemistry (Merrill and Logan 2009), and operation modes (Lee et al. 2009) can affect hydrogen production rate in MECs and are the subject of current research and development advancements.

The hydrogen yield in MECs, Y_{H2} (mg-H2 mg-COD^{-1}), can be calculated based on the chemical oxygen demand (COD) removal as

$$Y_{H_2} = \frac{n_{H_2} M_{H_2}}{v_L \Delta COD} \tag{10.1}$$

where n_{H_2} is the moles of hydrogen recovered in the system, M_{H_2} is the molecular weight of hydrogen, and v_L the volume of liquid in the anode chamber. ΔCOD is the change in COD based on the concentrations in the reactor influent and effluent for continuous flow tests, or the starting and final CODs for batch tests (Logan 2008). The hydrogen yield in MECs can also be calculated in molar units (mol-H_2 mol-COD^{-1}) by

$$Y_{H_2} = \frac{n_{H_2}}{n_{\Delta COD}} \tag{10.2}$$

where $n\Delta_{COD}$ is the molar change in COD in the reactor (Logan 2008).

Compared to other biohydrogen processes, MECs have higher hydrogen yields and the ability to use both fermentable and non-fermentable organics. At an E_{ps} of 0.6 V, hydrogen yields using cellulose, glucose, or volatile acids including acetic acid, butyric acid, lactic acid, propionic acid, or valeric acid, ranged from 3.65 to 8.77 mol-H_2 mol-substrate^{-1} (Table 10.1) (Cheng and Logan 2007).

Table 10.1. Hydrogen production using cellulose, glucose, or different volatile acids at an applied voltage of 0.6 V (Cheng and Logan 2007).

Substrate	Y_{H2}, mol of H_2 per mol of substrate	Production rate $m^3\ m^{-3}\ d^{-1}$
Glucose	8.55	1.23
Cellulose	8.20[a]	0.11
Acetic acid	3.65	1.10
Butyric acid	8.01	0.45
Lactic acid	5.45	1.04
Propionic acid	6.25	0.72
Valeric acid	8.77	0.14

[a] Calculated per mole of hexose equivalent

In MECs, the direct energy input includes both the substrate degraded by the bacteria and the power source for poising the cell voltage. The total energy recovery of the MEC process, η_{W+S}, based on both the substrate consumption and electricity input is calculated as

$$\eta_{W+S} = \frac{W_{H_2}}{W_{in} + W_S} \tag{10.3}$$

where W_{H_2} (kWh) is the energy value of the recovered hydrogen, calculated as follows based on the moles of hydrogen recovered, n_{H_2}, and the heat of combustion for hydrogen, H_{H_2} (kJ mol^{-1})

$$W_{H_2} = n_{H_2} \Delta H_{H_2}. \tag{10.4}$$

The energy value of the substrate, W_s (kWh), is similarly calculated based on the moles the substrate, n_S, and the heat of combustion for the substrate, as

$$W_S = n_S \Delta H_S. \tag{10.5}$$

The energy of electricity input, W_{in} (kWh), is calculated by subtracting energy lost by the inclusion of the external resistor, W_R, from the total energy from the power source, W_{ps}, as

$$W_{in} = W_{ps} - W_R = \int_{t=0}^{t} \left(IE_{ps} - I^2 R_{ex} \right) dt \tag{10.6}$$

where $I = E/R_{ex}$ is the current calculated in the MEC circuit based on measuring the cell voltage (E) across an external resistor (R_{ex}), and dt the time increment. The subtraction of W_R is to adjust the power lost that does not go into hydrogen production. Sometimes, energy recovery is calculated based on only electricity input as

$$\eta_W = \frac{W_{H_2}}{W_{in}}, \tag{10.7}$$

or substrate as

$$\eta_s = \frac{W_{H_2}}{W_{in}} \quad (Logan\ 2008). \tag{10.8}$$

The theoretical limits of the energy recovery for hydrogen production from acetate in MECs under standard biological conditions, based on electricity (η_W), substrate (η_s), or both (η_{W+S}), are 1,094 %, 131 %, and 117 %, respectively (Logan et al. 2008a). In practical operation of a membrane-less MEC system with increased anode surface area and decreased electrode spacing, the total energy efficiency based on both electricity input and substrate degraded could reach up to 86 % at E_{ps}=0.2 V (Call and Logan 2008).

II. Microbial Catalysts at the Anode

While there is considerable variability in MEC designs and features, an essential component of all MECs is an exoelectrogenic microbial catalyst at the anode that converts electron donor substrate into electrical current. There are several known mechanisms for this extracellular electron transfer, including the use of redox mediators, which may be self-produced (Rabaey et al. 2005), produced by another microbial community member (Pham et al. 2008a, b), or exogenously added (Sund et al. 2007); outer membrane cytochromes (Bretschger et al. 2007); or conductive pili (Reguera et al. 2005; Gorby et al. 2006). Significant advancements have been made in understanding the physiology of these mechanisms through the study of model organisms such as Geobacter spp. (Lovley et al. 2011) and Shewanella spp. (Biffinger et al. 2011). Molecular community analyses of mixed-culture systems have characterized the exoelectrogenic communities in a variety of conditions (Kiely et al. 2011; Lu et al. 2012). The results often show a predominance of Geobacter spp., but some communities have other dominant populations, sometimes of unknown exoelectrogenic potential. Isolation efforts have broadened the collection of known exoelectrogens to include species such as *Comamonas denitrificans* (Xing et al. 2010), *Ochrobactrum anthropi* (Zuo et al. 2008), and *Enterobacter chloacae* (Rezaei et al. 2009), but this is still very much an emerging area in microbiology.

There are two noteworthy nonexoelectrogenic metabolisms that can play a significant role in MEC communities, homoacetogens and methanogens. The former group (Parameswaran et al. 2011) can affect the recycling of electrons in the system by converting hydrogen that leaks to the anode back into acetate, which is a suitable substrate for exoelectrogens. Methanogens, on the other hand, can cause a loss of hydrogen productivity in MECs by converting acetate and hydrogen into methane (Jung and Regan 2011).

III. Cathode Reaction

A. Inorganic Catalysts

Electrochemical hydrogen production on the cathode of MECs can be greatly accelerated by metal catalysts, since they can effectively decrease the cathode overpotential. The most used metal catalyst for this process is platinum. However, this noble catalyst is not an ideal choice for the practical large-scale application of MECs due to its high cost and poisoning by chemicals such as sulfide, which is a common constituent of wastewater (Zhang et al. 2010). Therefore, non-precious inorganic catalysts such as cobalt and iron cobalt-based compounds (Cheng and Logan 2008), nickel oxides and alloy (Hu et al. 2009; Selembo et al. 2009a, 2010), tungsten carbide powder (Harnisch et al. 2009), and a combination of palladium and platinum (Tartakovsky et al. 2008) have been studied to replace Pt for hydrogen production in MECs.

In addition to evaluating alternate catalysts for proton reduction, studies using stainless steel (SS) instead of platinized carbon cloth or carbon fiber as cathode material have also been reported (Selembo et al. 2009a; Zhang et al. 2010). MECs with nickel, SS, or their alloys and related compounds as alternative cathode materials and catalyst showed comparable performance to platinum in terms of hydrogen production rate, hydrogen yield, and energy recovery (Table 10.2), while these non-precious catalysts only cost less than 20 % of platinum. However, both nickel and SS were found prone to corrosion, which was not detected on a platinized carbon electrode.

B. Biocathodes

An alternative approach to catalyzing proton reduction is the use of a biocathode, which offers a great alternative to platinum in that biocathodes are low cost and self-generating. To catalyze the cathodic reaction, microorganisms on biocathodes take up electrons from the electrode material and use these

Table 10.2. Performance of acetate-fed MECs with different cathode materials and catalysts.

E_{ps} (V)	Catalyst mg/cm^2	Q_{H2} (m^3 m^{-3} d^{-1})	Y_{H2} (mol/mol)	η_w (%)	η_{w+s} (%)	Cost relative to Pt	References
0.9	SS[a] alloys	1.5	NA	107	–	<1 %	Selembo et al. (2009a)
0.9	Ni alloys	0.79	NA	68	29	<1 %	Selembo et al. (2009a)
0.9	SS mesh	2.1	NA		74	<3 %	Zhang et al. (2010)
0.6	Ni alloys	1.5–2	2.2–2.6	114–182	–	<10 %	Hu et al. (2009)
0.6	Ni powder	1.3	–	210	65	<1 %	Selembo et al. (2010)
0.6	Ni powder/CB[a]	1.2	–	252	73	–	Selembo et al. (2010)
0.6	NiOx	0.9	–	215	67	–	Selembo et al. (2010)
0.6	SS brush	1.7	–	221	–	<20 %	Call et al. (2009)
0.6	Pt, 0.5	0.5–2.3	1.9–2.5	151–204	–	100 %	Call and Logan (2008), Cheng and Logan (2007), and Hu et al. (2008, 2009)
0.4	Ni alloys	1.1–1.6	1.4–2.2	200–240	–	<10 %	Hu et al. (2009)
0.4	Pt, 0.5	0.2–1.6	1.0–3.3	205–430	–	100 %	Call and Logan (2008), Cheng and Logan (2007), and Hu et al. (2008, 2009)

[a]*SS* stainless steel, *CB* carbon black

electrons to produce hydrogen. Metal-oxidizing microorganisms are known for taking up electrons from extracellular solid material and using the electrons from this reaction for metabolic process. On the other hand, microorganisms that contain hydrogenases catalyzing the reversible reaction of hydrogen oxidation and reduction are found in various environments.

The microbial uptake of electrons from a cathode for the production of hydrogen in an MEC was shown for the first time by Rozendal et al. (2008). The biocathode they developed was obtained by enriching an anodic biofilm of hydrogen-oxidizing and electrochemically active microorganisms. When stable anodic current was reached, the polarities of anode and cathode were reversed, and the anode-enriched biofilm was shifted to a biocathode. With the cathode potential poised at −0.7 V vs. standard hydrogen electrode (SHE), the biocathode achieved a hydrogen production rate of 0.63 m^3 m^{-3} d^{-1}, compared to 0.08 m^3 m^{-3} d^{-1} in a negative control. Further analysis of the microbial community indicated that the dominant microorganism was Desulfovibrio spp., which can metabolize hydrogen and has the potential for exocellular electron transfer. This was confirmed by the pure culture study of *Desulfovibrio vulgaris* strain G11 on a cathode for current uptake and hydrogen production (Croese et al. 2011). Later, Desulfitobacterium-enriched culture also showed the capability of catalyzing hydrogen production without mediators at a cathode potential of −750 mV, with a hydrogen production rate of 13.5 µeq mg VSS^{-1} d^{-1} (Villano et al. 2011). These studies of MEC biocathodes show promising potentials of improving MEC cost-effectiveness. However, for mixed-culture MEC biocathodes, the presence of methanogenesis might introduce competitive reactions to hydrogen production, resulting in decreased hydrogen content and increased methane ratio in the produced gas.

Acknowledgements

This work was supported by Award KUS-I1-003-13 from the King Abdullah University of Science and Technology (KAUST) and Grant Number W911NF-11-1-0531 from the U.S. Department of the Army – Army Research Office.

References

Aulenta F, Canosa A, Reale P, Rossetti S, Panero S, Majone M (2009) Microbial reductive dechlorination of trichloroethene to ethene with electrodes serving as electron donors without the external addition of redox mediators. Biotechnol Bioeng 103:85–91

Biffinger JC, Fitzgerald LA, Ray R, Little BJ, Lizewski SE, Petersen ER, Ringeisen BR, Sanders WC, Sheehan PE, Pietron JJ, Baldwin JW, Nadeau LJ, Johnson GR, Ribbens M, Finkel SE, Nealson KH (2011) The utility of *Shewanella japonica* for microbial fuel cells. Bioresour Technol 102:290–297

Bond DR, Holmes DE, Tender LM, Lovley DR (2002) Electrode-reducing microorganisms that harvest energy from marine sediments. Science 295:483–485

Bretschger O, Obraztsova A, Sturm CA, Chang IS, Gorby YA, Reed SB, Culley DE, Reardon CL, Barua S, Romine MF, Zhou J, Beliaev AS, Bouhenni R, Saffarini D, Mansfeld F, Kim BH, Fredrickson JK, Nealson KH (2007) Current production and metal oxide reduction by *Shewanella oneidensis* MR-1 wild type and mutants. Appl Environ Microbiol 73:7003–7012

Call D, Logan BE (2008) Hydrogen production in a single chamber microbial electrolysis cell lacking a membrane. Environ Sci Technol 42:3401–3406

Call DF, Merrill MD, Logan BE (2009) High surface area stainless steel brushes as cathodes in microbial electrolysis cells. Environ Sci Technol 43:2179–2183

Cao XX, Huang X, Liang P, Xiao K, Zhou YJ, Zhang XY, Logan BE (2009) A new method for water desalination using microbial desalination cells. Environ Sci Technol 43:7148–7152

Cheng S, Logan BE (2007) Sustainable and efficient biohydrogen production via electrohydrogenesis. Proc Natl Acad Sci U S A 104:18871–18873

Cheng SA, Logan BE (2008) Evaluation of catalysts and membranes for high yield biohydrogen production via electrohydrogenesis in microbial electrolysis cells (MECs). Water Sci Technol 58:853–857

Cheng SA, Logan BE (2011) High hydrogen production rate of microbial electrolysis cell (MEC) with reduced electrode spacing. Bioresour Technol 102:3571–3574

Cheng SA, Xing DF, Call DF, Logan BE (2009) Direct biological conversion of electrical current into methane by electromethanogenesis. Environ Sci Technol 43:3953–3958

Clauwaert P, Rabaey K, Aelterman P, De Schamphelaire L, Ham TH, Boeckx P, Boon N, Verstraete W (2007) Biological denitrification in microbial fuel cells. Environ Sci Technol 41:3354–3360

Croese E, Pereira MA, Euverink GJW, Stams AJM, Geelhoed JS (2011) Analysis of the microbial community of the biocathode of a hydrogen-producing microbial electrolysis cell. Appl Microbiol Biotechnol 92:1083–1093

Feng Y, Wang X, Lee BELH (2008) Brewery wastewater treatment using air-cathode microbial fuel cells. Appl Microbiol Biotechnol 78:873–880

Gorby YA, Yanina S, McLean JS, Rosso KM, Moyles D, Dohnalkova A, Beveridge TJ, Chang IS, Kim BH, Kim KS, Culley DE, Reed SB, Romine MF, Saffarini DA, Hill EA, Shi L, Elias DA, Kennedy DW, Pinchuk G, Watanabe K, Ishii S, Logan B, Nealson KH, Fredrickson JK (2006) Electrically conductive bacterial nanowires produced by *Shewanella oneidensis* MR-1 and other microorganisms. Proc Natl Acad Sci U S A 103:11358–11363

Harnisch F, Sievers G, Schroder U (2009) Tungsten carbide as electrocatalyst for the hydrogen evolution reaction in ph neutral electrolyte solutions. Appl Catal B-Environ 89:455–458

Hays S, Zhang F, Logan BE (2011) Performance of two different types of anodes in membrane electrode assembly microbial fuel cells for power generation from domestic wastewater. J Power Sources 196:8293–8300

Heilmann J, Logan BE (2006) Production of electricity from proteins using a microbial fuel cell. Water Environ Res 78:531–537

Hsu L, Masuda SA, Nealson KH, Pirbazari M (2012) Evaluation of microbial fuel cell Shewanella biocathodes for treatment of chromate contamination. RSC Adv 2:5844–5855

Hu HQ, Fan YZ, Liu H (2008) Hydrogen production using single-chamber membrane-free microbial electrolysis cells. Water Res 42:4172–4178

Hu HQ, Fan YZ, Liu H (2009) Hydrogen production in single-chamber tubular microbial electrolysis cells using non-precious-metal catalysts. Int J Hydrog Energy 34:8535–8542

Jung S, Regan JM (2011) Influence of external resistance on electrogenesis, methanogenesis, and anode prokaryotic communities in microbial fuel cells. Appl Environ Microbiol 77:564–571

Kiely PD, Cusick R, Call DF, Selembo PA, Regan JM, Logan BE (2011) Anode microbial communities produced by changing from microbial fuel cell to microbial electrolysis cell operation using two different wastewaters. Bioresour Technol 102: 388–394

Kim Y, Logan BE (2011) Series assembly of microbial desalination cells containing stacked electrodialysis cells for partial or complete seawater desalination. Environ Sci Technol 45:5840–5845

Kim JR, Jung SH, Regan JM, Logan BE (2007) Electricity generation and microbial community analysis of alcohol powered microbial fuel cells. Bioresour Technol 98:2568–2577

Lee HS, Rittmann BE (2010) Significance of biological hydrogen oxidation in a continuous single-chamber microbial electrolysis cell. Environ Sci Technol 44:948–954

Lee HS, Torres CI, Parameswaran P, Rittmann BE (2009) Fate of h-2 in an upflow single-chamber microbial electrolysis cell using a metal-catalyst-free cathode. Environ Sci Technol 43:7971–7976

Li CL, Fang HHP (2007) Fermentative hydrogen production from wastewater and solid wastes by mixed cultures. Crit Rev Environ Sci Technol 37:1–39

Liu H, Cheng S, Logan BE (2005a) Production of electricity from acetate or butyrate in a single chamber microbial fuel cell. Environ Sci Technol 39: 658–662

Liu H, Grot S, Logan BE (2005b) Electrochemically assisted microbial production of hydrogen from acetate. Environ Sci Technol 39:4317–4320

Logan BE (2008) Microbial fuel cells. Wiley-Interscience, Hoboken

Logan BE, Call D, Cheng S, Hamelers HVM, Sleutels T, Jeremiasse AW, Rozendal RA (2008a) Microbial electrolysis cells for high yield hydrogen gas production from organic matter. Environ Sci Technol 42:8630–8640

Logan BE, Call D, Cheng S, Hamelers HVM, Sleutels THJA, Jeremiasse AW, Rozendal RA (2008b) Microbial electrolysis cells for high yield hydrogen gas production from organic matter. Environ Sci Technol 42:8630–8640

Lovley DR (2011) Live wires: direct extracellular electron exchange for bioenergy and the bioremediation of energy-related contamination. Energy Environ Sci 4:4896–4906

Lovley DR, Ueki T, Zhang T, Malvankar NS, Shrestha PM, Flanagan KA, Aklujkar M, Butler JE, Giloteaux L, Rotaru AE, Holmes DE, Franks AE, Orellana R, Risso C, Nevin KP (2011) Geobacter: the microbe electric's physiology, ecology, and practical applications. Adv Microb Physiol 59:1–100

Lowy DA, Tender LM, Zeikus JG, Park DH, Lovley DR (2006) Harvesting energy from the marine sediment-water interface ii – kinetic activity of anode materials. Biosens Bioelectron 21: 2058–2063

Lu L, Xing D, Ren N (2012) Pyrosequencing reveals highly diverse microbial communities in microbial electrolysis cells involved in enhanced H_2 production from waste activated sludge. Water Res 46: 2425–2434

Luo HP, Jenkins PE, Ren ZY (2011) Concurrent desalination and hydrogen generation using microbial electrolysis and desalination cells. Environ Sci Technol 45:340–344

Merrill MD, Logan BE (2009) Electrolyte effects on hydrogen evolution and solution resistance in microbial electrolysis cells. J Power Sources 191: 203–208

Nevin KP, Hensley SA, Franks AE, Summers ZM, Ou JH, Woodard TL, Snoeyenbos-West OL, Lovley DR (2011) Electrosynthesis of organic compounds from carbon dioxide is catalyzed by a diversity of acetogenic microorganisms. Appl Environ Microbiol 77: 2882–2886

Parameswaran P, Torres CI, Lee HS, Rittmann BE, Krajmalnik-Brown R (2011) Hydrogen consumption in microbial electrochemical systems (MXCs): the role of homo-acetogenic bacteria. Bioresour Technol 102:263–271

Pham TH, Boon N, Aelterman P, Clauwaert P, De Schamphelaire L, Vanhaecke L, De Maeyer K, Hofte M, Verstraete W, Rabaey K (2008a) Metabolites produced by Pseudomonas sp enable a gram-positive bacterium to achieve extracellular electron transfer. Appl Microbiol Biotechnol 77: 1119–1129

Pham TH, Boon N, De Maeyer K, Hofte M, Rabaey K, Verstraete W (2008b) Use of Pseudomonas species producing phenazine-based metabolites in the anodes of microbial fuel cells to improve electricity generation. Appl Microbiol Biotechnol 80:985–993

Pikaar I, Rozendal RA, Yuan ZG, Rabaey K (2011) Electrochemical caustic generation from sewage. Electrochem Commun 13:1202–1204

Rabaey K, Boon N, Hofte M, Verstraete W (2005) Microbial phenazine production enhances electron transfer in biofuel cells. Environ Sci Technol 39:3401–3408

Rabaey K, Butzer S, Brown S, Keller J, Rozendal RA (2010) High current generation coupled to caustic production using a lamellar bioelectrochemical system. Environ Sci Technol 44:4315–4321

Reguera G, McCarthy KD, Mehta T, Nicoll JS, Tuominen MT, Lovley DR (2005) Extracellular electron transfer via microbial nanowires. Nature 435:1098–1101

Ren Z, Ward TE, Regan JM (2007) Electricity production from cellulose in a microbial fuel cell using a defined binary culture. Environ Sci Technol 41:4781–4786

Rezaei F, Xing D, Wagner R, Regan JM, Richard TL, Logan BE (2009) Simultaneous cellulose degradation and electricity production by *Enterobacter cloacae* in an MFC. Appl Environ Microbiol 75:3673–3678

Rozendal RA, Hamelers HVM, Euverink GJW, Metz SJ, Buisman CJN (2006) Principle and perspectives of hydrogen production through biocatalyzed electrolysis. Int J Hydrog Energy 31:1632–1640

Rozendal RA, Jeremiasse AW, Hamelers HVM, Buisman CJN (2008) Hydrogen production with a microbial biocathode. Environ Sci Technol 42: 629–634

Rozendal RA, Leone E, Keller J, Rabaey K (2009) Efficient hydrogen peroxide generation from organic matter in a bioelectrochemical system. Electrochem Commun 11:1752–1755

Selembo PA, Merrill MD, Logan BE (2009a) The use of stainless steel and nickel alloys as low-cost cathodes in microbial electrolysis cells. J Power Sources 190:271–278

Selembo PA, Perez JM, Lloyd WA, Logan BE (2009b) High hydrogen production from glycerol or glucose by electrohydrogenesis using microbial electrolysis cells. Int J Hydrog Energy 34:5373–5381

Selembo PA, Merrill MD, Logan BE (2010) Hydrogen production with nickel powder cathode catalysts in microbial electrolysis cells. Int J Hydrog Energy 35:428–437

Sund CJ, McMasters S, Crittenden SR, Harrell LE, Sumner JJ (2007) Effect of electron mediators on current generation and fermentation in a microbial fuel cell. Appl Microbiol Biotechnol 76:561–568

Tartakovsky B, Manuel MF, Neburchilov V, Wang H, Guiot SR (2008) Biocatalyzed hydrogen production in a continuous flow microbial fuel cell with a gas phase cathode. J Power Sources 182:291–297

Tartakovsky B, Manuel MF, Wang H, Guiot SR (2009) High rate membrane-less microbial electrolysis cell for continuous hydrogen production. Int J Hydrog Energy 34:672–677

Tender LM, Gray SA, Groveman E, Lowy DA, Kauffman P, Melhado J, Tyce RC, Flynn D, Petrecca R, Dobarro J (2008) The first demonstration of a microbial fuel cell as a viable power supply: powering a meteorological buoy. J Power Sources 179: 571–575

Villano M, De Bonis L, Rossetti S, Aulenta F, Majone M (2011) Bioelectrochemical hydrogen production with hydrogenophilic dechlorinating bacteria as electrocatalytic agents. Bioresour Technol 102: 3193–3199

Wagner RC, Regan JM, Oh SE, Zuo Y, Logan BE (2009) Hydrogen and methane production from swine wastewater using microbial electrolysis cells. Water Res 43:1480–1488

Xing D, Cheng S, Logan BE, Regan JM (2010) Isolation of the exoelectrogenic denitrifying bacterium *Comamonas denitrificans* based on dilution-to-extinction of the microbial community. Appl Microbiol Biotechnol 85:1575–1587

Zhang YM, Merrill MD, Logan BE (2010) The use and optimization of stainless steel mesh cathodes in microbial electrolysis cells. Int J Hydrog Energy 35:12020–12028

Zuo Y, Xing D, Regan JM, Logan BE (2008) An exoelectrogenic bacterium *Ochrobactrum anthropi* YZ-1 isolated using a u-tube microbial fuel cell. Appl Environ Microbiol 74:3130–3137

Part II

Applied Aspects in Biohydrogen Production

Chapter 11

Applications of Photofermentative Hydrogen Production

Inci Eroglu*
Department of Chemical Engineering, Middle East Technical University, Ankara, Turkey
Department of Biotechnology, Middle East Technical University, Ankara, Turkey

Ebru Özgür
MEMS Research and Application Center, Middle East Technical University, Ankara, Turkey

Ela Eroglu
School of Chemistry and Biochemistry, Centre for Strategic Nano-Fabrication, The University of Western Australia, Perth, Australia

Meral Yücel and Ufuk Gündüz
Department of Biotechnology, Middle East Technical University, Ankara, Turkey
Department of Biology, Middle East Technical University, Ankara, Turkey

*Author for correspondence, e-mail: ieroglu@metu.edu.tr

D. Zannoni and R. De Philippis (eds.), Microbial BioEnergy: Hydrogen Production, Advances in Photosynthesis and Respiration 38, pp. 237–267, DOI 10.1007/978-94-017-8554-9_11,

Summary

Scientific and market strategy is essential in developing biological hydrogen production processes. Plans for future research should be based on current knowledge, experience and techniques. This chapter focuses on the applied issues of photofermentative H_2 production using purple non sulfur bacteria (PNSB) in combined systems, and in particular, the optimization of the process on real feedstock such as olive mill wastewater and dark fermenter effluents (DFEs) of thick juice, molasses, and potato steam peels. Based on the current state of the knowledge in the field, the future applicability and prospects of these systems are evaluated. Strategies to overcome the problems are outlined.

I. Introduction

Laboratory scale biohydrogen studies have mostly been carried out with synthetic culture media. High production costs associated with these media are prohibitive for large scale processing, and as a result, the utilization of waste materials as renewable microbial substrate sources is increasingly being considered to address the economic restrictions of biological hydrogen production. Several researchers working on photofermentative hydrogen production have based their studies on the utilization of food and agricultural waste materials with high levels of organic compounds as feedstock (Rocha et al. 2001). This approach can potentially connect the benefits of energy production with waste management.

Availability, biodegradability, cost, organic acid content and the carbon to nitrogen ratio

---

Abbreviations: BCI – Biomass cost index; BOD – Biological oxygen demand; COD – Chemical oxygen demand; CSTR – Continuously stirred tank reactor; DFE – Dark fermenter effluent; FVW – Fruit and vegetable wastes; $g_{dcw}\ l_c^{-1}$ – Gram cell dry weight per liter of culture; HRT – Hydraulic retention time; LDPE – Low density polyethylene; l_{H2} – Liters of hydrogen; l_{OMW} – Liters of olive mill wastewater; M – Molar; OMW – Olive mill wastewater; PBR – Photobioreactor; PNSB – Purple non-sulfur bacteria; SRW – Sugar refinery wastewater; TOC – Total organic carbon; VFA – Volatile fatty acids of feedstocks are some of the critical selection measures for finding the right waste material for photofermentative hydrogen production (Kapdan and Kargi 2006). The choice of waste material depends not only on its hydrogen production performance, but also on its abundance and environmental impact, which vary regionally. For example, olive mill wastewater, with a total annual production of 30 million m^3, is an environmental concern in the Mediterranean region, (Ergüder et al. 2000; Sabbah et al. 2004) whereas tofu wastewater is a significant food waste for countries in Eastern Asia. As it can be expected, the optimum conditions and the economics for these widely varying applications can be completely different from one another; hence, biological H_2 production technology must be developed specifically for the selected waste. To form an overall framework for a sustainable biohydrogen economy, it is crucial to exchange and accumulate the knowledge and expertise obtained from these diverse case studies.

The use of integrated processes, which combine multiple organisms and metabolic modes for hydrogen production, is a rather recent development in the biohydrogen field with the intent to reduce the amount of feedstock utilization and its related costs while improving the hydrogen production (Eroglu and Melis 2011). Thus, an increasing number of studies focused on the combination of dark and photo fermentation processes either towards sequential two-step or combined single-step processes (Argun et al. 2009; Zong et al. 2009; Özgür et al. 2010a, b; Yang et al. 2010; Avcioglu et al. 2011; Boran et al. 2012a; Rai et al. 2012). Sequential two-step processes are based on the utilization of dark fermenter effluents (DFEs) as a substrate source for photofermentative hydrogen production. Dark-fermentation can be used either as a pretreatment stage for improving the physicochemical properties of the waste material, or to produce hydrogen itself. On the other hand, the integration of dark-fermentation and photofermentation into a single stage has also been proposed as an alternative and less labor-intensive process option. Such an arrangement may also reduce or eliminate the need for external pH adjustment, as the acidic environment caused by the dark fermentation could be neutralized via photofermentation (Keskin et al. 2011).

The purpose of the present chapter is to highlight photofermentative hydrogen production using waste materials for future large-scale applications. Here, we would like to provide hints on how to choose photofermentable feedstocks and emphasize their key specifications. There is no general methodology to be followed for the development of photofermentative H_2 production technology; however, we would like to share the experience we have gained in this area during a 3 year COST project (COST Action 841) "Biological and biochemical diversity of hydrogen metabolism" followed by a 5 year European Union 6th Frame Integrated Project, HYVOLUTION, "Nonthermal production of pure hydrogen from biomass". The aim of the latter project was to prepare a blueprint for integrated biological hydrogen production processes composed of a dark fermentation step followed by photofermentation. The overall objective of our workpackage was the utilization of the effluent of a thermo-bioreactor, for highly efficient and sustainable hydrogen production in photobioreactors (PBRs) by photosynthetic purple non sulfur bacteria (PNSB). Hence, the scope of this chapter is limited to a review of photofermentation and integrated dark and photofermentation processes, based on reported results from our work and from those of other researchers in this field.

II. Guidelines for Effective Photofermentative Hydrogen Production

What makes a waste material a good candidate as a resource for large-scale photofermentative hydrogen production? Advances reported in the relevant literature are primarily focused on finding ways to lower the operational costs of hydrogen production through the utilization of economical and regionally available feedstocks. A sustainable hydrogen economy cannot be solely based on wastes, hence, industrial and agricultural side products, crops, and biomass that do not strain the food supply must be also considered.

As mentioned previously, the optimal set of parameters for photofermentative hydrogen production from wastewater or other complex feedstock depend strongly on the individual application and therefore should be analyzed on a case-by-case basis. Nevertheless, based on accumulated insight gained from prior studies, a generally applicable set of guidelines to consider prior to developing such a waste-based photofermentative process can be proposed. The following properties are the most crucial parameters that need to be assessed before making any further plans about using that material as a substrate source.

A. Selection of the Microorganism

Photosynthetic purple non sulfur bacteria (PNSB) e.g. members of the *Rhodobacter* (*Rb.*) species are the most preferred ones for photofermentative hydrogen production. However, it should be emphasized that the hydrogen productivity on different feedstocks depends on the strain. Hence, screening of different PNSB strains for a specific feedstock is necessary. The absorption spectra of the feedstock and that of the PNSB should be compared to check if there is interference between the absorption spectrum of the medium and the bacteria. In PNSB a single photosystem, located in the intracellular membrane, is responsible for the absorption of the light energy. In practice, antenna pigments, namely carotenoids and bacteriochlorophylls, absorb the energy of a photon. These molecules show specific absorption spectra related to their characteristic color. Chlorophyll a, commonly available in all photosynthetic organisms, absorbs light at 375 nm, 590 nm, 800–810 nm and 830–890 nm, while the main peak is located at 850–890 nm (Weaver et al. 1975). Carotenoids, which are additionally located in the intercellular membrane of PNSB, are known to absorb light energy in the range of 400–550 nm (Uyar et al. 2007). It should also be noted that feedstocks with high absorbance values at wavelength ranges of 400–800 nm might interfere with the H_2 production (Boran 2011). It has been reported that the percent penetration of light intensity for each 1 cm of depth is 70 % for thick juice DFE and 51 % for molasses DFE, which is significantly lower than that for artificial medium (89 %) (Boran et al. 2012a).

During photofermentation it is observed that H_2 production starts above a certain critical cell concentration. However, when cell concentration exceeds a certain limit, hydrogen evolution stops. An optimum cell concentration of 0.5–0.7 g_{dcw} per l_c for the employed microorganisms has been reported (Gebicki et al. 2010; Androga et al. 2011a, b).

Outdoor production of hydrogen with photosynthetic bacteria is also strongly affected by fluctuations in temperature due to the day-night cycle and other seasonal, geographic and climatic conditions. The optimum temperature for hydrogen production by PNSB ranges between 30 °C and 35 °C depending on the strain used (Stevens et al. 1984). Although production at lower temperatures is possible, the significantly lower growth rate of the microorganisms below 20 °C renders the process very inefficient. It is also known that the PNSB may slow down or even turn off their hydrogen production at high temperatures. An actual maximum is not known yet, but temperatures higher than 45 °C should be avoided.

In outdoor applications, the light absorbed and the heat generated by bacteria cause the bioreactor temperature to rise up to 55 °C (18–20 °C higher compared to the air temperature) during the day. Reactor temperatures can be decreased by adapting a suitable cooling system or by using a proper shading material. Water spraying is an effective method for cooling solar glass bioreactors but not recommended for acrylic flat plate reactors. The temperature difference on the surface may cause cracking.

Thermo-resistant strains of PNSB (e.g. *Rhodospirillum centenum*) capable of growing at higher temperatures (optimum at 40–45 °C) are also known (Favinger et al. 1989). These microorganisms might be more interesting for regions of higher global radiation intensities if they could be used for biohydrogen production.

B. Composition of the Feedstock and Its Possible Adjustments

Purple non-sulfur bacteria are able to consume a wide variety of organic substrates including short-chain organic acids such as acetate, butyrate, propionate, and lactate, sugars such as glucose and sucrose, and mixtures of these. Short-chain organic acids are preferred for H_2 production; however, H_2 production performances of different strains vary depending on the type and the concentration of the substrate. The initial organic acid concentrations are known to have an impact on lag time, biomass growth rate, and H_2 production, as will be discussed in Sect. IV. Feedstocks usually have very high organic acid concentrations that bring about the dilution requirement, which increases the water consumption rate. Recirculation of treated wastewater may decrease the fresh water consumption rate. The concentration of the nitrogen source, and especially the ammonium ion, NH_4^+, also has a primary importance. It strongly influences the photo-fermentation process, as the nitrogenase enzyme is inhibited by the presence of ammonium. For this reason, the NH_4^+ content of the feedstock should be lowered by dilution or by pretreatment. It is known that the C/N ratio of substrate is one of the most important parameters affecting hydrogen productivity and yield in photo-fermentation. An optimum C/N ratio of 25 was reported when acetate and glutamate were used as C and N sources, respectively (Androga et al. 2011a).

During fermentation with artificial media, the addition of supplementary nutrients (iron, molybdenum, trace elements and vitamins) is necessary. Photo-fermentation experiments using various types of real feedstocks showed that most components necessary for growth and hydrogen production are readily available in the raw substrates. Yet, it was found that real feedstocks usually lack Fe and Mo, and supplementation of the media with these minerals was shown to enhance H_2 production by PNSB.

Another parameter affecting the photo-fermentation process is the pH. pH adjustment requires either acid or base addition. However, due to large reactor areas, effective pH control is difficult. To keep the pH level between 6.5 and 8.0, phosphate buffer is used at a concentration range of 4–20 mM. Buffering improves H_2 productivity, but if the buffer contains phosphate, its addition highly increases the environmental impact. More environmentally benign buffers like carbonate may be used to decrease this impact.

Is sterilization of the feedstock necessary? Small-scale experiments are usually carried out under sterile conditions. Yet, in large scale, sterilization would increase the cost substantially. However, contamination may halt the process during long-term operations, and should be avoided to obtain longer processes periods. Hence, a cheap sterilization method need to be applied in large scale.

Pretreatment may improve yield but brings additional cost to the process. Pretreatments include filtration, discoloration, purification, adsorption and chemical treatment. Discoloration significantly affects the absorption spectra of the feedstocks, thus the medium does not interfere with the absorption spectrum of the PNSB.

An anaerobic atmosphere is mandatory for hydrogen production by photosynthetic PNSB. Presence of N_2, O_2 and CO_2 adversely affects the hydrogen production pathway (Koku et al. 2002); hence, these gases must be eliminated from the photobioreactor. An argon atmosphere is especially preferred in laboratory experiments. Argon sparging at the start-up decreases the lag-time to produce hydrogen. In large-scale applications, the system is completely filled with culture to eliminate air.

Dissolved air could be consumed during the growth phase. A continuous diffusion of air into the reactor due to diffusion of air through reactor material or air leakage at the fittings will result in the complete cessation of hydrogen production. In many PNSB, a visual indication of such drastic air leakage is the gradual transformation of the culture medium color into a deep-red hue.

III. Utilization of Waste Materials for Photofermentative Hydrogen Production

A large variety of waste materials have been evaluated with respect to their hydrogen conversion via fermentative processes, as listed in Table 11.1. These studies also differ from each other in terms of the microorganism used. Hydrogen productivities of waste materials are mostly given in the literature as hydrogen production yield (units of H_2 volume per volume of waste liquid, l_{H2} per l_{ww}), and hydrogen production rate (either units of H_2 volume per culture volume over time, $ml_{H2}l_c^{-1}h^{-1}$; or units of H_2 volume per bacterial cell dry weight over time, $ml_{H2}g_{cell}^{-1}h^{-1}$). It should also be pointed out that the variations in the process parameters such as reactor geometry, waste material, microorganism, light source and illumination period make the direct productivity comparison quite difficult.

Fruit and vegetable, diary, and sugar wastes can be classified as the main groups of organic wastes used for hydrogen production. Some of the recent developments for each group will be explained in the subsections below, including an extended one for the "olive mill wastewater" as being one of the key research topics explored by the current chapter authors.

Table 11.1. Summary of photofermentative hydrogen production studies using various waste materials as substrate sources.

Waste material	Microorganism	Process details	H_2 Production potential	Reference
Brewery wastewater	*Rb. sphaeroides* O.U.001	10 % ww & Biebl and Pfennig medium, indoor, batch	0.22 l_{H2} l^{-1}_{ww}	Seifert et al. (2010b)
Tofu wastewater	*Rb.sphaeroides* RV	100 % ww, immobilized cultures in agar gel, indoor, batch	1.9 l_{H2} l^{-1}_{ww}	Zhu et al. (1999)
Milk industry wastewater	*Rb.sphaeroides* O.U.001	30 % ww & malate, indoor, batch	2.0 l_{H2} l^{-1}_{ww}	Türkarslan et al. (1998)
Sugar refinery wastewater	*Rb.sphaeroides* O.U.001	20 % ww & L-malic acid and sodium glutamate, indoor, batch	4.63 l_{H2} l^{-1}_{ww}	Yetis et al. (2000)
		20 % ww & L-malic acid and sodium glutamate, indoor, fed-batch	8.61 l_{H2} l^{-1}_{ww}	
Olive mill wastewater	*Rb.sphaeroides* O.U.001	4 % ww, outdoor, batch	11.4 l_{H2} l^{-1}_{ww}	Eroglu et al. (2008)
		4 % ww, indoor, batch	10.1 l_{H2} l^{-1}_{ww}	
		2 % ww, indoor, batch	13.9 l_{H2} l^{-1}_{ww}	Eroglu et al. (2004)
		Pretreatment with dark fermentation, 4 % OMW dark effluent, indoor, batch	29 l_{H2} l^{-1}_{ww}	Eroglu et al. (2006)
		Pretreatment with clay, 4 % of pretreated OMW, indoor, batch	31.5 l_{H2} l^{-1}_{ww}	Eroglu et al. (2008a)
Dairy waste	*Rb.sphaeroides* O.U.001	5 % ww & Biebl and Pfennig medium, indoor, batch	16.9 l H_2 l^{-1}_{ww}	Seifert et al. (2010a)
Yogurt waste	*Rhodospirillum rubrum* S-1	100 % ww, indoor, fed-batch	45 l_{H2} l^{-1}_{ww}	Zürrer and Bachofen (1979)
Whey waste			47 l_{H2} l^{-1}_{ww}	

A. Olive Mill Wastewater (OMW)

Olives are one of the important agricultural crops across the Mediterranean area. During the extraction of olive oil, significant quantities of water are used especially for the continuous washing of the olive paste with warm water before the separation of oil from its olive paste (Kiritsakis 1991; Visioli et al. 1999). The olive oil manufacturing process generates a dark-colored, oily wastewater phase, the so called olive mill wastewater (OMW), which contains olive fruit juice, residual pulp, and process water in a moderately stable emulsion of oil (Tsagaraki et al. 2006; Eroglu et al. 2008a). OMW is generated in large amounts throughout the world, with over 30 million m^3 per year from Mediterranean countries alone (Ergüder et al. 2000). Discarding this waste effluent to the environment requires critical precautions as it can create significant problems due to its high phenolic content (Eroglu et al. 2006, 2008a, 2009a).

Olive mill wastewater is a dark colored effluent with high organic content, with chemical oxygen demand (COD) values usually in between 50 and 200 g per l (Eroglu et al. 2009b; Yesilada et al. 1999). It is mostly composed of water (83–94 %), organic matter (4–16 %) and mineral salts (0.4–2.5 %) (Ramos-Cormenzana et al. 1996). The main organic constituents can be listed as oils, polysaccharides, proteins, organic acids, polyalcohols, and polyphenols (Cabrera et al. 1996). The characteristic dark color of OMW is mainly related to these phenolic constituents and lignin derivatives (Gonzalez et al. 1990). OMW also contains significant amounts of K, Ca, Na, Mg and Fe elements (Eroglu et al. 2004, 2009b).

Dilutions were optimized to obtain higher yields (Eroglu et al. 2004, 2009b). Various pretreatment stages enhanced hydrogen productivity (Eroglu et al. 2006, 2008a, 2009a). Different illumination regimes applied under indoor (Eroglu et al. 2010) or outdoor conditions (Eroglu et al. 2008b) also influenced the productivity. Iron and molybdenum are the other key components to supplement OMW for improving the yield (Eroglu et al. 2011).

Olive mill wastewater has been utilized as the sole substrate source for photofermentative hydrogen production (Eroglu et al. 2004) using the PNSB *Rb. sphaeroides* O.U.001, with glass column-photobioreactors (400 ml) under artificial illumination. Due to the dark color of the raw material, dilutions were needed for efficient hydrogen production. Dilutions of 1, 1.5, 2, 2.5, 3, 4, 5, 10, 20 % (v/v) in H_2O, were investigated. Although bacterial growth could be achieved at all of these concentrations, hydrogen production could only be observed for dilutions at or below 4 %. The highest bacterial mass (0.553 $g_{dcw}l_c^{-1}$) and hydrogen productivity were achieved with 4 % diluted OMW. On the other hand, the maximum hydrogen yield (13.9 l_{H2} per l_{OMW}) was obtained with 2 % diluted OMW. In addition to the hydrogen production, significant reduction of the initial biochemical oxygen demand (58 %), chemical oxygen demand (35 %), and total phenolic content (60 %) were observed when 2 % diluted OMW was used (Eroglu et al. 2004). The physicochemical characteristics of the OMW, collected from different regions of Western Anatolia, were slightly variable. Four of these were compared on the basis of their photofermentative hydrogen production efficiencies. It was observed that the hydrogen productivity was directly related with the organic acid content and the carbon-to-nitrogen (C/N) molar ratio (Eroglu et al. 2009b). The highest hydrogen yield (19.9 l_{H2} per l_{OMW}) was obtained with the OMW having the highest C/N molar ratio (73.8 MM^{-1}), and the highest organic acid content, particularly acetic acid.

Figure 11.1 illustrates hydrogen production in a flat plate solar bioreactor (8 l) operating under outdoor conditions, utilizing 4 % diluted OMW in batch mode. Diurnal periods were 14 h light and 10 h dark. Bacterial growth stopped during the periods of darkness. Hydrogen yield was 11.4 l_{H2} per l_{OMW} (Eroglu et al. 2008b).

Various two-stage hydrogen production processes with OMW have been reported

Fig. 11.1. Flat-plate solar bioreactor filled with 4 % (v/v) OMW (Olive Mill Wastewaters) (Middle East Technical University, Ankara).

(Eroglu et al. 2006, 2008a). In Eroglu et al. (2006), the first stage was based on the dark-fermentation of raw OMW by the mixed-cultures of activated sludge, which is followed by photofermentation via *Rb. sphaeroides* O.U.001. This approach converted the raw OMW to an effluent with favorable physicochemical properties for the photofermentation stage. The 50 % (v/v) diluted DFE achieved the highest photofermentative hydrogen production rate (8 mll^{-1} h^{-1}), while the 4 % (v/v) diluted DFE achieved the highest hydrogen production potential (29 l_{H2} per l_{OMW}). This two-stage process was mainly investigated for its potential to utilize raw OMW for photofermentative hydrogen production even at very high concentrations.

We proposed another two-stage process for enhancing the physicochemical properties of the raw OMW, which was based on a clay pretreatment step followed by photofermentation (Eroglu et al. 2006, 2008a). Clay pretreatment was investigated in detail to gain further insight into its overall effect on photofermentative hydrogen production (Eroglu et al. 2008a). The fundamental organic compounds of the clay pretreatment effluent were found as acetic, lactic, propionic, and butyric acids, which were favored during the photofermentative hydrogen production by *Rb. sphaeroides* O.U. 001. Hydrogen production with the effluent of the clay pretreatment process nearly doubled (31.5 l_{H2} per l_{OMW}), in comparison with the raw olive mill wastewater (16 l_{H2} per l_{OMW}) under the same experimental conditions. Clay pretreatment was mostly effective on the removal of unwanted compounds such as phenols, while the removal of desired substrates including organics acids, amino acids and sugars was minimal. The decrease in the phenolic content was also effective for color removal, which enhances the photosynthetic efficiency of the system by increasing the transmittance of light within the photobioreactor. Pretreatment methods applied to OMW include (i) oxidation with ozone, (ii) oxidation with Fenton's reagent, (iii) photodegradation via UV radiation, (iv) adsorption with clay, and (v) adsorption with zeolite (Eroglu et al. 2009a). Among these pretreatment processes, adsorption with clay was found to be the most applicable one for the photofermentation process. The effluents of the pretreatment stage by strong chemical oxidants were found to be unsuitable for either hydrogen production or bacterial growth, despite having the highest color removal (90 %).

B. Sugar Manufacturing Wastes

Sugar manufacture constitutes an important, widespread industry throughout the world. During the manufacturing of sugar, various wastes such as molasses, bagasse, and wastewater are produced. These wastes are known to have a high BOD content, which could be suitable for the growth of the photosynthetic microorganisms (Hampannavar and Shivayogimath 2010).

Singh et al. (1994) produced hydrogen via the utilization of potato starch, sugarcane juice and cheese whey by free and Ca-alginate immobilized cells of *Rhodopseudomonas* sp. Each feedstock was diluted with the Biebl and Pfennig medium (Biebl and Pfennig 1981). Among these three substrates, the maximum amount of hydrogen production was obtained with sugarcane juice, followed by potato starch and whey.

Almost no hydrogen gas was produced on sucrose by *Rb. sphaeroides* O.U.001, while very low hydrogen production rate (1 $ml_{H2}l_c^{-1}h^{-1}$at 20 % (v/v) dilution) was obtained on diluted sugar refinery wastewater (SRW). Enrichment of the SRW with L-malic acid and sodium glutamate gave a higher hydrogen production rate and yield values as 5 $ml_{H2}l_c^{-1}h^{-1}$and 4.63 l_{H2} per l_{SRW}, respectively (Yetis et al. 2000). Extra nutrients were added to achieve a significantly higher carbon to nitrogen ratio (C/N: 70/2 MM^{-1}). Continuous hydrogen production was achieved for 100 days with 20 % SRW supplemented with malate and glutamate mixture operating at fed-batch mode. The maximum hydrogen production yield of 8.61 l_{H2} per l_{SRW} was observed at a dilution rate of 0.0013 h^{-1}.

C. Dairy Product Wastes

Dairy product wastes, including milk or cheese residues and whey, are known to usually have high amounts of organic material content with a COD value varying between 5 and 50 g/l (Seifert et al. 2010a).

Zürrer and Bachofen (1979) used lactic acid containing whey and yogurt wastes as carbon sources for *Rhodospirillum rubrum* S-1. They reported very high hydrogen production yields of 47 and 45 l_{H2} per l_{ww} with whey and yogurt wastes, respectively. Sasikala and Ramana (1991) used the wastewater of a lactic acid fermentation plant for the production of hydrogen by *Rb. sphaeroides* O.U.001. Hydrogen production was observed for different dilutions of wastewater, ranging from 5 % to 100 %. The maximum hydrogen production rate was obtained as 5 $ml_{H2}h^{-1}\ l^{-1}$ while the hydrogen production yield was found to be 4.5 l_{H2} per l_{ww}.

Türkarslan et al. (1998) also investigated the hydrogen production potential of the dairy plant wastewater by *Rb. sphaeroides* O.U. 001. Milk factory waste was not sufficient for growth and hydrogen production by the photosynthetic bacteria. The dairy plant wastewater 30 % (v/v) with added malate produced hydrogen at a rate and yield of 5.5 $ml_{H2}l^{-1}\ h^{-1}$ and 2.0 l_{H2} per l_{ww}, respectively. Seifert et al. (2010a) reported the highest hydrogen production rate for 40 % (v/v) dairy wastewater after mixing with the Biebl and Pfennig medium. No hydrogen was produced from more concentrated dairy water due to the inhibitory effect of the high ammonium content. The highest hydrogen production yield (16.9 l_{H2} per l_{ww}) was achieved by 5 % (v/v) sterile dairy wastewater mixed with the Biebl and Pfennig medium (Biebl and Pfenning 1981).

D. Other Types of Waste Materials

Zhu et al. (1999) achieved high hydrogen production yields using immobilized cells of *Rb. sphaeroides* RV for hydrogen production from the wastewater of a tofu factory. They reported that 100 % (v/v) wastewater had hydrogen production rate and yield values of 59.0 $ml_{H2}l_c^{-1}h^{-1}$ and 1.9 l_{H2} per l_{ww}, respectively. Several factors, such as the use of a cellular immobilization technique, the utilization of a *Rb. sphaeroides* RV strain capable of using significant amounts of glucose, and the feeding of the tofu wastewater only after the cells had begun to evolve hydrogen from a pre-culture of lactate medium, were given as the possible reasons for a higher hydrogen production rate. Mitsui and his colleagues isolated marine photosynthetic bacteria for photofermentative hydrogen production from orange processing wastes under outdoor conditions (Mitsui et al. 1983). These wastes were diluted with seawater until a total organic carbon (TOC) value of 430 ppm was obtained. They used 4 l outdoor reactors while marine photosynthetic bacteria were immobilized in agar plates. In addition to hydrogen production, they also observed a 90 % decrease in the initial BOD and 37 % decrease in TOC content of the wastewater.

Vatsala and Ramasamy (1987) investigated photofermentative H_2 production from distillery waste by *Rhodospirillum rubrum* 11170 under both indoor and outdoor conditions. For the indoor conditions, 100 % (v/v) distiller waste resulted in a hydrogen production at a rate of 3 $ml_{H2}l_c^{-1}h^{-1}$, while this rate was 0.8 $ml_{H2}l_c^{-1}h^{-1}$ when 5 % distillery waste was used under outdoor conditions.

Brewery wastewater is also reported to be rich in organic content including organic acids, sugars, amino acids and etc., which would be favored during photofermentative hydrogen production. Seifert et al. (2010b) investigated the photofermentative hydrogen production potential of brewery wastewaters by using *Rb. sphaeroides*. They found the optimal dilution rate as 10 % (v/v), after mixing with the Biebl and Pfennig medium (Biebl and Pfenning 1981), which is mainly based on the high nitrogen content of the brewery wastewater. Diluted (10 %) brewery wastewater resulted in a hydrogen production yield of 0.22 l_{H2} per l_{ww} (2.24 l_{H2} per l_{media}) (Seifert et al. 2010b).

IV. Photofermentative Hydrogen Production with Dark Fermenter Effluents

One of the advantages of using photofermentative hydrogen producing bacteria is that they can utilize organic acids produced by dark fermentative bacteria. Biohydrogen production by combination of dark and photofermentation has been considered as a promising route to increase the hydrogen yield, due to the potential of complete substrate oxidation. Dark fermentative biohydrogen production liberates reduced organic compounds like acetate, butyrate, lactate and propionate, which can be readily fermented by photofermentative bacteria to produce more hydrogen. This can be realized in a single-step (co-culture of dark and photofermentative bacteria) or in two steps (dark fermentation followed by photofermentation) (Fig. 11.2). Single-step processes require optimization of process conditions (i.e., pH, temperature, light intensity and biomass concentration) for both types of microorganisms grown in the same reactor at the same time. Selection of suitable microorganisms that can coexist is important.

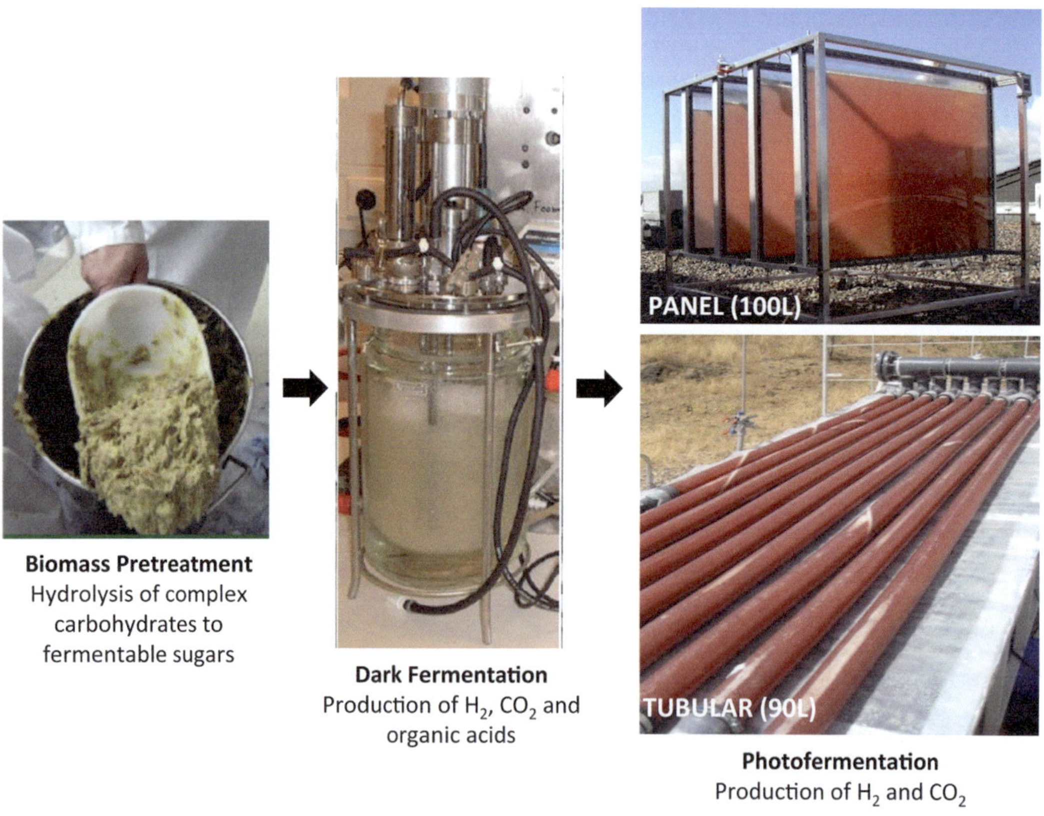

Fig. 11.2. Sequential operation of dark and photo-fermentative hydrogen production from biomass.

In two-step processes, suitable microorganisms for different feedstocks can be selected independently for each step. Moreover, process conditions can be adjusted separately for higher efficiency. For example, effluents of dark fermentation can be supplemented with necessary nutrients or, they can be pre-treated to remove toxic or inhibitory compounds. It was reported that the supplementation of DFEs with appropriate amounts of iron and molybdenum, the co-factors of nitrogenase, increases the photofermentative hydrogen production significantly (Özgür et al. 2010a, b; Afşar et al. 2011; Afşar 2012). Ammonium, an additive for dark fermentation process, strongly inhibits the nitrogenase enzyme of PNSB, leading to cessation of photofermentative hydrogen production at concentrations higher than 2 mM (Yakunin and Hallenbeck 1998; Akköse et al. 2009). For successful process integration, the ammonium content of DFEs should be lowered. This can be achieved by: (i) lowering the amount of ammonium added during dark fermentation, (ii) diluting the DFEs with water, (iii) using/developing bacterial strains with ammonium tolerance or, (iv) pretreating the DFEs so that ammonium is removed. Androga et al. (2012) suggested a pretreatment method for high ammonium containing DFEs using clinoptilolite, a natural zeolite, to reduce the ammonium concentration. Pretreatment resulted in an 80 % decrease of the ammonium concentration of molasses DFE. After using the clinoptilolite-pretreated effluent, *Rb. capsulatus* produced hydrogen at a high yield (90 %) and productivity (1.16 mmol $l_c^{-1} h^{-1}$), while no hydrogen production was observed with the untreated one. If DFEs are too dark, as is the case for molasses DFE, such pretreatment techniques can also be useful for color reduction to increase light penetration through the PBR. Another important parameter for high efficiency photofermentative hydrogen production using DFEs is the concentration of the carbon (C) source. DFEs mainly contain acetate, lactate and butyrate, which are utilized by PNSB during photofermentation. The concentration of the carbon source is important for optimum metabolic processes that result in high hydrogen yields. Acetate at 30–40 mM (Özgür et al. 2010c), butyrate at 20 mM (Shi and Yu 2006) and lactate at 10–20 mM (Lo et al. 2011) were reported to be optimum for photobiological hydrogen production with different PNSB species. Dark fermenter effluents usually contain significant concentrations of acetate, which may decrease hydrogen production efficiency due to the diversion of the cell resources towards the synthesis of PHB. The effects of acetate concentration on hydrogen and PHB production in *Rb. capsulatus* were studied at 10–65 mM acetate and gene expression analysis of related genes (*NifD* for nitrogenase and *PhaC* for PHB synthase) were carried out (Özsoy 2012). Optimum acetate concentration for photofermentation with high hydrogen yield and low PHB amount was around 25–50 mM, where *NifD* expression was high. *PhaC* expression was highest at 65 mM meaning that high concentrations of acetate (above 50 mM) lead to direction of PNSB metabolism to PHB biosynthesis, decreasing the hydrogen yield. Adjustment of the carbon source concentration can be achieved through dilution with water. Buffering of the DFEs is also necessary to adjust and keep the pH at a tolerable range (6.0–8.0) for photofermentation. It was reported that 20 mM of potassium phosphate (pH 6.4) was required as a buffer supplement for molasses, thick juice, barley straw and potato steam peels DFEs (Özgür et al. 2010a, b; Afşar et al. 2011).

In the remainder of this section, photobiological hydrogen production studies on DFEs of a variety of agro-industrial wastes/wastewaters are reviewed and compared. Some of the applications are summarized in Table 11.2. Feedstocks used in sequential dark-photo fermentation processes are classified as lignocellulosic, starchy or sugar containing, based on their primary carbohydrate constituent.

Table 11.2. Summary of the studies of photofermentative hydrogen production on dark fermenter effluent of various feedstocks.

Feedstock	Microorganism: Dark fermentation (DF)	Microorganism: Photofermentation (PF)	PF process details	Yield (PF)	Reference
Barley straw hydrolysate	*Caldicellulosiruptor saccharolyticus*	*Rb. capsulatus hup-* (YO3)	Indoor, batch	49 %	Özgür and Peksel (2013)
Miscanthus hydrolysate	*Thermotoga neopolitana*	*Rb. capsulatus* DSM155	Indoor, batch	28 %[a]	Uyar et al. (2009)
Thick juice	*C. saccharolyticus*	*Rb. capsulatus hup-*	Indoor, batch	50 %	Özgür et al. (2010d)
Thick juice	*C. saccharolyticus*	*Rb. capsulatus hup-*	Outdoor panel PBR (4 l), fed-batch, 15 days	77 %	Özkan et al. (2012)
Thick juice	*C. saccharolyticus* *Caldicellulosiruptor owensensis*	*Rb. capsulatus* DSM1710	Outdoor tubular PBR (90 l), fed-batch, 21 days	11 %	Boran et al. (2012a)
Molasses	*C. saccharolyticus*	*Rb. capsulatus hup-*	Indoor, batch	58 %	Özgür et al. (2010a)
Molasses	*C. saccharolyticus*	*Rb. capsulatus* DSM1710	Outdoor, panel PBR (4 l), fed-batch, 55 days	50 %	Avcioglu et al. (2011)
Molasses	*C. saccharolyticus*	*Rb. capsulatus hup-*	Outdoor, panel PBR (4 l), fed-batch, 75 days	78 %	Avcioglu et al. (2011)
Potato steam peels hydrolysate	*C. saccharolyticus*	Different PNSB strains	Indoor, batch	<25 %	Afşar et al. (2011)
Sweet potato starch residue + corn steep liquor	*Clostridium butyricum & Enterobacter aerogenes*	*Rhodobacter* sp. M-19	Indoor, batch	56 %[a]	Yokoi et al. (2002)
Cassava starch	Preheated activated sludge	*Rp. palustris*	Indoor, batch	18 mmol $H_2.g^{-1}$ starch[b]	Su et al. (2009)
Ground wheat starch	Heat treated anaerobic sludge	*Rb. sphaeroides* (NRRL- B1727)	Indoor batch	185 ml $H_2.g^{-1}$ VFA	Ozmihci and Kargi (2010)
Cheese whey	Mixed thermophilic dark fermentative consortia	*Rp. palustris*	Indoor, batch	349 ml $H_2.g^{-1}$ COD	Azbar and Dokgöz (2010)
Cheese whey	*Enterobacter aerogenes*	*Rp.* BHU01	Indoor, immobilized batch	67 %[a]	Rai et al. (2012)

[a]Calculated from given data
[b]Overall hydrogen yield of two stages

A. DFEs Obtained from Lignocellulosic Feedstocks

Selection of cheap, renewable and nonfood resources is necessary for sustainable biohydrogen production. Lignocellulosic biomass, which constitutes a major portion of agricultural and forest wastes, and of industrial effluents from the pulp and paper industry, is of prime interest due to its abundance and nonfood nature. The main challenge in using lignocellulosic feedstock is that it is composed mainly of cellulose, hemicellulose, and lignin, which cannot be degraded by most of the fermentative bacteria utilized for biohydrogen production. Hence, pre-hydrolysis of such biomasses through physical, chemical or biological processes is necessary to produce monomeric carbohydrates from cellulose and hemicellulose (Ren et al. 2009).

1. Barley Straw DFE

Barley straw is a lignocellulosic agricultural residue of barley production. It has a dry matter content of 91.1 %: 38.9 % being glucan, 23.7 % xylan, 3.5 % other hemicelluloses and 22.8 % lignin (Foglia et al. 2011). The annual barley straw production was estimated to be 114 million tons in Europe, which corresponds to 511 TWh, assuming a lower heating value of 16.3 MJ per kg dry matter (Ljunggren et al. 2011). Worldwide annual yields of lignocellulosic biomass residues were estimated to exceed 220 billion tons, equivalent to 60–80 billion tons of crude oil (Ren et al. 2009). It has been considered as a feedstock for biofuel production through biochemical processes in several studies (Qureshi et al. 2010; Sigurbjornsdottir and Orlygsson 2012).

Biohydrogen production from barley straw DFE was carried out (Özgür and Peksel 2013). Barley straw hydrolysate, obtained after chemical and biological pretreatment, was fermented with a hydrogen-producing thermophilic dark fermentative bacterium, *Caldicellulosiruptor saccharolyticus*, and the effluent was used for photofermentative hydrogen production, in indoor batch cultures of *Rb. capsulatus* DSM1710 and *Rb. capsulatus* YO3 (uptake hydrogenase deleted strain of *Rb. capsulatus* MT1131). The effluent was diluted to have an initial acetate concentration of 30–35 mM and supplemented with potassium phosphate buffer. Good bacterial growth and hydrogen production with yields and productivities as high as 50 % and 0.58 mmol $l_c^{-1} h^{-1}$, respectively, were obtained with the Hup^- strain. Addition of iron and molybdenum resulted in 30–50 % increase in productivity.

2. Miscanthus DFE

Miscanthus is a lignocellulosic energy crop mostly grown in USA and Europe as biomass feedstock for biofuel production, especially bioethanol. It is a perennial C_4 grass that grows rapidly with dry weight yields of around 8–15 t ha^{-1}, and requires only low inputs of nutrients for cultivation. *Miscanthus* hydrolysate, obtained after alkaline and enzymatic pretreatment, was used for biohydrogen production through sequential dark and photofermentation (Uyar et al. 2009; de Vrije et al. 2009). Dark fermentation, carried out using the thermophilic bacterium *Thermotoga neopolitana* on *Miscanthus* hydrolysate containing 10 g per l monomeric sugars in batch cultures, yielded 3.3 mol H_2 per mole hexose with a productivity of 13 mmol $H_2 l_c^{-1} h^{-1}$. The effluent of dark fermentation, which mainly contained acetate (94 mM) together with some lactate (8.9 mM), was fed to PBRs operated at batch mode using *Rb. capsulatus* DSM155. Before being fed to the PBR, the effluent was centrifuged, diluted with water (1:1) to decrease the acetate concentration, supplemented with buffer to control pH, and sterilized. Effluents supplemented with iron and vitamins resulted in a hydrogen productivity of 19 $ml_{H2} l_c^{-1} h^{-1}$.

B. DFEs Obtained from Sugar Processing Industry Intermediates

Intermediates from sugar beet processing industries, like molasses and thick juice, are very good sources for biohydrogen production due to their high content of fermentable sugars that can be readily utilized by microorganisms without any pretreatment. Molasses is the final effluent obtained in the manufacturing of sucrose by repeated evaporation, crystallization and centrifugation of juices from sugar cane and sugar beets. It contains more than 46 % total sugars (mostly sucrose), as well as nitrogenous compounds like amino acids and inorganic compounds, all of which make it an excellent substrate for fermentation. It is mainly used as animal feed, and as raw material for alcohol, yeast, and citric acid production. Thick juice is another by-product of sugar industry obtained after evaporation of sugar beet juice. It also has a high sugar content that can be utilized by microorganisms without any pretreatment. Unlike molasses, which can be stored for long times at room temperature without contamination, thick juice is contaminated easily, hence, it can only be used for biohydrogen production during sugar factory processing periods of the year.

1. Thick Juice DFE

Biohydrogen production on thick juice through dark and photofermentation has been addressed within the context of the FP6 HYVOLUTION Project (Claassen and de Vrije 2006). The thermophilic dark fermentative bacterium, *C. saccharolyticus,* was utilized in a continuously operated 30 l CSTR dark fermenter, with an initial sucrose concentration of 10 g per l. The effluent of dark fermentation, containing acetate as the main product, was fed to photofermenters after necessary adjustments were made, including dilution to reduce acetate concentration to around 30 mM and ammonium concentration below 2 mM, and additions of buffer and iron. In batch operations using *Rb. capsulatus* DSM1710 and *Rb. capsulatus hup-* (YO3) strains, a 40 % molar yield with productivities of 0.4 mmol $l_c^{-1} h^{-1}$ and 0.6 mmol $l_c^{-1} h^{-1}$ was achieved with DSM1710 and YO3 strains, respectively. Promising results with continuously operated indoor 4 l panel PBRs were also reported on thick juice DFEs with a feeding rate of 10 % (vv^{-1}) daily. During 25 days of continuous operation with the YO3 strain, stable biomass concentration of around 1 g_{dcw} per l_c was obtained with average hydrogen productivity and molar yield of 0.9 mmol $l_c^{-1} h^{-1}$ and 50 %, respectively (Özgür et al. 2010d). Continuous photofermentative hydrogen production under natural sunlight on thick juice DFEs was also reported (Özkan et al. 2012; Boran et al. 2012a). The DFEs were diluted and supplemented with buffer, Fe and Mo before feeding to the PBRs. It was reported that *Rb. capsulatus* YO3 strain grows and produces hydrogen successfully on thick juice DFE in outdoor conditions, unless the temperature of the reactor is kept above 35 °C. Temperature control was necessary as the reactor temperature exceeds 50 °C during summer, in Ankara, Turkey. To control the reactor temperature, chilled water was recirculated through the internal cooling coils. During 15 days of continuous operation inside panel PBRs with a daily feeding rate of 10 %, stable biomass concentration of around 1 g_{dcw} per l_c was achieved with a daily average hydrogen productivity of 1.12 mmol $l_c^{-1} h^{-1}$ and molar yield of 77 % (Fig. 11.3).

Thick juice DFEs obtained with co-cultures of thermophilic dark fermentative bacteria, *C. saccharolyticus* and *C. owensensis,* were also utilized for continuous photofermentative hydrogen production by *Rb. capsulatus* DSM1710 using a pilot scale (90 l) solar tubular PBR with internal cooling coils. After a long lag phase (7 days), which was attributed to adaptation of bacteria to outdoor conditions, rapid bacterial growth with a growth rate of 0.025 h^{-1} was observed. Hydrogen production started after 9 days with daily average productivity and molar yield of 0.27 mmol $l_c^{-1} h^{-1}$ and 10 %, respectively, which are significantly lower compared to those obtained with panel PBRs. This was attributed to the tube diameter (6 cm), which is larger than the panel PBR thickness (2 cm), substantially lowering the light penetration through the deeper sites of

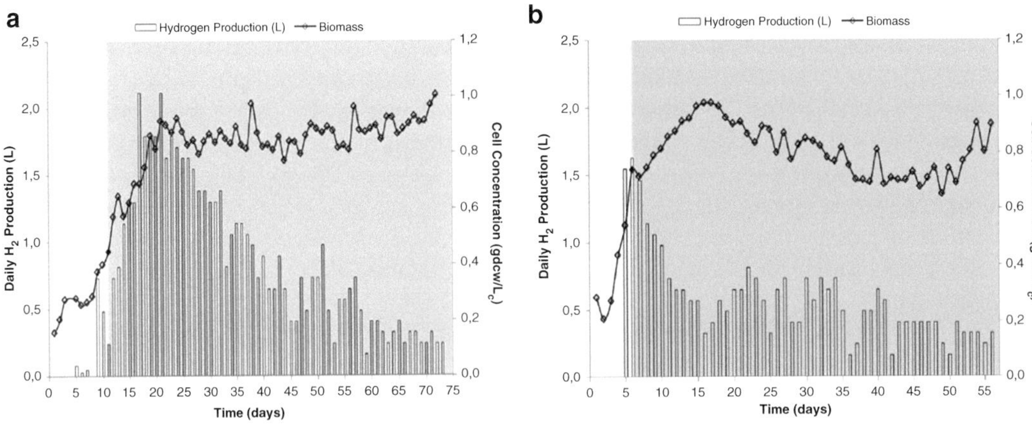

Fig. 11.3. Continuous photofermentative hydrogen production and biomass growth in solar panel PBRs (4L) on molasses DFEs outdoors, during summer 2009 in Ankara, Turkey. (**a**) *Rb. capsulatus hup-*, (**b**) *Rb. capsulatus* DSM1710 (Reproduced with permission from Avcioglu et al. 2011).

the reactor (Boran et al. 2012b). It was also noted that the hydrogen permeability of the wall material used for the tubular reactor (low density polyethylene, LDPE) was around ten times higher than that of Plexiglass® used in panel PBR, with the added effect of wall thickness (150 µm for tubular PBR, 6 mm for panel PBR). Hence, both low light penetration and hydrogen leakage problems in tubular PBR resulted in lower hydrogen productivities. To increase the hydrogen production, lower tube diameters and continuous circulation to decrease the residence time of hydrogen within the reactor, were suggested, taking into consideration the energy requirements for circulation and land area needed. A direct correlation with the hydrogen yield factor (moles of hydrogen per gram cell dry weight) and daily-received light energy was reported, suggesting that hydrogen production efficiency is directly related to the light exposure of cells.

2. Molasses DFE

Due to its high fermentable sugar and nutritional content, molasses is one of the most promising substrates for sustainable fermentative biohydrogen production. There are a number of studies carried out on biohydrogen production from molasses through dark fermentation (Ren et al. 2006, 2009; Aceves-Lara et al. 2008; Guo et al. 2008; Wang and Jin 2009).

Özgür et al. (2010a) reported sequential operation of dark and photofermentation for biological hydrogen production from sugar beet molasses. An extreme thermophile *C. saccharolyticus* was used for the dark fermentation, and several photosynthetic bacteria (*Rb. capsulatus* DSM1710, *Rb. capsulatus hup-* mutant (YO3), and *Rp. palustris*) were used for the photofermentation. *C. saccharolyticus* was grown in a pH-controlled bioreactor, in batch mode, on molasses with an initial sucrose concentration of 15 g per l. The effects of addition of ammonium on dark and photofermentative processes were also determined. *C. saccharolyticus* fermentation yielded acetate and lactate as the main organic acids, with a hydrogen yield of 4.2 mol per mole sucrose in the absence of ammonium addition. Molasses DFE obtained without ammonium addition, which yielded highest hydrogen production in dark fermentation, was utilized for photobiological hydrogen production, under continuous illumination, in batch mode. Adjustments, including dilution, addition of buffer, iron and molybdenum, were carried out on DFE to improve the photofermentative hydrogen production. The highest hydrogen yield

(58 % of theoretical hydrogen yield over consumed organic acids) and productivity (1.37 $mmoll_c^{-1} h^{-1}$) was attained using a Hup- mutant of *Rb. capsulatus*. The overall yield increased from 4.2 mol H_2 per mole sucrose in dark fermentation to 13.7 mol H_2 per mole sucrose by sequential dark and photofermentation (corresponding to 57 % of the theoretical yield of 24 mol of H_2 per mole sucrose).

Long term continuous hydrogen production under outdoor conditions from sugar beet molasses DFE in solar panel PBR using *Rb. capsulatus* DSM1710 and *Rb. capsulatus hup-* (YO3) was also reported (Avcioglu et al. 2011). The DFE obtained by a continuous dark fermentation process using thermophilic dark fermentative bacteria, *C. saccharolyticus*, contained mainly acetate as a carbon source, together with some lactate and formate. This DFE was fed to a PBR equipped with cooling coils to control the temperature below 35 °C using chilled water. The PBRs were successfully operated with a daily feeding rate of 10 %, for 55 days using *Rb. capsulatus* DSM1710, and 75 days for *Rb. capsulatus hup-*. An average biomass concentration of 0.9 g_{dcw} per l_c was achieved over the continuous operation, in outdoor conditions during summer 2009, in Ankara, Turkey (Fig. 11.3). The maximum hydrogen yield obtained using *Rb. capsulatus hup-* was 78 % (of the theoretical maximum) and the maximum hydrogen productivity was 0.67 mmol $l_c^{-1} h^{-1}$. The maximum hydrogen productivity and yield of the wild type strain on the molasses DFE were 0.50 mmol $l_c^{-1} h^{-1}$ and 50 %, respectively.

C. DFEs Obtained from Starch-Based Biomass

Starch-based biomass, like vegetable raw materials including potato and cereals and food wastes from the industry and household, contain high levels of carbohydrate and protein. Biohydrogen production studies from starch containing biomasses were reported to have promising hydrogen yields. Like lignocellulosic biomass, starchy biomass also needs pretreatment in order to hydrolyze starch to fermentable sugars, which is generally carried out through thermal and enzymatic pretreatment processes. However, some thermophilic dark fermentative bacteria can digest starch, eliminating the need for a pretreatment (Mars et al. 2010).

1. Potato Steam Peels (PSP) Hydrolysate DFE

Potato steam peel (PSP) waste is a co-product from the potato processing industry that is rich in starch, and available in large quantities. It is currently used as animal feed, but studies have shown that it is also a good source for fermentative hydrogen production (Claassen et al. 2005). A life cycle assessment (LCA) analysis made to evaluate the main environmental benefits and burdens of using PSP to produce hydrogen in a two-step fermentation process showed that it is more beneficial to use PSP for hydrogen production, and the protein-rich residue of this process for animal feed (Djomo and Humbert 2008). Results obtained from a dark fermentation process revealed that either PSP hydrolysate or untreated PSP are very suitable substrates for efficient fermentative hydrogen production at moderate substrate loadings (ca. 10 g per l of glucose), using thermophilic dark fermentative bacteria, *C. saccharolyticus* (Mars et al. 2010).

PSP hydrolysate, obtained after enzymatic liquefaction and saccharification steps, where starch is converted into glucose, was utilized for sequential dark and photofermentation (Afşar et al. 2011). Diluted hydrolysate (10 g C_6 sugar per l) supplemented with yeast extract was fed to bioreactor with thermophilic dark fermentative bacteria, *C. saccharolyticus*. The DFE of PSP containing acetate (102 mM), lactate (28 mM) and NH_4Cl (4.0 mM) was sterilized and diluted by three times with sterile dH_2O. The effluent was also supplemented with 20 mM potassium phosphate buffer (pH 6.4), iron (Fe-citrate, 0.1 mM) and molybdenum ($NaMO_4.2H_2O$, 0.16 mM) before being fed to PBRs. Photobiological hydrogen production has been carried out in indoor, batch cultures of different PNSB strains (namely, *Rb. capsulatus* DSM1710, *Rb. capsulatus hup-* (YO3), *Rb. sphaeroides* O.U.001 (DSM5864), *Rb. sphaeroides hup-* (uptake hydrogenase deleted mutant of O.U.001) and *Rp. palustris*) under continuous illumination. Although the process was highly efficient in terms of acetate consumption (100 %), lower hydrogen yields (<25 %) and productivities (<0.55 mmol $l_c^{-1} h^{-1}$) were reported compared to the results obtained on molasses, thick juice or barley straw effluents.

Yokoi et al. (2002) used sequential hydrogen production from a media consisting of sweet potato starch residue as carbon source and corn steep liquor as a nitrogen source. The initial stage was composed of dark-fermentation of repeated batch cultures of *Clostridium butyricum* and *Enterobacter aerogenes;* this was followed by photofermentation using *Rhodobacter* sp. M-19. The overall hydrogen yield was observed to be the highest (4.5 mol H_2 per mol glucose) when the DFE was supplemented with molybdenum and ethylene-diamine-tetra-acetic acid (EDTA).

Laurinavichene and her colleagues (2008) also applied a sequential two-step process to produce hydrogen from potato starch. Dark fermentation was performed using natural microbial consortia. Thereafter, the volatile fatty acid rich dark-fermentation effluent was used for photofermentative hydrogen production by *Rb. capsulatus* B-10. They observed that butyrate could not be consumed until the complete consumption of acetate, propionate, and lactate, after providing equimolar mixture of volatile fatty acids.

2. Cassava Starch DFE

Cassava is a starch containing root crop grown as a staple food and animal feed in subtropical and tropical regions of Africa, Latin America and Asia, with a total cultivated area over 18 million hectare. A high starch content (up to 90 %) together with its low agro-chemical requirements, high drought and heat tolerance make it an attractive plant for biofuel production (Jansson et al. 2009). Biohydrogen production from cassava starch through combination of dark and photofermentation has been reported (Su et al. 2009; Zong et al. 2009). Dark fermentation of raw cassava starch, after its gelatinization and enzymatic hydrolysis, was carried out at different initial starch concentrations (10–25 g per l) using hydrogen producing bacteria from preheated activated sludge (Su et al. 2009). Effluents of dark fermentation, which mainly consisted of acetate and butyrate, were used for photofermentative hydrogen production in batch cultures of *Rp. palustris,* after being diluted to optimal concentrations. The maximum hydrogen yield was reported to increase from 9.3 to 10.7 mmol H_2 per g starch only in the dark fermentation to 17.5–18.0 mmol H_2 per g starch in the combined dark and photo fermentation, which corresponds to 63.1–93.7 % improvement.

3. Ground Wheat Starch DFE

Waste ground wheat starch constitutes a reliable and renewable resource for biohydrogen production due to its high starch and gluten content (Kargi and Pamukoglu 2009; Argun et al. 2009). Ground wheat starch is reported to be rich in carbohydrates but deficient in nitrogen and phosphorous. Hence, supplementation with N and P maximizes the hydrogen yield by dark fermentation (Kargi and Pamukoglu 2009). The DFE of ground wheat starch containing 1950 ± 50 mg per l total volatile fatty acid (VFA), obtained by heat treated anaerobic sludge as the bacterial culture, were utilized for photofermentative hydrogen production by *Rb. sphaeroides* (NRRL-B1727) (Ozmihci and Kargi 2010). Continuous photofermentative hydrogen production at varying hydraulic retention times (HRT) (24–120 h) was carried out on the DFEs supplemented with several nutrients, such as $MgSO_4$, EDTA, $(NH_4)_2SO_4$ and KH_2PO_4 buffer to adjust the pH to 7.0 and C/N/P ratio to 100/1/0.3. The highest steady-state daily hydrogen production (55 ml per day) and hydrogen yield (185 ml H_2 per g VFA) were obtained at 72 h HRT (3 days).

D. DFEs from Other Feedstocks

Cheese whey is a lactose-rich (about 5 %) byproduct of the cheese manufacturing industry and its removal represents a significant problem in the dairy industry. Utilization of cheese whey wastewater to produce biohydrogen through mesophilic and thermophilic dark fermentation processes using mixed microbial communities has been reported (Azbar et al. 2009a, b, c). Biohydrogen production through sequential operation of thermophilic dark fermentation and photo fermentation using cheese whey wastewater has been investigated (Azbar and Cetinkaya-Dokgöz 2010). After centrifugation and dilution at different degrees, the effluent containing mainly acetate, isobutyrate, and lactate was used for photofermentation by *Rp. palustris*. The highest hydrogen production (349 ml H_2 per g COD) was obtained with the five-times diluted effluent. Overall hydrogen production performance of two-stage system was found to vary between 2 and 10 mol H_2 per mole of lactose consumed. It was concluded that the dilution of anaerobic effluent helps to reduce the nitrogen and the volatile fatty acid content in the feeding, which can otherwise be inhibitory. It was also reported that addition of malate significantly improves the hydrogen production, hence, mixing of cheese whey effluent with malate containing material such as apple juice processing effluent, might be useful.

Rai et al. (2012) investigated sequential two-stage biohydrogen production on cheese whey using *Enterobacter aerogenes* in dark fermentation step, and *Rhodopseudomonas* BHU 01 in photofermentation steps. They showed that the cumulative H_2 yield (dark and photo-fermentation) obtained by cultures immobilized by alginate entrapment was better (5.88 mol H_2 per mole lactose) compared to the suspension cultures (3.40 mol H_2 per mole lactose).

Fruit and vegetable wastes (FVW) are produced in large amounts mainly from the domestic households, restaurants and market places. Due to their high biodegradability, they can cause several irritations in municipal landfills, which bring a need for their recycling (Bouallagui et al. 2005). FVW is mainly composed of easily biodegradable constituents such as sugars and hemicellulose (75 %) (Verrier et al. 1987). On the other hand, some vegetable processing effluents contain non-biodegradable organic constituents that should be minimized via pretreatment processes before being used for any bioprocesses (Bouallagui et al. 2005). Fascetti et al. (1998) used organic-acid rich (mostly lactic-acid and acetic-acid) DFE of fruit and vegetable wastes for photofermentative hydrogen production by *Rb. sphaeroides* RV. They achieved significant hydrogen production rates of around 100 $ml_{H2}\,g^{-1}.h^{-1}$ after the continuous processing of a 1 liter chemostat.

Abd-Alla et al. (2011) studied biohydrogen production from rotten date palm fruits in a three-stage process including dark fermentation by the facultative anaerobe *E. coli* EGY, followed by the strict anaerobe *Clostridium acetobutylicum* ATCC 824 and lastly by the PNSB *Rb. capsulatus* DSM1710, within the same reactor. They reported maximum hydrogen yield of 7.8 mol H_2 per mole of sucrose corresponding to 162 l H_2 per kg fresh rotten dates.

V. Optimization of Hydrogen Yield

A. Genetic Modifications

Genetic engineering is a promising tool to increase the yield and productivity of photofermentative hydrogen production (Vignais et al. 2006). Considering the H_2 metabolism of PNSB, genetic modifications can be done to (i) inhibit H_2 utilization by deletion of uptake hydrogenase, (ii) eliminate the light-dependency of hydrogen evolution by recombinant expression of hydrogen-evolving hydrogenases, (iii) optimize the flow of reducing equivalents to nitrogenase by the inhibition of PHB and CO_2 fixation, (iii) eliminate/decrease the effect of environmental factors (e.g. NH_4^+, temperature) by producing resistant mutants, (iv) reduce the size of antenna pigments to increase light utilization efficiency.

1. Deletion of Uptake Hydrogenase

The physiological function of uptake hydrogenase in most PNSB is to catalyze the conversion of molecular hydrogen to electrons and protons, which decreases the yield of H_2 production. The uptake hydrogenase helps to maintain a redox balance. It was shown that the inactivation of uptake hydrogenase results in significant increase in total hydrogen production in these bacteria. *Rb. sphaeroides* O.U.001 is a purple non-sulfur bacterium producing hydrogen under photoheterotrophic conditions. Hydrogen is produced by Mo-nitrogenase enzyme and a substantial amount of H_2 is re-oxidized by a membrane-bound uptake hydrogenase. To improve the hydrogen producing capacity of the cells, a suicide vector containing a gentamicin cassette in the *hupSL* genes was introduced into *Rb. sphaeroiodes* O.U.001 and the uptake hydrogenase genes were destroyed by site directed mutagenesis. The wild type and the mutant cells showed similar growth patterns but the total volume of hydrogen gas evolved by the mutant was higher than that of the wild type strain (Kars et al. 2008). A similar approach was adopted by Öztürk et al. (2006) to develop an uptake hydrogenase lacking strain of *Rb. capsulatus* MT1131. The mutant strain produced around 30 % more hydrogen than the wild type MT 1131. The *hup-* strain was also tested in large-scale photobioreactor outdoor conditions on acetate as carbon source and enhancement in H_2 production rates and yields were observed (Androga et al. 2011a, b). The *Rb. capsulatus hup-* strain was also used for hydrogen production from real DFEs in a panel photobioreactor in outdoor conditions (Avcioglu et al. 2011). Similar studies involving the genetic manipulations of PNSB were reviewed by Kars and Gündüz (2010).

2. Expression of Hydrogen-Evolving Hydrogenases

Light is essential for efficient photobiological hydrogen production by PNSB but it is discouraging to know the fact that continuous H_2 production in the outdoor using sunlight is impossible or rather inefficient during night. Therefore, one promising approach is to relieve the dependency of PNSB from light for hydrogen production by recombinant expression of hydrogen-evolving hydrogenases in these bacteria. Thus, continuous hydrogen production under light and dark conditions would be possible and certainly improve the efficiency of biological hydrogen production process. In a study by Kim et al. (2008), constitutive hydrogen evolution under both photoheterotrophic and dark fermentative conditions by recombinant *Rb. sphaeroides* was reported. They developed a recombinant *Rb. sphaeroides* KCTC 12085 strain that harbor, with all the accessory genes necessary, formate hydrogen lyase and Fe-only hydrogenase from *Rs. rubrum*, to enable dark fermentative hydrogen production from *Rb. sphaeroides*. The strain produced hydrogen during dark fermentative growth, and photofermentative hydrogen production increased by twofolds.

3. Redirecting the Electron Flow to Nitrogenase

The charge or redox status of the cells is very important for biohydrogen production, since nitrogenase needs electrons and ATPs in order to reduce protons to hydrogen. To optimize the flow of reducing equivalents to nitrogenase, genetic modifications were carried out targeting the CO_2 fixation and PHB synthesis pathways, which compete for reducing equivalents. The Calvin-Benson-Bassham (CBB) pathway is not only the heart of photoautotrophic metabolism but also important in maintenance of redox homeostasis via CO_2 assimilation. Öztürk et al. (2012) investigated the relationship between the redox balancing system and hydrogen production in various *Rb. capsulatus* strains whose CBB pathway was inactivated by deleting one of the key enzyme phosphoribulokinase (PRK). The results indicated that in the absence of the functional CBB pathway, the excess reducing equivalents were dissipated mainly through the nitrogenase in the form of hydrogen under nitrogen limiting conditions. The rate of hydrogen production was enhanced slightly for *Rb. capsulatus* Hup-, PRK- and *Rb. capsulatus* Hup-, PRK-, cbb3- mutants.

Spontaneous variants of *Rb. capsulatus* strains deficient in the CBB pathway have been shown to express nitrogenase structural genes to dissipate excess reducing equivalents, even in the presence of high concentrations of ammonium that is sufficient to repress nitrogenase expression in wild type (Tichi and Tabita 2000). In *Rb. sphaeroides* KD131, inactivation of PHB synthase resulted in a two-fold increase in hydrogen production on acetate and butyrate, in spite of depressed cellular growth and lower substrate utilization (Kim et al. 2011). It was shown that overexpression of *rnf* operon, which is thought to be dedicated to electron transport to nitrogenase, in *Rb. capsulatus* enhanced *in vivo* nitrogenase activity (Jeong and Jouanneau 2000). In another study done by Öztürk et al. (2006), a loss of function in the electron carriers in the membrane of *Rb. capsulatus* resulted in significant decrease in H_2 production. These results suggest that the electron flow to nitrogenase is critical in the H_2 production process and needs to be manipulated for the enhanced H_2 production.

4. Improving the Ammonium Ion Tolerance

Ammonium is one of the substances regulating nitrogenase activity. Removal of ammonium inhibition on nitrogenase activity is important especially for integrated dark and photofermentation studies, in which the dark fermenter effluent is usually rich in ammonium. To develop mutants with ammonium insensitive-nitrogenase activity, Pekgöz et al. (2011) deleted genes expressing two regulatory proteins of ammonium-dependent nitrogenase regulation, GlnB and GlnK in *Rb. capsulatus* DSM1710. However, *glnB* mutants showed lower hydrogen production, while *glnK* mutants were unviable. This observation suggests that the GlnB/GlnK two component regulatory system most probably has roles in other metabolic pathways as well. An ammonia tolerant mutant strain of *Rp. palustris* has been developed through a mutation in *nifA* gene (NH_4^+-dependent a transcriptional regulator of nitrogenase expression). The mutant strain constitutively expressed nitrogenase even in the presence of nitrogen (McKinley and Harwood 2010). Hydrogen production by NifA mutant *Rp. palustris* on undiluted vegetable waste derived medium has shown significant improvements in productivity, start-up time and substrate conversion efficiency at NH_4^+ concentrations as high as 6.1 mM (Adessi et al. 2012).

5. Proteomics and Microarray Analysis of Rb. capsulatus

For further improvement of biological hydrogen production, the understanding of whole genome expression profile of *Rb. capsulatus* under different stress conditions is required. Although the genome database is available, there is not enough information about the proteomics and transcriptomics of *Rb. capsulatus*. The first proteomic study, reported by Daldal's group (Önder et al. 2010) revealed more than 450 proteins including the localization information. The data obtained by the proteome study of *Rb. capsulatus* SB1003 under different growth conditions, namely, those leading to aerobic respiratory, anaerobic photofermentative and anaerobic respiratory modes, contributed to the extent of the protein database and highlighted the proteins associated with these growth modes (Peksel 2012). A total 460 proteins were identified with 17 proteins being unique to particular growth conditions.

Understanding the whole genome expression profile under different stress conditions, such as temperature, UV, and high light intensity, is important for outdoor PBR applications. A custom designed Affymetrix Gene Chip for *Rb. capsulatus* DSM1710 (GEO Accession number: GPL18063) was constructed. The effects of temperature stress on transcriptome of *Rb. capsulatus* were investigated by comparing expression profiles under optimum hydrogen production condition (30 °C), heat (42 °C) and cold (4 °C) stress conditions (GSE53477) (Gürgan 2011). The influence of different nitrogen sources on transcriptome of *Rb. capsulatus* was investigated by comparing expression profile on 5 mM ammonium chloride and 2 mM glutamate. Carbon source was 40 mM acetate on both conditions. To study the effect of different acetate concentrations, 40 mM and 80 mM acetate were used with 2 mM glutamate as nitrogen source (GSE53303).

The data obtained from physiological, biochemical, proteomics and microarray studies has to be combined to further understand the metabolism of the microorganism and the genetic manipulation of the bacterial strains.

B. Photobioreactor Design

Photobioreactors (PBRs) are reactors that accommodate microbial culture systems to carry out light dependent biological reactions (Tredici 2004). In these reactors, light has to pass through the transparent reactor walls to reach the cells. A fundamental parameter for the evaluation of the photofermentation process is its photochemical efficiency (PE). This is defined as the conversion efficiency of the energy of the incident light to chemically bound energy in the hydrogen produced. In contrast to the performance of a fermenter, which is usually reported as a volumetric productivity, the performance of a PBR is usually given as an area related productivity – as hydrogen productivity per illuminated area or per land area occupied by the bioreactor. When designing an industrial process, however, not only the input energy of the incident light, but also the input energy for operation of the installation should be taken into consideration when evaluating the efficiency of the process in terms of energy conversion. When evaluating the economic efficiency of the process, costs of construction material, construction and maintenance as well as resources necessary for operation should also be taken into consideration. The ideal PBR requires low investment costs and low operational costs, but still should have a high productivity and be easily scalable. In practice it is not easy to provide a large surface-to-volume ratio for sufficient supply with solar light and enable at the same time a good mass and heat transfer.

In terms of reactor geometry, two main categories of PBRs can be distinguished as:

1. Tubular reactors
2. Flat panel reactors

1. Tubular Reactors

For hydrogen producing bacteria, residence time in the tubular part should be optimized since organic acids are already dissolved in the medium, but the gases produced (H_2 and CO_2) need to be removed. Therefore, a manifold configuration of the photo-bioreactor is preferred (Fig. 11.4).

Gas collection in a photo-bioreactor with a manifold configuration could be accomplished by tilting of the surface. Attachment of small gas bubbles to the wall of the tubes might be solved by large gas bubbles moving in the upward direction, as was done by Tredici and co-workers in the Near Horizontal Tubular Reactor (NHTR) (Chini Zittelli et al. 1999). In the NHTR designed by Tredici, gas is injected to create an airlift. In the case of hydrogen production, gas bubbles are only required to collect the small bubbles of hydrogen (and CO_2) produced in the medium. The energy input for pumping liquid should be limited to enable a positive energy balance of the photofermentation process (Fig. 11.4).

The tubular PBR finally constructed and used for continuous outdoor experiments by our group (Boran et al. 2012a, b) had a volume of 85 l (Fig. 11.5). It is currently equipped with new manifolds with a tube length of 3 m. For up-scaling the reactor size, the tube length can be easily increased. However, it should be pointed out that the maximum allowable pressure for the tubes depends on the used tubing material and its wall thickness.

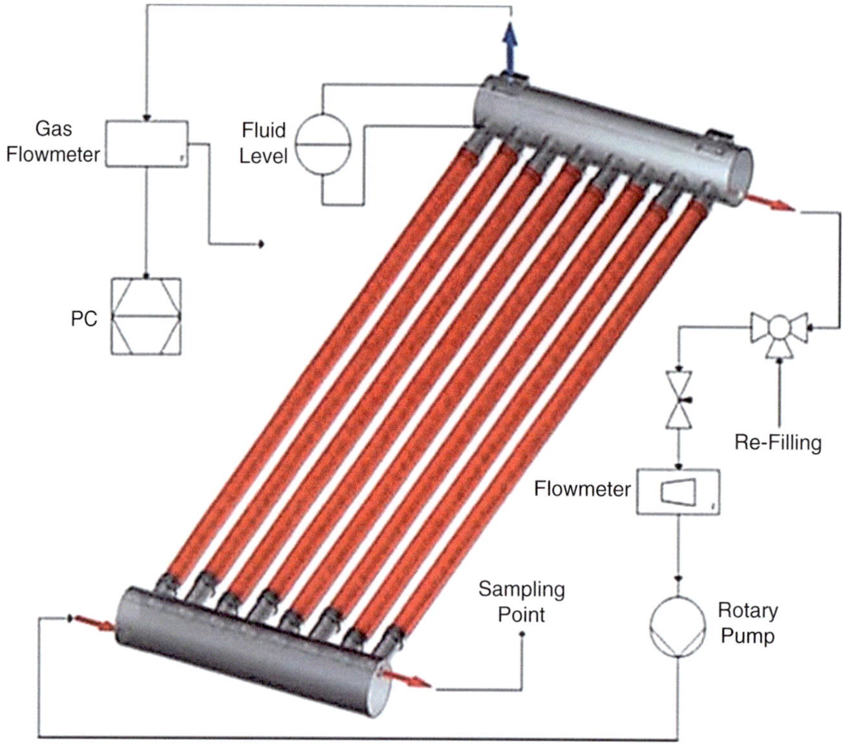

Fig. 11.4. Principle scheme of a manifold based tubular photo-bioreactor for the production of biohydrogen.

Fig. 11.5. Tubular photo-bioreactor (85 L) operated at METU, Ankara (Reproduced with permission from Boran et al. 2012a).

Fig. 11.6. Modular arrangement of flat panel reactor (112 L) operated at RWTH, Aachen (Reproduced with permission from Gebicki et al. 2010).

2. Flat Panel Reactor

In flat panel reactors it is observed that the bacteria only settle in the dark. As soon as hydrogen is produced, the bacterial cells are 'self-suspended' due to the upward movement of the H_2-bubblets onto which the cells tend to attach. To design a PBR as compact as possible, the design must aim to achieve a maximal ratio of illuminated surface area to land space covered by the reactors. The illuminated surface per ground space is higher in vertical reactors than in horizontal systems.

The performance of a photobioreactor is strongly dependent on the light availability for each single cell of the dense microorganism suspension within the panels. Self-shading due to light absorption by the pigments causes an exponential decay in light intensity with culture depth. If a fixed density of the culture is assumed, the spatial light distribution in the suspension can be regulated by means of the depth of the panels.

Since the bacteria are able to use diffuse light as well, the intensity of the incident light can be controlled by the orientation of the photo-bioreactor. If the illuminated surfaces of the panels are facing the east–west direction the heat input during noontime is reduced and the utilization of the sunlight is improved because purple bacteria mainly use long wave radiation, which dominate the morning and evening sky. Direct radiation of comparably low intensity enters the reactor in the mornings and evenings. At noontime, when the sun emits light with a high content of UV rays, which might damage the culture, mainly diffuse light impinges on the reactor surfaces. The optimal spacing between the panels depends on light distribution and temperature. One panel of the photobioreactor consists of a frame covered by a transparent plate on both sides. The height of the panels is limited to 1 m in order to reduce the deflection of the transparent plates and to guarantee the gas tightness of the enclosed volume. For this reactor PMMA plates of 2 mm in thickness were used since they are able to withstand the hydrostatic pressure at the bottom with a fairly low deflection.

The flat panel reactor used in the study (Gebicki et al. 2010) consists of four parallel panels, which are arranged vertically as shown in Fig. 11.6. Each panel has an illuminated area of 2 m^2 and the plate thickness varies between 20 and 30 mm. This results in a total reactor volume of 112 l (4 × 28 l).

VI. Efficiency Analysis

Cost efficiency of any biofuel production from biomass depends primarily on the feedstock cost. According to a biomass cost index (BCI) analysis for biohydrogen production by sequential dark and photofermentation using a variety of feedstock, where production, logistics and pretreatment costs were compared, the biomass cost for sugar and starch containing feedstocks are lower than lignocellulosic feedstocks, due primarily to higher pretreatment costs associated with lignocellulosic biomass (Diamantopoulou et al. 2011). In selecting the biomass for biofuel production, one should also consider the local feedstock availability to decrease the logistics cost.

An exergy analysis to determine the efficiency of the integrated dark and photofermentation has been carried out for different feedstocks based on the HYVOLUTION process described by Claassen and de Vrije (2006). It was shown that exergy efficiency is influenced by the used feedstock, applied process parameters, and process and heat integration (Modaressi et al. 2010). It was demonstrated by Foglia and colleagues (Foglia et al. 2010, 2011) that with proper heat integration of pretreatment, thermophilic dark fermentation, and photofermentation steps, and recirculation of process effluents, biohydrogen production from variety of feedstocks, including PSP, barley straw, and thick juice, through the HYVOLUTION process is technically feasible. However, techno-economical analysis of two-stage biohydrogen production process using barley straw as feedstock revealed that with the current technologies it is around 20 times more expensive compared to the bioethanol production process, due mainly to the low productivity, low energy efficiency, and high cost of buffer and base required to control the pH (Ljunggren et al. 2011). Improvements in bioreactor design and process conditions to decrease the chemical and water requirements are essential for the transition of two-stage biohydrogen production process from lab-scale to industrial scales.

VII. Future Prospects

The aim of photofermentation is the optimal conversion of organic acids present in wastewater or fermentation effluents to hydrogen and carbon dioxide, in a prolonged stable operation. A yield above 75 % is targeted with a high productivity. A long-term stable operation is limited by environmental factors, as photobiological hydrogen production has to be carried out in outdoor conditions relying on natural sunlight for an energy-efficient process. The application of the process in technical scale demands:

- Stable microorganisms working under non sterile conditions with a high yield of capturing sunlight and with broad range of operational temperatures
- Productive feedstocks with higher yields
- Strain improvement to enhance the efficiency of the process under high/low light intensity and temperature variations
- Reactors with a high illuminated surface per ground space, cheap and mechanically stable construction material
- Plant design for the photofermentation, which guarantees well-defined operational conditions in all parts of the plant through development of a process control system. Design should include process control for pH, substrate concentration, temperature and liquid input/output control in each module.

The following items have to be investigated in the future to transfer the process to technical application:

(a) *Improvement of hydrogen productivity of the micro-organism(s)*: The specific production rate of the microorganism has to be further increased to decrease the demand on reactor area. Their adaptability to fluctuating environmental conditions, like light intensity and temperature, has to be improved to increase the stability of the process. It is now clear that changes in metabolism alone, e.g. nitrogen-fixing conditions to induce nitrogenase biosynthesis, is not enough to force the phototrophs towards approaching the theoretical limits of hydrogen production. It is necessary to create mutants for high, stable and long-lasting hydrogen productivity. The PNSB could be genetically improved to produce hydrogen during night time as well by dark fermentation, which would increase the daily productivity.

(b) *Designed co-cultures*: Use of designed co-cultures should also be studied in photofermentation. By this approach, the effect of environmental conditions on photo-biohydrogen production can be partially avoided by using selected strains with different properties such as thermo-tolerant versus mesophilic strains, or strains adapted to high/low light intensity.

(c) *Nutrient requirements*: There are certain nutrients such as Mo and Fe, which are necessary to activate the nitrogenase of PNSB and there are certain compounds like ammonium that are inhibitors of nitrogenase activity. However, more attention should be paid to find out the possible other inhibitors of hydrogen production metabolism of PNSB, especially those that may potentially be present in feedstocks of different biomass sources. These may include alcohols (like ethanol), polyphenols and aldehydes.

(d) *Use of immobilized microorganisms*: Entrapment of the photosynthetic bacteria into a solid support improves the productivity as a higher cell density can be used compared to the suspension cultures (Elkahlout 2011). This also increases the hydrogen yield, as there is only limited bacterial growth in immobilized systems. A large-scale outdoor operation requires a special reactor design, which should be investigated in future studies. The immobilized systems will also require a novel feeding protocol.

(e) *Modeling*: Development of predictive models for photofermentation for design and process control purposes that should include new bioreactors, microorganisms, and changing environmental conditions is needed.

(f) *Optimization of reactor design*: The geometry of the PBRs should meet the requirements for locations with different process conditions: low temperature combined with low sunlight intensity (Central/Eastern Europe) versus high temperature combined with high sunlight intensity (Southern Europe). Outdoor tests have to be performed at larger scale (some 10 m^2 reactors) to demonstrate their functional performance in the long run. Additionally, an improvement of the concept for the panel reactor is necessary to reduce the demand on construction material, to improve internal mixing and to allow for cooling. New construction materials have to be applied to minimize the Net Energy Ratio (NER). NER gives a monetary independent analysis for the viability of an energy conversion process. It is the relationship between the energy output and the energy content of all the materials of which the plant is constructed plus the energy needed for all operations, calculated for the lifetime of the system.

(g) *Optimization of the design of the plant*: Due to the inherent low productivity of the PNSB, a large illuminated area is necessary to produce an acceptable amount of hydrogen. Larger volumes of tubular and panel reactors may be obtained by repeating the original unit reactor operated in parallel. The concept for the arrangement of the reactors, the piping, and distribution of input and recirculation flow has to be demonstrated at large scale and has to be optimized as well. The optimal mixing of the reactor volume has to be ensured in the large-scale plant. This is simplified by dividing the whole reactor volume into smaller compartments, which are fed and mixed separately. The size of the compartments has to be defined. Smaller compartments are also favorable if a contamination emerges.

(h) *Concept for handling contamination*: Contamination may result in a breakdown of the hydrogen production. Since competing microorganisms will not stay in one module but spread out in the whole system a solution to handle contamination must be developed. The separation of the reactor into smaller units with independent feed inlets and effluent outlets might be a solution.

(i) *Heat economizing*: Operation of PBRs in outdoors is an energy requiring process due to the need of temperature control and recirculation. Exploitation of other renewable energy sources (sunlight, wind, geothermal energy, etc.) to supply energy to the PBR for recirculation or temperature control can be explored and implemented in the design of a biohydrogen plant. Heat economizing is necessary for the plant with the integration of cooling and heating streams in a heat exchange network.

(j) *Development of process control system*: Development of a complex process control system is inevitable. The variables that affect the PBR performance are pH, temperature, C/N ratio, and ammonium and acetate concentration in the effluent. In the lab-scale operation of the photobioreactor, the analysis of the feed and effluent may be carried out by means of an on-line HPLC. However, for a prototype PBR, practical devices and methods have to be developed to detect the control variables.

(k) *Stability in long-term operation*: Stability in long term operation has to be investigated at the molecular level. Metabolic understanding of the photo-sensing mechanism of bacteria may be helpful in the design of large-scale PBRs and process conditions. There are evidences that especially in complex media like real DFEs, photosynthetic PNSB conduct both dark and light fermentation in a competitive manner. If the light intensity, biomass concentration and temperature are not within the favorable intervals, bacteria switch off the photofermentative hydrogen production and proceed with dark fermentation alone. The understanding of hydrogen production switch-on/off mechanism may be beneficial for the development of long-term stable operations.

Acknowledgements

This work was supported by European Cooperation in Science and Technology (COST Action 841), European Commission – Research: The Sixth Framework Program for Research and Technological Development Sustainable Energy Systems EU FP6-SES IP HYVOLUTION (contract no. 019825), Turkish State Planning Organization (DPT) (BAP-08-11-DPT.2005K120600), Turkish Scientific Research Council (TÜBİTAK) (108T455) and Middle East Technical University Research Fund.

References

Abd-Alla MH, Morsy FM, El-Enany AWE (2011) Hydrogen production from rotten dates by sequential three stages fermentation. Int J Hydrog Energy 36:13518–13527

Aceves-Lara CA, Latrille E, Bernet N, Buffière P, Steyer JP (2008) A pseudo-stoichiometric dynamic model of anaerobic hydrogen production from molasses. Water Res 42:2539–2550

Adessi A, McKinlay JB, Harwood CS, De Philippis R (2012) A *Rhodopseudomonas palustris nifA** mutant produces H_2 from NH_4^+-containing vegetable waste. Int J Hydrog Energy 37:15893–15900

Afşar N (2012) A Global approach to the hydrogen production, carbon assimilation and nitrogen metabolism of Rhodobacter capsulatus by physiological and microarray analyses. Ph.D. thesis. Graduate School of Natural and Applied Sciences, METU, Ankara, Turkey. http://etd.lib.metu.edu.tr/upload/12615030/index.pdf

Afşar N, Özgür E, Gürgan M, Akkose S, Yücel M, Gündüz U, Eroglu I (2011) Hydrogen productivity of photosynthetic bacteria on dark fermenter effluent of potato steam peels hydrolysate. Int J Hydrog Energy 36:432–438

Akköse S, Gündüz U, Yücel M, Eroglu I (2009) Effects of ammonium ion, acetate and aerobic conditions on hydrogen production and expression levels of nitrogenase genes in *Rhodobacter sphaeroides* OU001. Int J Hydrog Energy 34:8818–8827

Androga DD, Özgür E, Eroglu I, Gündüz U, Yücel M (2011a) Significance of carbon to nitrogen ratio on the long-term stability of continuous photofermentative hydrogen production. Int J Hydrog Energy 36:15583–15594

Androga DD, Özgür E, Gündüz U, Yücel M, Eroglu I (2011b) Factors affecting the long-term stability of biomass and hydrogen productivity in outdoor photofermentation. Int J Hydrog Energy 36:11369–11378

Androga DD, Özgür E, Eroglu I, Gündüz U, Yücel M (2012) Amelioration of photofermentative hydrogen production from molasses dark fermenter effluent by zeolite-based removal of ammonium ion. Int J Hydrog Energy 37:16421–16429

Argun H, Kargi F, Kapdan IK (2009) Hydrogen production by combined dark and light fermentation of ground wheat solution. Int J Hydrog Energy 34: 4304–4311

Avcioglu SG, Özgür E, Eroglu I, Yücel M, Gündüz U (2011) Biohydrogen production in an outdoor panel photobioreactor on dark fermentation effluent of molasses. Int J Hydrog Energy 36:11360–11368

Azbar N, Cetinkaya-Dokgöz FT (2010) The effect of dilution and L-malic acid addition on bio-hydrogen production with *Rhodopseudomonas palustris* from effluent of an acidogenic anaerobic reactor. Int J Hydrog Energy 35:5028–5033

Azbar N, Çetinkaya Dokgöz FT, Keskin T, Korkmaz KS, Syed HM (2009a) Continuous fermentative hydrogen production from cheese whey wastewater under thermophilic anaerobic conditions. Int J Hydrog Energy 34:7441–7447

Azbar N, Dokgöz FT, Keskin T, Eltem R, Korkmaz KS, Gezgin Y, Akbal Z, Öncel S, Dalay MC, Gönen Ç, Tutuk F (2009b) Comparative evaluation of biohydrogen production from cheese whey wastewater under thermophilic and mesophilic anaerobic conditions. Int J Green Energy 6:192–200

Azbar N, Dokgöz FTÇ, Peker Z (2009c) Optimization of basal medium for fermentative hydrogen production from cheese whey wastewater. Int J Green Energy 6:371–380

Biebl H, Pfennig N (1981) Isolation of members of the family Rhodosprillaceae. In: Starr MP, Stolp H, Trüper HG, Balows A, Schlegel HG (eds) The prokaryotes, vol 1. Springer-Verlag, New York, pp 267–273

Boran E (2011) Process development for continuous photofermentative hydrogen production. Thesis, Graduate School of Natural and Applied Sciences, METU, Ankara, Turkey. (http://etd.lib.metu.edu.tr/upload/12612955/index.pdf)

Boran E, Özgür E, Yücel M, Gündüz U, Eroglu I (2012a) Biohydrogen production by *Rhodobacter capsulatus* in solar tubular photobioreactor on thick juice dark fermenter effluent. J Clean Prod 31:150–157

Boran E, Özgür E, Yücel M, Gündüz U, Eroglu I (2012b) Biohydrogen production by *Rhodobacter capsulatus* Hup− mutant in pilot solar tubular photobioreactor. Int J Hydrog Energy 37:16437–16445

Bouallagui H, Touhami Y, Cheikh RB, Hamdi M (2005) Bioreactor performance in anaerobic digestion of fruit and vegetable wastes. Process Biochem 40:989–995

Cabrera F, Lopez R, Martinez-Bordiu A, Dupuy de Lome E, Murillo JM (1996) Land treatment of olive oil mill wastewater. Int Biodeter Biodegrad 38:215–255

Chini Zittelli G, Lavista F, Bastianini A, Rodolfi L, Vincenzini M, Tredici MR (1999) Production of eicosapentaenoic acid by *Nannochloropsis sp.* cultures in outdoor tubular photobioreactors. J Biotechnol 70:299–312

Claassen PAM, de Vrije T (2006) Non-thermal production of pure hydrogen from biomass: HYVOLUTION. Int J Hydrog Energy 31:1416–1423

Claassen PAM, Budde MAW, Niel EWJV, de Vrije T (2005) Utilization of biomass for hydrogen fermentation. In: Lens P, Westermann P, Haberbauer M, Monero A (eds) Biofuels for fuel cells: biomass fermentation towards usage in fuel cells. IWA Publishing, London, pp 221–230

de Vrije T, Bakker RR, Budde MA, Lai MH, Mars AE, Claassen PAM (2009) Efficient hydrogen production from the lignocellulosic energy crop *Miscanthus* by the extreme thermophilic bacteria *Caldicellulosiruptor saccharolyticus* and *Thermotoga neapolitana*. Biotechnol Biofuels 2:12

Diamantopoulou LK, Karaoglanoglou LS, Koukios EG (2011) Biomass cost index: mapping biomass-to-biohydrogen feedstock costs by a new approach. Bioresour Technol 102:2641–2650

Djomo S, Humbert S (2008) Life cycle assessment of hydrogen produced from potato steam peels. Int J Hydrog Energy 33:3067–3072

Elkahlout KE (2011) Phototrophic hydrogen production by agar-immobilized *Rhodobacter capsulatus*. Ph.D. thesis, Graduate School of Natural and Applied Sciences, METU, Ankara, Turkey. https://etd.lib.metu.edu.tr/upload/12613033/index.pdf

Ergüder TH, Güven E, Demirer GN (2000) Anaerobic treatment of olive mill wastes in batch reactors. Process Biochem 36:243–248

Eroglu E, Melis A (2011) Photobiological hydrogen production: recent advances and state of the art. Bioresour Technol 102:8403–8413

Eroglu E, Gündüz U, Yücel M, Turker L, Eroglu I (2004) Photobiological hydrogen production by using olive mill wastewater as a sole substrate source. Int J Hydrog Energy 29:163–171

Eroglu E, Gündüz U, Yücel M, Turker L, Eroglu I (2006) Biological hydrogen production from olive mill wastewater with two stage processes. Int J Hydrog Energy 31:1527–1535

Eroglu E, Eroglu I, Gündüz U, Yücel M (2008a) Effect of clay pretreatment on photofermentative hydrogen production from olive mill wastewater. Bioresour Technol 99:6799–6808

Eroglu I, Tabanoglu A, Gündüz U, Eroglu E, Yücel M (2008b) Hydrogen production by *Rhodobacter sphaeroides* O.U.001 in a flat plate solar bioreactor. Int J Hydrog Energy 33:531–541

Eroglu E, Eroglu I, Gündüz U, Yücel M (2009a) Treatment of olive mill wastewater by different physicochemical methods and the utilization of their liquid effluents for biological hydrogen production. Biomass Bioenergy 33:701–705

Eroglu E, Eroglu I, Gündüz U, Yücel M (2009b) Comparison of the physicochemical characteristics and photofermentative hydrogen production potential of wastewaters produced from different olive-oil mills in Western-Anatolia Turkey. Biomass Bioenergy 33(4):706–711

Eroglu E, Gündüz U, Yücel M, Eroglu I (2010) Photosynthetic bacterial growth and productivity under continuous illumination or diurnal cycles with olive mill wastewater as feedstock. Int J Hydrog Energy 35:5293–5300

Eroglu E, Gündüz U, Yücel M, Eroglu I (2011) Effect of Fe and Mo addition on the photofermentative hydrogen production from olive mill wastewater. Int J Hydrog Energy 36:5895–5903

Fascetti E, D'Addario E, Todini O, Robertiello A (1998) Photosynthetic hydrogen evolution with volatile organic acids derived from the fermentation of source selected municipal solid wastes. Int J Hydrog Energy 23:753–760

Favinger J, Stadtwald R, Gest H (1989) *Rhodospirillum centenum*, sp. nov., a thermotolerant cyst-forming anoxygenic photosynthetic bacterium. A Van Leeuw J Microbiol 55:291–296

Foglia D, Ljunggren M, Wukovits W, Friedl A, Zacchi G, Urbaniec K, Markowski M (2010) Integration studies on a two-stage fermentation process for the production of biohydrogen. J Clean Prod 18:S72–S80

Foglia D, Wukovits W, Friedl A, Ljunggren M, Zacchi G, Urbaniec K, Markowski M (2011) Effects of feedstocks on the process integration of biohydrogen production. Clean Technol Environ 13:547–558

Gebicki J, Modigell M, Schumacher M, van der Burg J, Roebroeck E (2010) Comparison of two reactor concepts for anoxygenic H_2 production by *Rhodobacter capsulatus*. J Clean Prod 18:S36–S42

Gonzalez MD, Moreno E, Quvedo-Sarmiento F, Ramos- Cormenzana A (1990) Studies on the antibacterial activity of wastewaters from olive oil mills (alpechin): inhibitory activity of phenolic and fatty acids. Chemosphere 20:423–432

Guo WQ, Ren NQ, Wang XJ, Xiang WS, Meng ZH, Ding J, Qu YY, Zhang LS (2008) Biohydrogen production from ethanol-type fermentation of molasses in an expanded granular sludge bed (EGBS) reactor. Int J Hydrog Energy 33:4981–4988

Gürgan Doğan M (2011) Microarray analysis of the effects of heat and cold stress on hydrogen production metabolism of Rhodobacter capsulatus. M. Sc. thesis Graduate School of Natural and Applied Sciences, METU, Ankara, Turkey. http://etd.lib.metu.edu.tr/upload/12613668/index.pdf

Hampannavar US, Shivayogimath CB (2010) Anaerobic treatment of sugar industry wastewater by upflow anaerobic sludge blanket reactor at ambient temperature. Int J Environ Sci 1:631–639

Jansson C, Westerbergh A, Zhang J, Hu X, Sun C (2009) A potential biofuel crop in (the) People's Republic of China. Appl Energy 86:S95–S99

Jeong HS, Jouanneau Y (2000) Enhanced nitrogenase activity in strains of *Rhodobacter capsulatus* that overexpress the *rnf* genes. J Bacteriol 182:1208–1214

Kapdan IK, Kargi F (2006) Bio-hydrogen production from waste materials. Enzyme Microbiol Technol 38:569–582

Kargi F, Pamukoglu MY (2009) Dark fermentation of ground wheat starch for bio-hydrogen production by fed-batch operation. Int J Hydrog Energy 34:2940–2946

Kars G, Gündüz U (2010) Towards a super H_2 producer: improvements in photofermentative biohydrogen production by genetic manipulations. Int J Hydrog Energy 35:6646–6656

Kars G, Gündüz U, Rakhely G, Yücel M, Eroglu I, Kovacs KL (2008) Improved hydrogen production by uptake hydrogenase deficient mutant strain of *Rhodobacter sphaeroides* O.U.001. Int J Hydrog Energy 33:3056–3060

Keskin T, Abo-Hashesh M, Hallenbeck PC (2011) Photofermentative hydrogen production from wastes. Bioresour Technol 102:8557–8568

Kim EJ, Kim MS, Lee JK (2008) Hydrogen evolution under photoheterotrophic and dark fermentative conditions by recombinant *Rhodobacter sphaeroides* containing the genes for fermentative pyruvate metabolism of *Rhodospirillum rubrum*. Int J Hydrog Energy 33:5131–5136

Kim MS, Kim DH, Son HN, Ten LN, Lee JK (2011) Enhancing photo-fermentative hydrogen production by *Rhodobacter sphaeroides* KD131 and its PHB synthase deleted-mutant from acetate and butyrate. Int J Hydrog Energy 36:13964–13971

Kiritsakis A (1991) Olive oil. American Oil Chemists' Society, Champaign

Koku H, Eroglu I, Gündüz U, Yücel M, Türker L (2002) Aspects of the metabolism of hydrogen production by *Rhodobacter sphaeroides*. Int J Hydrog Energy 27:1315–1329

Laurinavichene TV, Tekucheva DN, Laurinavichius KS, Ghirardi ML, Seibert M, Tsygankov AA (2008) Towards the integration of dark and photo fermentative waste treatment. 1. Hydrogen photoproduction by purple bacterium *Rhodobacter capsulatus* using potential products of starch fermentation. Int J Hydrog Energy 33:7020–7026

Ljunggren M, Wallberg O, Zacchi G (2011) Techno-economic comparison of a biological hydrogen process and a 2nd generation ethanol process using barley straw as feedstock. Bioresour Technol 102:9524–9531

Lo YC, Chen CY, Lee CM, Chang JS (2011) Photo fermentative hydrogen production using dominant components (acetate, lactate, and butyrate) in dark fermentation effluents. Int J Hydrog Energy 36:14059–14068

Mars AE, Veuskens T, Budde MAW, van Doeveren PFNM, Lips SJ, Bakker RR, de Vrije T, Claassen PAM (2010) Biohydrogen production from untreated and hydrolyzed potato steam peels by the extreme thermophiles *Caldicellulosiruptor saccharolyticus* and *Thermotoga neapolitana*. Int J Hydrog Energy 35:13206–13213

McKinley LB, Harwood CS (2010) Carbon dioxide as a central redox cofactor recycling mechanism in bacteria. PNAS 107:11669–11675

Mitsui A, Philips EJ, Kumazawa S, Reddy KJ, Ramachandran S, Matsunaga T, Haynes L, Ikemoto H (1983) Progress in research toward outdoor biological hydrogen production using solar energy, sea water, and marine photosynthetic microorganisms. Ann N Y Acad Sci 413:514–530

Modaressi A, Wukovits W, Foglia D, Friedl A (2010) Effect of process integration on the exergy balance of a two-stage process for fermentative hydrogen production. J Clean Prod 18:S63–S71

Önder O, Sunar SA, Selamoglu N, Daldal F (2010) A glimpse into the proteome of phototrophic bacterium *Rhodobacter capsulatus*. Adv Exp Med Biol 675:179–209

Özgür E, Peksel B (2013) Biohydrogen production from barley straw hydrolysate through sequential dark and photofermentation. J Clean Prod 52:14–20

Özgür E, Mars A, Peksel B, Lowerse A, Afşar N, Vrije T, Yücel M, Gündüz U, Claassen PAM, Eroglu I (2010a) Biohydrogen production from beet molasses by sequential dark and photofermentation. Int J Hydrog Energy 35:511–517

Özgür E, Afşar N, Vrije T, Yücel M, Gündüz U, Claassen PAM, Eroglu I (2010b) Potential use of thermophilic dark fermentation effluents in photofermentative hydrogen production by *Rhodobacter capsulatus*. J Clean Prod 18:S23–S28

Özgür E, Uyar B, Öztürk Y, Yücel M, Gündüz U, Eroglu I (2010c) Biohydrogen production by *Rhodobacter capsulatus* on acetate at fluctuating temperatures. Resour Conserv Recy 54:310–314

Özgür E, Uyar B, Gürgan M, Yücel M (2010d) Hydrogen production by Hup- mutant and wild type strains of *Rhodobacter capsulatus* on dark fermenter effluent of sugar beet thick juice in batch and continuous photobioreactors. In: Detlef S, Grube T (eds) Hydrogen production technologies, part-1, Proceedings of WHEC 2010, Essen. Forschunszentrum Jülich GmbH, Verlag, GmbH, pp 228–233

Özkan E, Uyar B, Özgür E, Yücel M, Eroglu I, Gündüz U (2012) Photofermentative hydrogen production using dark fermentation effluent of sugar beet thick juice in outdoor conditions. Int J Hydrog Energy 37:2044–2049

Ozmihci S, Kargi F (2010) Bio-hydrogen production by photo-fermentation of dark fermentation effluent with intermittent feeding and effluent removal. Int J Hydrog Energy 35:6674–6680

Özsoy B (2012) Hydrogen and poly-hydroxy butyric acid production and expression analyses of related genes in *Rhodobacter capsulatus* at different acetate concentrations. M.Sc. thesis, Graduate School of Natural and Applied Sciences, METU, Ankara, Turkey. http://etd.lib.metu.edu.tr/upload/12614076/index.pdf

Öztürk Y, Yücel M, Daldal F, Mandaci S, Gündüz U, Turker L, Eroglu I (2006) Hydrogen production by using *Rhodobacter capsulatus* mutants with genetically modified electron transfer chains. Int J Hydrog Energy 31:1545–1552

Öztürk Y, Gökce A, Peksel B, Gürgan M, Özgür E, Gündüz U, Eroglu I, Yücel M (2012) Hydrogen production properties of *Rhodobacter capsulatus* with genetically modified redox balancing pathways. Int J Hydrog Energy 37:2014–2020

Pekgoz G, Gündüz U, Eroglu I, Yücel M, Kovacs K, Rakhley G (2011) Effect of inactivation of genes involved in ammonium regulation on the biohydrogen production of *Rhodobacter capsulatus*. Int J Hydrog Energy 36:13536–13546

Peksel B (2012) Proteome analysis of hydrogen production mechanism of *Rhodobacter capsulatus* grown on different growth conditions. M.Sc. thesis, Graduate School of Natural and Applied Sciences, METU, Ankara, Turkey (http://etd.lib.metu.edu.tr/upload/12614133/index.pdf)

Qureshi N, Saha BC, Dien B, Hector RE, Cotta MA (2010) Production of butanol (a biofuel) from agricultural residues: part I – Use of barley straw hydrolysate. Biomass Bioenergy 34:559–565

Rai PK, Singh SP, Asthana RK (2012) Biohydrogen production from cheese whey wastewater in a two-step anaerobic process. Appl Biochem Biotechnol 167:1540–1549

Ramos-Cormenzana A, Jimenez B, Pareja G (1996) Antimicrobial activity of olive mill wastewaters (alpechin) and biotransformed olive oil mill wastewater. Int Biodeter Biodegrad 38:283–290

Ren N, Li J, Li B, Wang Y, Liu S (2006) Biohydrogen production from molasses by anaerobic fermentation with a pilot-scale bioreactor system. Int J Hydrog Energy 31:2147–2157

Ren N, Wang A, Cao G, Xu J, Gao L (2009) Bioconversion of lignocellulosic biomass to hydrogen: potential and challenges. Biotechnol Adv 27:1051–1060

Rocha JS, Barbosa MJ, Wijffels RH (2001) Hydrogen production by photosynthetic bacteria: culture media, yields and efficiencies. In: Miyake J, Matsunaga T, San Pietro A (eds) Biohydrogen II – an approach to environmentally acceptable technology. Elsevier Science Ltd., London, pp 3–32

Sabbah I, Marsook T, Basheer S (2004) The effect of pretreatment on anaerobic activity of olive mill wastewater using batch and continuous systems. Process Biochem 39:1947–1951

Sasikala K, Ramana CV (1991) Photoproduction of hydrogen from waste water of a lactic acid fermentation plant by a purple non-sulfur photosynthetic bacterium, *Rhodobacter sphaeroides* O.U. 001. Indian J Exp Biol 29:74–75

Seifert K, Waligorska M, Laniecki M (2010a) Hydrogen generation in photobiological process from diary wastewater. Int J Hydrog Energy 35:9624–9629

Seifert K, Waligorska M, Laniecki M (2010b) Brewery wastewaters in photobiological hydrogen generation in presence of *Rhodobacter sphaeroides* O.U.001. Int J Hydrog Energy 35:4085–4091

Shi X, Yu H (2006) Continuous production of hydrogen from mixed volatile fatty acids with *Rhodopseudomonas capsulata*. Int J Hydrog Energy 31:1641–1647

Sigurbjornsdottir MA, Orlygsson J (2012) Combined hydrogen and ethanol production from sugars and lignocellulosic biomass by *Thermoanaerobacterium* AK_{54}, isolated from hot spring. Appl Energy 97:785–791

Singh SP, Srivastava SC, Pandley KD (1994) Hydrogen production by *Rhodopseudomonas* at the expense of vegetable starch, sugarcane juice and whey. Int J Hydrog Energy 19:437–440

Stevens P, Vertonghen C, De Vos P, De Ley J (1984) The effect of temperature and light intensity on hydrogen production by different *Rhodopseudomonas capsulata* strains. Biotechnol Lett 6:277–282

Su H, Cheng J, Zhou J, Song W, Cen K (2009) Improving hydrogen production from cassava starch by combination of dark and photo fermentation. Int J Hydrog Energy 34:1780–1786

Tichi MA, Tabita FR (2000) Maintenance and control of redox poise in *Rhodobacter capsulatus* strains deficient in the Calvin-Benson-Bassham pathway. Arch Microbiol 174:322–333

Tredici MR (2004) Mass production of microalgae: photobioreactors. In: Richmond A (ed) Handbook of microalgal culture: biotechnology and applied phycology. Blackwell Publishing, Oxford, pp 178–214

Tsagaraki E, Lazarides HN, Petrotos KB (2006) Olive mill wastewater treatment. In: Oreopoulou V, Russ W (eds) Utilization of by-products and treatment of waste in the food industry. Springer, New York, pp 133–157

Türkarslan S, Yigit DO, Aslan K, Eroglu I, Gündüz U (1998) Photobiological hydrogen production by *Rhodobacter sphaeroides* O.U.001 by utilization of waste water from milk industry. In: Zaborsky OR (ed) Biohydrogen. Plenum Press, London, pp 151–156

Uyar B, Eroglu I, Yücel M, Gündüz U, Turker L (2007) Effect of light intensity, wavelength and illumination protocol on hydrogen production in photobioreactors. Int J Hydrog Energy 32:4670–4677

Uyar B, Schumacher M, Gebicki J, Modigell M (2009) Photoproduction of hydrogen by *Rhodobacter capsulatus* from thermophilic fermentation effluent. Bioprocess Biosyst Eng 32:603–606

Vatsala TM, Ramasamy V (1987) Photo-hydrogen production from distillery waste. In: Veziroğlu TN (ed) Proceedings of the 8th Miami international conference on alternative energy sources, vol 2. Miami, Florida, pp 618–623

Verrier D, Ray F, Albagnac G (1987) Two-phase methanization of solid vegetable wastes. Biol Waste 22:163–177

Vignais PM, Magnin J-P, Willison JC (2006) Increasing biohydrogen production by metabolic engineering. Int J Hydrog Energy 31:1478–1483

Visioli F, Romani A, Mulinacci N, Zarini S, Conte D, Vincieri FF, Galli C (1999) Antioxidant and other biological activities of olive mill waste waters. J Agric Food Chem 47:3397–3401

Wang X, Jin B (2009) Process optimization of biological hydrogen production from molasses by a newly isolated *Clostridium butyricum* w5. J Biosci Bioeng 107:138–144

Weaver PF, Wall JD, Gest H (1975) Characterization of *Rhodopseudomonas capsulata*. Arch Microbiol 105:207–216

Yakunin AF, Hallenbeck PC (1998) Short-term regulation of nitrogenase activity by NH_4^+ in *Rhodobacter capsulatus*: multiple in vivo nitrogenase responses to NH_4^+ addition. J Bacteriol 180:6392–6395

Yang H, Guo L, Liu F (2010) Enhanced bio-hydrogen production from corncob by a two-step process: dark- and photo-fermentation. Bioresour Technol 101:2049–2052

Yesilada Ö, Sık S, Sam M (1999) Treatment of olive oil mill wastewater with fungi. Turk J Biol 23:231–240

Yetis M, Gündüz U, Eroglu I, Yücel M, Turker L (2000) Photoproduction of hydrogen from sugar refinery wastewater by *Rhodobacter sphaeroides* O.U.001. Int J Hydrog Energy 25:1035–1041

Yokoi H, Maki R, Hirose J, Hayashi S (2002) Microbial production of hydrogen from starch-manufacturing wastes. Biomass Bioenergy 22:389–395

Zhu H, Suzuki T, Tsygankov AA, Asada Y, Miyake J (1999) Hydrogen production from tofu wastewater by *Rhodobacter sphaeroides* immobilized in agar gels. Int J Hydrog Energy 24:305–310

Zong W, Yu R, Zhang P, Fan M, Zhou Z (2009) Efficient hydrogen gas production from cassava and food waste by a two-step process of dark fermentation and photo-fermentation. Biomass Bioenergy 33:1458–1463

Zürrer H, Bachofen R (1979) Hydrogen production by the photosynthetic bacterium *Rhodospirillum rubrum*. Appl Environ Microbiol 37:789–793

Chapter 12

Photosynthesis and Hydrogen Production in Purple Non Sulfur Bacteria: Fundamental and Applied Aspects

Alessandra Adessi
Department of Agrifood Production and Environmental Sciences, University of Florence, 50144 Florence, Italy

and

Roberto De Philippis*
Department of Agrifood Production and Environmental Sciences, University of Florence, 50144 Florence, Italy

Institute of Chemistry of Organometallic Compounds (ICCOM), CNR, 50019 Sesto Fiorentino, Florence, Italy

Summary

Light-dependent hydrogen production by purple non sulfur bacteria (PNSB) has been studied for several decades. However the exact route that energy takes from the moment a photon is absorbed to the formation of a molecule of hydrogen is quite complex. The aim of this

*Author for correspondence, e-mail: roberto.dephilippis@unifi.it

D. Zannoni and R. De Philippis (eds.), Microbial BioEnergy: Hydrogen Production, Advances in Photosynthesis and Respiration 38, pp. 269–290, DOI 10.1007/978-94-017-8554-9_12,

chapter is to review the researches carried out on the metabolic processes related to hydrogen production in PNSB, in particular stressing the issues related with the efficiency in the conversion of the energy deriving from the light in the energy-rich H_2 molecule produced. The metabolic processes that bring form the light capturing to hydrogen production are described, with the relative bottlenecks and hurdles.

The information currently available on the light distribution in various kind of photobioreactors are also reviewed, mainly focusing on the photosynthetic efficiency and on the efficiency in substrate conversion to H_2 obtained in laboratory and outdoor experiments.

From these data, it comes out how many different cellular processes can interact and affect photosynthetic efficiency and how complex is the route that brings from light energy to hydrogen energy.

I. Introduction

In 1949, Howard Gest and Martin Kamen first observed light dependent hydrogen production by the purple non sulfur bacterium (PNSB) *Rhodospirillum rubrum* (Gest and Kamen 1949). The dependence of H_2 production on nitrogenase activity was also observed, opening the way for research in this field. Since 1949, the metabolic routes that bring to the production of a molecule of hydrogen gas by PNSB have been thoroughly investigated and elucidated.

The process will be analyzed step by step in this chapter, stressing the complexity of the different metabolic pathways that interact with and influence the production of hydrogen.

II. The H_2 Production Process in Purple Bacteria

PNSB produce hydrogen via nitrogenase (see also Bothe et al., Chap. 6 of this volume). Although the major role of nitrogenase is to fix molecular nitrogen to ammonia, giving molecular hydrogen as a by-product (Eq. 12.1), the enzyme can also work in absence of molecular nitrogen and give hydrogen as the sole product (Eq. 12.2).

$$N_2 + 8H^+ + 8e^- + 16ATP \rightarrow 2NH_3 + H_2 + 16ADP + 16P_i \quad (12.1)$$

$$8H^+ + 8e^- + 16ATP \rightarrow 4H_2 + 16ADP + 16P_i \quad (12.2)$$

Nitrogenase is a two-protein complex consisting of a dinitrogenase containing Fe and Mo as cofactors and having a molecular weight of 250 kDa, and of a dinitrogenase reductase (containing Fe) of about 70 kDa. Mo-nitrogenase is the most common and the most efficient nitrogenase for converting N_2 to NH_3 (12.1), but other two isozymes have been described (Eady 1996) that contain Fe or V as cofactors and *Rhodopseudomonas palustris* is the only purple bacterium that encodes all three of them (Larimer et al. 2004). Those alternative nitrogenases are less efficient in reducing N_2 and more efficient in producing H_2 in nitrogen fixing conditions; however, they are produced only in case of lack of Mo and presence of Fe or V (Oda et al. 2005).

PNSB synthesize large amounts of nitrogenase (up to 2 % of cellular protein), not only because of its crucial role in cellular metabolism, but also because the enzyme is known as a slow catalyst (its turnover time per electron is ~5 s^{-1}), so a larger amount of enzyme provides a larger amount of nitrogen fixed. Although the mechanism by which nitrogenase reduces its substrates has been deeply

Abbreviations: BChl – Bacteriochlorophyll; Cyt bc_1 – Cytochrome bc_1 complex; Cyt c_2 – Cytochrome c_2; Fd – Ferredoxin; LH – Light harvesting; PE – Photosynthetic efficiency; PHB – Poly-β-hydroxybutyrate; PNSB – Purple non sulfur bacterium/bacteria; PSU – Photosynthetic unit; RC – Reaction center; SC – Substrate conversion studied during the last decades -the first comprehensive scheme was proposed by Thorneley and Lowe in 1985- it proves to be very complex so that it still has not been completely clarified (Seefeldt et al. 2009).

H_2 production via nitrogenase has a specific activity one order of magnitude lower than Ni–Fe hydrogenase typical of oxygenic photosynthetic organisms (e.g. 1.3 mmol mg protein^{-1} min^{-1} for Mo-nitrogenase). Even so, in vivo H_2 production rates by nitrogenase-utilizing PNSB are comparable with those by hydrogenase-utilizing oxygenic phototrophs (Harwood 2008).

The H_2 produced during N_2-fixation by nitrogenase can be reused by uptake-hydrogenases (Vignais and Billoud 2007). High uptake-hydrogenase activities have been observed in cells possessing an active nitrogenase; the hydrogen produced by the nitrogenase stimulated the activity of hydrogenase in growing cells even though the synthesis of hydrogenase is not closely linked to the synthesis of nitrogenase (Colbeau et al. 1980).

Uptake-hydrogenases usually need to be genetically deleted to accumulate larger amounts of H_2. However, a few H_2-producing hydrogenases in PNSB have been described. *Rs. rubrum* and *Rp. palustris* BisB18 have a Ni–Fe hydrogenase that couples H_2 production to CO or formate oxidation (Fox et al. 1996; Oda et al. 2008).

A. Electron Transport to Nitrogenase

Ferredoxins are small proteins that function as electron carriers and take part in various metabolic processes, such as photosynthesis, nitrogen fixation, steroid hydroxylation, and degradation of aromatic compounds (Bruschi and Guerlesquin 1988). *Rhodobacter capsulatus* has been shown to synthesize six soluble ferredoxins which can be divided into two groups according to the number of [2Fe-2S] clusters present: FdI, FdII, and FdIII constitute the group of dicluster ferredoxins; FdIV, FdV, and FdVI contain a single [2Fe-2S] cluster. The genes encoding FdI, FdIII, FdIV, and FdV were localized within three *nif*-regulated operons (Moreno-Vivian et al. 1989; Schatt et al. 1989; Grabau et al. 1991; Willison et al. 1993), indicating that these ferredoxins participate in nitrogen fixation. FdI has been shown to serve as the physiological electron donor to nitrogenase (Jouanneau et al. 1995; Naud et al. 1996).

Göbel (1978) calculated that 1.5 photons are needed to synthesize 1 ATP molecule at 860 nm. Moreover, Miyake (1998) calculated that at 860 nm 11 photons are required for the production of a single molecule of H_2. Considering reaction (12.2), 4 ATP molecules are consumed for every H_2 molecule produced (i.e. 6 photons at 860 nm, according to Göbel 1978); the remaining 5 photons are most probably needed for the ferredoxin mediated (ATP consuming) electron transfer to nitrogenase.

B. The Role of ATP in Nitrogenase Activity

As it has been above described, nitrogenase catalyses a very expensive reaction in terms of ATP consumption to synthesize hydrogen (12.2), being the latter an endothermic reaction requiring external energy to overcome the positive free energy barrier ($G_0 = 75$ kJ). The balance is of 4 ATP molecules and 2 electrons consumed per molecule of H_2 produced. Besides these 4 ATP molecules, an additional number of ATP molecules is needed, through an energy requiring process, for reducing the ferredoxins that act as electron donors to nitrogenase (see Sect. II.A).

In detail, it has been schematically reported that 2 molecules of ATP are needed for every electron transfer from the Fe-protein nitrogenase subunit to the Mo-Fe protein subunit (in Mo-nitrogenase). However, experimental results demonstrated that the actual ratio of ATP hydrolyzed per couple of protons reduced is ~4.5 (Eady 1996).

Furthermore, it looks like the energy gained from the hydrolysis of those ATP molecules is used not only to overcome the thermodynamic barrier for N_2 reduction, but also has a role in the kinetic mechanism, not yet completely understood (Rees and Howard 2000).

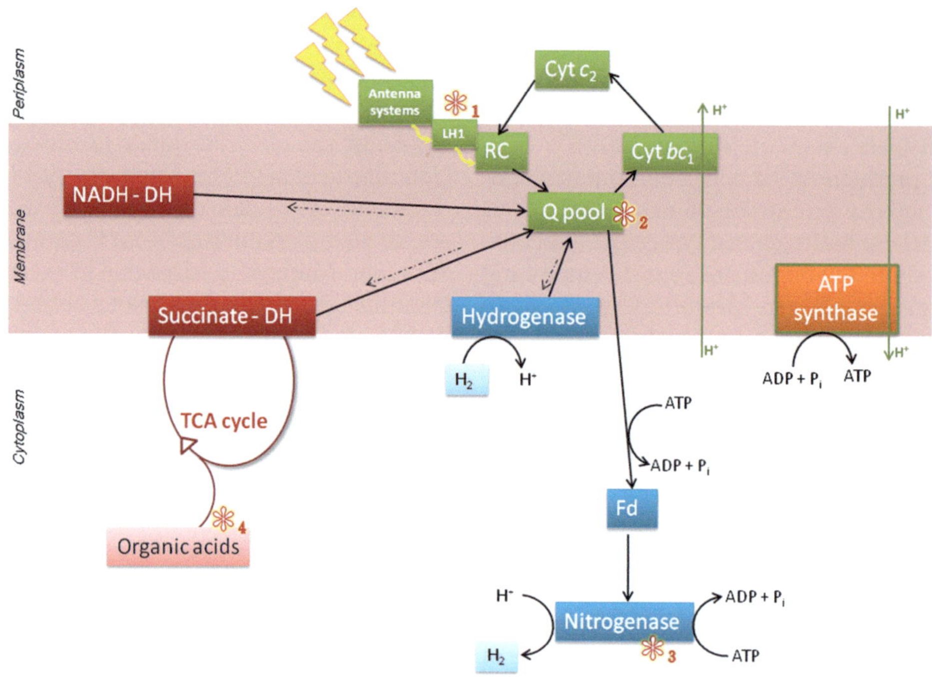

Fig. 12.1. Main processes related to hydrogen production, under photoheterotrophic growth in non-nitrogen fixing conditions: anoxygenic photosynthesis (in *green*), ATP synthesis (in *orange*), TCA cycle (in *red*), hydrogenase and nitrogenase activities (in *blue*). The straight *black arrows* indicate the electron flow. The *dotted lines* indicate minor electron flow. The lightning symbols indicate light excitation. The waving *arrows* indicate light energy transfers. The straight green lines indicate proton translocation. Abbreviations: *Cyt* bc_1, cytochrome bc_1 complex; *Cyt* c_2, cytochrome c_2, *Fd* ferredoxin, *RC*, Reaction center, *Succinate – DH* succinate dehydrogenase; *NADH-DH* NADH dehydrogenase. Numbered 1–4* signs will be discussed in Sect. VI of this chapter (Scheme modified from Adessi and De Philippis 2012, with kind permission from Springer Science+Business Media BV).

During photofermentation, cells are in photoheterotrophic conditions, i.e. the ATP is formed via anoxygenic photosynthesis, while the reducing power is derived by the catabolism of organic substrates, as it is schematically shown in Fig. 12.1.

The main energy carrier of a cell, ATP, is thus only synthesized by photosynthesis and many molecules are consumed while nitrogenase is active; this makes the parameters that refer to light and photosynthesis of crucial importance for the nitrogenase-mediated hydrogen production process in PNSB.

For this reason in recent years a large number of researches was aimed at optimizing illumination protocols in order the parameters capable of increasing photosynthetic efficiency (see Sect. IV).

III. Anoxygenic Photosynthesis

Generally speaking, in the reaction center of a photosynthetic organism, the excitation energy of photons is used to move one electron from one chemical compound (donor) towards another compound (acceptor). Thus, the reaction center is where the separation of charges actually occurs: the excitation energy is then stored into an energy-rich chemical bond.

The specificity of purple bacteria is given by their ability to form their energy carrier (ATP) in absence of oxygen by using sunlight as a source of energy (Imhoff 1995). Indeed, in anoxygenic photosynthesis, a special pair of bacteriochlorophylls, $[BChl]_2$, are both the primary electron donors and the final electron acceptors, as this photosynthetic process is operationally defined as cyclic (Fig. 12.1).

A photon is absorbed by the light harvesting (LH) complexes that funnel the excitation towards bacteriochlorophylls in the reaction center (RC) and charge separation occurs. This energy is used for the release of an electron which reduces the quinone into a semiquinone. Once the quinone is doubly reduced (i.e. after a second photon is captured) it picks up protons from the cytoplasmic space and translocates them through the membrane to reach the cytochrome bc_1 complex: here electrons are channeled to the cytochrome c_2 (Cyt c_2) while protons are released in the periplasmic space. Cyt c_2 is then able to reduce the oxidized primary electron donors in the RC, i.e. $[Bchl]_2$, thus closing the cycle. The protons accumulated in the periplasmic space form an electrochemical gradient ($\Delta\mu H^+$) which is used by the ATP-synthase to generate ATP.

The photo-electron cycle can be opened by the "Δp-driven reversed electron transport" (indicated by the dotted black arrows in Fig. 12.1), i.e. by the action of both the NADH dehydrogenase working in the "reversed" way to reduce NAD^+ to NADH, and the succinate dehydrogenase also working "backwards" reducing fumarate to succinate (Klamt et al. 2008). As those two enzymes are able to catalyze both the forward and reverse reactions, the force that drives the direction is the presence or absence of the products; in particular the reversed reactions are ways to get rid of the possible excess of the reduced quinol. The reversed NADH dehydrogenase reaction is also the way to refurnish the cell with NADH reducing equivalents.

A. The Photosynthetic Unit (PSU)

The absorption spectrum of purple bacteria (Fig. 12.2) is very wide as covers the two ends of the visible spectrum, and even a little wider. The 'three fingered' absorption bands between 450 and 550 nm are due to carotenoids. Bacteriochlorophylls (BChls) show two characteristic absorption bands: the Soret band in the near UV region, around

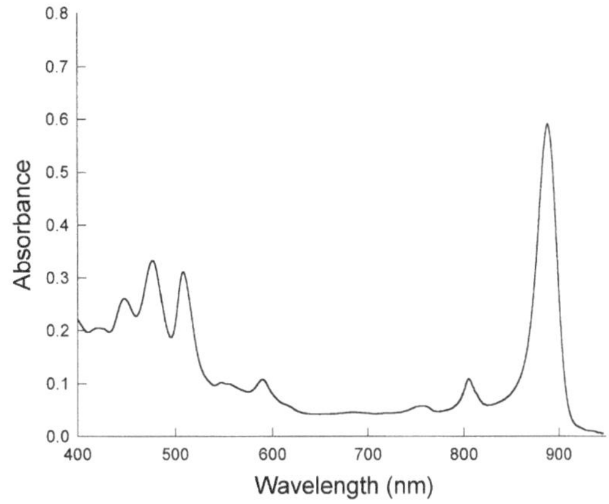

Fig. 12.2. Absorption spectrum of *Rb. sphaeroides* taken as an example for purple bacteria absorption spectrum. Peak wavelengths slightly vary among species.

390 nm (not shown in Fig. 12.2), and a band called Q band in the visible region of the spectrum. This band can usually be decomposed in two distinct bands called Q_x andQ_y according to their predominant polarization. The absorption peaks at 800, 850 and 880 nm in the near infrared region are due to the Q_y shift. The peaks at 850 and 880 often merge in one larger peak, having the maximum at an intermediate wavelength in between, but shoulders can be observed occasionally. The peak at 590 is due to the Q_x shift of BChl *a* (Hoff and Deisenhofer 1997; Blankenship 2002; Frank and Polívka 2008; Loach and Parker-Loach 2008; Robert 2008).

The pigment-protein complexes in the bacterial PSU are responsible for the absorption of light energy and its conversion to electronic excitation that drives the primary charge separation process. All the pigment-protein complexes bind both BChls and carotenoids; with a typical number of 10 LH-2 s, 1 LH-1 and 1 RC the PSU contains approximately 300 BChls. The observed stoichiometric ratio BChl:carotenoids of 3:2 implies the presence of 200 carotenoids (Hu et al. 2002). Out of all these pigments, only very few BChls in the RC directly take part in photochemical reactions; most BChls serve as light-harvesting antennae capturing the sunlight and channeling electronic excitation towards the RC. A wealth of evidence has accumu-

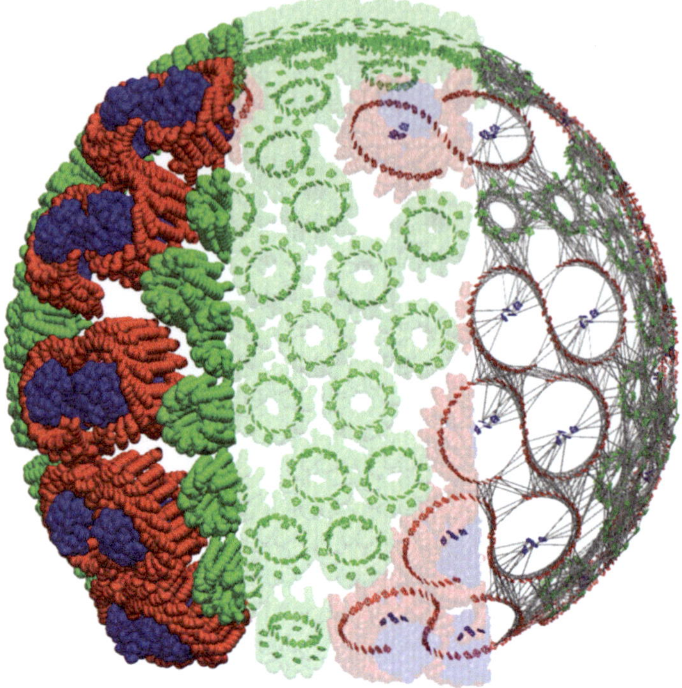

Fig. 12.3. Organization of the photosynthetic unit (PSU) from the supramolecular organization (*left hand side*) to individual bacteriochlorophylls and the electronic couplings (*right hand side*). Protein complexes and the BChls they respectively embed are colored as follows: LH2 in *green*, LH1 in *red* and RC in *blue*. On the *left* side, proteic subunits are represented with tubes; in the *middle* the proteic complexes are transparent to make the BChls visible; on the *right* side the strong electronic couplings of BChls are shown (Figure from Şener and Schulten 2008, with kind permission from Springer Science+Business Media BV).

lated now which proves that the organization of PSUs, to surround an RC with aggregates of chlorophylls and associated carotenoids, is universal in photosynthetic bacteria, higher plants and other photosynthetic organisms (Cogdell et al. 1996; Fromme 1996; Hankamer et al. 1997; Hu and Schulten 1997).

An overall scheme of the photosynthetic units is given in Fig. 12.3, where are represented the protein complexes, differently colored for LH2, LH1 and RCs, the pigment organization and the electronic couplings.

1. Carotenoids

An amount of about 50 distinct carotenoids have been described for purple anaerobic bacteria, and most of them have structures that significantly differ from those found in other photosynthetic or non-photosynthetic organisms.

The function of carotenoids in photosynthetic organisms, and in purple anaerobic bacteria as well, is both to harvest light energy and to protect the cell from photo-induced stress; also structural roles cannot be excluded (Takaichi 2008).

The pathways for carotenoid-genesis in purple bacteria have been classified in two distinct main pathways that contain many variations on their inside: the spirilloxantin pathway (that contain normal spirilloxantin, unusual spirilloxantin, sphaeroidene and carotenal pathways) and the okenone pathway (comprising okenone and *R.g.*-keto carotenoids pathways).

The structure of a carotenoid molecule affects its ability to transfer energy to BChls: *Rb. sphaeroides* containing spheroidene showed a 80–100 % energy transfer efficiency; *Rp. acidophila* containing rhodopin glucoside showed a 35–70 % efficiency; even a lower efficiency has been detected (~30 %) for *Rs. rubrum* LH1 containing spirilloxantin (Frank and Polívka 2008).

2. Light Harvesting (LH) Complexes

LH complexes are able to harvest a great portion of photons hitting the cellular membrane and then to transfer the energy to the "open" (i.e. ready to accept photons) reaction center. LH1 and LH2 differ in structure and in the role covered: LH1 is intimately associated to the RC (for that reason is often referred to as the "core" or "primary" complex) and is found in all purple photosynthetic bacteria. LH2 is present in most species, but not all of them; this complex is distal from the RC (and is also referred to as the "secondary" complex) and channels the excitation to the RC only trough LH1 complexes. This is the reason why the LH2 complexes present absorption bands at higher energy levels (i.e. lower wavelengths) than those of LH1 (Gabrielsen et al. 2008).

Both antenna complexes are composed of repetitions of an elementary unit formed by α and β hydrophobic apoproteins (Zuber and Brunhisolz 1991). Those αβ subunits bind two BChl molecules and one or two carotenoids molecules (Cogdell et al. 2006); the subunits are organized in rings of different dimensions for the two kinds of complexes and the rings themselves can form different structures (Sheuring et al. 2004, 2005). "Basic" LH1 complexes are composed of at least 16 αβ subunits forming a ring that surrounds the RC (Karrash et al. 1995; Walz and Ghosh 1997; Walz et al. 1998; Jamieson et al. 2002); however, the structure varies significantly among bacterial species. In some cases LH1 complexes also contain an additional polypeptide named PufX (Cogdell et al. 1996).

LH2 is structured as circles as well, but composed of a different number of subunits compared to LH1; differences in the number of subunits and in their organization have been observed among different species. Generally speaking LH2 are smaller cycles than LH1 (Robert 2008).

These structural differences between LH1 and LH2, and the different environment and organization of the BChls they contain, give to the two complexes a difference in the absorption bands (Gabrielsen et al. 2008). In particular, LH2 contains B800 and B850 BChls, while LH1 contains B880 BChls (the numbers following the letter B indicate the wavelength corresponding to the Q_y transition of BChl; for details see Frank and Polívka 2008).

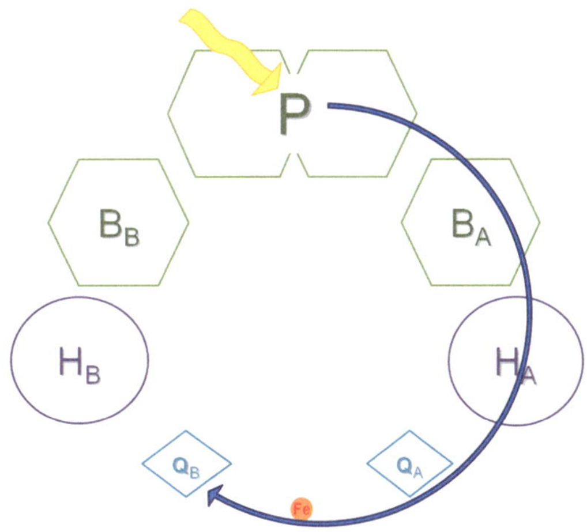

Fig. 12.4. Schematic representation of the reaction center (RC) of purple bacteria. RC is an integral membrane protein complex composed of L, M and H subunits. The H subunit contributes to the complex stability, while L and M compose the core of the RC and embed nine cofactors, symmetrically disposed in an A and a B branch: 2 BChl *a* molecules that form a dimer (P or $[BChl]_2$) connecting the two branches, 2 monomeric bacteriochlorophyll *a* molecules (B_A and B_B), 2 bacteriopheophytin *a* molecules (H_A and H_B), 2 ubiquinone molecules (Q_A and Q_B), and an iron (Fe) atom. The *round arrow* shows the electron path. The *waving arrow* shows excitation energy.

3. The Reaction Center (RC)

The reaction center (RC) of purple bacteria is an integral membrane protein complex composed of L, M and H subunits. The H subunit contributes to the complex stability and is involved in proton binding and transfer, while L and M compose the core of the RC and embed 9 cofactors, symmetrically disposed in an A and a B branch as shown in Fig. 12.4: 2 BChl *a* molecules that form a dimer (P or $[BChl]_2$) connecting the two branches, 2 monomeric bacteriochlorophyll *a* molecules (B_A and B_B), 2 bacteriopheophy-

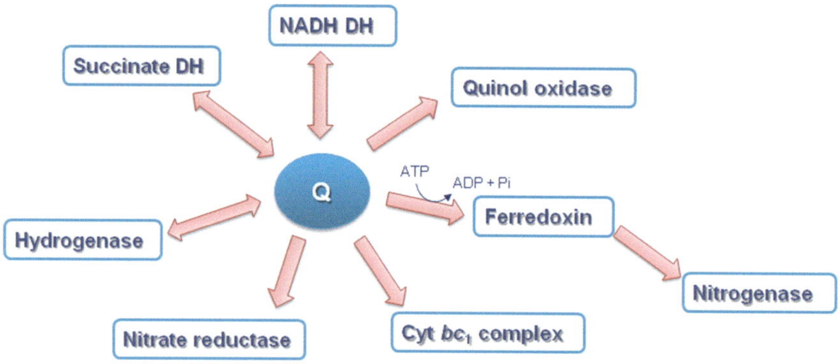

Fig. 12.5. The quinone pool functions. *Q* quinones, *Cyt bc*$_1$ complex: cytochrome bc_1 complex, *NADH DH* NADH dehydrogenase, *Succinate DH* succinate dehydrogenase.

tin *a* molecules (H_A and H_B), 2 ubiquinone molecules (Q_A and Q_B), and an iron (Fe) atom. Some authors suggested also the presence of a carotenoid molecule located next to the B_B monomer. Bacteriochlorophylls and bacteriopheophytins can be substituted by some other molecules of the same class (Williams and Allen 2008).

When a photon is conveyed to the P dimer an electron is raised to a higher energy level; the excited dimer P* transfers the electron, selectively through the A branch, to B_A (Fig. 12.4). Then, the electron is transferred to H_A; the overall transfer from P* to H_A takes from 3–5 ps, depending on bacterial species. Then the electron is transferred from H_A^- to Q_A and finally to Q_B, reducing it to QH_2. The oxidized P^+ is reduced back to P by an electron donated by Cyt c_2 (Fig. 12.1).

B. The Role of Quinones

Purple bacteria export reducing equivalents in pair from the RC, and the molecule migrating from the RC to the other membrane proteins is the reduced quinol (QH_2).

Two electrons are needed to reduce the quinone to quinol, so, as the RC transfers electrons one by one, two quinones (Q_A and Q_B) are needed and act in a cycle in order to accumulate two reducing equivalents; after two RC turnovers a QH_2 is released from the Q_B site (Okamura and Feher 1992, 1995; Shinkarev and Wraight 1993; Okamura et al. 2000; Paddock et al. 2003; Wraight 2004, 2005). During this cycle, protons are captured from the cytoplasm through the acceptor quinone cycle, and then released in the periplasm by the cytochrome bc_1 complex to form the proton gradient needed for ATP synthesis.

QH_2 is enough lipophilic to freely move throughout the membrane and to transport its reducing power to different membrane-bound enzymes. Indeed, quinones hold a crucial role for the energetic processes in the cell, as it is schematically shown in Fig. 12.5. They not only take part in photosynthesis, as it has been above reported, but they also function as electron carriers for all of the redox processes taking place in the membrane (Adessi and De Philippis 2013); for the sake of brevity not all redox processes are treated in this chapter, but specific references are given for further information.

During photosynthesis, they acquire electrons from the RC and reach the cytochrome bc_1 complex to donate them; they also can accept or donate electrons from/to NADH dehydrogenase and succinate dehydrogenase, as described earlier.

During respiration, they receive electrons from NADH-dehydrogenase and succinate dehydrogenase and carry the electrons to both the quinol oxidase and the Cyt bc_1 complex (Zannoni 1995; Zannoni et al. 2008).

During denitrification, they carry the electrons to nitrite reductase (Shapleigh 2008).

In anaerobic conditions and under H_2 containing atmosphere, quinones can accept electrons from hydrogenase (either donate them, but rarely and depending on the abundance of substrates and products of the hydrogenase catalyzed reaction).

Again, in anaerobic conditions electrons can also be transferred to ferredoxins, with the expense of a molecule of ATP, that will then carry the reducing power to nitrogenase in order to fix nitrogen and/or reduce protons to hydrogen.

Moreover, they are not only involved in many essential redox reactions, but also are the mediators of the integrative cell-redox state signal to the RegA/RegB regulon (Swem et al. 2006) which regulates the major processes taking place in purple non sulfur bacteria: photosynthesis, respiration, nitrogen and carbon fixation (Elsen et al. 2004).

IV. Photosynthetic Efficiency (PE)

Generally speaking, the photosynthetic efficiency (PE) is defined as the energy stored as biomass produced per unit of light energy absorbed (Gadhamshetty et al. 2011). It is also called light conversion efficiency. The calculation of the light energy absorbed can be based either on the full solar irradiance or on the photosynthetically active radiation (PAR) range. As a measure for efficiency, when working with photosynthetic organisms, biomass yield (calculated as protein content or dry weight) on light energy is often used. However, in the case of hydrogen production, the product of the process is energy in the form of H_2 and the limiting factor is light; therefore, in this specific case, expressing the efficiency on the basis of the hydrogen-related energy produced per unit of light energy absorbed is more accurate (Eq. 12.3).

$$PE = \frac{\text{Free energy of the total amount of } H_2}{\text{Total energy of the light incident on the culture}} \% \quad (12.3)$$

It has to be stressed that in this calculation only the energy input of light is considered, and not the energetic contribution of organic substrates consumed by the cell during hydrogen production (Hallenbeck and Benemann 2002).

The efficiency by which the light energy can be transformed into hydrogen gas energy is dependent not only on the part of the energy that is absorbed by the antenna systems of the organism studied but also on the energy loss during the several steps of excitation and electron transfers that follow. In addition to that, PE is a measure derived from the amount of hydrogen produced, thus all cellular processes that deviate from the routes that bring to hydrogen production negatively affect PE. To summarize, hydrogen production depends both on photoparameters such as quality and quantity of light and on biological factors such as pigment composition, quantum requirements and the metabolism of the different PNSB strains. Therefore, qualitative and quantitative understanding of each of those factors is important to optimize PE. Moreover, only the 65.8 % of the spectrum is actually part of the PAR for purple bacteria; the energy associated to one mole of photons depends on the wavelengths inside this 65.8 %, so PE should be calculated for each wavelength. However, quantum yields are only known for 522 and 860 nm wavelenghts (Miyake 1998), and lead to wavelength specific PEs of 8.4 % and 19 % respectively. From these data it was calculated that the theoretical maximum PE for PNSB, based on the natural sunlight spectrum, is at least 10 % (Akkerman et al. 2002).

PE is a crucial factor for optimum hydrogen production and it is the most important aspect to be taken into account designing a photobioreactor (Akkerman et al. 2002; Gadhamshetty et al. 2011). Indeed, high surface to volume ratios are necessary to capture sufficient light (Dasgupta et al. 2010); this means that operating at a scarce PE implicates the need to occupy a large surface area with the photobioreactor for producing satisfactory amounts of hydrogen. Consequently, photobioreactors that optimize light distribution and capturing by cells are required for efficient hydrogen production processes.

Moving from theory to practice, low PE has always constituted a very hard obstacle to overcome, not only using natural sunlight but also using artificial irradiation. Indeed, up to 2010, high photosynthetic efficiencies have been reached using such low light intensities that the production rates were not enough to be considered interesting for an H_2 production process. Barbosa et al. (2001) observed that higher light intensities may decrease PE, but usually increase hydrogen productivity. Miyake and Kawamura (1987) reported light conversion efficiencies of 7.9 and 6.2 %: they illuminated the cultures with a xenon lamp at 50 W m^{-2} and by a solar simulator at 75 Wm^{-2}, respectively. Those light intensities allowed to reach a relevant value of PE, but are too low to reach gas evolution rates that can be considered interesting for production processes. However, recently, impressive high PEs have been obtained (Tian et al. 2010; Wang et al. 2013) associated to high hydrogen production rates, by paying attention to the quality of the incident light. These data will be discussed thoroughly in this paragraph, Sect. IV.B.

A. Light Intensity

In *Rb. capsulatus* it has been observed that light strongly stimulates not only the activity but also the amount of nitrogenase synthesized (up to 25 % of all soluble proteins); as a consequence, a larger amount of hydrogen under high light conditions was observed as well (Jouanneau et al. 1985; Vignais et al. 1985). The photophosphorylation capacity is also slightly greater in cells grown under high light-intensity than in cells grown under low-light intensities (Steinborn and Oelze 1989). Therefore, Kars and Gündüz (2010) proposed that high ATP production rates under well-illuminated conditions result in higher hydrogen production activity.

Uyar et al. (2007) indicated, for *Rb. sphaeroides*, a minimum light intensity of 270 W m^{-2} to obtain high hydrogen production rates, this value being equivalent to 4,000 lx and 1,370 μmol(photons) m^{-2} s^{-1}.

However, it must be underlined that the intensity of light impinging on the photobioreactor, which is the parameter usually reported in the literature, is not equivalent to the intensity of light actually faced by the single cells. Indeed, due to the self-shading of cells, a phenomenon always present in the cultivation systems used for growing phototrophic microorganisms (Vonshak and Richmond 1985; Tredici 1999), the amount of light reaching the single cell may be hundreds of times lower than the light reaching the photobioreactor, or even zero (Gadhamshetty et al. 2011). This well known phenomenon depends on the concentration of the cells, the light path in the photobioreactor and the amount of light absorbed by the single cell. This amount, on its side, depends on cell diameter, pigment composition and concentration, antenna size.

From these considerations, it is quite evident that it is a problematic task to evaluate the exact amount of light that is absorbed by a cell, thus making the photosynthetic efficiency a parameter that relies on a few approximations.

B. Light Quality and Sources

Purple bacteria are able to absorb light at the very extremities of the visible spectrum (Fig. 12.2) with one main absorption band at 300–500 nm and the other above 800 nm in the near infrared region.

When using artificial light, the most frequently used light sources are incandescent lamps. Among them, tungsten lamps have an emission spectrum that covers the whole absorption spectrum of PNSB (Adessi and De Philippis 2012). Particularly important is the high near infrared emission, where is located the absorption maximum of bacteriochlorophylls. Halogen lamps are quite frequently used as well.

As incandescent lamps are energy-expensive light sources, an interesting alternative is offered by Light Emitting Diodes (LEDs). Kawagoshi et al. (2010) prefigured a reduction of energy cost by 98 % using LEDs instead of tungsten lamps. In a cost effective scaled-up system, though, the best solution would appear the use of natural solar light.

Table 12.1 shows the results deriving from different kinds of photobioreactors illuminated by different light sources: incandescent lamps, LEDs and solar light. Summarizing the experimental results reported in Table 12.1, PE is <10 % in the experiments carried out using incandescent lamps, it stands around 1 % when using solar light but reaches very high values when using LEDs.

Apparently, specific LED illumination granted the highest ever obtained PE values. Those data were obtained using immobilized systems: glass beads biofilm (Tian et al. 2010) or a mix of PVA, carrageenan and alginate (Wang et al. 2013). These outstanding efficiencies, respectively 56 and 82 % (see Table 12.1), were to a great extent due to the use of LEDs illuminating at a selected specific wavelength (590 nm). This specific excitation allowed the cells to utilize a large part of the incident energy, as it was at the exact wavelength that could be absorbed by the culture.

590 nm as an emission wavelength was the best among other wavelength specific LEDs (470, 520, 590, 620 nm), tested by Zhang et al. (2010); comparing those emission wavelengths with the PNSB absorption spectrum (Fig. 12.2) it emerges that 590 nm is very close to one of the absorption peaks of bacteriochlorophylls. Also long-wavelength emitting LEDs have been used for hydrogen production processes with PNSB (Kawagoshi et al. 2010); those LEDs provided an emission spectrum having a maximum at 850 nm and were used for illuminating a halotolerant *Rhodobacter* sp., but poor hydrogen production was obtained. However, the scarce hydrogen production was not due to the light source, but most probably to sub- optimal growth conditions.

The use of natural sunlight opens a series of more complex issues, not all related to light itself. Indeed, most sunlight illuminated photobioreactors are outdoor large-scale systems, whose management complicates the processes comparing it with a lab-scale process (Chen et al. 2011). However, focusing exclusively on light-related issues, not all of the solar spectrum is part of the PAR for PNSB, as mentioned earlier. Moreover, Miyake et al. (1999) pointed out how the intrinsic variability of solar light makes the rates vary along with light intensity during the day, giving an endemic inconstancy to the process. Furthermore, they observed a probable photoinhibition under the highest irradiation of the day (0.9 kW m^{-2}).

Generally speaking, when H_2 production is carried out outdoors using solar irradiation, PE is around 1 %. The highest PE reported with a solar photobioreactor containing purple bacteria (1.4 %) was obtained with a flat plate reactor provided with light shading bands (Wakayama and Miyake 2002). The special feature of the system used by Wakayama and Miyake (2002) was the reduction of the total amount of light irradiation reaching the cells through the use of shading bands.

Photoinhibition in outdoors purple bacteria culturing systems has not been studied thoroughly yet, as a few studies on PNSB pay a specific attention to it. Miyake et al. (1999) observed a delay of 2–4 h of the maximum H_2 production rate after the maximum irradiation at noon. At the same regard, Adessi et al. (2012a) tried to avoid photoinhibition cutting sunlight intensity by 50 % using a light-shield. Hydrogen production rates were not negatively affected by this light intensity decrease; on the contrary, the maximum rate (27.2 ml l^{-1} h^{-1}), reached 2 h after noon, was comparable to artificial lab-conditions results. However, this delay in achieving the maximum production rate compared to the peak of irradiation might still indicate photoinhibition by the highest irradiation of the day.

Although the photoinhibition might affect hydrogen production during the peak irradiation of the day, Adessi et al. (2012a) observed a good physiological long-term fitting of a *Rp. palustris* culture, as confirmed by the BChl *a* fluorescence analysis. Twenty-one days of exposure to solar irradiation was

Table 12.1. Light sources and intensities and types of photobioreactors (PBR) with the corresponding hydrogen production rates (HPR), photosynthetic efficiencies (PE) and substrate conversions (SC).

Light source and intensity	Type of PBR	Organism	HPR (ml l^{-1} h^{-1})	PE (%)	SC (%)	Ref.
Traditional lamps						
Tungsten lamp 200 W m^{-2}	Tubular, vertical	*Rb. sphaeroides*	20.0	1.1[a]	n.a.	Eroglu et al. (1998)
Tungsten lamp 10.25 W m^{-2}	Flat, rocking	*Rb. sphaeroides*	11.0	3.31	45	Gilbert et al. (2011)
Tungsten lamp 500 W m^{-2}	Double-layer	*Rb. sphaeroides* (wild type + reduced pigment mutant)	3640[b]	2.18	n.a.	Kondo et al. (2002a, b)
Tungsten lamps 500 lx	Fermentor type	*Rs. rubrum*	107.5 [c]	8.67	n.a.	Ismail et al. (2008)
Halogen lamp	Flat, floating type	*Rp. palustris*	10.0–12.0	0.3	n.a.	Otzuki et al. (1998)
Halogen lamp	Induced diffuse light	*Rb. sphaeroides*	7577[b]	6.12 Peak value: 9.23	61	El-Shishtawy et al. (1997)
Luminine tubular light 15 W m^{-2}	Annular triple jacketed	*Rb. sphaeroides*	6.5	3.7	75	Basak and Das (2009)
Halogen +Tungsten lamps 95 W m^{-2} (with optical fibers)	Immobilized, clay	*Rp. palustris*	43.8	2.34	90.8	Chen and Chang (2006)
Incandescent lamp 150 W m^{-2}	Immobilized, activated carbon fibers	*Rp. faecalis*	32.9	0.96	77.0	Xie et al. (2012)
Metal–halide lamp 12 W m^{-2} (with optical fibers)	Biofilm, rough surface	*Rp. palustris*	39.2 [c]	9.3	75	Guo et al. (2011)

LED lights							
LED (590 nm) 6.75 W m^{-2}	Biofilm, glass beads	*Rp. palustris*	38.9		56	1.7	Tian et al. (2010)
LED (590 nm) 6.75 W m^{-2}	Biofilm, groove-type	*Rp. palustris*	86.46[b,c]		3.8	6.25	Zhang et al. (2010)
LED (590 nm) 6,000 lx	Immobilized, PVA + carrageenan + alginate	*Rp. palustris*	58.4		82.3	62.3	Wang et al. (2013)
LED (590 nm)	Flat, 30° inclined	*Rp. palustris*	26.9		8.9	2.1	Liao et al. (2010)
Solar light							
Sulight (+ initial artificial light)	Tubular, near horizontal	*Rb. capsulatus*	Avg:6.9[a]	Max: 16.6 [a]	1.0	16	Boran et al. (2010)
Sulight (+ initial artificial light)	Tubular, near horizontal	*Rb. capsulatus hup*- mutant	Avg:4.5[a]	Max: 8.9[a]	0.2	12	Boran et al. (2012)
Sunlight (shielded)	Tubular, horizontal	*Rp. palustris*	Avg:10.7	Max: 27.2	Avg 0.63 Max 0.92	10.3	Adessi et al. (2012a)
Sunlight	Flat, 30° inclined	*Rb. sphaeroides*	10.0		n.a.	76.7	Eroglu et al. (2008)
Sunlight	Flat with light shading bands	*Rb. sphaeroides*	0.87[b]		1.4	n.a.	Wakayama and Miyake (2002)
Sunlight + tungsten lamp (with optical fibers)	Fermentor type	*Rp. palustris*	22.7		n.a.	62.3	Chen et al. (2008)

n.a not available
[a]Calculated by Akkerman et al. (2002)
[b]Volume expressed as ml m^{-2} h^{-1}
[c]Value in mmoles of H_2 converted to ml of H_2 by 22.4 multiplying factor

not affecting the functionality of the photosystem, and it was not degraded by the excess of solar radiation. This indicates a very high capability of purple bacteria to acclimate to conditions that would potentially be a source of stress, as an excessive light irradiance. Structural and functional homology between the purple bacterial RC and the RC of the photosystem II (PSII) of oxygenic photosynthesis allows the use of variable bacteriochlorophyll (Bchl *a*) fluorescence to investigate the energy transfer and electron transport within the photosynthetic apparatus. There is evidence that confirms the applicability of chlorophyll fluorescence analysis (usually used for oxygenic photosynthetic organisms) to photosynthetic bacteria (Koblízek et al. 2005; Maróti 2008; Asztalos et al. 2010; Adessi et al. 2012a).

V. Substrate to Hydrogen Conversion (SC)

As it has been described earlier (see Fig. 12.1), under photoheterotrophic growth conditions the ATP is formed via anoxygenic photosynthesis, while the reducing power is derived by the catabolism of organic substrates. Thus, the efficiency of the carbon related metabolic processes has a role in determining the amount of electrons accumulated as quinols and then, if that is the case, transferred to nitrogenase through ATP-dependent ferredoxin reduction. Therefore, even going back to the role that quinones have in the cell (Fig. 12.5), it looks clear how hydrogen production in purple bacteria is related to many metabolic processes that deal with ATP generation (photosynthesis), carbon metabolism and nitrogen fixation. Usually, all the processes involved in energy generation, as photosynthesis and H_2 oxidation, and energy consumption, as N_2 and CO_2 fixation, are globally regulated by the two component system RegB-RegA (Elsen et al. 2000).

The preferred substrates for hydrogen production are the low-molecular weight organic acids that can easily enter the TCA cycle, which is very active during anaerobic photosynthetic growth.

Carbon metabolism in purple non-sulfur bacteria has been schematically compiled by Koku et al. (2002). That scheme describes the metabolism of *Rb. sphaeroides*, but not all species and genera follow the same scheme: for example *Rp. palustris* does not possess the Entner-Doudoroff pathway (Larimer et al. 2004). A well documented description of C metabolism in PNSB in relation with hydrogen production is reported in Chap. 7 of this volume (McKinlay).

An important parameter to evaluate the yield of a hydrogen production process is the substrate conversion efficiency, calculated as the ratio between the moles of hydrogen produced and the moles theoretically obtainable if all the substrate consumed was converted to CO_2 and H_2. Thus, considering the most common organic acids utilized in photofermentation processes (Barbosa et al. 2001), the conversion yields can be calculated from the following reactions:

$$\text{Lactate}: C_3H_6O_3 + 3H_2O \rightarrow 6H_2 + 3CO_2 \quad (12.4)$$

$$\text{Acetate}: C_2H_4O_2 + 2H_2O \rightarrow 4H_2 + 2CO_2 \quad (12.5)$$

$$\text{Malate}: C_4H_6O_5 + 3H_2O \rightarrow 6H_2 + 4CO_2 \quad (12.6)$$

It has to be stressed that these reactions are theoretical, because they are neither considering the utilization of the substrate for the growth neither limiting factors occurring in a culture. On the basis of these reactions, the gas expected should be composed of a 66.7 % of H_2 and a 33.3 % of CO_2 when growing on lactate and on acetate; a 60 % of H_2 and a 40 % of CO_2 when growing on malate. Actually, the gas phase above the culture is much richer in H_2 then in CO_2, due to a partial solubilization of CO_2 in the culture medium and also to a partial fixation to CO_2 for anabolic reactions (McKinlay, Chap. 7 of this volume).

The presence of Calvin cycle in carbon metabolism has been described by Joshi and Tabita (1996). They demonstrated that the absence of the reductive pentose phosphate CO_2 fixation pathway enhances the synthesis of nitrogenase also in presence of ammonium ions, as the reduction of CO_2, in photoheterotrophy, is another way to dissipate the excessive reducing power deriving from organic carbon compounds. McKinlay and Harwood (2010, 2011) followed the exact pathway of CO_2 molecules, getting to the conclusion that CO_2 fixation actually deprives the cells from a part of the electrons that would either be useful for hydrogen production, but that it is not true that a more reduced substrate generates a larger quantity of electrons available for H_2 production.

A 100 % substrate conversion efficiency has been reported by Sasikala et al. (1990), but in a limited culture volume (2 ml); conversion efficiencies (reported in Table 12.1) mainly range between 60 % and 90 %.

The substrate conversion efficiency is strongly affected by the C/N ratio in the culture. Indeed, a high C/N ratio in the culture media usually leads to higher hydrogen production compared with low C/N ratio, where a higher cell growth occurs (Kapdan and Kargi 2006; Redwood et al. 2009; Keskin et al. 2011). In the latter case, the conversion efficiency decreases due to the consumption of the organic acids for cell growth instead that for hydrogen production. This problem becomes a very relevant matter when wastewaters or liquors deriving from other fermentation processes are utilized for the production of H_2 by means of photofermentation (Eroglu et al., Chap. 11 of this volume).

As regarding the processes that compete with the conversion of the carbon substrate to hydrogen, PNSB are also capable of producing poly-β-hydroxybutyrate (PHB) a valuable by-product which is a biodegradable thermoplastic having industrial and medical interest (De Philippis et al. 1992; Sasikala and Ramana 1995; Reddy et al. 2003; Franchi et al. 2004). PHB synthesis not only consumes the carbon substrate itself (acetate molecules are needed as building blocks for PHB), but also the reducing power, as PHB synthesis utilizes NADH. Both those aspects are in competition with hydrogen production during photofermentation (De Philippis et al. 1992; Hustede et al. 1993; Vincenzini et al. 1997; Koku et al. 2003; Franchi et al. 2004).

Other factors affecting substrate conversion are the mixing and the availability of the substrate for the cells in terms of exchanging surface between cells and medium. In particular, immobilized or biofilm reactors show a higher conversion efficiencies for that reason (Table 12.1).

VI. Process Bottlenecks – Conclusions

In this chapter, an overview of the process leading from light energy to the production of a molecule of hydrogen was given. From this description, it appears evident that this process encounters many hurdles and deviations from the desired route.

In Fig. 12.1, the most critical steps, that can be bottlenecks of the process are indicated by star signs:

$*_1$ Energy transfer through the photosynthetic unit;
$*_2$ Reducing power deviation;
$*_3$ Nitrogenase activity;
$*_4$ Substrate deviation.

Recently, Harwood (2008) and Kars and Gündüz (2010) reviewed the research on mutagenesis of strains in order to overcome the hurdles described. Below are listed the cultivation strategies that can be adopted for each of the previous points.

$*_1$ Strategies for improving the energy transfer through the photosynthetic unit:

Reduced pigment strains: reducing pigment content by genetic manipulations is aimed at reducing the self-shading effect of the cells. As a consequence, the light penetration inside the bioreactor increases, causing an increase in H_2 production (Vasilyeva et al. 1999; Kondo et al. 2002a, b).

Protection from photoinhibition: light shading bands (Wakayama and Miyake 2002) or light shields (Adessi et al. 2012a) prevent the cells to be burnt by the excess of sunlight.

Use of specific light sources: the use of specific-wavelength LEDs brought to higher PEs (Tian et al. 2010; Wang et al. 2013).

*₂ Strategies for preventing the of deviation reducing power:

Hydrogenase inhibition: since uptake-hydrogenase decreases the efficiency of H_2 production, it was targeted to be eliminated in many PNSB either by antibiotic resistance gene insertion into the *hup* genes or by deletion of *hup* genes (Kern et al. 1994; Ooshima et al. 1998; Franchi et al. 2004; Kim et al. 2006; Öztürk et al. 2006; Kars et al. 2008; Kars et al. 2009); other deviations are avoided by supplementing the cultures with reduced carbon substrates in order not to let the cells run out of reducing equivalents.

CBB defective strains: PNSB fix CO_2 via the Calvin Benson Bassham (CBB) cycle; the overall pathway consumes both NADPH and ATP to synthesize cell material; carbon dioxide fixation defective strains have been constructed in order to save electrons for hydrogen production (Joshi and Tabita 1996; Qian and Tabita 1996; McKinlay and Harwood 2010, 2011).

$*_3$ Strategies for enhancing nitrogenase activity:

Enhanced electron flow to nitrogenase: the *rnf* operon has been identified as related to the electron transport to nitrogenase. Overexpression of this operon led to higher nitrogenase activity (Jeong and Jouanneau 2000).

Enhanced nitrogenase activity: the limiting step for nitrogenase mediated catalysis is difficult to overcome due to the complexity of the mechanism. However, a possible strategy could be acting on the substrate selectivity of nitrogenase in presence of dinitrogen (Harwood 2008), as it has already been proved to be successful in *Azotobacter vinelandii* (Barney et al. 2004).

Deregulation of nitrogenase: nitrogenase is strongly inhibited by the presence of fixed nitrogen in the cell (Rey et al. 2007; Heiniger et al. 2012); deregulating nitrogenase appears to have a stabilizing effect on hydrogen production even on ammonia containing substrates (Adessi et al. 2012b).

$*_4$ Strategies for avoiding substrate deviation:

PHB route inactivation: polyhydroxyalkanoic acids biosynthesis constitutes a way for disposing of the excess of reducing power in the cell; it competes with hydrogen production and consumes acetate. Deletion of the PHB synthesizing route resulted in enhanced hydrogen production (Hustede et al. 1993); following studies reported an enhanced hydrogen production only when PHB deficiency was coupled with uptake hydrogenase inactivation (Franchi et al. 2004; Kim et al. 2006).

It looks clear how complex is the itinerary that brings from light to hydrogen and how many different routes have been taken to optimize the process. A "supermutant" bringing all the mutations above reported is not a solution that the authors consider feasible, due to the consequent fragility of an organism that would bring all those kind of genetic alterations. A fragile organism, even if the hydrogen production route might be stabilized in its metabolism, would not be a good choice for an up-scaled process in particular if operating on complex substrates derived from other fermentation or degradative processes. On the contrary the authors believe that a limited number of genetic manipulations carefully combined with the more appropriate culture conditions and photobioreactor geometry should be designed for each specific process in order to obtain the maximum hydrogen production without affecting process stability.

Acknowledgements

The authors gratefully acknowledge the Italian Ministry of Agricultural, Food and Forest Politics (MIPAAF; project IMERA), the Italian Ministry of the Environment (MATTM; project PIRODE), the Italian Ministry of the University and Research (MIUR) and the Italian National Research Council (CNR) (EFOR project) that partially supported the researches carried out in their lab and mentioned in this chapter. The Authors would also like to mention the contribution to the development of their researches on biological hydrogen given by the activities carried out by RDP in the frame of the IEA-HIA (International Energy Agency – Hydrogen Implementation Agreement), Annex 21 "Bioinspired and biological hydrogen".

References

Adessi A, De Philippis R (2012) Hydrogen production: photofermentation. In: Hallenbeck PC (ed) Microbial technologies in advanced biofuels production. Springer Science+Business Media, New York, pp 53–75

Adessi A, De Philippis R (2013) Purple bacteria: electron acceptors and donors. In: Lennarz WJ, Lane MD (eds) The encyclopedia of biological chemistry, vol 3. Academic, Waltham, pp 693–699

Adessi A, Torzillo G, Baccetti E, De Philippis R (2012a) Sustained outdoor H_2 production with *Rhodopseudomonas palustris* cultures in a 50 L tubular photobioreactor. Int J Hydrog Energy 37:8840–8849

Adessi A, McKinlay JB, Harwood CS, De Philippis R (2012b) A *Rhodopseudomonas palustris nifA** mutant produces H_2 from NH_4^+-containing vegetable wastes. Int J Hydrog Energy 37:15893–15900

Akkerman I, Janssen M, Rocha J, Wijffels RH (2002) Photobiological hydrogen production: photochemical efficiency and bioreactor design. Int J Hydrog Energy 27:1195–1208

Asztalos E, Italiano F, Milano F, Maróti P, Trotta M (2010) Early detection of mercury contamination by fluorescence induction of photosynthetic bacteria. Photochem Photobiol Sci 9:1218–1223

Barbosa MJ, Rocha JMS, Tramper J, Wijffels RH (2001) Acetate as a carbon source for hydrogen production by photosynthetic bacteria. J Biotechnol 85:25–33

Barney BM, Igarashi RY, Dos Santos PC, Dean DR, Seefeldt LC (2004) Substrate interaction at an Iron-Sulfur face of the FeMo-cofactor during nitrogenase catalysis. J Biol Chem 279:53621–53624

Basak N, Das D (2009) Photofermentative hydrogen production using purple non-sulfur bacteria *Rhodobacter sphaeroides* O.U.001 in an annular photobioreactor: a case study. Biomass Bioenergy 33:911–919

Blankenship RE (2002) Molecular mechanisms of photosynthesis. Blackwell Science Ltd, Oxford

Boran E, Özgür E, Van der Burg J, Yücel M, Gündüz U, Eroğlu I (2010) Biological hydrogen production by *Rhodobacter capsulatus* in solar tubular photo bioreactor. J Clean Prod 18:S29–S35

Boran E, Özgür E, Yücel M, Gündüz U, Eroğlu I (2012) Biohydrogen production by *Rhodobacter capsulatus* Hup- mutant in pilot solar tubular photobioreactor. Int J Hydrog Energy 37:16437–16445

Bruschi M, Guerlesquin F (1988) Structure, function and evolution of bacterial ferredoxins. FEMS Microbiol Rev 54:155–176

Chen CY, Chang JS (2006) Enhancing phototrophic hydrogen production by solid-carrier arrested fermentation and internal optical-fiber illumination. Process Biochem 41:2041–2049

Chen CY, Saratale GD, Lee CM, Chen PC, Chang JS (2008) Phototrophic hydrogen production in photobioreactors coupled with solar energy excited optical fibers. Int J Hydrog Energy 33:6686–6695

Chen CY, Liu CH, Lo YC, Chang JS (2011) Perspectives on cultivation strategies and photobioreactor designs for photo-fermentative hydrogen production. Bioresour Technol 102:8484–8492

Cogdell R, Fyfe P, Barrett S, Prince S, Freer A, Isaacs N, McGlynn P, Hunter C (1996) The purple bacterial photosynthetic unit. Photosynth Res 48:55–63

Cogdell RJ, Southall J, Gardiner AT, Law CJ, Gall A, Roszak AW, Isaacs NW (2006) How purple photosynthetic bacteria harvest solar energy. CR Chim 9:201–206

Colbeau A, Kelley BC, Vignais PM (1980) Hydrogenase activity in *Rhodopseudomonas capsulata*: relationship with nitrogenase activity. J Bacteriol 144:141–148

Dasgupta CN, Gilbert JJ, Lindblad P, Heidorn T, Borgvang SA, Skjanes K, Das D (2010) Recent trends on the development of photobiological processes and photobioreactors for the improvement of hydrogen production. Int J Hydrog Energy 35:10218–10238

De Philippis R, Ena A, Guastini M, Sili C, Vincenzini M (1992) Factors affecting poly-β-hydroxybutyrate accumulation in cyanobacteria and in purple non-sulfur bacteria. FEMS Microbiol Rev 103:187–194

Eady RR (1996) Structure-function relationships of alternative nitrogenases. Chem Rev 96:3013–3030

Elsen S, Dischert W, Colbeau A, Bauer CE (2000) Expression of uptake hydrogenase and molybdenum nitrogenase in *Rhodobacter capsulatus* is coregulated by the RegB-RegA two-component regulatory system. J Biotechnol 182:2831–2837

Elsen S, Swem LR, Swem DL, Bauer CE (2004) RegB/RegA, a highly conserved redox-responding global two-component regulatory system. Microbiol Mol Biol Rev 68:263–279

El-Shishtawy RMA, Kawasaki S, Morimoto M (1997) Biological H_2 production using a novel light-induced and diffused photoreactor. Biotechnol Tech 11:403–407

Eroglu I, Aslan K, Gunduz U, Yucel M, Turker L (1998) Continuous hydrogen production by Rhodobacter sphaeroides O.U.001. In: Zaborsky OR (ed) Biohydrogen. Plenum Press, London, pp 143–151

Eroğlu I, Tabanoğlu A, Gündüz U, Eroğlu E, Yücel M (2008) Hydrogen production by *Rhodobacter sphaeroides* O.U.001 in a flat plate solar bioreactor. Int J Hydrog Energy 33:531–541

Fox JD, He Y, Shelver D, Roberts GP, Ludden PW (1996) Characterization of the region encoding the CO-induced hydrogenase of *Rhodospirillum rubrum*. J Bacteriol 178:6200–6208

Franchi E, Tosi C, Scolla G, Penna DG, Rodriguez F, Pedroni MP (2004) Metabolically engineered *Rhodobacter sphaeroides* RV strains for improved biohydrogen photoproduction combined with disposal of food wastes. Mar Biotechnol 6:552–565

Frank HA, Polívka T (2008) Energy transfer from carotenoids to bacteriochlorophylls. In: Hunter CD, Daldal F, Thurnauer MC, Beatty JT (eds) The purple phototrophic bacteria, vol 28, Advances in photosynthesis and respiration. Springer Science+Business Media, Dordrecht, pp 213–230

Fromme P (1996) Structure and function of photosystem I. Curr Opin Struct Biol 6:473–484

Gabrielsen M, Gardiner AT, Cogdell RJ (2008) Peripheral complexes of purple bacteria. In: Hunter CD, Daldal F, Thurnauer MC, Beatty JT (eds) The purple phototrophic bacteria, vol 28, Advances in photosynthesis and respiration. Springer Science+Business Media, Dordrecht, pp 135–153

Gadhamshetty V, Sukumaran A, Nirmalakhandan N (2011) Photoparameters in photofermentative biohydrogen production. Environ Sci Technol 41:1–51

Gest H, Kamen MD (1949) Photoproduction of molecular hydrogen by *Rhodospirillum rubrum*. Science 109:558–559

Gilbert JJ, Ray S, Das D (2011) Hydrogen production using *Rhodobacter sphaeroides* (O.U.001) in a flat panel rocking photobioreactor. Int J Hydrog Energy 36:3434–3441

Göbel F (1978) Quantum efficiencies of growth. In: Clayton RK, Sistrom WR (eds) Photosynthetic bacteria. Plenum Press, New York, pp 907–925

Grabau C, Schatt E, Jouanneau Y, Vignais PM (1991) A new [2Fe-2S] ferredoxin from *Rhodobacter capsulatus*. Coexpression with a 2[4Fe-4S] ferredoxin in *Escherichia coli*. J Biol Chem 266:3294–3299

Guo CL, Zhu X, Liao Q, Wang YZ, Chen R, Lee DJ (2011) Enhancement of photo-hydrogen production in a biofilm photobioreactor using optical fiber with additional rough surface. Bioresour Technol 102:8507–8513

Hallenbeck PC, Benemann JR (2002) Biological hydrogen production; fundamentals and limiting processes. Int J Hydrog Energy 27:1185–1193

Hankamer B, Barber J, Boekema EJ (1997) Structure and membrane organization of photosystem II in green plants. Annu Rev Plant Physiol 48:641–671

Harwood CS (2008) Nitrogenase-catalyzed hydrogen production by purple nonsulfur photosynthetic bacteria. In: Demain AL, Wall JD, Harwood CS (eds) Bioenergy. Springer Science+Business Media, Dordrecht, pp 259–271

Heiniger EK, Oda Y, Samanta SK, Harwood CS (2012) How posttranslational modification of nitrogenase is circumvented in *Rhodopseudomonas palustris* strains that produce hydrogen gas constitutively. Appl Environ Microbiol 78:1023–1032

Hoff AJ, Deisenhofer J (1997) Photophysics of photosynthesis. Structure and spectroscopy of reaction centers of purple bacteria. Phys Rep 287:1–247

Hu X, Schulten K (1997) How nature harvests sunlight. Phys Today 50:28–34

Hu X, Ritz T, Damjanović A, Autenrieth F, Shulten K (2002) Photosynthetic apparatus of purple bacteria. Q Rev Biophys 35:1–62

Hustede E, Steinbüchel A, Schlegel HG (1993) Relationship between the photoproduction of hydrogen and the accumulation of PHB in non-sulfur purple bacteria. Appl Microbiol Biotechnol 39:87–93

Imhoff JF (1995) The anoxygenic phototrophic purple bacteria. In: Boone DR, Castenholz RW, Garrity GM (eds) Bergey's manual of systematic bacteriology, 2nd edn. Springer, New York, pp 631–637

Ismail KSK, Najafpour G, Younesi H, Mohamed AR, Kamaruddin AH (2008) Biological hydrogen production from CO: bioreactor performance. Biochem Eng J 39:468–477

Jamieson SJ, Wang P, Qian P, Kirkland JY, Conroy MJ, Hunter CN, Bullough PA (2002) Projection structures of the photosynthetic reaction centre-antenna complex from *Rhodospirillum rubrum* at 8.5 Å resolution. EMBO 21:3927–3935

Jeong HS, Jouanneau Y (2000) Enhanced nitrogenase activity in strains of *Rhodobacter capsulatus* that overexpress the *rnf* genes. J Bacteriol 182:1208–1214

Joshi HM, Tabita FR (1996) A global two component signal transduction system that integrates the control of photosynthesis, carbon dioxide assimilation, and nitrogen fixation. Proc Natl Acad Sci U S A 93:14515–14520

Jouanneau Y, Wong B, Vignais PM (1985) Stimulation by light of nitrogenase synthesis in cells of *Rhodopseudomonas capsulata* growing in N-limited continuous cultures. BBA-Bioenerg 808:149–155

Jouanneau Y, Meyer C, Naud I, Klipp W (1995) Characterization of an *fdxN* mutant of *Rhodobacter capsulatus* indicates that ferredoxin I serves as electron donor to nitrogenase. Biochim Biophys Acta 1232:33–42

Kapdan IK, Kargi F (2006) Bio-hydrogen production from waste material. Enzyme Microb Technol 38:569–582

Karrash S, Bullough PA, Ghosh R (1995) The 8.5 Å projection map of the light-harvesting complex I from *Rhodospirillum rubrum* reveals a ring composed of 16 subunits. EMBO J 14:631–638

Kars G, Gündüz U (2010) Towards a super H_2 producer: improvements in photofermentative biohydrogen production by genetic manipulations. Int J Hydrog Energy 35:6646–6656

Kars G, Gündüz U, Rakhely G, Yücel M, Eroğlu I, Kovacs LK (2008) Improved hydrogen production by hydrogenase deficient mutant strain of *Rhodobacter sphaeroides* O.U.001. Int J Hydrog Energy 33:3056–3060

Kars G, Gündüz U, Yücel M, Rakhely G, Kovacs K, Eroğlu I (2009) Evaluation of hydrogen production by *Rhodobacter sphaeroides* O.U.001 and its *hup*SL deficient mutant using acetate and malate as carbon sources. Int J Hydrog Energy 34:2184–2190

Kawagoshi Y, Oki Y, Nakano I, Fujimoto A, Takahashi H (2010) Biohydrogen production by isolated halotolerant photosynthetic bacteria using long-wavelength light-emitting diode (LW-LED). Int J Hydrog Energy 35:13365–13369

Kern M, Klipp W, Klemme HJ (1994) Increased nitrogenase dependent H_2 photoproduction by *hup* mutants of *Rhodospirillum rubrum*. Appl Environ Microbiol 60:1768–1774

Keskin T, Abo-Hashesh M, Hallenbeck PC (2011) Photofermentative hydrogen production from wastes. Bioresour Technol 102:8557–8568

Kim MS, Baek JS, Lee JK (2006) Comparison of H_2 accumulation by *Rhodobacter sphaeroides* KD131 and its uptake hydrogenase and PHB synthase deficient mutant. Int J Hydrog Energy 31:121–127

Klamt S, Grammel H, Straube R, Ghosh R, Gilles ED (2008) Modeling the electron transport chain of purple non-sulfur bacteria. Mol Syst Biol 4:156–174

Koblízek M, Shih JD, Breitbart SI, Ratcliffe EC, Kolber ZS, Hunter CN, Niederman RA (2005) Sequential assembly of photosynthetic units in *Rhodobacter sphaeroides* as revealed by fast repetition rate analysis of variable bacteriochlorophyll a fluorescence. Biochim Biophys Acta 1706:220–231

Koku H, Eroğlu I, Gündüz U, Yücel M, Türker L (2002) Aspects of metabolism of hydrogen production by *Rhodobacter sphaeroides*. Int J Hydrog Energy 27:1315–1329

Koku H, Eroğlu I, Gündüz U, Yücel M, Türker L (2003) Kinetics of biological hydrogen production by the photosynthetic bacterium *Rhodobacter sphaeroides* O.U. 001. Int J Hydrog Energy 28:381–388

Kondo T, Arakawa M, Hirai T, Wakayama T, Hara M, Miyake J (2002a) Enhancement of hydrogen production by a photosynthetic bacterium mutant with reduced pigment. J Biosci Bioeng 93:145–150

Kondo T, Arakawa M, Wakayama T, Miyake J (2002b) Hydrogen production by combining two types of photosynthetic bacteria with different characteristics. Int J Hydrog Energy 27:1303–1308

Larimer FW, Chain P, Hauser L, Lamerdin J, Malfatti S, Do L, Land ML, Pelletier DA, Beatty JT, Lang AS, Tabita FR, Gibson JL, Hanson TE, Bobst C, Torres y Torres JL, Peres C, Harrison FH, Gibson J, Harwood CS (2004) Complete genome sequence of the metabolic versatile photosynthetic bacterium *Rhodopseudomonas palustris*. Nat Biotechnol 22:55–61

Liao Q, Wamg YJ, Wang YZ, Zhu X, Tian X, Li J (2010) Formation and hydrogen production of photosynthetic bacterial biofilm under various illumination conditions. Bioresour Technol 101:5315–5324

Loach PA, Parker-Loach PS (2008) Structure-function relationships in bacterial light-harvesting complexes investigated by reconstitution techniques. In: Hunter CD, Daldal F, Thurnauer MC, Beatty JT (eds) The purple phototrophic bacteria, vol 28, Advances in photosynthesis and respiration. Springer Science+Business Media, Dordrecht, pp 181–198

Maróti P (2008) Kinetics and yields of bacteriochlorophyll fluorescence: redox and conformation changes in reaction center of *Rhodobacter sphaeroides*. Eur Biophys J 37:1175–1184

McKinlay JB, Harwood CS (2010) Carbon dioxide fixation as a central redox cofactor recycling mechanism in bacteria. Proc Natl Acad Sci U S A 107:11669–11675

McKinlay JB, Harwood CS (2011) Calvin cycle flux, pathway constraints, and substrate oxidation state together determine the H_2 biofuel yield in photoheterotrophic bacteria. mBio 2:e00323-10

Miyake J (1998) The science of biohydrogen. In: Zaborsky OR (ed) Biohydrogen. Plenum Press, London, pp 7–18

Miyake J, Kawamura S (1987) Efficiency of light energy conversion to hydrogen by the photosynthetic bacterium *Rhodobacter sphaeroides*. Int J Hydrog Energy 12:147–149

Miyake J, Wakayama T, Schnackenberg J, Arai T, Asada Y (1999) Simulation of the daily sunlight illumination pattern for bacterial photo-hydrogen production. J Biosci Bioeng 88:659–663

Moreno-Vivian C, Hennecke S, Pühler S, Klipp W (1989) Open reading frame 5 (ORF5), encoding a ferredoxin-like protein, and *nif*Q are cotranscribed with *nif*E, *nif*N, *nif*X, and ORF4 in *Rhodobacter capsulatus*. J Bacteriol 171:2591–2598

Naud I, Meyer C, David L, Breton J, Gaillard J, Jouanneau Y (1996) Identification of residues of *Rhodobacter capsulatus* ferredoxin I important for its interaction with nitrogenase. Eur J Biochem 237:399–405

Oda Y, Samanta SK, Rey FE, Wu L, Liu X, Yan T, Zhou J, Harwood CS (2005) Functional genomic analysis of three nitrogenase isozymes in the photosynthetic bacterium *Rhodopseudomonas palustris*. J Bacteriol 187:7784–7794

Oda Y, Larimer FW, Chain PS, Malfatti S, Shin MV, Vergez LM, Hauser L, Land ML, Braatsch S, Beatty JT, Pelletier DA, Schaefer AL, Harwood CS (2008) Multiple genome sequences reveal adaptations of a phototrophic bacterium to sediment microenvironments. Proc Natl Acad Sci U S A 105:18543–18548

Okamura MY, Feher G (1992) Proton transfer in reaction centers from photosynthetic bacteria. Annu Rev Biochem 61:861–896

Okamura MY, Feher G (1995) Proton-coupled electron transfer reactions of QB in reaction centers from photosynthetic bacteria. In: Blankenship RE, Madigan MT, Bauer CE (eds) Anoxygenic photosynthetic bacteria, vol 2, Advances in photosynthesis and respiration. Springer Science+Business Media, Dordrecht, pp 577–593

Okamura MY, Paddock ML, Graige MS, Feher G (2000) Proton and electron transfer in bacterial reaction centers. Biochim Biophys Acta 1458:148–163

Ooshima H, Takakuwa S, Katsuda T, Okuda M, Shirasawa T, Azuma M, Kato J (1998) Production of hydrogen by a hydrogenase deficient mutant of *Rhodobacter capsulatus*. J Ferment Bioeng 85:470–475

Otzuki T, Uchiyama S, Fujiki K, Fukunaga S (1998) Hydrogen production by a floating-type photobioreactor. In: Zaborsky OR (ed) Biohydrogen. Plenum Press, London, pp 369–374

Öztürk Y, Yücel M, Daldal F, Mandacı S, Gündüz U, Türker L, Eroğlu I (2006) Hydrogen production by using *Rhodobacter capsulatus* mutants with genetically modified electron transfer chains. Int J Hydrog Energy 31:1545–1552

Paddock ML, Feher G, Okamura MY (2003) Proton transfer pathways and mechanism in bacterial reaction centers. FEBS Lett 555:45–50

Qian Y, Tabita FR (1996) A global signal transduction system regulates aerobic and anaerobic CO2 fixation in *Rhodobacter sphaeroides*. J Bacteriol 178:12–18

Reddy CSK, Ghai R, Rashmi V, Kalia C (2003) Polyhydroxyalkanoates: an overview. Bioresour Technol 87:137–146

Redwood MD, Paterson-Beedle M, Macaskie LE (2009) Integrating dark and light bio-hydrogen production strategies: towards the hydrogen economy. Rev Environ Sci Biotechnol 8:149–185

Rees DC, Howard JB (2000) Nitrogenase: standing at the crossroads. Curr Opin Chem Biol 4:559–566

Rey FE, Heiniger EK, Harwood CS (2007) Redirection of metabolism for biological hydrogen production. Appl Environ Microbiol 73:1665–1671

Robert B (2008) Spectroscopic properties of antenna complexes from purple bacteria. In: Hunter CD, Daldal F, Thurnauer MC, Beatty JT (eds) The purple phototrophic bacteria, vol 28, Advances in photosynthesis and respiration. Springer, Dordrecht, pp 199–212

Sasikala C, Ramana CV (1995) Biotechnological potentials of anoxygenic phototrophic bacteria: II. Biopolyesters, biopesticide, biofuel and biofertilizer. Adv Appl Microbiol 41:227–278

Sasikala K, Ramana CV, Raghuveer Rao P, Subrahmanyam M (1990) Effect of gas phase on the photoproduction of hydrogen and substrate conversion efficiency in the photosynthetic bacterium *Rhodobacter sphaeroides* O.U. 001. Int J Hydrog Energy 15:795–797

Schatt E, Jouanneau Y, Vignais PM (1989) Molecular cloning and sequence analysis of the structural gene of ferredoxin I from the photosynthetic bacterium *Rhodobacter capsulatus*. J Bacteriol 171:6218–6226

Seefeldt LC, Hoffman BM, Dean DR (2009) Mechanism of Mo-dependent nitrogenase. Annu Rev Biochem 78:701–722

Sener MK, Schulten K (2008) From atomic-level structure to supramolecular organization in the photosynthetic unit of purple bacteria. In: Hunter CD, Daldal F, Thurnauer MC, Beatty JT (eds) The purple phototrophic bacteria, vol 28, Advances in photosynthesis and respiration. Springer, Dordrecht, pp 275–294

Shapleigh JP (2008) Dissimilatory and assimilatory nitrate reduction in the purple photosynthetic bacteria. In: Hunter CD, Daldal F, Thurnauer MC, Beatty JT (eds) The purple phototrophic bacteria, vol 28,

Advances in photosynthesis and respiration. Springer, Dordrecht/London, pp 623–642
Sheuring S, Rigaud JL, Sturgis JN (2004) Variable LH2 stoichiometry and core clustering in native membranes of *Rhodospirillum photometricum*. EMBO J 23:4127–4413
Sheuring S, Lévy D, Rigaud JL (2005) Watching the components of photosynthetic bacterial membranes and their in situ organization by atomic force microscopy. Biochim Biophys Acta 1712:109–127
Shinkarev VP, Wraight CA (1993) Electron and proton transfer in the acceptor quinone complex of reaction centers of phototrophic bacteria. In: Deisenhofer J, Norris JR (eds) The photosynthetic reaction center, vol 1. Academic, San Diego, pp 193–255
Steinborn B, Oelze J (1989) Nitrogenase and photosynthetic activities of chemostat cultures of *Rhodobacter capsulatus* 37b4 grown under different illuminations. Arch Microbiol 152:100–104
Swem LR, Gong X, Yu CA, Bauer CE (2006) Identification of a ubiquinone-binding site that affects autophosphorylation of the sensor kinase RegB. J Biol Chem 281:6768–6775
Takaichi S (2008) Distribution and biosynthesis of carotenoids. In: Hunter CD, Daldal F, Thurnauer MC, Beatty JT (eds) The purple phototrophic bacteria, vol 28, Advances in photosynthesis and respiration. Springer, Dordrecht/London, pp 97–117
Tian X, Liao Q, Zhu X, Wang Y, Zhang P, Li H, Wang H (2010) Characteristics of a biofilm photobioreactor as applied to photo-hydrogen production. Bioresour Technol 101:977–983
Tredici MR (1999) Bioreactors, photo. In: Flickinger MC, Drew SW (eds) Encyclopedia of bioprocess technology: fermentation, biocatalysis, and bioseparation, vol 1. Wiley, New York, pp 395–419
Uyar B, Eroğlu I, Yücel M, Gündüz U, Türker L (2007) Effect of light intensity, wavelength and illumination protocol on hydrogen production in photobioreactors. Int J Hydrog Energy 32:4670–4677
Vasilyeva L, Miyake M, Khatipov E, Wakayama T, Sekine M, Hara M, Nakada E, Asada Y, Miyake J (1999) Enhanced hydrogen production by a mutant of *Rhodobacter sphaeroides* having an altered light-harvesting system. J Biosci Bioeng 87:619–624
Vignais PM, Billoud B (2007) Occurrence, classification, and biological function of hydrogenases: an overview. Chem Rev 107:4206–4272
Vignais PM, Colbeau A, Willison JC, Jouanneau Y (1985) Hydrogenase, nitrogenase, and hydrogen metabolism in the photosynthetic bacteria. Adv Microb Physiol 26:155–234
Vincenzini M, Marchini A, Ena A, De Philippis R (1997) H_2 and poly-β-hydroxybutyrate, two alternative chemicals from purple non sulfur bacteria. Biotechnol Lett 19:759–762
Vonshak A, Richmond A (1985) Problems in developing the biotechonology of algal biomass production. Plant Soil 89:129–135
Wakayama T, Miyake J (2002) Light shade bands for the improvement of solar hydrogen production efficiency by *Rhodobacter sphaeroides* RV. Int J Hydrog Energy 27:1495–1500
Walz T, Ghosh R (1997) Two-dimensional crystallization of the light-harvesting I-reaction center photounit from *Rhodospirillum rubrum*. J Mol Biol 256:107–111
Walz T, Jamieson SJ, Bowers CM, Bullogh PA, Hunter CN (1998) Projection structures of three photosynthetic complexes from *Rhodobacter sphaeroides*: LH2 at 6 Å, LH1 and RC-LH1 at 25 Å. J Mol Biol 282:833–845
Wang YZ, Liao Q, Zhu X, Chen R, Guo CL, Zhou J (2013) Bioconversion characteristics of *R. palustris* CQK 01 entrapped in a photobioreactor for hydrogen production. Bioresour Technol 135:331–338
Williams JC, Allen JP (2008) Directed modification of reaction centers from purple bacteria. In: Hunter CD, Daldal F, Thurnauer MC, Beatty JT (eds) The purple phototrophic bacteria, vol 28, Advances in photosynthesis and respiration. Springer, Dordrecht/London, pp 337–353
Willison JC, Pierrard J, Hübner P (1993) Sequence and transcript analysis of the nitrogenase structural gene operon *nif*HDK of *Rhodobacter capsulatus*: evidence of an intramolecular processing of *nif*HDK mRNA. Gene 133:39–46
Wraight CA (2004) Proton and electron transfer in the acceptor quinone complex of bacterial photosynthetic reaction centers. Front Biosci 9:309–327
Wraight CA (2005) Intraprotein proton transfer – Concept and realities from the bacterial photosynthetic reaction center. In: Wilkström M (ed) Biophysical and structural aspects of bioenergetics. Royal Society of Chemistry, Cambridge, pp 273–313
Xie GJ, Liu BF, Ding J, Xing DF, Ren HY, Guo WG, Ren NQ (2012) Enhanced photo-H_2 production by *Rhodopseudomonas faecalis* RLD-53 immobilization on activated carbon fibers. Biomass Bioenergy 44:122–129
Zannoni D (1995) Aerobic and anaerobic electron transport chains in anoxygenic phototrophic bacteria. In: Blankenship RE, Madigan MT, Bauer CE (eds) Anoxygenic photosynthetic bacteria, vol 2,

Advances in photosynthesis and respiration. Springer Science+Business Media, Dordrecht, pp 949–971

Zannoni D, Schoepp-Cothenet B, Hosler J (2008) Respiration and respiratory complexes. In: Hunter CD, Daldal F, Thurnauer MC, Beatty JT (eds) The purple phototrophic bacteria, vol 28, Advances in photosynthesis and respiration. Springer, Springer Science+Business Media, Dordrecht, pp 537–561

Zhang C, Zhu X, Liao Q, Wang Y, Li J, Ding Y, Wang H (2010) Performance of a groove-type photobioreactor for hydrogen production by immobilized photosynthetic bacteria. Int J Hydrog Energy 35:5284–5292

Zuber H, Brunhisolz RA (1991) Structure and function of antenna polypeptides and chlorophyll-protein complexes: principles and variability. In: Sheer H (ed) Chlorophylls. CRC Press, Boca Raton, pp 627–703

Chapter 13

Photobioreactors Design for Hydrogen Production

José Maria Fernández-Sevilla,
Francisco Gabriel Acién-Fernández, and Emilio Molina-Grima*
Department of Chemical Engineering, University of Almería, Almería 04071, Spain

Summary

The photobiological production of H_2 is a subject that has been studied with great intensity over the past 50 years using different approaches; direct or indirect biophotolysis (green algae and cyanobacteria) and photo-fermentations (photosynthetic bacteria). The number of publications on the subject is impressive. However, hardly any of the production methods proposed have progressed beyond the laboratory, and the photobioreactors (PBR) used to carry out the processes are still bench-top scale laboratory devices. The scale up of some of the proposed PBR to carry out the process outdoor using full solar radiation is just beginning and the existing data are too scarce.

This chapter is mainly addressing the major issues in the design and scale up of photobioreactors (PBR) for the eventual photobiological production of H_2 when using an envisaged two-stage scheme. A first one in which microalgae are cultivated in large open ponds to produce microalgae biomass with a high C/N ratio; then, by changing the physiological conditions, a second anoxygenic step to produce hydrogen in closed PBRs. The different

*Author for correspondence, e-mail: emolina@ual.es

D. Zannoni and R. De Philippis (eds.), Microbial BioEnergy: Hydrogen Production, Advances in Photosynthesis and Respiration 38, pp. 291–320, DOI 10.1007/978-94-017-8554-9_13,

designs currently used for practical microalgae mass culture are reviewed, identifying their characteristic parameters. The major operational variables impacting on PBR performances are also highlighted, as well as the challenges associated with the PBR design and scale up. Finally, the bottlenecks for the scaling up of the different technologies and thus of the photobiological H_2 production are discussed.

I. Introduction

The current hydrogen production is based on thermochemical processes (mainly steam reforming of methane) that use fossil fuel, thus being great producers of CO_2. Only a small fraction of current world production comes from water electrolysis (Berberoglu et al. 2008). This latter approach would be even more sustainable if the electricity required for electrolysis came from photovoltaic cells. However, this technology is quite expensive and the cost remains a major drawback. Therefore, the world's growing energy needs will place much greater reliance on a combination of fossil fuel-free energy sources and new technologies for capturing sunlight and converting atmospheric CO_2. Within these technologies falls the photosynthetic hydrogen production from sunlight and water with the possible advantage of CO_2 capture, which has been investigated with great dedication over the past 50 years (Levin et al. 2004; Laurinavichene et al. 2006; Das and Veziroglu 2008). The advantages most highlighted in these studies, with respect to the current industrial technologies for producing H_2 are: (a) the biological H_2 production occurs under mild temperature and pressure conditions, (b) the reaction specificity is typically higher than that of inorganic catalysts used in thermochemical processes and (c) there is a diverse collection of raw materials, including waste, that can serve as feedstock for the production of photobiological H_2.

In theory, the photobiological production of hydrogen from water requires an efficient microalgae capable of converting protons to H_2 and low cost PBRs. The selected microalgal strains should exhibit a high hydrogen production rate and high light to hydrogen conversion efficiencies when using dense cultures outdoors (Benemann 1997, 2000). The PBRs must expose the H_2-producing cultures to sunlight and at the same time allow the recovery of the gas produced (Berberoglu et al. 2008). Most of the studies on photobiological H_2 production so far available have been carried out in well-equipped lab scale photobioreactors having various geometries (Levin et al. 2004; Berberoglu et al. 2008; Oncel and Sabankay 2012). In addition, due to differences in (i) the design of the PBRs, (ii) the light sources, (iii) the temperature of operation, (iv) the microorganism used and (v) the media used, the results reported in the literature for H_2 production show large variations of over a hundred fold (Eroglu and Melis 2011). For example: the maximum reported H_2 production rate for *C. reinhardtii* outdoors was 0.61 ml L^{-1} h^{-1} in a process which used a second phase with sulphur deprivation to induce H_2 production (Giannelli and Torzillo 2012). However, for the cyanobacterium *Anabaena variabilis,* some of the reported data are: 167.6 mmol H_2 g chl a^{-1} h^{-1} in a process using indirect biophotolysis (Sveshnikov et al. 1997); or the 150 mmol H_2 g chl a^{-1} h^{-1} reported for the marine green algae *Scenedesmus obliquus* (Florin et al. 2001). In contrast, the maximum reported H_2 production rates when using purple-non sulphur (PNS) bacteria in a outdoor photofermentative process was 27.2 ml L^{-1} h^{-1} by using a horizontal tubular PBR (Adessi et al. 2012), and 5.9 mol H_2 kg^{-1} h^{-1} for *Rhodobacter spharoides* (Sasikala et al. 1991).

Although some advances have been recently made, there are still many problems in the different routes proposed that should have to be solved before a practical photobiological production process could be set up. The major drawbacks so far reported are related to the very low yields attained, well below those achievable for the production of other biofuels from the same feedstocks (Hallenbeck and Benemann 2002). Moreover, low yields may also lead to the generation of side products which being produced in large volumes, would generate a significant disposal problem. On the other hand, another major challenge when scaling up the photobiological H_2 production, is connected with the use of economically viable PBRs, and the sparse data available on outdoor production of H_2 using a fully sealed PBR (Hallenbeck et al. 2012).

II. Major Routes for the Photobiological H_2 Production

There are several pathways for photobiological hydrogen production (note that dark fermentation for producing H_2 is not addressed in this chapter). Photobiological production of H_2 includes three distinct routes such as (1) direct biophotolysis, (2) indirect biophotolysis and (3) photo-fermentation. It is also possible to design integrated systems incorporating both photosynthetic and fermentative processes (Levin et al. 2004).

Direct Biophotolysis. When microalgae use this pathway, the energy source is the sunlight in the spectral range from 400–700 nm. The catalyst producing hydrogen is the enzyme hydrogenase, extremely sensitive to oxygen. This mechanism can be considered to be the photobiological electrolysis of water and it is, theoretically, the most energy efficient for H_2 production (Prince and Kheshgi 2005). However, the oxygen produced during water splitting, irreversibly inhibits the functioning of the [Fe-Fe]-hydrogenase and the practical light-to-hydrogen conversion efficiency at full solar radiation is well below 0.1 %. This makes the process impractical for industrial applications (Hallenbeck and Benemann 2002). Green algae such as *Chlamydomonas reinhardtii*, *Scenedesmus obliquus*, and *Chlorococcum littorale* are capable of producing H_2 via direct biophotolysis (Das and Veziroglu 2001) with H_2 production rates varying between 2.5 and 13 ml of H_2 L^{-1} h^{-1} (Tsygankov et al. 1998; Laurinavichene et al. 2006).

Indirect Biophotolysis. In this mechanism, the source of electrons is also water. The electrons are first used to reduce CO_2 to form organic compounds during photosynthesis and O_2 is simultaneously generated and released. Then, in a second step, that can be carried out in the same reactor if it is closed or in another closed reactor if the first step was carried out in an open reactor, electrons are recovered from the oxidation of the organic compounds and used in generating H_2 through the action of nitrogenase (Hallenbeck and Benemann 2002). Thus, no O_2 is generated during H_2 production. Cyanobacteria such as *Anabaena variabilis*, are capable of indirect biophotolysis (Sveshnikov et al. 1997; Tsygankov et al. 1999; Pinto et al. 2002). The maximum theoretical light to H_2 energy conversion efficiency of indirect biophotolysis is about 60 % lower than that of direct biophotolysis (Prince and Kheshgi 2005) due to the fact that (i) multiple steps are involved in converting solar energy to H_2 and (ii) the use of nitrogenase enzyme requires ATP. However, the problems associated to O_2 that happen in direct biophotolysis make its actual efficiency lower than in the indirect biophotolysis. Production rates of 12.5 ml kg dry $cells^{-1}$ h^{-1} were obtained (Markov et al. 1997) by using a partial vacuum in the second step; similar to those achieved by Tsygankov et al. (2002). Yoon et al. (2006) incorporated nitrate to the culture medium to enhance the biomass productivity of *Anabaena* in the first step and increased the H_2 production rate in the second step. The authors reported maximum specific H_2 production rates of 4.1 and 0.45 L kg dry $cells^{-1}$ h^{-1} for flat panel reactors of 2 and 4 cm thick respectively. They flushed with argon in order to force the removal of the dissolved H_2 and to increase the H_2 production in the second step, as an alternative to the partial vacuum used by Markov et al. (1997). Even higher productivities were obtained by Berberoglu et al. (2008): 5.6 L kg dry cells^{-1} h^{-1}, at 1 atm and 30 °C, when using Allen-Arnon medium in comparison with those achieved when using BG 11 medium (1 L kg dry cells^{-1} h^{-1}). However again, the reality is that the conversion efficiencies achieved by indirect biophotolysis are also below 1 %.

Photo-Fermentation. This mechanism is similar to indirect biophotolysis with the distinction that the organic compounds used are produced outside the cells via the photosynthesis of other organisms. These extracellular organic materials, such as organic acids, carbohydrates, starch, and cellulose (Kapdan and Kargi 2006), are used as electron source and sunlight is used as energy source to produce H_2 by the enzyme nitrogenase (Das and Veziroglu 2001). Since this enzyme is repressed by nitrogen fixation, photofermentation needs a high C/N ratio biomass. Since cells do not carry out oxygenic photosynthesis, no O_2 is generated and all the solar energy can be used to produce H_2. Thus, this mechanism is viewed as the most promising microbial system to produce H_2 (Das and Veziroglu 2001). The major advantages of this route are (i) the absence of O_2 evolution, which inhibits the H_2-producing enzymes, and (ii) the ability to consume a wide variety of organic substrates found in waste waters. Due to their ability to harvest a wider light spectrum, from 300 to 1,000 nm, Purple non-sulphur (PNS) bacteria such as *Rhodobacter sphaeroides, Rhodobacter capsulatus* and *Rhodopseudomonas palustris* hold promise as photo-fermentative H_2 producers (Sasikala et al. 1991; Das and Veziroglu 2001; Adessi et al. 2012).

As stated before, the [Fe-Fe] hydrogenase involved in H_2 generation during biophotolysis is irreversibly inactivated by the O_2 produced during water splitting. A milestone in solving this problem when using the green algae *C. reinhardtii* was achieved by Melis and co-workers in 2000. The key of this process lies on the second phase of microalgae growth in which Photosystem II (PSII), that plays a vital role in oxygen generation, is deactivated the by sulphur deprivation (Melis et al. 2000). They demonstrated H_2 production in a sulphur-depleted, sealed, illuminated *C. reinhardtii* culture for several days, starting one day after sulphur deprivation (Melis et al. 2000). In a first step the algae was grown in a medium containing acetate until a high biomass concentration was obtained. Upon harvesting, the biomass slurry was transferred to a sulfur-deprived medium in a second step. Sulphur deprivation causes a progressive decrease in the photosynthetic O_2-evolving capacity of the cells, because the PSII repair function is slowed nearly to a halt (Wykoff et al. 1998; Melis et al. 2000). When the photosynthesis rate drops below the level of respiration, the culture becomes anaerobic in a short period time as long as no oxygen is present in the photobioreactor (Melis et al. 2000). Under these conditions *C. reinhardtii* is able to synthesize an [Fe-Fe]-hydrogenase which combines electrons and protons from a remaining PSII activity and especially from the degradation of the starch (that has been accumulated in first stage 1 of the whole process) to produce significant amounts of H_2 for several days (Antal et al. 2003; Posewitz et al. 2004; Rupprecht et al. 2006; Hemshemeier et al. 2008; Chochois et al. 2009). In all the green microalgal species analyzed so far hydrogenases are coupled to the photosynthetic electron transport chain via ferredoxin, which after being reduced by PS1 donates its electrons to the hydrogenase, as has been shown for *C. reinhardtii* (Winkler et al. 2009).

This chapter addresses the main aspects to be tackled in the design of photobioreactors for the photobiological production of H_2 from green microalgae when using the route proposed by Melis and co-workers. For economic reasons, the growth phase should be performed first in raceways open ponds. Those systems are not appropriate for the H_2 production phase since the collection of the desorbing hydrogen would pose serious difficulties unless fully sealed closed photobioreactors were used. Figure 13.1 schematically illustrates a typical process flow envisioned for continuous photobiological hydrogen production at industrial scale.

The photobioreactor configuration proposed in Fig. 13.1 could also be used, with slight operational modifications and the establishing the proper physiological conditions, for the production of H_2 by indirect biophotolysis with cyanobacteria. Thus, in case the selected strain for producing H_2 is a cyanobacteria, the biomass production step could be carried out in the open photobioreactor and the subsequent H_2 evolving step in the closed tubular system (Fig. 13.1a). To that end, if tubular technology were used in the H_2 production step not even a O_2-free airstream can be used for degassing as both enzymes [Ni-Fe] hydrogenase and nitrogenase (depending on the cyanobacteria used) are inhibited by any nitrogen source. Any N_2 could be assimilated through the heterocysts of the cyanobacteria. Therefore, when using the tubular technology with cyanobacteria in this step, the centrifugal pump employed to drive the culture through the circuit would also have to be used to transport, coalesce and separate the H_2 bubbles that would gather in the upper part of the tubes and collect them in the degasser. H_2 removal in these conditions can be further favoured by a partial vacuum established to this purpose in the headspace of the degasser. On the other hand, the schemes proposed in Fig. 13.1a, b, could be used for producing H_2 by a photo-fermentative process of the biomass produced in the open raceway by using PNS bacteria to degrade the organic material (i.e., biomass coming from the open raceway) in the enclosed photobioreactors. Therefore, according with the photosynthetic microorganism used, the scheme of Fig. 13.1 might be used for producing H_2 by means of both photolytic and photofermentative routes. The number of modules required will depend on the biomass productivity of the first step. Details about how estimate them depending on the microorganism and technology used in this second step are described later (see scale up section for each type of technology). Finally, the depleted biomass can be used as a feedstock for biofuel. Depending on the downstream process chosen for this biomass (thermochemical, anaerobic digestion, direct transesterification, etc.) the characteristics of the produced biofuel will be different (bio-oil, biogas, biodiesel, etc.).

Regardless the photobioreactor used in either the first step of biomass production or in the hydrogen production phase, the photobioreactor must be designed and operated to meet the optimal conditions required for the microalgal strain in each phase of the process, making clear how important is to know what the main factor determining growth rate either phase are.

III. Major Factors Impacting on Photobioreactor Performance

Microalgae, like all living microorganisms, require the proper conditions to grow. The best the culture conditions the higher the growth rate and productivity obtained. Thus, the pH, temperature and culture medium must be those appropriate for the microalgae. The mineral culture medium can be easily supplied so as to not limit the growth; however, the correct supply of light requires a deep analysis due to its different nature with respect to mineral nutrients. In the same way, CO_2 must be supplied to meet the carbon demand and maintain the proper pH in the culture. At the same time, the assimilation of CO_2 provokes the consumption of water and the generation of oxygen that must be removed. This consequently impacts the required mass transfer capacity of the culture system. In addition, the power supply to maintain the cells in suspension and reduce light and nutrient gradients within the culture must be considered, keeping in mind that inadequate fluid dynamics can have deleterious effects on fragile microalgal cells. Finally the control of temperature is mandatory to ensure the optimal temperature to grow and also to prevent overheating.

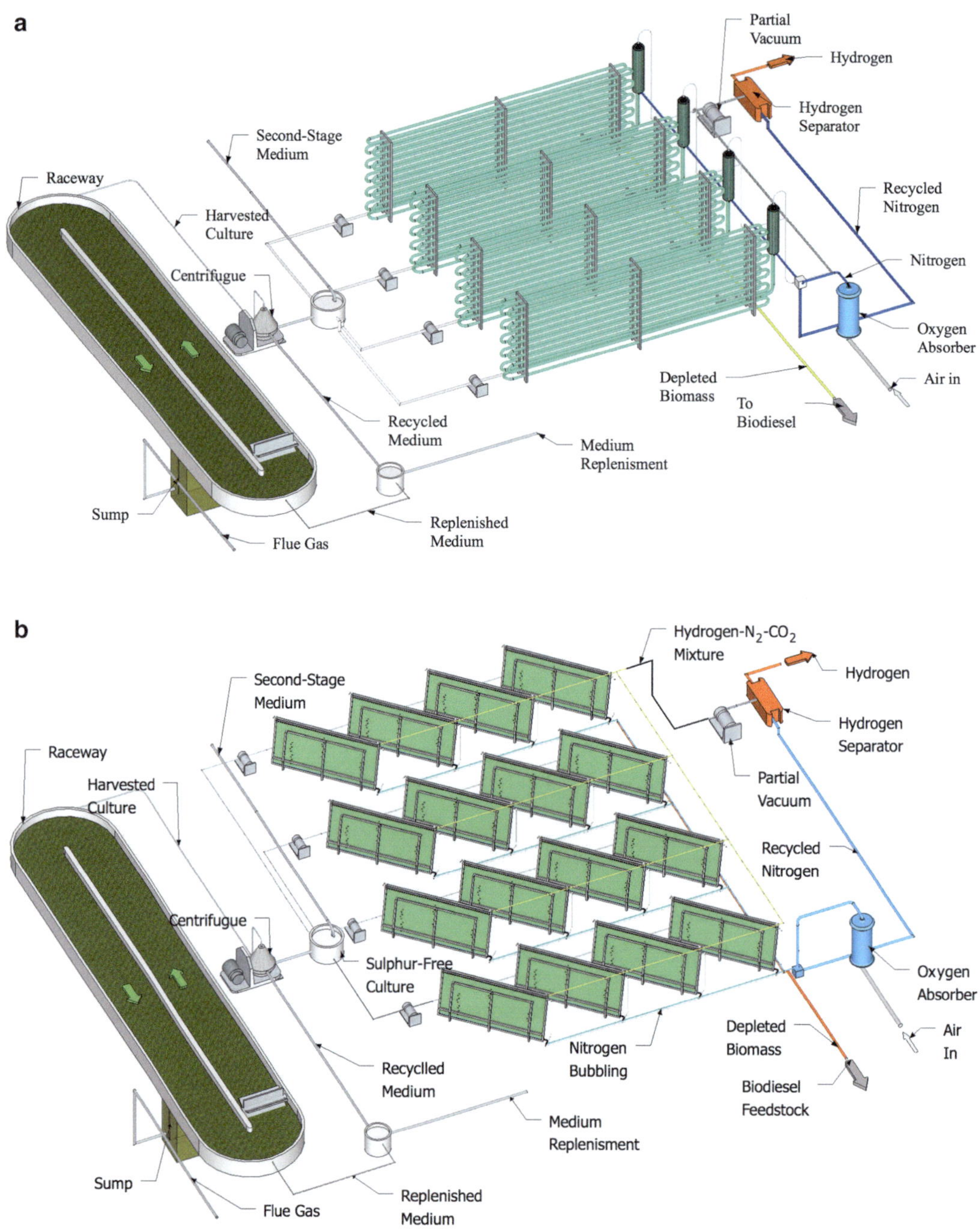

Fig. 13.1. Conceptual scheme of an integrated, microalgae-based, continuous two-step process for H_2 production with tubular PBR (**a**) or flat plate reactor (**b**). The biomass from the raceway reactor of the first step is harvested by centrifugation. The daily harvested slurry is then transferred to the appropriate culture medium and drove to the closed reactors by centrifugal pumps. The H_2 produced is stripped by bubbling in the degasser an *in situ* prepared O_2-free airflow (bubble column or the entire culture, for tubular PBR or flat plate reactor, respectively). Upon separated the H_2 from the rest of gases, mainly N_2 and a small amount of CO_2, the O_2-free gas is recycled. Finally the depleted biomass can be used for additional biofuels. The most appropriate technology to use in the H_2 production step (flat plate or tubular) and culture medium are depending of the microorganism used for producing H_2. For details about the selection of technology, culture media, procedures for producing the O_2-free airflow and downstream processing possibilities for the depleted biomass (see text).

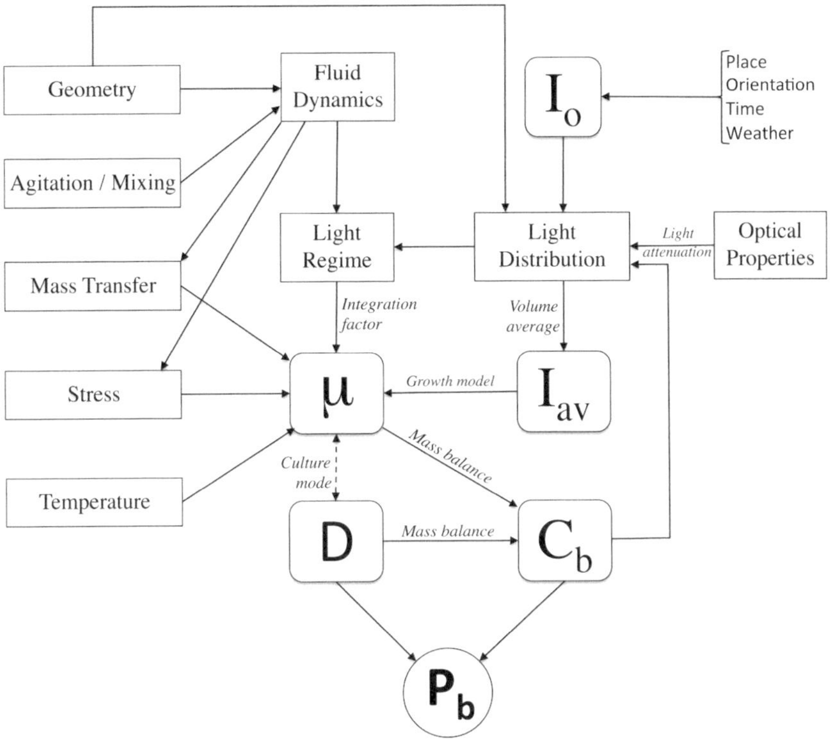

Fig. 13.2. Relationships between the major factors influencing biomass productivity of microalgal mass cultures. Incident irradiance (I_o) promotes microalgal growth. The biomass productivity (P_b) is the result of multiplying biomass concentration (C_b) by the specific growth rate (μ), which is a function of light availability (calculated as average irradiance I_{av}). In continuous cultures, the dilution rate (D) equals μ in steady state (Adapted from Molina 1999 and Molina et al. 2010).

Figure 13.2 shows the main factors impacting on biomass productivity and their relationships, whatever the outdoor photobioreactor used (open or closed). Light availability inside the photobioreactor and temperature have been found to be the main factors determining optimum system performances (Molina Grima 1999). Thus, when the temperature of the culture is kept within an appropriate interval, light availability is the only factor limiting growth. The productivity is determined by the growth rate (μ) and the biomass concentration (C_b), which is a function of the light distribution inside the photobioreactor and the light regime at which the cells are subjected. Once this function is known, it is possible to obtain a correlation between biomass productivity and the average irradiance within the reactor (Acién-Fernández et al. 1998). On the other hand, average irradiance within the reactor is a function of the irradiance impinging on the reactor surface (I_o), which comes as a consequence of the geographic and environmental factors (Acién-Fernández et al. 1997, 2012). The geographic location and day of the year determine the solar incident radiation and therefore the temperature in the culture. While temperature can be kept within a narrow interval by using suitable thermostatic systems, solar radiation cannot be controlled.

The incident solar radiation, which is a function of climatic and geographic parameters of the facility location (Incropera and Thomas 1978), as well as the design and orientation of the photobioreactor (Lee and Low 1992; Qiang and Richmond 1996; Sierra et al. 2008), determines the maximum energy available for growth. The incident solar radiation is attenuated inside the photobioreactor in a way that depends on the geometry, the biomass concentration and the optical properties of the biomass. Thus, a heterogeneous light distribution always takes place inside dense microalgal cultures (Molina Grima et al. 1994; Acién-Fernández et al. 1997). The light availability or average irradiance inside the culture can be calculated by a volumetric integration of the irradiance profile. The cell metabolism adapts to this light availability and so does the biochemical composition and growth rate (Acién-Fernández et al. 1998). On the other hand, the photobioreactor fluid dynamics also determine the mass transfer and the light regime of the cells. The latter is the result of the distribution of residence times between light and dark zones (Phillips and Myers 1954; Terry 1986; Grobbelaar 1994). This light regime affects the photosynthetic efficiency of the cultures, modulating the use of the available light calculated as averaged irradiance (I_{av}). High mixing favours an adequate light regime and thus an efficient use of light although the mixing power supplied to the cultures must not reach intensities that can damage the individual cells by hydrodynamic stress.

The biomass concentration is influenced by the growth rate. Biomass accumulation is the result of a mass balance between biomass generation (μ) and the output rate that can be expressed as a dilution rate (D) in continuous or semicontinuous cultures. Both, growth rate and biomass concentration, determine the final biomass productivity of the system. Thus, in an optimum system where there are no limitations other than light, a direct interrelationship between light availability, rate of photosynthesis and productivity may be expected. In fact, it seems that other limitations do not only limit growth through their direct effect, but also impose a limitation on the ability to utilize the absorbed solar energy. Therefore, in the end the most important design criterion will be to enhance the light availability per cell and consequently high efficiency in transforming the sunlight reaching the culture.

IV. Principles for Photobioreactors Design and Scale Up

According with the envisioned scheme of Fig. 13.1, for the hydrogen production step, a PBR with high surface to volume ratio is mandatory to obtain high cell densities, maximum interception of light per unit of occupied area, a good light distribution within the reactor and an efficient H_2 removal to overcome product inhibition (Akkerman et al. 2002; Posten 2009). Flat plate reactors (FPRs), alveolar plate reactors and tubular reactors may provide such a high surface to volume ratios. These culture systems may provide the requested mass transfer capabilities for stripping the dissolved H_2 from the culture and maintain the temperature of the culture within acceptable limits. This section describes the basic principles for designing both the open raceway (RWs), to be used for the preparation of the inocula, and closed reactors needed in the scale up of a process for photobiological production of hydrogen with microalgae.

A. Open Raceway (RW) Reactor

The principles for the design and construction of shallow paddle-stirred raceways for large microalgal production (Oswald and Golueke 1968) were reviewed by Oswald (1988) and recently have been updated by Chisti (2013) and Acién-Fernández et al. (2013). A theoretical approach to modelling microalgae growth in raceway reactors, taking into account the biological and hydrodynamic phenomena occurring in the reactor, has recently been reported (James and Boriah 2010). A photograph of an open 100 m^2 set-up at the authors' facility is showing in Fig. 13.3.

Selection of a suitable bottom lining and wall construction are important to the success of the open pond. The lining may be made of concrete, sheets of plastic or rubber material. The stirring is accomplished with one paddle wheel per pond to match the designed liquid depth for the raceway. The presence of deflectors in the bends also improves the performance of the RW by enhancing the liquid velocity and reducing the power consumption for the same liquid velocity. Recently the use of asymmetric islands to minimize existence of dead zones in the bend areas and maximize the energy yield has also been reported (Chisti 2013). A wheel of large diameter (ca. 2.0 m in

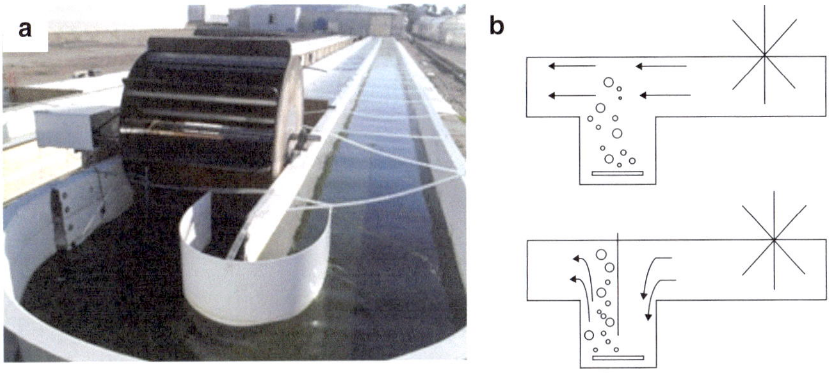

Fig. 13.3. Image of an open 100 m² raceway set-up at the authors' facilities (**a**), and scheme of the sump configurations tested (**b**). The channel length is the distance travelled by the culture from the discharge side of the paddle wheel to the entering side.

diameter) revolving slowly (e.g. 10 rpm) is preferable to smaller diameter wheels that have to rotate faster and produce excessive shear damage and foam. Under these conditions biomass concentrations of up to $1.0\ g \cdot L^{-1}$ and productivities of $0.1\ g \cdot L^{-1} \cdot d^{-1}$ are possible. The two major technical issues to keep in mind at the time of designing an open RW are related to the poor gas-liquid mass transfer and mixing capability of the current industrial scale reactors which have a great impact in RW performance.

1. Mass Transfer. The Supply of Carbon Dioxide and the Removal of Oxygen in an Open RW

Several systems have been developed to supply CO_2 efficiently to shallow suspensions. In most cases the gas is supplied in form of fine bubbles. Due to the shallowness of the suspension the residence time of the bubbles is not sufficient for the CO_2 to dissolve and a significant part of the CO_2 supplied is lost to the atmosphere. For this reason, raceway reactors are frequently equipped with devices such as sumps or mixing columns to increase the gas/liquid contact time (Azov and Shelef 1982; Weissman and Goebel 1987; Weissman et al. 1988; Doucha et al. 2005; Moheimani and Borowitzka 2007; Park and Craggs 2010; Park et al. 2011; Putt et al. 2011). An arrangement frequently used to attain this is to install a baffle dividing vertically the sump in two sections of ascending and descending liquid (Fig. 13.3b). In this sump configuration the culture velocity in the descending section can be adjusted to match the ascension of the small CO_2 bubbles. Using this technique, Laws et al. (1986) reported a 70 % efficiency in CO_2 transfer, a large improvement compared with the 13–20 % efficiency obtained when gas is injected in the shallow channel. The use of sumps is the simplest way to improve carbonation in raceways because they can be easily incorporated in the channels and do not need an external energy supply. However, we have carried out experiments with and without baffle with the raceway pond shown in Fig. 13.3a, and the usefulness of introducing a baffle into the sump is questionable. There was a slight increase of the mass transfer capacity in the sump in the experiment at the expense of increased power consumption and a reduction in the culture velocity and the mixing degree in the system. In the authors' opinion the sump configuration with baffle for the counter-current injection of CO_2 should be considered as a serious disadvantage. The CO_2 transfer under both sump configurations was tested in the same experimental set up and similar CO_2 uptake efficiencies in the sump were achieved (>95 %).

On the other hand, oxygen desorption has been usually disregarded in the design of these type of photobioreactors. Thus, oxygen removal is performed through the culture surface the mass transfer coefficient being extremely low and oxygen accumulating into the culture (Jiménez et al. 2003). It has been previously reported that, in large ponds with small water circulation and turbulence, O_2 concentration may reach concentrations as high as 500 % saturation, inhibiting photosynthesis and growth, and eventually leading to culture death (Vonshak 1997).

The mass transfer capacity of the system (i.e., *Kla* coefficient) and mixing (given by both mixing time and dispersion coefficient) are key parameters to be considered in the proper design of a raceway system for CO_2 supplying and for removal of the photosynthetic oxygen generated. As the oxygen generation rate is about 1.3–1.4 g O_2 per gram of biomass and that the CO_2 consumption rate is roughly 2 g per gram of biomass produced, the minimum *Kla* values which would be requested for satisfying an appropriate mass transfer capacity in a raceway are much higher (one order of magnitude greater) than those currently provided in the existing industrial raceway reactors. Taking into account that, it makes no sense to supply CO_2 in the channel because in this zone the mass transfer capacity is virtually zero, the only zone available for CO_2 supplying is the sump. On the other hand, for O_2 degassing the available zones are the sump and the zone of the paddle wheel. Nonetheless, the only practical way to increase the O_2 removal, for a determined paddle wheel configuration rotating at a specific frequency, is by increasing the relative volume of the sump with respect the total volume of the raceway and by increasing the depth of the sump.

2. *Mixing and Power Consumption*

The evolution of the open culture technology is the consequence of the development of the mixing systems that have been designed in the time. Mixing is necessary in order to prevent the cells from settling and sticking to the bottom and to avoid thermal stratification of the culture. Mixing is of paramount importance since it is directly linked to other key parameters (Fig. 13.2). Mixing determines the light-dark cycling frequency, improves the mass transfer capability of the culture system, reduces the mutual shading between cells and decreases the potential photoinhibition effect at the pond surface. Properly designed paddle wheels are by far the most efficient and durable pond mixers. They discharge all of the culture entering the system and are thus highly efficient. With reference to Fig. 13.3a, the design is based in a culture flowing at depth *d* in a channel with finite width, *w*, and unspecified length, *L*. Water depth (*d*) is maximum just after the discharge side of paddlewheel and minimum in the entering side. This depth reduction (Δd), termed head loss or depth change, determines the rate of energy that must be provided to maintain circulation at the chosen velocity. The head losses (energy dissipation) depend on: (i) flow around the two 180° curves (bend losses) and (ii) the friction with the surface (side wall and bottom). The head losses as water flows in bends is calculated by (Acién-Fernández et al. 2013):

$$\Delta d_b = \frac{k \cdot v^2}{2g} \tag{13.1}$$

where v is the mean velocity (ms^{-1}); g the acceleration of gravity (9.8 ms^{-2}) and k is the kinetic loss coefficient for each bend. Similarly, the channel and wall friction loss, Δd_c, can be calculated by (Acién-Fernández et al. 2013):

$$\Delta d_c = v^2 n^2 \frac{L}{R^{4/3}} \tag{13.2}$$

where n is the roughness factor; R (m) is the channel hydraulic radius: $R=4w{\cdot}d/(w+2d)$ and L the length of the channel (m). The total head loss or change in depth is $\Delta d=\Delta d_b+\Delta d_c$. The channel length, L, that corresponds to the calculated head losses for a given friction factor and a culture velocity (v) is given as Eq. (13.3):

$$L = \frac{\Delta d\left(dw/\left(w+2d\right)\right)^{4/3}}{v^2 \cdot n^2} \tag{13.3}$$

where n is the Manning friction factor ($s \cdot m^{-1/3}$), L is the channel length that corresponds to the head loss (Δd) and w is the channel width. The value of n varies according to the relative roughness of the channel. Experimentally determined n values in algae growth channels vary from 0.008 to 0.030, the former for smooth plastic-lined channels and the latter for relatively rough earth. The channel velocity, v, impacts on the paddle wheel's power requirements, calculated as:

$$P = \frac{Q \cdot \rho \cdot g \cdot \Delta d}{\eta} \quad (13.4)$$

where P is the power (kW); Q the culture flow rate in motion ($m^3\ s^{-1}$); ρ is the specific weight of culture ($kg \cdot m^{-3}$); g is the gravitational acceleration ($m\ s^{-2}$); Δd is the change in depth generated in the paddle wheel (m) and η is the efficiency of the paddle wheel, which usually is about 0.2. Because Δd is a function of v^2, the power consumption, P, increases as the cube of velocity. It is therefore worthwhile to minimize velocity whenever energy is a major cost factor. Typical values of flow rates range between 15 and 30 $cm \cdot s^{-1}$, whereas the power supply is around 2–4 $W \cdot m^{-3}$. Velocities greater than 30 $cm \cdot s^{-1}$ will result in large values of Δd in long channels and may require high channel walls and higher divider walls.

The mixing time reduces when liquid velocity through the system increases and the channel length-to-width ratio (L/w) decreases. For the system shown in Fig. 13.3a this L/w ratio is about 100 and the mixing time is about 2 h, when no baffle is inserted, or about 5 h when the baffle is inserted in the sump (Acién-Fernández et al. 2013). A more accurate quantification of mixing in the different zones of the raceway can be achieve by measuring the Bodenstein number (Bo) which is related to the dispersion coefficient (Dz) by (Verlaan et al. 1989):

$$Dz = \frac{v L_{section}}{Bo} \quad (13.5)$$

where $L_{section}$ represents the length of each zone within the reactor (channel, sump, paddlewheel and bends). As a rule of thumb, when Bo is ≤ 20 (i.e., high dispersion coefficient) the mixing pattern in that raceway zone is a perfect mixing and when Bo is ≥ 100 the pattern corresponds to a plug flow. Overall, in an industrial raceway pond with a length to width ratio over 50, the sump, paddle wheel and bends shows a complete mix pattern and in the channel however the pattern corresponds to a completely plug flow.

3. Scale-Up of Raceway Reactors

Reactor scale-up is based on reactor surface area rather than volume. An open RW facility capable of providing the necessary biomass for the H_2 production step (closed PBR described in Fig. 13.1) will require the RW facility running in continuous or semicontinuous mode (the biomass productivity in continuous mode is at least 2.3 times greater than in batch mode, and generally about five times greater). The question to solve in the scale up is to calculate the land demand of a RW facility and the number of raceway units needed for producing M tonnes of biomass (dry weight) per year of a specific microalgae. The tools for designing this facility are described next.

The growth rate of the strain can be calculated by Eq. (13.6) (Molina Grima et al. 1994).

$$\mu = \frac{\mu_{max} I_{av}^{n}}{I_k^n + I_{av}^{n}} \quad (13.6)$$

In Eq. (13.6), the maximum specific growth rate, μ_{max}, the light saturation constant, I_k, and the shape parameter, n, are kinetic parameters which are species specific and must be determined experimentally. In Eq. (13.6), I_{av} represents the average irradiance within the raceway that can be estimated by:

$$I_{av} = \frac{I_o}{K_a \cdot d \cdot C_b}\left(1 - \exp\left(K_a \cdot d \cdot C_b\right)\right) \quad (13.7)$$

Equation (13.7) is a simplified model to calculate I_{av}. This model is suitable for any combination of disperse and direct light as long as it is impinging uniformly on the reactor surface (Molina Grima et al. 1996; Acién-Fernández et al. 1997). According to this model, the average irradiance, I_{av}, is a function of the irradiance measured on the reactor surface, I_o, the extinction coefficient of the biomass, K_a, the optical light path (depth of the culture), d, and the biomass concentration in the culture, C_b.

The average volumetric biomass productivity (g L^{-1} d^{-1}) all year long in a continuous, or semicontinuous, culture is determined by:

$$P_{bv} = D \bullet C_b \tag{13.8}$$

where D is the average dilution rate all through the year (d^{-1}) and C_b the average biomass concentration during the year (g L^{-1}). As a rule of thumb, D is about 40 % the maximum specific growth rate of the strain and ranges between 0.2 and 0.5 d^{-1} for winter and summer time, respectively (Acién-Fernández et al. 2013). From Eq. (13.8), it is possible to calculate the areal productivity P_{ba}, taking into account the volume to surface ratio of the culture system.

$$P_{ba} = P_{bv} \frac{V}{S} \tag{13.9}$$

The V/S ratio is a function strongly dependent on the culture depth, d, and ranges between 150 and 250 L m^{-2} for depths of the culture in the raceway fluctuating between 15 and 25 cm, respectively. On the other hand, the land demand S (m^2) i.e., the mixable area of raceway, for producing the M tonnes of biomass a year is related to P_{ba} by means of:

$$S = 2.74.10^3 \frac{M}{P_{ba}} \tag{13.10}$$

Note that, for a finite value of the channel width, w, the permissible mixing channel length, L, and thus the mixable area, $S = L\,w$, is a function strongly dependent on depth, d. From Eq. (13.10), we can determine the number of raceway units needed, taking into account that the optimal mixable surface of raceway should not exceed 0.5 ha and that, as a rule, the typical permissible length (Eq. 13.3) to raceway width ratio (L/w) is about 30–50.

B. Closed Photobioreactors

A completely sealed photobioreactor is needed in the H_2 production phase. The two technologies already tested at lab scale, which may be useful in this step are the flat panel and the tubular systems. These reactors meet a set of conditions: (1) low optical path, i.e., a high surface to volume ratio (S/V), so that the average irradiance within the culture is high; (2) high biomass concentrations (around 1 g L^{-1}), which may provide a chlorophyll content of about 20–30 mg chl *a* L^{-1}, which facilitates the use of the maximum number of photons (direct and diffuse light) impinging on the reactor surface; and (3) these reactors can be easily implemented with a proper degassing system in order to remove all the produced H_2 and, at the same time, to maintain the appropriate fluid-dynamic conditions to enhance the mass transfer and light distribution within the reactor.

In both photobioreactors the H_2 produced can be stripped out of the reactor by maintain a partial vacuum (roughly −4 kPa) (Giannelli et al. 2009) in the headspace (upper part) of the flat panel or the headspace of the degasser (bubble column) in the tubular photobioreactor, respectively. In these conditions, the culture is oversaturated and close to cause bubbles of pure H_2 forming in the culture. However, the H_2 can be kept under saturating concentrations by stripping it with O_2-free-air flow bubbled through the reactor in the case of flat panel, which also provides mixing and allows working at the proper fluid dynamics conditions within the culture; or through the bubble column (degasser system) in the case of the tubular technology. This O_2-free gas can be obtained by bubbling air through a sulphite solution before entering to the PBR. The sodium sulphite is oxidized to sodium sulphate by the O_2. The concentration of sodium sulphite in the absorber should never be below 0.5 M. This guarantees a O_2-free airflow in the gas leaving the O_2 absorption column (Camacho Rubio et al. 1991, 1999), that can be used for removal the H_2 produced in the closed photobioreactor.

Table 13.1. Mass transfer coefficient required for the stripping out of the H_2 produced as a function of hydrogen production rates (r_{H2}) reported.

Microorganism	PBR type	Reported H_2 production rate	Calculated production rate (r_{H2}) ($molH_2/L \cdot h$)	Kla (1/h)	References
Green alga *C. reinhardtii*	Tubular 110 L	0.62 ml/L·h	2.77E-05	0.90	Giannelli and Torzillo (2012)
Green alga *C. reinhardtii*	Benchtop n.a.	200 mmol/gchla·h	2.00E-03	2.35	Winkler et al. (2002)
Green alga *C. reinhardtii*	Benchtop n.a.	13 ml/L·h	5.80E-04	0.68	Laurinavichene et al. (2006)
Green alga *S. obliquus*	Benchtop n.a.	150 mmol/gchla·h	1.50E-03	1.76	Florin et al. (2001)
Cyanobacterium *A. variabilis*	Benchtop helical reactor ~ 2 L	167.6 mmol/gchla·h	1.68E-03	1.97	Sveshnikov et al. (1997)
Cyanobacterium *A. variabilis*	Benchtop flat plate, 2 cm	4.1 L/kg·h	1.83E-04	0.22	Yoon et al. (2006)
Cyanobacterium *A. variabilis*	Benchtop flat plate, 4 cm width	0.45 L/kg·h	2.01E-05	0.02	Yoon et al. (2006)
Cyanobacterium *A. variabilis*	Benchtop Flat plate, 2 cm width	5.6 L/kg·h	2.50E-04	0.29	Berberoglu et al. (2008)
PNS bacterium *Rp. palustris*	Tubular 50 L	27.2 ml/L·h	1.21E-03	35.72	Adessi et al. (2012)
PNS bacterium *Rb. spharoides*	Benchtop Flat plate	5.9 mol/kg·h	5.90E-03	6.94	Sasikala et al. (1991)

Mass transfer capacity required has been calculated considering that O_2-free air is bubbled into the culture to remove hydrogen ($[H_2^*] = 1.06 \cdot 10^{-7}$ mol/L) and that the culture become saturated with pure hydrogen ($[H_2]_{sat} = 8.50 \cdot 10^{-4}$ mol/L)

The hydrogen desorption capacity, N_{H2}, is calculated as a function of the volume of the mass transfer unit, V_{mt}, (i.e., the entire culture volume in the case of flat panel, or the culture volume contained inside the degassing bubble column in the case of tubular reactors), the volumetric mass transfer coefficient, $K_l a_{l,H2}$, and the driving force for hydrogen desorption. The latter is calculated as the difference between dissolved hydrogen in the culture, $[H_2]$ and the dissolved hydrogen in equilibrium with the gas phase which is in contact with the liquid, $[H_2^*]$, i.e., oxygen-free air (Eq. 13.11). The dissolved hydrogen concentration in equilibrium with gas phase is calculated according to Henry's law for diluted gases under ideal conditions, as a function of Henry's constant H_{H2}; total pressure, P_T, and the molar fraction of hydrogen in the gas phase, y_{H2} (Eq. 13.12).

$$N_{H2} = K_l a_{1,H2} \left([H_2] - H_2^* \right) V_{mt,H2} \quad (13.11)$$

$$H_2^* = H_{H2} P_T y_{H2} \quad (13.12)$$

The molar fraction of H_2 in air is $1.0 \cdot 10^{-4}$, then the hydrogen molar fraction in O_2-free air, y_{H2}, is $1.25 \cdot 10^{-4}$, whereas the molar fraction of hydrogen in pure hydrogen gas is 1.00. Considering that the Henry's law constant for H_{H2} is 8.50 10^{-4} mol $L^{-1} \cdot atm^{-1}$, at 25 °C, it is possible to determine the required value of the mass transfer coefficient to strip the dissolved H_2 accumulated in the culture as a function of hydrogen productivity and hydrogen concentrations in the liquid and gas phase (Eq. 13.11). Table 13.1 shows the calculated $K_l a_l$ values needed for the desorption of H_2 by using an O_2-free airflow (y_{H2}–$1.25 \cdot 10^{-4}$), for selected H_2 production rates, r_{H2}, reported, considering that the liquid become saturated in equilibrium with pure hydrogen. These values are easy to achieve in flat panels, and feasible with more difficulty in tubular reactors

$$K_l a_{l,H2} = \frac{P_{H2} V}{\left(\left[H_2\right]-\left[H_2^*\right]\right) V_{mt,H2}} \quad (13.13)$$

1. Flat Panels

Flat panels are transparent plates joined close together so that they contain the culture, Thus, they can be illuminated by one or both sides and stirred by bubbling the O_2-free airflow. The dimensions of flat panels are variable but a height under 1.5 m and a separation between plates shorter than 0.10 m are preferred, to avoid the use of high mechanical resistance materials (Fig. 13.4).

Pulz and Scheibenbogen (1998) used flat panels with inner walls arranged to promote an ordered horizontal culture flow that was forced by a mechanical pump (Fig. 13.4). The most innovative aspect of the commercially available Pulz's reactor was that several parallel plates were packed together; close enough to attain up to 6 m^3 of culture volume on 100 m^2 of land, which at the same time improves the dilution of light, with a total illuminated culture surface of ca. 500 m^2. This flat alveolar reactor has no gas headspace and the entire volume of the panel is filled with the culture. The research of Hu et al. (1996) resulted in a type of flat plate reactor made of glass sheets, glued together with silicon rubber to make flat vessels. This simple methodology for the construction of glass reactors provided the opportunity to easily build up reactors with any desired light-path. Doucha et al. (1996) described an optimized large-scale flat plate photobioreactor module of 1,000 m^2. Recently, a new design of vertical flat panel photobioreactor consisting of a plastic bag located between two iron frames has been proposed (Rodolfi et al. 2009); this brings a substantial cost reduction to this type of reactors but the system is not completely sealed as it is required for H_2 production. The O_2-free aeration is done through a PVC plastic tube perforated to provide minute holes of about 1 mm. A separate degasser must be used in the case of the alveolar flat plate reactor. The cooling system is a heat exchanger inserted in the reactor. The mass transfer, mixing and heat transport capacities in flat panel reactors are usually very good. The main advantages of this reactor type are the low power consumption (roughly 50 W m^{-3}) and the high mass transfer capacity ($Kl_l a_l = 25\ h^{-1}$). The major technical issues in designing and building up this type of reactors are (1) the panel orientation and light path depth, (2) the O_2-free aeration rate for both maintaining the proper fluid-dynamic and removal of H_2, and its impact on power supply, mass transfer and mixing; and (3) the temperature control.

a. Panel Orientation and Light Path Depth

The solar radiation intercepted may vary significantly with orientation and position. For latitudes above 35°N the east-faced/west-faced orientation are favourable to north/south, the higher the latitude the higher the increase in the solar radiation intercepted. On the contrary, for latitudes under 35°N the north/south-orientated reactors intercept more radiation and the difference is more pronounced the closer to the equator (Acién-Fernández et al. 2013). The position of the reactor also influenced the type of radiation intercepted. In vertical panel PBR the proportion of disperse radiation is dominant (Qiang et al. 1996; García Camacho et al. 1999). Vertically arranged flat panels intercept less solar radiation than inclined flat panels but have the advantage of less cost and overheating. The vertical arrangement allows reducing solar radiation peaks at noon increasing the interception of solar radiation in the morning and afternoon. Moreover, the vertical arrangement also shows an improved interception of solar radiation in winter with respect to summer (Sierra et al. 2008). Disperse radiation has been consistently

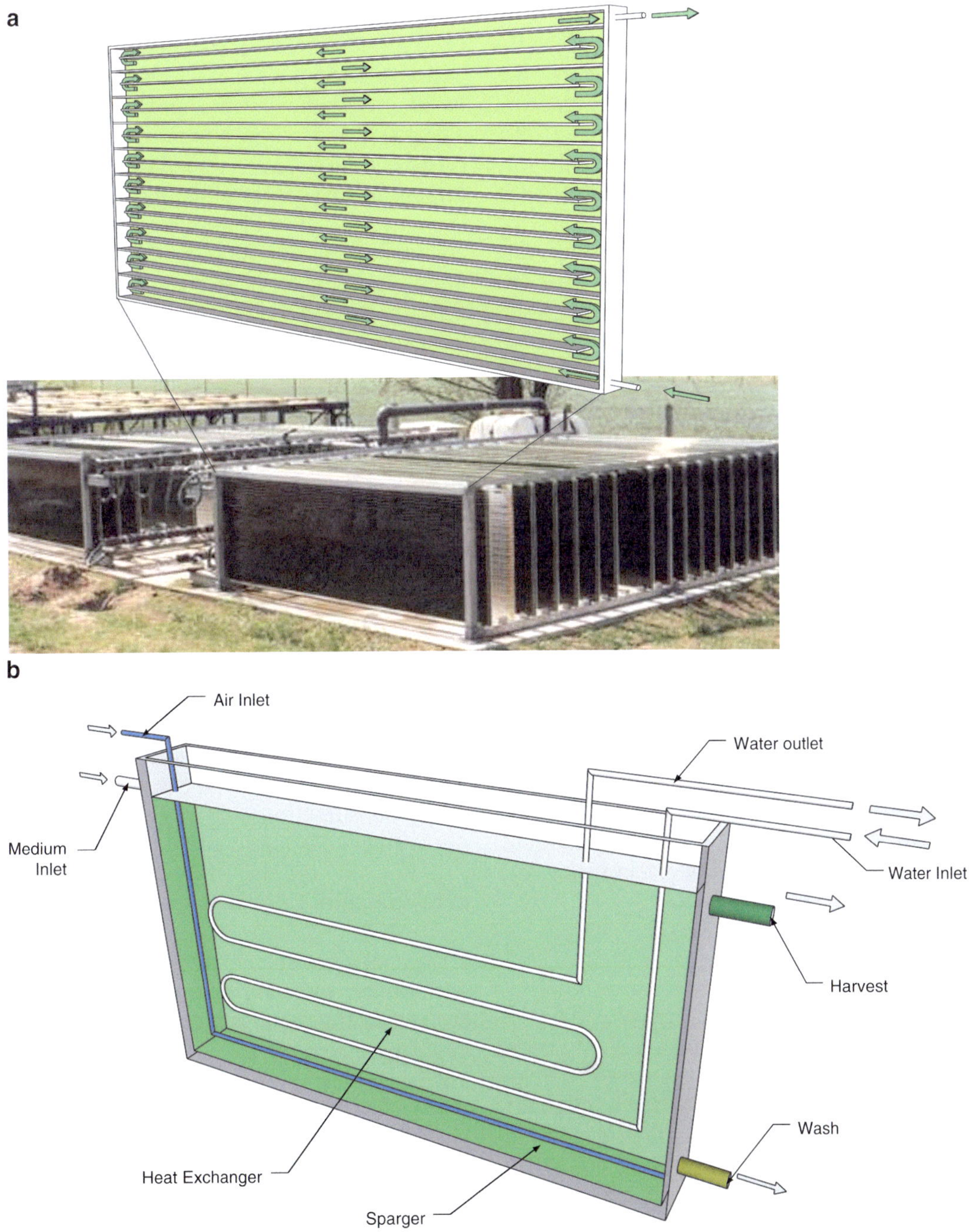

Fig. 13.4. Image of an alveolar flat plate photobioreactor field (**a**) (Courtesy of Dr. O. Pulz, IGV GmbH, Nuthetal, Germany). Detail shows the horizontal channels through which the culture is circulated. Below. (**b**) Is a scheme of a non-alveolar flat plate reactor in which the appropriate fluid-dynamic conditions are provided by bubbling O_2-free airflow at the bottom contributing, at the same time, to strip out the dissolved hydrogen produced.

reported to be more efficient for microalgal cultures. Indeed, the photosynthetic efficiency of vertical photobioreactors has resulted higher than optimal-tilt reactors, reaching values of 20 % (Qiang et al. 1996). This is due to the fact that low irradiance levels normally result in higher photosynthetic efficiencies; this is, when cells are growing under irradiance levels far from saturating light assimilation is more efficient. This can be accomplished by increasing the light-receiving surface of photobioreactor per square meter of occupied land, a technique usually referred to as "dilution" of light.

With respect to the panel depth, to maintain average irradiances over 100 $\mu E \cdot m^{-2} \cdot s^{-1}$ in flat panels while maintaining a cell density about 1 g L^{-1} the panel depth should be below 7 cm (typically 4–6 cm) in order to achieve chlorophyll concentrations in the range of 20–30 mg chl *a* L^{-1} for an increased light-to-H2 conversion efficiency.

b. O_2-Free Aeration Rate and Its Impact on Power Supply, Mass Transfer and Mixing

The power input per volume unit due to aeration, P_G/V_L, in a flat panel reactor is a function of aeration rate, the density of the liquid, ρ_L and the gravitational acceleration, g and can be calculated:

$$\frac{P_G}{V_L} = \rho_L g U_G \tag{13.14}$$

where U_G is the superficial gas velocity in the O_2-free aerated zone. U_G is easily derived from the O_2-free airflow rate, in v/v/m, multiplying this by the total volume of the culture and dividing by the cross-sectional area of the aerated zone. The power supply also impacts on the mass transfer capacity of the flat panel reactor according to the following equation:

$$K_L a_L = 2.39 \bullet 10^{-4} \left(\frac{P_G}{V_L} \right)^{0.86} \tag{13.15}$$

Note that, in spite of the low power supply used, the volumetric mass transfer coefficient $K_l a_l$ reached values of about 25 h^{-1}, enough to avoid dissolved hydrogen accumulation with O_2-free air (Table 13.1). This mass transfer capacity in the flat panel photobioreactor can be attained with a power supply of 50 W m^{-3}. The low power supply requirement and the relatively high mass transfer capacity are important advantages of the flat panel photobioreactor because of the sensitivity to stress caused by intense turbulence that show many microalgal strains. In addition to mass transfer capacity, the power supply also determines the mixing time inside the rector. In the range of typical aeration rates: 0.05–0.35 v/v/min (i.e., power supply between 5 and 55 W m^{-3}) the complete mixing in the flat panel photobioreactor ranged between 150 and 100 s, much lower than those obtained in tubular systems and open raceway.

In the case of alveolar panel reactor, the culture is forced to circulate through the internal channels by means of a centrifugal pump. In these systems, thus, there is no headspace in the reactor body and the supply of O_2-free gas has to be carried out in an auxiliary degasser connected to the reactor body, in a similar implementation as in tubular technology. Therefore the assessment of the power consumption, mass transfer and mixing is analogous to tubular technology.

c. Temperature Control

Flat panels can be cooled by water spray or alternatively by using internal heat exchangers. According to the authors' experience, the cooling capacity of spray systems is limited and its application is only possible under certain environmental conditions (temperature, humidity, etc.). Most times the use of a heat exchanger is needed. The reduction of solar radiation interception in the vertical arrangement also allows reducing the overheating of the cultures at noon, thus reducing the requirements of cooling. In any case, the heat transport capacities in flat panel reactors are usually very good. The values of the heat transfer coefficient in these systems range from 300 to 1,000 $W \cdot m^{-2} \cdot K$ (Sierra et al. 2008).

The design of the tubular heat exchanger can be done from the following heat balance:

$$m_{water} \bullet Cp_{water} \bullet \left(T_{outlet} - T_{inlet}\right) = U \bullet A \bullet \frac{\left(T_{culture} - T_{inlet}\right) - \left(T_{culture} - T_{outlet}\right)}{\ln\left(\frac{T_{culture} - T_{inlet}}{T_{culture} - T_{outlet}}\right)} \quad (13.16)$$

where the left hand side of the equation represents the heat flow gained by the cooling water passing through the internal side of the heat exchanger, and the right hand side represents the heat lost by the culture and transferred to the cooling water. U is the global heat exchanger coefficient ($W\ m^{-2}\,°C^{-1}$) and A is the external surface of the heat exchanger.

d. Scale-Up of Flat Panels

According to the conceptual scheme of Fig. 13.1, the scale-up of the number of flat panel modules for the second step of H_2 production, is based on the M Tn $year^{-1}$ produced in the scale up performed for the first step (biomass production). Thus, the daily M/365 tonnes of biomass produced in the open RW must be re-suspended in the proper medium so that the resulting biomass concentration is over 1 g L^{-1} (i.e., sulphate free-Tris-acetate-phosphate medium (TPA-S) for *C. reinhardtii*; Nitrogen free-Allen-Arnon medium (and no aeration) for the cyanobacteria *Anabaena variabilis*). This dense culture (about 1 g L^{-1}), along with the large surface to volume ratio (optical path of about 3–7 cm) would allow a good use of the solar radiation. This means that the volume of flat panel reactor (m^3) should be about $2.74 \cdot M$. By taking into account that the volume of one module is 0.1875 m^3 (2.5 m length, 1.5 m high, 0.05 m width), the number of modules needed for culturing the daily biomass produced in the first step is $14.6M$. ($\sim 15 \cdot M$).

Bearing in mind that the H_2 production phase last about 4 days in the case of *C. reinhardtii*, and 7 days in the case of cyanobacteria or PNS bacteria, the total amount of modules needed are $4 \times 14.6M$ or $7 \times 14.6M$ (rounding up to $60 \cdot M$ and $100 \cdot M$ respectively). Finally, the land demand required can be estimated by using, as rule of thumb, that the distance between a row of modules and others should be about 0-75-1.00 m.

Note that the use of O_2-free aeration for flat panel reactors, according to the scheme shown in Fig. 13.1, is restricted to the H_2 production by using green microalgae that requires anoxygenic conditions in the second step, i.e., the absence of oxygen from the algal environment. Green algae use the iron [Fe-Fe]-hydrogenase that catalyzes both the production and the consumption of hydrogen through the reversible reaction:

$$2H^+ + 2e^- \leftrightarrow H_2.$$

The rate at which the [Fe-Fe]-hydrogenase catalyzes the production of H_2 decreases significantly with increasing partial pressure of H_2, and hence the need of removing the dissolved H_2 by bubbling a stream of O_2-free gas and to maintain a slightly negative pressure (about −4kPa). On the other hand, aerated flat panel technology is not the technology of choice when using cyanobacteria or PNS bacteria in the H_2 production step. Cyanobacteria use either [Ni-Fe]-hydrogenase or nitrogenase (the latter is the enzyme used by PNS bacteria) and both, nitrogen-fixing cyanobacteria and PNS bacteria, in order to produce hydrogen require the absence of nitrogen sources (N_2, NO_3^- or NH_4^+) in addition to anaerobic conditions. Therefore, since in the flat panel technology is mixed by a O_2-free air stream to keep the proper fluid dynamics conditions, the presence of N_2 would inhibit the nitrogenase (if using PNS bacteria), or both [Ni-Fe]-hydrogenase or nitrogenase, depending on the cyanobacterium used in the second step. In the case of a flat panel with internal channels through which the culture is circulated by means of a centrifugal pump, it would be possible the use of cyanobacteria or PNS bacteria in them, as is done with the tubular reactors.

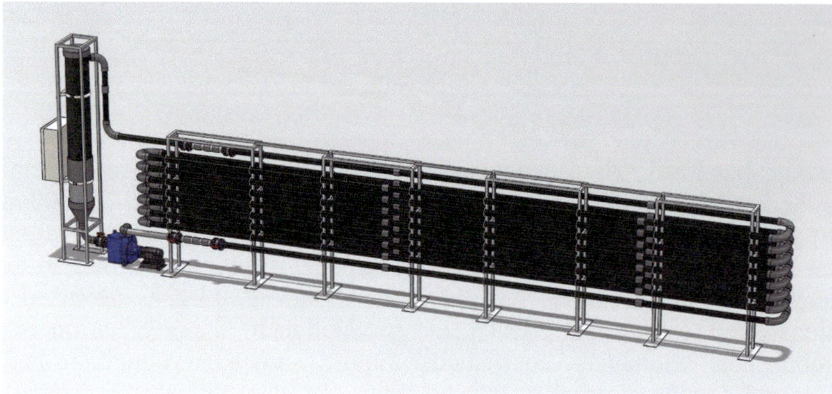

Fig. 13.5. Fence-type tubular photobioreactor.

2. Tubular Photobioreactors

Tubular photobioreactors are the most widely used closed systems for the production of microalgae. They consist of a solar collector made of tubes, a degasser unit which usually is a bubble column and a pump for the culture impulsion. A conceptual fence configuration tubular photobioreactor, as those existing in authors' facilities, is shown in Fig. 13.5.

The pump circulates the culture through the solar collector tubing where most of the photosynthesis, or the photofermentative process, occurs. The H_2 produced by photofermentation or photosynthesis is accumulated in the broth (as well as pure H_2 bubbles in the upper part of the tube when the dissolved H_2 concentration exceeds the saturation value) until the fluid returns to the degasser zone (bubble column), where the accumulated hydrogen is stripped by a counter current O_2-free airflow. A gas-liquid separator in the upper part of the bubble column prevents the H_2 bubbles from being recirculated into the solar collector. The major drawback of this technology is the high power consumption used for the impulsion of the liquid through the tube, roughly 500 W m^{-3} for the pump-driven fence configuration (Acién-Fernández et al. 2012). The solar loop is designed to be efficient in collecting the solar radiation and in promoting the light dilution effect, to minimize the resistance to flow and to occupy as little land as possible. Similarly to the flat panels, the diameter of the solar tubing is selected so that the volume of the dark zone (i.e., the zone with light intensity below saturation) is kept to a minimum. Also, the movement of fluid between the light and the dark zones in the solar collector should be rapid enough to prevent an excessive residence time of any element of fluid in the dark zone. Increasing the culture velocity in order to enhance the turbulent mixing (and therefore the light to dark cycle frequencies) and reducing the mixing time appear to be crucial in the photobioreactor scale-up for the hydrogen production stage (Oncel and Sabankay 2012). The length of the tube is limited by the H_2 build-up. As a rule of thumb, the maximum tube length is determined by the maximum dissolved hydrogen that the specific strain can withstand with an acceptable drop in the H_2 production rate. The H_2 stripping capacity of the culture broth is a key factor for designing a tubular reactor for hydrogen production. Increasing the gas phase to the liquid phase by a factor of 4 resulted in a 100 % increase in the H_2 output (Giannelli and Torzillo 2012). These findings are important for a rational design of PBR. However, in the case of the tubular PBR characterized by perfect plug flow behaviour inside the tubes, the increase of the liquid free surface in the headspace of the degasser has little impact on the gas removal due to the extremely high ratio between the residence time inside the tube circuit compared to that in the degasser. Increasing the flow rate can help to reduce both the mixing and the residence time, thus helping to reduce the contact between the hydrogen gas and the culture. However, replacing the curves with proper designed manifolds conveying the gas toward the degasser could be the best solution to the gas removal problem in a tubular rector (Giannelli and Torzillo 2012). Therefore these reactors are usually modular. The relevant design aspects are discussed next.

a. The Liquid Velocity and Length of the Solar Tube

The design of tubular photobioreactor must guarantee that the flow in the solar tube is turbulent (i.e., Reynolds number should exceed 10,000) so that the cells do not stagnate in the dark interior of the tube. At the same time, the dimensions of the fluid micro eddies should always exceed those of the algal cells; so that turbulence-associated damage is avoided (Acién-Fernández et al. 2001; Camacho et al. 2001). The need to control eddy size places an upper limit on the flow rate through the solar tubing. The length scale of the microeddies may be estimated by applying Kolmogorof's theory of local isotropic turbulence (Kawase and Moo-Young 1990):

$$\lambda = \left(\frac{\mu_L}{\rho}\right)^{3/4} \xi^{-1/4} \qquad (13.17)$$

Where λ is the microeddy length, ξ is the energy dissipation per unit mass, μ_L is the viscosity of the fluid, and ρ is the fluid density. The specific energy dissipation rate within the tube depends on the pressure drop and the liquid velocity, U_L,

$$\xi = \frac{2C_f U_L^3}{d_t} \qquad (13.18)$$

Where C_f is the Fanning friction factor that can be estimated by using the Blausius equation (Eq. 13.19). Thus, for any selected strain the cell size is known. Using this size as the microeddie length, allows calculating the maximum energy dissipation rate per unit mass (Eq. 13.15), and from this, the maximum liquid velocity, U_l, that makes the microeddie length similar to cell size (Eq. 13.18).

$$C_f = 0.0791 Re^{-0.25} \qquad (13.19)$$

Another restriction on the design of the solar collector is imposed by the acceptable upper limit on the concentration of dissolved hydrogen. The length of the solar collector must not be long enough as to achieve an inhibiting hydrogen concentration in the culture. The maximum length, L, is a function of the liquid velocity, dissolved hydrogen concentration and the hydrogen production rate, r_{H2}, and can be calculated as follows:

$$L = \frac{U_L\left([H_2]_{out} - [H_2]_{in}\right)}{r_{H2}} \qquad (13.20)$$

Where U_L is the liquid velocity (ratio between the liquid flow rate and the cross sectional area of the tube). Note that U_L is always lower than the maximum velocity imposed by microeddies length. $[H_2]_{in}$ is the hydrogen concentration at the beginning of the solar collector (i.e., the saturation value when the fluid is in equilibrium with the atmosphere), $[H_2]_{out}$, is the hydrogen concentration at the outlet of the solar collector (i.e., maximum acceptable value that does not inhibit hydrogen production), and r_{H2} is the volumetric rate of hydrogen generation reported for the strain in well-controlled laboratory experiments. If the culture is circulated by pumps (fence type configuration), thc typc and power of the pump determines the liquid velocity. The above Eqs. (13.17), (13.18), (13.19) and (13.20) can also be applied for alveolar flat panel reactors.

Considering the reported data of hydrogen production rates it is possible to determine the maximal length of tubular photobioreactors as a function of the tolerable hydrogen concentration at the beginning and end of the loop. Assuming that the hydrogen production rate is not inhibited at concentrations close to saturation with pure hydrogen, this value can be used for the H_2 concentration at the end of the loop i.e., this is a scenario in which no H_2 bubbles are produced within the tube circuit. For the concentration at the beginning of the loop, a 40 % of the saturation level with pure hydrogen is accepted. The higher the initial hydrogen concentration is, the shorter the loop has to be to prevent oversaturation. On the other hand, a high hydrogen concentration at the beginning of the loop, and hence in the degasser, implies a high driving force for the desorption process thus a lower mass transfer capacity requirement. The mass transfer capacity needed to achieve a stable operation of the system can be calculated from these values if the volumes of the total reactor and of mass transfer unit (V_{mt}) are known. As a rule of thumb, we can take V_{mt} as a 10 % of total tubular photobioreactor volume.

b. Combining Flow and Gas-Liquid Mass Transfer Within the Tube

In the previous scenario, we have assumed that the dissolved hydrogen concentration does not surpass the saturation level in any point of the loop. It is also possible to design for a situation in which the level of dissolved hydrogen be over the saturation value in a part or in the whole loop, giving rise to coexisting gas and liquid phases in the loop. This is highly likely, above all, when working with PNS bacteria because these microorganisms have H_2 production rates substantially higher than rates green algae have. Design of tubular photobioreactors in this scenario must also consider gas-liquid mass transfer and hydrodynamics within the tube. By applying mass balances to the different zones of the tube circuit for which the fluid-dynamic conditions remain constant, the hydrogen transfer between the liquid and gas phase can be modelled. For the liquid phase, the changes in concentrations of dissolved hydrogen along the circuit can be related to the gas-liquid mass transfer rates and the generation rates by mass balances as follows:

$$Q_L d[H_2] = K_l a_{H_2}\left([H_2]^* - [H_2]\right) Sdx + r_{H_2}(1-\varepsilon) Sdx \qquad (13.21)$$

In this equation, $K_l a_{H2}$ is the volumetric gas-liquid mass transfer coefficient for hydrogen in the solar collector (i.e., within the tube circuit), dx is the differential distance along the direction of flow in the solar tube, $[H_2]$ is the liquid phase concentration of hydrogen, ε is the gas holdup; S is the cross-sectional area of the tube; r_{H2} is the volumetric generation of hydrogen and Q_L is the volumetric flow rate of the liquid. Note that the concentration values marked with an asterisk correspond to the equilibrium concentration, i.e., the maximum possible liquid-phase concentration of hydrogen in contact with the gas phase of a given composition. This term only exists if there is a gas phase from which mass is transferred. The gas phase exists if the culture becomes oversaturated (pure hydrogen bubbles) or gas is artificially injected into the tubes.

As for the liquid phase, a component mass balance can be established also for the gas; hence,

$$dF_{H_2} = -K_l a_{H2}\left([H_2]^* - [H_2]\right) Sdx \qquad (13.22)$$

Here F_{H2} is the molar flow rate of the hydrogen in the gas phase. Note that because of the change in molar flow rate, the volumetric flow rate of the gas phase may change along the tube. The equilibrium concentrations of the hydrogen in the liquid can be calculated by using the Henry's law:

$$[H_2]^* = H_{H_2} P_{H_2} = H_{H_2}(P_T - P_v) \qquad (13.23)$$

where H_{H2} is the Henry's constant for hydrogen, P_{H2} is the partial pressure of hydrogen in the gas phase existing in the upper part of the tube (i.e., roughly 1 atm, because gas phase is almost pure hydrogen); the partial pressures can be calculated knowing the total pressure (P_T) and the vapour pressure (P_v). The previous equations and the initial conditions, allow numerical integration and consequently, the determination of the H_2 axial profiles in the liquid phase and the molar flow rates of H_2 in the gas phase. The model is simple and can be adapted to any photobioreactor and H_2 producing strain in the second phase (note the same rationale and Eqs. (13.21), (13.22) and (13.23) can be used for alveolar flat pannel reactors). Moreover, since the model can simulate the H_2 profile along the tube as a function of the tube length and operational variables, it would allow the rational design and scale-up of tubular PBR for H_2 production.

c. Hydrogen Removal and Temperature Control

Once the solar collector has been designed, it is necessary to calculate the degasser unit used for the removal of hydrogen and temperature control. For this purpose the use of bubble columns (generally used in the fence configuration reactor, Fig. 13.5) is preferred because these systems are simple, well-known and widely used at industrial scale. The mass transfer coefficient can be calculated as a function of O_2-free aeration rate (Eqs. 13.13 and 13.14) and from this the volume of bubble column required to remove the hydrogen is calculated as:

$$V = \frac{Q_L\left([H_2]_{out} - [H_2]_{in}\right)}{K_l a_l \cdot \left([H_2^*] - [H_2]\right)(1-\varepsilon)} \tag{13.24}$$

where Q_L is the liquid flow rate entering to the bubble column, $[H_2]_{in}$ is the hydrogen concentration at the inlet of the bubble column, $[H_2]_{out}$ is the hydrogen concentration at the outlet of the bubble column, K_La_L is the volumetric mass transfer coefficient for the bubble column, and $([H_2^*]-[H_2])$ is the driving force for the transport of hydrogen from the liquid to the gas phase, calculated as the logarithmic mean of the concentration difference at the inlet and the outlet. To avoid the recirculation of bubbles from the bubble column to the solar collector, the superficial liquid velocity downstream must be lower than the velocity of the rising bubbles, U_b of the O_2-free stream. Thus, the minimum diameter of bubble column, dc, can be calculated (Eq. 13.25), as well as the minimum column height, hc, (Eq. 13.26).

$$dc = \sqrt{\frac{4Q_L}{\pi U_b}} \tag{13.25}$$

$$hc = \frac{4V}{\pi dc^2} \tag{13.26}$$

The heat transfer equipment must be designed analogously to mass transfer. The equipment must be able to remove the heat absorbed by radiation. This is a function of the solar radiation received by the solar collector, $Qrad$, and the thermal absorptivity of the culture, a_{rad}. Finally, the area of heat exchanger necessary, $A_{exchanger}$ is calculated as a function of the overall heat transfer coefficient, $U_{exchanger}$, and the temperature of cooling water by:

$$A_{exchanger} = \frac{Q_{rad}\, a_{rad}}{U_{exchanger}\left(T_{culture} - T_{water}\right)} \tag{13.27}$$

d. Examples of Pilot Scale (>50 L) Sealed PBR for H_2 Production

Although algal H_2 production has been extensively investigated, there are almost no publications on the subject carried out with PBR volumes anything beyond lab-scale. Nonetheless, some publications have started to appear recently with the objective of assessing the production of H_2 in tubular reactors with volume capacities over 50 L using sunlight.

In this sense, Torzillo and co-workers (Scoma et al. 2012) built a 50 L horizontal tubular PBR for producing H_2 with the microalga *Chlamydomonas reinhardtii*. The device was made up of ten parallel glass tubes connected by PVC U-bends (Fig. 13.6). The illuminated area was 1.5 m^2 with a surface-to-volume ratio of 30.5 m^{-1}.

Fig. 13.6. Overview of the 50-L horizontal tubular photobioreactor used for outdoor experiments with *C. reinhardtii*. *Inset*: details of the PVC pump and the degasser (Courtesy of Dr. G. Torzillo, ISE, CNR, Florence, Italy).

The circulation of the fluid within the loop was done with a PVC centrifugal pump with three stainless-steel flat blades placed at an angle of 120° with respect to each other on the propeller shaft. It is interesting to note that the distance between the blades and the box of the rotor is 0.5 cm while the height of the casing is 6.5 cm (see detail in Fig. 13.6). This pump provides adequate fluid dynamics, liquid velocity (range 0.2–0.5 m s^{-1} and mixing time (around 1 min). At the end of the loop (with a total length of 23 m) the culture flows through a 2.2 L degasser (PVC tube, 10 cm internal diameter and 28 cm height). The degasser contains several hose fittings to feed culture medium, flushing gas, and for the withdrawal of culture and gas samples.

During the hydrogen production phase, the headspace of the PBR (i.e., the volume above the culture level) was about 0.2 L (0.4 % of the total volume) and the degasser was flushed with a N_2 gas stream. In our opinion, the free G-L surface in the degasser is very little and so will be the mass transfer capacity of this unit; nonetheless the experiment was performed with a partial vacuum of −4.03 kPa to facilitate the degassing of the dissolved H_2 entering the degasser. The maximum H_2 production rate reported was 0.3 ml L^{-1} h^{-1}, production that was 18 % lower than that obtained with the same microalga in 1 L flat panel reactor. The major differences are in the mixing time (15.5 s) in lab-scale *vs*. 60 s in the 50 L reactor, and in the illumination pattern (both sides illuminated in the 1 L flat reactor while in the 50 L outdoor reactor about a 30 % of the culture is permanently in dark). The illumination and mixing degree have both been proved to be key features to take into account in the scale-up of the H_2 production phase; although as commented before, care must be taken with the potential damage of cells caused by an excessive turbulence level (Oncel and Sabankay 2012). Therefore, the scaling up of this tubular configuration has serious difficulties since the increase of liquid velocity needed to achieve the proper light-dark cycle frequencies may conflict with the cell damage that this can cause.

The same reactor has also been used by Adessi et al. (2012) for H_2 production by the purple non-sulphur bacterium *Rhodopseudomonas palustris*. The average H_2 production rate was 10.7 ml L^{-1} h^{-1} with a reported maximum of 27.2 ml L^{-1} h^{-1}. These productivities were two orders of magnitude higher that that obtained with *C. reinhardtii* with the same reactor and similar environmental conditions. A micro sensor for measuring the dissolved H_2 concentration in the range 0.0–1.5 mg L^{-1} (i.e., 0.0–0.75 mM) revealed that the elapsed time from the sunrise until dissolved hydrogen appeared in the culture was about 1.3 h and that H_2 gas was effectively collected between 11:00 and 19:00 h local time (Adessi et al. 2012). The light to H_2 conversion efficiency with the PNS bacteria was 0.63 *vs.* the 0.21 % for *C. reinhardtii*. This shows how the processes based on PNS bacteria are more promising than those using green algae.

A 80 L tubular reactor for the H_2 production with the PNS bacterium *Rhodobacter capsulatus* by the photofermentation of thick juice effluents was designed by Boran et al. (2010, 2012). The reactor was placed with an inclination of 10º. This reactor is conceptually similar to the near horizontal tubular reactor designed by Tredici et al. (1998). The tubular reactor consisted of 9 tubes, 6 cm diameter and 2.35 m length. The total illuminated surface was 2 m^2 and the occupied ground area was 2.88 m^2. During the H_2 production phase, the culture was bubbled with an argon stream. The reactor temperature was kept below 35 °C by internal cooling coils. The culture was carried out in a semicontinuous mode with a dilution rate of 0.11 d^{-1}, trying to maintain cell concentration at around 1 g L^{-1} during the H_2 production phase. The average H_2 production rate was 0.15 mol $H_2 m^{-3}$ h^{-1}during 9 days, obtaining a light conversion efficiency of roughly 0.2 %, and demonstrating that the H_2 yield of the culture (mmol H_2 to cell dry weight ratio) was a potential function of the total light energy, *E*, received ($W \cdot h\ m^{-2}$), with a 1.4 exponent.

Possibly, the work carried out at the largest scale so far has been performed by Torzillo and co-workers by using a 110 L tubular PBR immersed in a light-scattering nanoparticle suspension (Fig. 13.7) (Giannelli and Torzillo 2012). The PBR was made up of 64 tubes (i.d., 2.75 cm; length 2 m) connected by 64 U-bends, with a total 133 m length. The tubes were immersed in a light scattering suspension of silica nanoparticles that increases the H_2 production rate up to 0.62 ml L^{-1} h^{-1} *vs.* the 0.42 ml L^{-1} h^{-1} achieved without immersing them into the nanoparticles bath (note that these productivities almost doubled those obtained in the 50 L reactor shown in Fig. 13.6). This demonstrates the positive effect of the dilution of light by nanoparticles, which enhances the H_2 production. In our opinion, the drawbacks already existing in the 50 L reactor (Fig. 13.6) are still present in this new design. The degasser again is hardly 2 L volume. During the H_2 production phase, the headspace of the degasser was 0.35 L (i.e., 15 % of the total volume of the degasser). Apparently, the mass transfer capacity of this reactor, similar to the 50 L reactor, is clearly insufficient. The light to H_2 conversion efficiency was only 0.21 %. Ignoring the economical considerations, the new method presented in this design for the dilution of light shows at least two major advancements over the state of the art of sealed PBR for H_2 production: (1) it not only improves the light conversion efficiency, but this new design also allows an efficient way for PBR scaling up by reducing the distance between the tubes, and consequently increasing the number of tubes, without altering the foot print; and (2) it is possible to modify the irradiance impinging on the surface of the tubes by varying the concentration of nanoparticles. This is a significant advantage in outdoor cultures where the intense direct light radiation may cause over saturation with loss of efficiency. With the PBR concept designed by Torzillo´s group, even direct incident sunlight can be diluted to levels of photosynthesis saturation (about 200 μE m^{-2} s^{-1}). Nonetheless, this new PBR design has only been operated so far with artificial light and it needs to be demonstrated operating with sunlight to actually test the above commented advancements.

e. Scale-Up of Tubular PBR

For practical purposes, the scaling up of a tubular photobioreactor requires the scaling

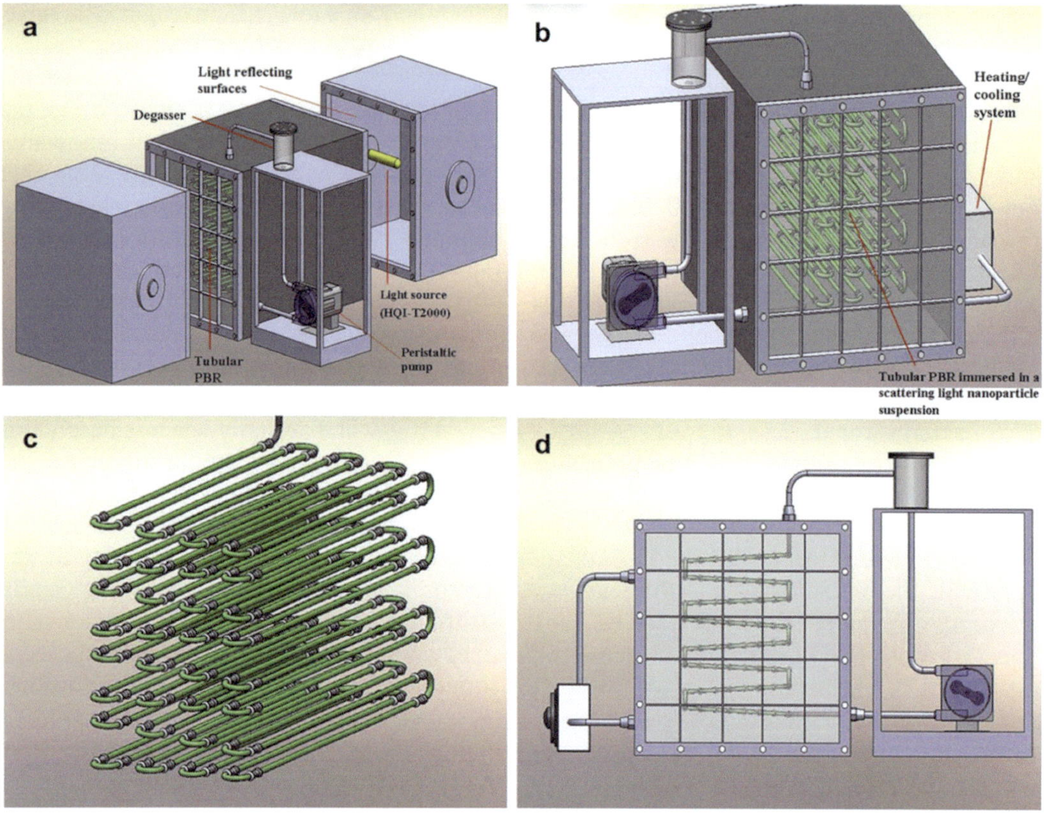

Fig. 13.7. Enclosed tubular photobioreactor used for f H_2 production with *C. reinhardtii* (**a**) general view of the 110-L PBR (**b**) frontal view of the tubular PBR set in a container filled with a light scattering nanoparticle suspension (**c**) detail of the PBR made up eight tube layers and connected each other by U-bends to form a 133 m long circuit (**d**) frontal view of the reactors showing the tube layers with opposite inclination to facilitate culture draining (Photography and description, courtesy of Dr. G. Torzillo, ISE, CNR, Florence, Italy).

up of both the solar receiver and the degasser system (i.e., the bubble column in the case of fence type configuration). Scaling of the degasser does not pose a limitation for any realistic size of the photobioreactor. However, there are limitations in the scaling up of a continuous run solar loop. In principle, the volume of the loop may be increased by increasing the diameter and the length of the tube. In practice, the maximum tube length is limited by gas buildup and the diameter should not exceed 0.10 m (Molina Grima et al. 2000; Brindley et al. 2004). Figure 13.8 shows an industrial size photobioreactor operated in a greenhouse in Almería, Spain, following Eqs. (13.7) to (13.8) and (13.17), (13.18), (13.19), (13.20), (13.21), (13.22), (13.23), (13.24), (13.25), (13.26) and (13.27) (Acién-Fernández et al. 2013). This reactor was scaled up to industrial size with the objective of producing a metabolite associated to growth. The production of biomass involves the removal of O_2 and the supply of CO_2, and therefore there are some differences between Eqs. (13.20), (13.21), (13.22), (13.23) and (13.24) and those shown for O_2 removal and CO_2 supply in Acién-Fernández et al. (2013). This reactor is running in the authors' facilities for producing high value products and may be adapted with some small modifications (head space of the degasser and type of centrifugal pump for liquid impulsion) to produce H_2.

For the scaling up of tubular reactors, first the light availability in PBR location should be calculated according to solar radiation

Fig. 13.8. Photograph of an industrial size fence configuration tubular photobioreactor 30 m^3 plant for the production of lutein from *Scenedesmus almeriensis* Almería. Fundación CAJAMAR (With permission of Fundación CAJAMAR, Almería, SPAIN).

knowledge. Then, simulations should be performed to determine the optimal tube diameter. The selection should be done taking into account the characteristic parameters of the microalga, cyanobacterium or PNS bacterium to be used (μ_{max}, Ik, n, Eq. 13.6). A practical tube diameter range of 6–9 cm is convenient (Brindley et al. 2004). This range allows the reactor operation at the proper fluid-dynamics conditions, promoting adequate light to dark cycle frequencies and a limited energy consumption. According to the target H_2 productivity r_{H2}, the maximum length of the solar collector can calculated with Eq. (13.20). Table 13.2 shows the tube length for a tubular reactor, similar to that presented in Fig. 13.8, for H_2 production in a scenario in which, during the time spent by the culture in the circuit tube no bubbles are generated. The dissolved hydrogen concentration at the beginning of the loop is 40 % of the saturation with pure hydrogen and 100 % saturation at the exit. This means that in the degasser of the tubular system, the counter current O_2-free airflow removes the 60 % of the dissolved hydrogen of the culture, which enters in degasser saturated. As can be seen in Table 13.2, provided that the hydrogen production rate is over 10^{-3} mol L^{-1} h^{-1} the length of the tube needed in order to have the culture hydrogen saturated is in the reasonable range 100–400 m. For lower hydrogen production rates, the culture will never reach saturation and tube length needed is much lower. Table 13.3 shows the tube length needed for the different scenario in which hydrogen bubbles are formed within the tubes (visible in the upper part of the tube), which have to be dragged to the degasser by the culture motion. The calculations have been made considering that the dissolved oxygen concentration at the entrance and exit of the solar collector are 40 % with respect to saturation with pure hydrogen and twice the saturation, respectively. In these conditions there is a large proportion of tube showing bubbles in the upper part. This scenario is more reasonable when using PNS bacteria in the H_2 production step, since their reported H_2 productivities are almost double in comparison with those obtained with green microalgae. Finally, a bubble column can be designed to be coupled to the solar collector to remove hydrogen as well as to control the temperature of the culture, Eqs. (13.24), (13.25), (13.26) and (13.27).

Table 13.2. Permissible length of tubular reactors as a function of hydrogen production rate, r_{H2}, reported.

Productivity (mol/L·h)	Tube length (m)	Volume (L)	Vmt (L)	Kla (1/h)
2.77E-05	19,900	11,820	1,182	0.47
2.00E-03	275	1,752	175	33.62
5.80E-04	949	6,038	604	9.76
1.50E-03	367	2,336	234	25.21
1.68E-03	329	2,091	209	28.17
1.83E-04	3,009	19,144	1,914	3.08
2.01E-05	27,418	174,423	17,442	0.34
2.50E-04	2,203	14,016	1,402	4.20
1.21E-03	454	2,886	289	20.41
5.90E-03	93	594	59	99.18

Values were obtained considering a liquid velocity of 0.3 m/s, and hydrogen dissolved concentrations at the beginning and end of the tube circuit equal to 40 % of saturation with pure hydrogen ($[H_2]=3.4\cdot10^{-4}$ mol/L) and saturation with pure hydrogen ($[H_2]_{sat}=8.50\cdot10^{-4}$ mol/L), respectively
The references for these values are the same as in Table 13.1

Table 13.3. Tube length of tubular photobioreactor as a function of hydrogen production rate values reported.

Productivity (mol/L·h)	Tube length (m)	Volume (L)	Vmt (L)	Kla (1/h)
2.77E-05	53,066	31,519	3,152	0.27
2.00E-03	734	4,672	467	19.61
5.80E-04	2,531	16,101	1,610	5.69
1.50E-03	979	6,229	623	14.71
1.68E-03	876	5,575	558	16.43
1.83E-04	8,025	51,051	5,105	1.79
2.01E-05	73,114	465,129	46,513	0.20
2.50E-04	5,875	37,376	3,738	2.45
1.21E-03	1,210	7,695	770	11.91
5.90E-03	249	1,584	158	57.85

Values obtained considering a liquid velocity of 0.3 m/s, and hydrogen concentrations at the beginning and end of the tube loop equal to 40 % of saturation with pure hydrogen ($[H_2]=3.4\cdot10^{-4}$ mol/L) and double than saturation with pure hydrogen ($2x[H_2]_{sat}=1.7\cdot10^{-3}$ mol/L), respectively

V. Concluding Remarks

This chapter shows the fundamental principles of photobioreactor design to be used in a hypothetical facility (Fig. 13.1) to produce hydrogen from green microalgae, cyanobacteria or PNS bacteria. This has been envisaged from the pioneering work of Melis et al. (2000), showing the possibility to separate the biomass production stage and the H_2 production stage for a culture of *C. reinhardtii*. The challenges associated with the two closed technologies capable of producing H_2were discussed followed with strategies to overcome the major technical issues. The previous experience and tools for the design and scaling up of industrial reactors has been adapted for the photobiological production of hydrogen in completely sealed photobioreactors. However the data available on hydrogen production is scarce for systems over 50 L capacity, and the light to H_2 conversion efficiencies, whatever the route used for producing H_2, are well below 1 %. Therefore the calculations made are subjected to many uncertainties. The majority of data on hydrogen production rate used have been taken from laboratory scale photobioreactors, which rarely exceed the 2 L, given the scarcity of outdoor pilot H_2 production. The length of 9 cm tube needed for the two production scenarios covered in Tables 13.2 (no H_2 bubbles in the tube circuit) and 13.3 (presence of H_2 bubbles) gives rise to volume reactors able to be fed with the harvested biomass produced, in a first step, in an open raceway facility producing about 2 and 4 tonnes of biomass (d.w) per year, respectively. In brief, photobiological hydrogen production is at an early stage of development that requires much more pilot experience.

Acknowledgements

The authors wish to acknowledge the contribution of all our colleagues of the Marine Microalgae Biotechnology Group of the University of Almería who have worked with us in the design and assessment of photobioreactors in the last 15 years. Special acknowledgement to Cajamar Foundation and the financial support from projects granted by EU (EnerBioAlgae. SOE2/P2/E374. SUDOE INTERREG IVB), Secretatia de Estado de Investigación, Ministerio de Economía y Competitividad (Project DPI2011-27818-C02-01) as well as by FEDER funds, PlanE for microalgae, ACCIONA S.A., ENDESAS.A. and Junta de Andalucía (CVI 131 &173).

References

Acién-Fernández FG, García Camacho F, Sánchez Pérez JA, Fernández Sevilla JM, Molina Grima E (1997) A model for light distribution and average solar irradiance inside outdoor tubular photobioreactors for the microalgal mass culture. Biotechnol Bioeng 55:701–714

Acién-Fernández FG, García Camacho F, Sánchez Pérez JA, Fernández Sevilla JM, Molina Grima E (1998) Modeling of biomass productivity in tubular photobioreactors for microalgal cultures: effects of dilution rate, tube diameter, and solar irradiance. Biotechnol Bioeng 58:605–616

Acién-Fernández FG, Fernández Sevilla JM, Sánchez Perez JA, Molina Grima E, Chisti Y (2001) Airlift-driven external-loop tubular photobioreactors for outdoor production of microalgae: assessment of design and performance. Chem Eng Sci 56:2721–2732

Acién-Fernández FG, Fernández Sevilla JM, Magán JJ, Molina Grima E (2012) Production cost of a real microalgae production plant and strategies to reduce it. Biotechnol Adv 30:1344–1353

Acién-Fernández FG, Fernández Sevilla JM, Molina Grima E (2013) Principles of photobioreactor design. In: Posten C, Walter C (eds) Microalgal biotechnology: potential and production. De Grutier, Göttingen, pp 151–179

Adessi A, Torzillo G, Baccetti E, De Philippis R (2012) Sustained outdoor H_2 production with *Rhodopseudomonas palustris* cultures in a 50 L tubular photobioreactor. Int J Hydrog Energy 37:8840–8849

Akkerman I, Jansen M, Rocha J, Wijjfels RH (2002) Photobiological hydrogen production: photochemical efficiency and bioreactor design. Int J Hydrog Energy 27:1195–1208

Antal T, Krendeleva T, Laurinavichene T, Makarova V, Ghirardi M, Rubin A, Tsygankov AA, Seibert M (2003) The dependence of algal H_2 production on photosystem II and O_2 consumption activities in sulfur-deprived *Chlamydomonas reinhardtii* cells. Biochim Biophys Acta 1607:153–160

Azov Y, Shelef G (1982) Operation of high-rate oxidation ponds: theory and experiments. Water Res 16:1153–1160

Benemann JR (1997) Feasibility analysis of photobiological hydrogen production. Int J Hydrog Energy 228:979–987

Benemann JR (2000) Hydrogen production by microalgae. J Appl Phycol 12:291–300

Berberoglu H, Jay J, Pilon L (2008) Effect of nutrient media on photobiological hydrogen production by *Anabaena variabilis* ATCC 29413. Int J Hydrog Energy 33:1172–1184

Boran E, Özgür E, van der Burg J, Yücel M, Gündüz U, Eroglu I (2010) Biological hydrogen production by *Rhodobacter capsulatus* in solar tubular photo bioreactor. J Clean Prod 18:S29–S35

Boran E, Özgür E, Yücel M, Gündüz U, Eroglu I (2012) Biohydrogen production by *Rhodobacter capsulatus* in solar tubular photobioreactor on thick juice dark fermenter effluent. J Clean Prod 3:150–157

Brindley C, Garcia-Malea MC, Acién-Fernández FG, Fernández Sevilla JM, García Sánchez JL, Molina Grima E (2004) Influence of power supply in the feasibility of *Phaeodactylum tricornutum* cultures. Biotechnol Bioeng 87:723–733

Camacho Rubio F, Molina Grima E, Valdés Sanz F, Andújar Peral JM (1991) Influence of operational and physical variables on interfacial area determination. AIChE J 37:1196–1204

Camacho Rubio F, Acién-Fernández FG, Sánchez Pérez JA, García Camacho F, Molina Grima E (1999)

Prediction of dissolved oxygen and carbon dioxide concentration profiles in tubular photobioreactors for microalgal culture. Biotechnol Bioeng 62:71–86

Camacho FG, Molina Grima EM, Mirón AS, Pascual VG, Chisti Y (2001) Carboxymethyl cellulose protects algal cells against hydrodynamic stress. Enzyme Microb Technol 29:602–610

Chisti Y (2013) Raceways-based production of algal crude oil. In: Posten C, Walter C (eds) Microalgal biotechnology: potential and production. De Grutier, Göttingen, pp 113–146

Chochois V, Dauvillée D, Beyly A, Tolleter D, Cuiné S, Timpano H, Ball S, Cournac L, Peltier G (2009) Hydrogen production in *Chlamydomonas:* photosystem II-dependent and independent pathways differ in their requirement for starch metabolism. Plant Physiol 151:631–640

Das D, Veziroglu TN (2001) Hydrogen production by biological processes: a survey of literature. Int J Hydrog Energy 26:13–28

Das D, Veziroglu TN (2008) Advances in biological hydrogen production processes. Int J Hydrog Energy 33:6046–6057

Doucha J, Livansky K, Kostelnik K (1996) Thin-layer microalgal culture technology. Abstracts of the 7th international conference of applied. Algology, Knysna, South Africa, p 32

Doucha J, Straka F, Lívanský K (2005) Utilization of flue gas for cultivation of microalgae (*Chlorella* sp.) in an outdoor open thin-layer photobioreactor. J Appl Phycol 17:403–412

Eroglu E, Melis A (2011) Photobiological hydrogen production: recent advances and state of the art. Bioresour Technol 102:8403–8413

Florin L, Tsokoglou A, Happe T (2001) A novel type of iron hydrogenase in the green alga *Scenedesmus obliquus* is linked to the photosynthetic electron transport chain. J Biol Chem 276:6125–6132

García Camacho F, Contreras A, Acién-Fernández FG, Fernández JM, Molina Grima E (1999) Use of concentric-tube airlift photobioreactors for microalgal outdoor mass cultures. Enzyme Microb Technol 24:164–172

Giannelli L, Torzillo G (2012) Hydrogen production with the microalga *Chlamydomonas reinhardtii* grown in a compact tubular photobioreactor immersed in a scattering light nanoparticle suspension. Int J Hydrog Energy 37:16951–16961

Giannelli L, Scoma A, Torzillo G (2009) Interplay between light intensity, chlorophyll concentration and culture mixing on the hydrogen production in sulfur-deprived *Chlamydomonas reinhardtii* cultures grown in laboratory photobioreactors. Biotechnol Bioeng 104:76–90

Grobbelaar JU (1994) Turbulence in mass algal cultures and the role of light/dark fluctuations. J Appl Phycol 6:331–335

Hallenbeck PC, Benemann JR (2002) Biological hydrogen production: fundamentals and limiting processes. Int J Hydrog Energy 27:1185–1193

Hallenbeck PC, Abo-Hashesh M, Ghosh D (2012) Strategies for improving biological hydrogen production. Bioresour Technol 110:1–9

Hemshemeier A, Fouchard S, Cournac L, Peltier G, Happe T (2008) Hydrogen production by *Chlamydomonas reinhardtii*: an elaborate interplay of electron sources and sinks. Planta 227:397–407

Hu Q, Guterman H, Richmond A (1996) A flat inclined modular photobioreactor for outdoor mass cultivation of photoautotrophs. Biotechnol Bioeng 51:51–60

Incropera FP, Thomas JF (1978) A model for solar radiation conversion to algae in a shallow pond. Sol Energy 20:157–165

James SC, Boriah V (2010) Modeling algae growth in an open-channel raceway. J Comp Biol 17: 895–906

Jiménez C, Cossío BR, Niell FX (2003) Relationship between physicochemical variables and productivity in open ponds for the production of *Spirulina*: a predictive model of algal yield. Aquaculture 221:331–345

Kapdan IK, Kargi F (2006) Bio-hydrogen production from waste materials. Enzyme Microb Technol 38:569–582

Kawase Y, Moo-Young M (1990) Mathematical models for design of bioreactors: applications of Kolmogoroff's theory of isotropic turbulence. Chem Eng J 43:B19–B41

Laurinavichene TV, Fedorov AS, Ghirardi ML, Seibert M, Tsygankov AA (2006) Demonstration of sustained hydrogen production by immobilized, sulfur-deprived *Chlamydomonas reinhardtii* cells. Int J Hydrog Energy 5:659–667

Laws EA, Taguchi S, Hirata J, Pang L (1986) High algal production rates achieved in a shallow outdoor flume. Biotechnol Bioeng 28:191–197

Lee YK, Low CS (1992) Productivity of outdoor algal cultures in enclosed tubular photobioreactor. Biotechnol Bioeng 40:1119–1122

Levin DB, Pitt L, Love M (2004) Biohydrogen production: prospects and limitations to practical application. Int J Hydrog Energy 29:173–185

Markov SA, Thomas AD, Bazin MJ, Hall DO (1997) Photoproduction of hydrogen by cyanobacteria under partial vacuum in batch culture or in a photobioreactor. Int J Hydrog Energy 22:521–524

Melis A, Zhang LP, Forestier M, Ghirardi ML, Seibert M (2000) Sustained photobiological hydrogen gas production upon reversible inactivation of oxygen evolution in the green alga *Chlamydomonas reinhardtii*. Plant Physiol 122:127–135

Moheimani NR, Borowitzka MA (2007) Limits to productivity of the alga *Pleurochrysis carterae* (haptophyta) grown in outdoor raceway ponds. Biotechnol Bioeng 96:27–36

Molina Grima E (1999) Microalgae, mass culture methods. In: Flickinger MC, Dew SW (eds) Encyclopedia of bioprocess technology: fermentation. Biocatalysis and bioseparations. Wiley, New York, pp 1753–1769

Molina Grima E, Garcia Camacho F, Sanchez Perez JA, Fernandez Sevilla JM, Acién-Fernández FG, Contreras Gomez A (1994) A mathematical model of microalgal growth in light-limited chemostat culture. J Chem Technol Biotechnol 61:167–173

Molina Grima E, Fernández Sevilla JM, Sánchez Pérez JA, García Camacho F (1996) A study on simultaneous photolimitation and photoinhibition in dense microalgal cultures taking into account incident and averaged irradiances. J Biotechnol 45:59–69

Molina Grima E, Acién-Fernández FG, García Camacho F, Camacho Rubio F, Chisti Y (2000) Scale-up of tubular photobioreactors. J Appl Phycol 12:355–368

Molina Grima E, Fernández Sevilla JM, Acién-Fernández FG (2010) Microalgae, mass culture methods. In: Flickinger MC (ed) Encyclopedia of industrial biotechnology: bioprocess, bioseparation, and cell technology. Willey, New York, pp 1–24

Oncel S, Sabankay M (2012) Microalgal biohydrogen production considering light energy and mixing time as the two key features for scale-up. Bioresour Technol 12:228–234

Oswald WJ (1988) Large scale algal culture systems. In: Borowitzka MA, Borowitzka LJ (eds) Microalgal biotechnology. Cambridge University Press, Cambridge, MA, pp 305–328

Oswald WJ, Golueke CG (1968) Large scale production of microalgae. In: Mateless RI, Tannenbaum SR (eds) Single cell protein. MIT Press, Cambridge, MA, pp 271–305

Park JBK, Craggs RJ (2010) Wastewater treatment and algal production in high rate algal ponds with carbon dioxide addition. Water Sci Technol 61:633–639

Park JBK, Craggs RJ, Shilton AN (2011) Wastewater treatment high rate algal ponds for biofuel production. Bioresour Technol 102:35–42

Phillips JN, Myers J (1954) Growth rate of Chlorella in flashing light. Plant Physiol 29:152–161

Pinto FA, Troshina LO, Lindblad P (2002) A brief look at three decades of research on cyanobacterial hydrogen evolution. Int J Hydrog Energy 27:1209–1215

Posewitz M, Smolinski S, Kanakagiri S, Melis A, Seibert M (2004) Hydrogen photoproduction is attenuated by disruption of an isoamylase gene in *Chlamydomonas reinhardtii*. Plant Cell 16:2151–2163

Posten C (2009) Design principles of photobioreactors for cultivation of microalgae. Eng Life Sci 9:165–177

Prince RC, Kheshgi HS (2005) The photobiological production of hydrogen: potential efficiency and effectiveness as a renewable fuel. Criti Rev Microbiol 31:9–31

Pulz O, Scheibenbogen K (1998) Photobioreactors: design and performance with respect to light energy input. Adv Biochem Eng Biotechnol 59:123–152

Putt R, Singh M, Chinnasamy S, Das KC (2011) An efficient system for carbonation of high-rate algae pond water to enhance CO_2 mass transfer. Bioresour Technol 102:3240–3245

Qiang H, Richmond A (1996) Productivity and photosynthetic efficiency of *Spirulina platensis* as affected by light intensity, algal density and rate of mixing in a flat plate photobioreactor. J Appl Phycol 8:139–145

Qiang H, Guterman H, Richmond A (1996) A flat inclined modular photobioreactor for outdoor mass cultivation of photoautotrophs. Biotechnol Bioeng 51:51–60

Rodolfi L, Chini Zittelli G, Bassi N, Padovani G, Biondi N, Bonini G, Tredici MR (2009) Microalgae for oil: strain selection, induction of lipid synthesis and outdoor mass cultivation in a low-cost photobioreactor. Biotechnol Bioeng 102:100–112

Rupprecht J, Hankamer B, Mussgnug J, Ananyev G, Dismukes C, Kruse O (2006) Perspectives and advances of biological H_2 production in microorganisms. Appl Microbiol Biotechnol 72:442–449

Sasikala K, Ramana CV, Rao PR (1991) Environmental regulation for optimal biomass yield and photoproduction of hydrogen by *Rhodobacter sphaeroides* O.U. 001. Int J Hydrog Energy 16:597–601

Scoma A, Giannelli L, Faraloni C, Torzillo G (2012) Outdoor H_2 production in a 50-L tubular photobioreactor by means of a sulfur-deprived culture of the microalga *Chlamydomonas reinhardtii*. J Biotechnol 157:620–627

Sierra E, Acién-Fernández FG, Fernández JM, García JL, González C, Molina Grima E (2008) Characterization of a flat plate photobioreactor for the production of microalgae. Chem Eng J 138:136–147

Sveshnikov DA, Sveshnikova NV, Rao KK, Hall DO (1997) Hydrogen metabolism of mutant forms of *Anabaena variabilis* in continuous cultures and under nutritional stress. FEMS Microbiol Lett 147:297–301

Terry KL (1986) Photosynthesis in modulated light: quantitative dependence of photosynthetic enhancement on flashing rate. Biotechnol Bioeng 28:988–995

Tredici MR, Chini Zittelli C, Benemann JR (1998) A tubular internal gas exchange photobioreactor for biological hydrogen production. In: Zaborsky OR (ed) BioHydrogen. Plenum Press, New York, pp 391–402

Tsygankov AA, Hall DO, Liu J, Rao KK (1998) An automated helical photobioreactor incorporating cyanobacteria for continuous hydrogen production. In: Zaborsky OR (ed) BioHydrogen. Plenum Press, New York, pp 431–440

Tsygankov AA, Borodin VB, Rao KK, Hall DO (1999) H_2 photoproduction by batch culture of *Anabaena variabilis* ATCC 29413 and its mutant PK84 in a photobioreactor. Biotechnol Bioeng 64:709–715

Tsygankov AA, Fedorov AS, Kosourov SN, Rao KK (2002) Hydrogen production by cyanobacteria in an automated outdoor photobioreactor under aerobic conditions. Biotechnol Bioeng 80:777–785

Verlaan P, Van Eijs AMM, Tramper J, Van't Riet K (1989) Estimation of axial dispersion in individual sections of an airlift-loop reactor. Chem Eng Sci 44:1139–1146

Vonshak A (1997) *Spirulina*: growth, physiology and biochemistry. In: Vonhask A (ed) *Spirulina platensis* (*Arthrospira*): physiology, cell-biology and biotechnology. Taylor & Francis, London, pp 43–65

Weissman JC, Goebel RP (1987) Design and analysis of microalgal open pond systems for the purpose of producing fuels: A subcontract report. United States: SERI/STR-231-2840

Weissman JC, Goebel RP, Benemann JR (1988) Photobioreactor design: mixing, carbon utilization, and oxygen accumulation. Biotechnol Bioeng 31:336–344

Winkler M, Hemschemeier A, Gotor C, Melis A, Happe T (2002) [Fe]-hydrogenases in green algae: photofermentation and hydrogen evolution under sulfur deprivation. Int J Hydrog Energy 27:1431–1439

Winkler M, Kuhlgert S, Hippler M, Happe T (2009) Characterization of the key step for light-driven hydrogen evolution in green algae. J Biol Chem 284:36620–36627

Wykoff DD, Davies JP, Melis A, Grossman AR (1998) The regulation of photosynthetic electron transport during nutrient deprivation in *Chlamydomonas reinhardtii*. Plant Physiol 117:129–135

Yoon JH, Shin JH, Kim MS, Sim SJ, Park TH (2006) Evaluation of conversion efficiency of light to hydrogen energy by *Anabaena variabilis*. Int J Hydrog Energy 31:721–727

Chapter 14

Immobilization of Photosynthetic Microorganisms for Efficient Hydrogen Production

Anatoly Tsygankov* and Sergey Kosourov
Institute of Basic Biological Problems, Russian Academy of Sciences, Pushchino, Moscow Region 142290, Russia

Summary

The immobilization is a process of catalyst (cells) attachment to the matrix. It separates effectively the cells from the liquid and gas phases, allowing a significant increase in the culture density. This review describes various approaches used for immobilization of photosynthetic cells. The main attention is focused on advantages and limitations of immobilized systems for hydrogen photoproduction.

*Author for correspondence, e-mail: ttt-00@mail.ru

D. Zannoni and R. De Philippis (eds.), Microbial BioEnergy: Hydrogen Production, Advances in Photosynthesis and Respiration 38, pp. 321–347, DOI 10.1007/978-94-017-8554-9_14,

I. Introduction

Molecular hydrogen is an ideal energy carrier for the future world. It can be produced from a wide range of sources and by a number of different technologies that includes: water electrolysis, reforming of natural gas, coal gasification, thermochemical production and biomass gasification. Hydrogen gas production by the cultures of phototrophic microorganisms is considered as one of the most promising and ecologically friendly approaches. The process occurs under ambient temperatures, requires sunlight, water and minimal amounts of macro and micronutrients. However, due to some limitations discussed in this and other chapters of the book it is still in the research and development stage. Recently obtained research data show promise in the practical application. However, there are a number of problems that should be solved before H_2 gas production by phototrophic microorganisms becomes a commercially competitive technology. Among the most important limitations are the low rates of hydrogen gas production and difficulties in maintaining and processing suspension cultures. These challenges might be addressed by immobilizing microbial cells.

When compared to suspensions, immobilized cultures have higher volumetric rates of hydrogen production; the separation of gas, liquid, and solid phases is natural; the catalyst (microbial cells) has higher stability due to a diffusion barrier of the matrix. These and some other advantages make immobilization a very suitable approach for biohydrogen applications. This chapter reviews the modern methods available for immobilization of photosynthetic microorganisms, describes criteria for selection of these methods, gives some recommendations for designing photobioreactors (PhBRs) with immobilized cells and presents examples of already described systems for hydrogen production by immobilized photosynthetic microorganisms.

Abbreviations: Chl – Chlorophyll; EPS – Extracellular polysaccharides; Fd – Ferredoxin; PAR – Photosynthetic active radiation; PhBR(s) – Photobioreactor(s); PSI – Photosystem I; PSII – Photosystem II; PVA – Polyvinyl alcohol

II. Methods of Immobilization

Immobilization in biotechnology is defined as "the confinement or localization of viable microbial cells to a certain defined region of space in such a way as to exhibit hydrodynamic characteristics, which differ from those of the surrounding environment" (Azbar and Kapdan 2011). Methods of immobilization can be divided into two large groups: artificial cell entrapments that assume application of matrices or substrates for attachment, entrapment within the matrix, or encapsulation of microorganisms, and natural cell entrapments, which allow microorganisms to form biofilm or granules. Independently on the nature, the ideal immobilization method should satisfy the following requirements:

- Neutrality to cell metabolism. The immobilization matrix should be non-toxic to the cells and stable against cellular activities.
- Stability in time and in space. Mechanical properties of the matrix should be strong enough to withstand against shear stress, but gentle to avoid mechanical inhibition of the cells.
- Porousness of the matrix or/and supporting substrate. Associated diffusion barriers should not limit the target reaction, but should be high enough to realize a high resistance of immobilized cells against toxic compounds and non-optimal environmental parameters like sharp pH changes.
- Absence of cell leakage outside of the immobilization space.
- Simplicity in production and operation.
- Low cost.
- Renewability of materials, if possible.

In addition to general requirements listed above, the immobilization method for photosynthetic microorganisms should allow the light penetration to the cells. Therefore, the matrix and support for the matrix should be transparent (translucent for foam and porous structures).

A. Artificial Immobilization

Microorganisms can be fixed in space using a variety of methods, which include: ionic adsorption on water-insoluble matrices, gel entrapments and encapsulations with solid or liquid membranes.

1. Ionic Adsorption on Water-Insoluble Matrices

Ionic adsorption of microorganisms can be non-specific and specific. An example of non-specific interactions is the cyanobacterial cells readily attached to hydrophilic polyvinyl or polyurethane foam (Hall and Rao 1988). Specific adsorption is developed due to surface charges. The surface of many bacteria has negative charge. When a substrate is charged positively, for example by activating the glass with aminosilane (Tsygankov et al. 1993), bacteria occupy up to 40 % of the substrate surface. This method is quick and simple, but produces monolayer of cells. If not followed by colonization, the process is reversible. Nevertheless, this approach is good for accelerating the natural immobilization (see below).

2. Gel Entrapments

A variety of natural and synthetic, organic and inorganic substances can be used for the entrapment of microorganisms.

a. Polysaccharides

Polysaccharides are excellent materials for entrapping microorganisms. They satisfy almost all requirements. Matrices with immobilized cells can be formed into any shape, including beads (Hallenbeck 1983), cubes (Saetang and Babel 2009) and fibers (Park et al. 1991). The main disadvantage of polysaccharides is a low chemical and mechanical stability of their matrices, which requires application of additional supporting substrates: glass plates, synthetic fibers, steel or plastic screens. The following groups of natural and modified polysaccharides are frequently used in immobilization (Stolarzewicz et al. 2011):

- polyuronides (polymers of uronic acids): alginate and pectins;
- galactans (galactose polymers): agar, agarose, carrageenan;
- glucans (polymers of D-glucose and its derivatives): chitosan, an amino polysaccharide derivative of chitin, starch, cellulose and its alkyl and carboxylic derivatives;
- some polysaccharides containing natural products like cashew apple bagasse, corn starch gel or orange peel.

The immobilization process using the above mentioned matrices is usually carried out by encapsulating the cells or entrapping them within gels of different shapes. Solidification (polymerization) of the gel is achieved by several physical and chemical treatments, largely depending on properties of the used polymer. In some cases, polymerization occurs in the presence of bivalent ions (alginic acid) and cationic compounds: metal ions, amines, amino acid derivates and organic solvents (carrageenan) or by changing the pH (alginic acid, chitosan, pectins). In other cases, gel thickening is driven by cooling the liquid polymer to the temperature of solidification (agar, agarose). The most commonly used polysaccharides for entrapping within gels are alginate, a linear unbranched polymer containing 1-4-linked-β-D-mannuronic acid and α-L-guluronic acid residues in different proportions and sequences, and carrageenan, a linear sulfated polysaccharide consisting of alternating 3-linked-β-D-galactopyranose and 4-linked-α-D-galactopiranose units.

b. Proteins

Similar to polysaccharides, proteins form hydrocolloids and may also be very effective for immobilization of enzymes and whole cells (Kourkoutas et al. 2004). The most often employed proteins are: albumin, gelatin, gluten and silk fibroin (Krastanov 1997; Gugerli et al. 2004). Encapsulation and entrapment of the cells inside fibers are commonly used for protein immobilization approaches. Protein hydrocolloids are transparent in visible part of spectrum and, therefore, can be used for immobilization of photosynthetic cells in biohydrogen applications. The main drawback of these matrices is their high cost.

c. Polyvinyl Alcohol

Polyvinyl alcohol (PVA) is another excellent, but synthetic material, for entrapping microorganisms due to its neutrality, stability, transparency, simplicity of operation, cost, and commercial availability. Two methods of matrix polymerization are often used for microbial immobilization. The first is "freezing-thawing", and the second is UV-light polymerization. PVA is not mechanically strong enough to create foam. For this purpose, the method of preparing macroporous PVA foam with improved stability was suggested, which involves adding calcium carbonate as a pore-forming agent and epichlorohydrin as a chemical cross-linking agent (Bai et al. 2010).

d. Polyacrylamide

Acrylic polymers obtained from acrylic or methacrylic acid and their derivatives, such as amides, esters and others, and cross-linked with *N,N'*-methylenebisacrylamide form transparent, mechanically and chemically stable matrices, which have a high diffusion rate regulated by cross-linking. These matrices are often used for immobilization of microorganisms (Stolarzewicz et al. 2011). The polymerization procedure (by irradiation or by chemical activation) is not neutral for cells and, therefore, should be tested for the influence on the target activity before any application for the particular microorganisms.

3. Thin-Layer Immobilization

The main limiting factor affecting the productivity of photosynthetic microorganisms is the light distribution within the culture (Torzillo et al. 2003). The immobilized cells are subject to the same rule. With increasing depth of the matrix, the lower cell layers experience shading, which can be significant in the dense culture. On the contrary, upper cell layers are subject to photoinhibition, especially under high light. As a result, overall light utilization efficiency by the culture is low. The problem can be addressed in part by immobilizing photosynthetic microorganisms in thin-layer matrices. Thin-layer immobilization allows a precise control of the matrix thickness and makes possible a more uniform light distribution to the cells. Although different transparent polymers can be applied for entrapping photosynthetic microorganisms within thin-layer films, the most frequently used polymers in this approach are sol-gels, latexes and alginates. Except for sol-gels, these materials are cheap and available in industrial quantities. In addition, alginates, as natural polymers, are renewable materials.

a. Sol-Gel Encapsulation

Starting from the first description of sol-gel application for the immobilization of microorganisms (Carturan et al. 1989) this technology is developed for encapsulation of different enzymes and cells, including photosynthetic microorganisms and plant cells (Rooke et al. 2008). Silica sol–gels are chemically inert, mechanically stable, transparent and can be used not only for encapsulation but also for entrapment (Kandimalla et al. 2006). One drawback is the procedure of evaporation during xerogel preparation since drying reduces the viability of cells immobilized within sol–gel matrices (Baca et al. 2007). Another drawback of sol-gel application is the formation of alcohols from alkosilane matrices (Meunier et al. 2010). Sol-gel materials based on alumina, titania and other compounds have similar drawbacks. To avoid cells drying during xerogel formation, several methods were applied. Fukushima et al. (1988) entrapped whole cells in alginate-silica gels with very promising results. Other authors employed biocompatible short chain phospholipids to direct the formation of an ordered silica mesophase during evaporative processes (Baca et al. 2007). To increase bacterial viability during sol-gel preparation, different authors used encapsulation of microorganisms by alginate (Pannier et al. 2011), PVA (Liu et al. 2009) with following coating by sol-gel. The addition of cryoprotectors like glycerol, quorum-sensing molecules or combination of them increased viability of bacteria in sol-gel (Meunier et al. 2010).

Different kinds of precursors have been developed to obtain the sol-gel matrices with improved properties like reduced brittle nature, high transparency, improved hydrophilicity, flexible and tuned porosity, etc. Nevertheless, the ideal matrix has not been developed. For example, inorganic sol-gels are good in transparency, but have low porosity in xerogels restricts their application. Organically modified sol–gels have good tunable porosity, but have limited optical transparency (Kandimalla et al. 2006) that limits their application for photosynthetic cells.

b. Latex Coatings

Although photosynthetic microbes were first immobilized in latex coatings nearly 20 years ago (Martens and Hall 1994), the latex polymers have been applied mainly for entrapping heterotrophic bacteria (Lyngberg et al. 2001; Flickinger et al. 2007). The entrapment of photosynthetic microorganisms into latex matrices has attracted more attention only recently with increasing demands of immobilizing H_2-photoproducing cultures. The main advantage of latex polymers compared to many other materials used for cell entrapments is their ability to form mechanically stable nanoporous coatings (Flickinger et al. 2007). The coating thickness can easily be controlled within 10–250 μm (Gosse et al. 2007). Using this advantage, Gosse et al. (2007) found the optimal thickness for H_2-producing phototrophic bacterium, *Rhodopseudomonas palustris* entrapped in latex coatings of around 50 μm (at 34 μE m^{-2} s^{-1} PAR). As expected, an increase in the thickness led to the loss in photoreactivity due to the self-shading effect. Immobilization of bacteria into the latex polymer also improved greatly their catalytic stability. The latex coatings of *Rp. palustris* produced H_2 gas for over 4,000 h (Gosse et al. 2010). They also remained active after hydrated storage for greater than 3 months in the dark and over 1 year when stored at −80° C (Gosse et al. 2007).

c. Alginate Films

Latex polymerization requires drying that leads to coalescence of latex particles. Commercial latex polymer formulations may also contain biocides and other toxic additives. As a result, not all microbial cultures can survive in latex coatings. In those cases, alginate and likely other hydrogels may replace latex polymers in the thin-layer immobilization approach. Recently, Kosourov and Seibert (2009) entrapped green alga, *Chlamydomonas reinhardtii* within thin Ca^{2+}-alginate films (Fig. 14.1). Since alginate is not mechanically stable polymer, authors introduced an additional polymeric screen for the mechanical support that improved the film longevity for up to 1 month. The immobilization of cells within alginate films has several advantages. The pH of a 4–6 % alginate solution is close to 7, and the polymerization process does not shift the pH inside the matrix, which is important for optimal survival of microbial cells. Furthermore, the polymerization process occurs at room temperature in the presence of added non-toxic divalent ions like Ca^{2+} and Ba^{2+}. The material is cheap and can be produced in industrial quantities from a renewable resource.

Fig. 14.1. The entrapment of *Chlamydomonas reinhardtii* cells in Ca^{2+}- alginate films is a convenient way for screening H_2 photoproduction activities in different mutants.

The mechanical stability of alginate films can further be improved by cross-linking alginate with other polymers, such as: chitosan, polyacrylic acid, polyurethane, polyvinyl alcohol and polyvinylamine or by coating the surface of alginate films with poly-L-lysine, polyethyleneimine or glutaraldehyde.

4. Covalent Attachment

Immobilization based on the formation of covalent bonds is widely used for enzymes. Covalent methods in general can be divided in two groups: activation of the matrix by addition of a reactive function to the polymer and modification of the polymer backbone to produce an activated group (Brena and Batista-Viera 2006). A wide variety of reactions have developed depending on the functional groups available on the matrix. Treatments with tresyl chloride, cyanogen bromide, epoxides, epichlorohydrin, glutaraldehyde, glycidol-glyoxyl are among them (Brena and Batista-Viera 2006). Covalent coupling of whole cells as immobilization method is often applied in scanning electron microscopy. This method and methods based on cross-linking and co-cross linking with neutral molecules are not applicable for long-term immobilization of whole cells producing valuable products due to negative influence on cell surface and leakage of cells into suspension due to growth and division. However, there is a possibility to apply mild covalent binding as a first step accelerating natural immobilization (see below).

All artificial immobilization methods have advantages and drawbacks. For example, adsorption is simple, cheap and effective but creates monolayers with limited applications and frequently reversible; covalent attachment and cross-linking are effective and durable, but expensive and often affect microorganisms viability; gel entrapment and encapsulation have inherent diffusion problems. Furthermore, the existence of many different immobilization methods indicates that no universal method exists for different microorganisms and various processes.

B. Natural Immobilization

Bacteria exhibit two different behavioral strategies: a planktonic state, in which individual cells move freely in liquid medium, or a benthic state, in which they are tightly clamp. If no support exists, benthic bacteria form mats, which can be seen in different ecological niches. If bacteria attach to the surface, they form biofilms. Most of physiological, biochemical, and genetic researches have been based on the assumption that bacterial populations consist of the individual cells with identical characteristics or of the colonies originating from a single cell. Development of microscopy and molecular genetic techniques promoted the understanding that natural bacterial populations exist mostly in the attached state, in which bacteria may exchange signals and exhibit coordinated activity with creation of subpopulations for specialized functions (Smirnova et al. 2010).

Cells in biofilms have enhanced resistance to solvents and toxins as compared to suspension cells (Dagher et al. 2010; Smirnova et al. 2010). Development of mat or biofilm communities is one of the main strategies for survival of bacteria in a certain ecological niche under different stresses. Due to a higher resistance, biofilms cause chronic infections and persistent diseases, food spoilage, metal corrosion that can wear out or block water pipes, and cause other common problems associated with surfaces exposed to water.

Apart from the nature and problems associated with formation of biofilms, they have found application in biotechnology as an immobilization method. Natural immobilization is widely used in industrial applications for the treatment of wastewater, gas desulfurization, and in food production (Dagher et al. 2010). Therefore, a number of publications dealing with different aspects of biofilm formation is growing.

1. Biofilm Formation

Different authors divide the process of biofilm formation into different stages, but all of them agree that three main stages exist: attachment, colonization with extracellular polymer production and growth with maturation of the biofilm (Dagher et al. 2010; Smirnova et al. 2010; Cheng et al. 2010).

a. Attachment

The first stage, named attachment, is greatly affected by the surface properties (i.e. roughness, porosity, hydrophobicity and charge) and the growth rate of microorganisms being transported to the surface. The hydrodynamic conditions of the medium can also affect adherence to the surface by increasing or decreasing cell shearing. Many microorganisms react to excessive turbulence and shearing forces by inducing a global genetic response that causes a complete modification of cell surface components including flagella, fimbriae, pili, capsule, and other cell-wall polysaccharides (Dagher et al. 2010). The attachment is the reversible process and cell surface structures like flagella, adhesins, fimbria, and pili participate in irreversible adsorption together with extracellular polymers synthesis (Smirnova et al. 2010).

b. Colonization

When the cell is irreversibly attached to the surface it continues growing and dividing. As a result, microcolonies of microorganisms appear on the support. Simultaneously, depending on the microorganism and environmental conditions cells continue synthesizing exopolymeric substances like polysaccharides, proteins, carbohydrates, DNA and lipids (Dagher et al. 2010). Alternatively, the monolayer of cells is formed by irreversible attachment of moving cells along the surface (Smirnova et al. 2010). Evidently, this event depends greatly on the environmental conditions, such as: availability of substrates and different gradients near the surface, turbulence of liquid, presence of toxins etc. When conditions favor biofilm formation, both events can take place simultaneously.

At the colonization stage, the polymer matrix between the surface and cells starts forming. The bulk of the matrix consists of extracellular polysaccharides (EPS). In some cases, the matured biofilm contains 85 % EPS (Romanova et al. 2011). Nevertheless, other constituents of the polymer matrix are also very important. For example, extracellular DNA participates in polymer matrix formation and addition of DNA-ase to the biofilm destroys the polymer matrix (Romanova et al. 2011).

EPS build bridges between negatively charged cells providing them with a natural matrix. The EPS matrix consists mainly of homo- and heteropolysaccharides. The EPS contain uronic acids, mostly glucuronic and aminosugars. For some bacteria the structure of EPS is already known. For example, *Pseudomonas* species produce alginate, *Escherichia coli* synthesizes colonic acid, and *Bacillus cepacia* – cepacian (Smirnova et al. 2010).

During the colonization stage of biofilm formation, concentration of the cells is enough for them to start interacting each other. The bacterial cell-to-cell communications occur on chemical and physical (via pili) levels. Cells can produce and sense molecules that allow the whole population initiating a concerted action of structural and metabolic changes once a critical concentration (which depends on the population density) of the signaling molecule is reached, a phenomenon known as quorum sensing. Autoinducer 2 (AI-2) is suggested to be a universal bacterial signaling molecule, which is synthesized by the luxS product (Dagher et al. 2010). The quorum sensing is a possibility for the cells to sense a combination of cell density, mass-transfer properties of the environment and spatial cell distribution, to estimate the efficiency of producing extracellular effectors and to respond only when the response is efficient (Carnes et al. 2010).

c. Biofilm Maturation

With increasing cell density in the matrix, the distribution of cells becomes non-homogeneous. In addition, the diffusion nature of interactions between the matrix and the environment creates local gradients of substrates, products, AI-2, pH, and other growth components along the biofilm. The last process results in changes of cell morphology (Smirnova et al. 2008) and leads to the differentiation of the microbial population inside the biofilm in subpopulations with specialized functions (Dagher et al. 2010). Continuous growth of the biofilm forms its 3-dimensional structure. The thickness of biofilm varies from a few microns to centimeters depending on microbial species, biofilm age, nutrient availability and environmental liquid hydrodynamic parameters (Cheng et al. 2010).

Simultaneously with biofilm formation, the dispersal of planktonic cells and the detachment of biofilm pieces happen. When the rate of biofilm formation is equal to the rate of the cell dispersal and the detachment of biofilm pieces, the biofilm reaches a steady state.

2. *Acceleration of Natural Immobilization*

Similar to artificial immobilization, natural immobilization or biofilm formation is the process of cell entrapment in polysaccharide matrix. At the start of the operation, artificial immobilization using appropriately adapted cells realizes higher rates of the target reactions. However, artificial immobilization has the long-term stability lower than natural biofilms. This conclusion comes from the consideration that cells in natural biofilms already have morphology and metabolism adapted to the immobilized state (Smirnova et al. 2010) and surrounded by the polymers of their own metabolism. The biofilms have optimal thickness adapted to the particular environment, substrates, toxins, and product concentrations. Therefore, natural immobilization is a preferable approach for long-term operation. Unfortunately, biofilm formation is a slow process. For example, the natural biofilm of cyanobacteria in polyurethane foam cubes was formed for 2 weeks (Park et al. 1991), the biofilm of microalgae on glass tissue was formed for 2–3 weeks (Laurinavichene

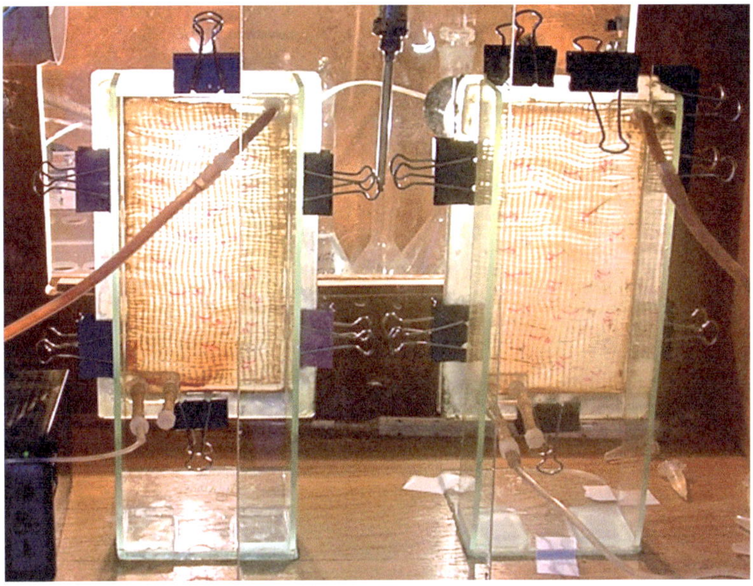

Fig. 14.2. The photobioreactors with *Rhodobacter sphaeroides* cells attached to the glass textile.

et al. 2006), the biofilm of purple bacteria on glass beads – 60 days (Tian et al. 2010). Therefore, acceleration of biofilm formation is required for adopting this process in practical applications.

Looking at different stages of biofilm formation, it becomes evident that the attachment of the cells is the most difficult and time-consuming stage. If the attached cells are kept in the appropriate medium, the rate of biofilm formation is determined by the growth rate of microorganisms. Several ways for accelerating the attachment exist. The simplest is to use the support, which is favorable for the cell attachment. Here, different polymers with the positively charged surface are used. For example, cyanobacteria attach readily to hydrophilic polyvinyl or polyurethane foams due to non-specific interaction (Hall and Rao 1988). The next way is modification of the substrate surface. For example, glass has the same, negative, charge as microbial cells. There is a possibility to modify it by aminosilanes. For example, glass surface activated by 3-(2-aminoethylaminopropyl)-trimetoxysilane was occupied by cells of purple bacterium, *Rhodobacter sphaeroides* after 2 h (Tsygankov et al. 1993). In contrast, pure glass surface did not contain bacteria during the same incubation. After 3 days of continuous medium flow, the biofilm on the glass surface was formed with stable rate of hydrogen production at least for 40 days (Tsygankov et al. 1994). It was shown later that not only purple bacteria but also green algae and cyanobacteria attach readily to the activated glass surface (Tsygankov et al. 1998b). Unfortunately, this procedure is not cheap and requires incubation of bacteria in distilled water for avoiding positive charges shading by different ions. Also, other methods of glass activation were studied: treatment of the glass surface with sulfuric acid, with hydrophobic silane reagent, and with sodium hydroxide (Tekucheva et al. 2011). After 4 days of continuous medium flow, glass textile treated with sodium hydroxide contained highest quantity of bacteria as measured by the content of bacteriochlorophyll *a* (Fig. 14.2). Other authors also used sodium hydroxide glass treatment (Tian et al. 2010). Therefore, simple and cheap procedure for accelerating the first stage of biofilm formation on glass surfaces is available.

III. Mechanical Support and Photobioreactors for Immobilized Photosynthetic Microorganisms

Ionic adsorption, gel entrapment, and biofilms need a mechanical support. For immobilization of photosynthetic microorganisms this support should not absorb light energy. In the case of smooth surfaces, it means that the surface should be transparent. When the support has complicated surface (foam, pores, cavities for the surface enhancement) even glass is not transparent but translucent. This mechanical support for biofilm formation in some cases is called as a matrix for immobilization (Tsygankov 2004). However, it is not correct: biofilm creates its own matrix on the support. A variety of inert light penetrable supports are available for immobilization: glass, polyurethane, polyvinyl chloride, other transparent polymers in different forms and shapes as sheets (Fedorov et al. 1998), hollow fibers (Park et al. 1991; Markov et al. 1993), polymeric screens (Kosourov and Seibert 2009), textiles made of glass fibers (Laurinavichene et al. 2006). Alternatively, in the case of entrapment of bacteria in gels, there is a possibility to form beads (Sasikala et al. 1992) or cubes (Kannaiyan et al. 1994) that can be placed directly in PhBRs.

The support with immobilized cultures should be placed in the vessel, which is called a photobioreactor. For immobilized heterotrophic bacteria a variety of bioreactors with high performance exists (Cheng et al. 2010). They can be divided in two categories: fixed bed and expanded bed bioreactors. Fixed bed reactors, where immobilized bacteria are fixed on static media, can be divided into submerged beds in which the biofilm particles are completely immersed in the liquid; trickling filters in which the liquid flows downward through the biofilm bed; rotating biological contactor, in which the biofilm develops on the surface of a partially submerged surfaces; membrane biofilm reactors in which the microbial layer is formed on a porous gas-permeable membrane. Expanded bed reactors contain biofilm with continuously moving media. They are divided into fluidized beds in which particles (bacterial granules or pieces of the support with immobilized bacteria) move up and down within the expanded bed and moving beds in which the whole expanded bed circulates throughout the reactors, such as air-lift reactor and circulating bed reactors.

Unfortunately, in the present state these reactors are not applicable for photosynthetic microorganisms due to the absence of light delivery. The problem of light delivery comes from the fact of light absorption by photosynthetic microorganisms. As a result, deeper layers of photosynthetic microorganisms receive less light energy. In practice, suspensions with approximately 1 g dry biomass of photosynthetic bacteria per liter decrease light intensity by 90 % after 1 cm of optical path. Therefore, PhBRs with immobilized photosynthetic microorganisms should be constructed in such a way that incident light goes through them as short as possible. The following types of PhBRs are already described in the literature:

- The column packed with beads or cubes of a support with immobilized microorganisms; light is delivered through transparent walls (Park et al. 1991).
- The column packed with hollow fibers; light is delivered through transparent walls while hydrogen (the product of cyanobacterial photosynthesis) is extracted to the inner space of hollow fibers by decreased pressure (Markov et al. 1995).
- The column packed with optical fibers; light is delivered through optical fibers and cells are immobilized around them (Matsunaga et al. 1991).
- The column packed with optical fibers; light is delivered through optical fibers and cells are immobilized around them. Optical fibers are covered by stainless steel mesh to increase the surface for biofilm formation (Guo et al. 2011).
- Plate type PhBRs with sheets of porous glass (Tsygankov et al. 1994) or glass textile (Laurinavichene et al. 2006) for biofilm and with partitions for directed medium flow.

Unfortunately, the geometry of the PhBR itself cannot define the efficiency of hydrogen production by immobilized microorganisms for several reasons:

- The efficiency of light delivery to the whole set of microorganisms is defined by interplay between the thickness of illuminated layer and cell density. For PhBRs with suspension cultures, particular criteria for estimating this parameter were created (Tsygankov 2001a). The simplest one is the ratio of illuminated surface to the volume. However, it is not very accurate for scaling up (Tsygankov 2001a). In practice, nobody applied this criterion for comparison of PhBRs with immobilized cells. It remains unclear how to compare efficiencies of light delivery to PhBRs with different shapes and sizes.
- The efficiency of substrate delivery, the product outflow, and protection of cultures against toxins and non-optimal environmental conditions depend on the PhBR geometry, mode of the process (batch or continuous), homogeneity of the liquid distribution, the rate of the process, as well as the concentration of immobilized cells. All these parameters depend on each other. Therefore, the interaction of liquid and solid phases in the PhBR should be analyzed separately for each PhBR.
- The concentration of immobilized cells should be as high as possible, taking into account the efficient light delivery and mass exchange between the cells and the liquid phase. This parameter defines the geometry of the support that should have as high surface as possible and the volume very close to the PhBR volume, taking into account light delivery and solid-liquid mass exchange. Otherwise, the volume of the PhBR will be used inefficiently and the volumetric rate will be lower than possible.
- As a matter of fact, different strains of microorganisms have different abilities for hydrogen production, distinct demands for the medium and the light supplement. If a wrong culture is used, the best PhBR satisfying all the other criteria above will not produce hydrogen with the highest rate.

Analyzing the above-listed considerations, one could suggest that the rate of hydrogen production by immobilized photosynthetic bacteria measured per unit of the volume is the result of very complicated interplay between the shape of the PhBR, the efficiency of its illumination, its mass transfer between solid, liquid, and gas phases, the shape of the support, concentration of cells in the support and in the PhBR, as well as the activity of cells.

IV. Hydrogen Production by Purple Bacteria

Purple non-sulfur bacteria are anoxygenic phototrophs. Due to the presence of only one photosystem they cannot use water as electron donor for photosynthesis and, therefore, do not produce oxygen. Instead of water, they utilize simple organics, reduced sulfur compounds, molecular hydrogen and some other reduced substrates.

Under combined nitrogen deficiency purple non-sulfur bacteria synthesize nitrogenase. This enzyme in reaction of nitrogen fixation produces 1 mol of H_2 per 1 mol of fixed N_2:

$$N_2 + 8e^- + 8H^+ + 16ATP \rightarrow 2NH_3 + H_2 + 16ADP + 16P_i \quad (14.1)$$

When N_2 is absent, all electron flows to nitrogenase are directed to H_2 synthesis:

$$2H^+ + 2e^- + 4ATP \rightarrow H_2 + 4ADP + 4P_i. \quad (14.2)$$

Purple non-sulfur bacteria belong to the most active nitrogen fixers (Vignais et al. 1985). Under the light and diazotrophic conditions, they produce H_2 at high rates. Coupling of nitrogen fixation to anoxygenic photosynthesis allows purple bacteria producing H_2 gas at high rates that are accelerated in the absence (or deficiency) of nitrogen. Simultaneously, they consume simple organics. These physiologic peculiarities allow researchers considering purple bacteria as biocatalysts in systems of wastewater treatment with simultaneous H_2 production. Extensive studies on mechanisms (Vignais et al. 1985) and major factors influencing the process (for review see Akkerman et al. 2002), as well as optimization of hydrogen photoproduction rates (Tsygankov et al. 1998a) were done using suspension cultures.

It was proved that immobilization of microorganisms is an efficient way of increasing the volumetric hydrogen production rates (Brodelius and Vandamme 1987). The first attempts of entrapping the purple bacteria were done more than 30 years ago (Vincenzini et al. 1981, 1982a, b, 1986; Zurrer and Bachofen 1985; Tsygankov et al. 1998a; Zhu et al. 1999a). These publications demonstrate that immobilized purple bacteria can be packed at significantly higher concentrations than suspension cultures. As a result, immobilized systems produce H_2 gas with higher volumetric rates than suspensions (Tsygankov 2001b). They also showed more stable H_2 production than suspensions and the time of operation close to 1,000 h (Tsygankov et al. 1994, 1998a; Tsygankov 2004).

At present, the list of available supports and matrices for immobilization of purple bacteria includes: gels like agar, agarose, carrageenan, alginate (Planchard et al. 1984; Fissler et al. 1995), PVA (Tian et al. 2009), agar with chitosan (Zhu et al. 1999b), PVA with carrageenan and alginate (Wang et al. 2010), clay (Chen and Chang 2006), latex (Gosse et al. 2007). The range of supports for biofilms of purple bacteria expands from porous glass and smooth glass surfaces (Tsygankov et al. 1993) to polyurethane foam (Fedorov et al. 1998), glass textile (Tekucheva et al. 2011), glass beads (Tian et al. 2010), plastic optical fibers with additional mesh support around them (Guo et al. 2011). As reported in the papers, all matrices and supports showed good characteristics. From a wide range of different methods used in immobilization of purple bacteria, it seems that the choice of the matrix or support is rather a preference of the authors than a necessity.

As a matter of fact, no any significant advances in the volumetric rate of hydrogen production by immobilized purple bacteria have been reported in the last two decades. In most cases, it is impossible to explain this phenomenon. In some cases, however, after a thorough analysis of published data one can conclude that authors did not take into account one or several interplaying parameters that influence the final process.

V. Hydrogen Production by Immobilized Microalgae

Despite a quite extensive list of different techniques and approaches described for immobilization of microalgae, they were mainly devised for water purification and production of some valuable metabolites, but not for H_2 photoproduction (Mallick 2002, 2006). As an example, the different species of *Chlorella, Scenedesmus, Chlamydomonas* and *Dunaliella* immobilized in alginate or carrageenan beads and screens, in polyvinyl or polyurethane foams and on hollow cellulose fibers were successfully applied for reduction of nitrogen and phosphorus contents in farm and industrial wastewater effluents (Travieso et al. 1992, 1996; Kaya and Picard 1995; Robinson 1998; Jimenez-Perez et al. 2004; Shi et al. 2007). In some cases, the immobilized cells removed up to 95 % of inorganic nitrogen and up to 99 % of phosphates (Lau et al. 1998). The immobilized microalgae were also applied for biosorption of heavy metals (Garnham et al. 1992; Moreno-Garrido et al. 2005) and biodegradation of industrial pollutants, including biocides, hydrocarbons and surfactants (Zhang et al. 1998; Semple et al. 1999). As biocatalysts, they showed promise in *de novo* biosynthesis of glycerol (Leon and Galvan 1995), hydrogen peroxide (Scholz et al. 1995) and (R)-1,2-propanediol (Hatanaka et al. 1999). More details on the environmental and industrial applications of immobilized microalgae can be found in recently published reviews (Moreno-Garrido 2008; de-Bashan and Bashan 2010).

The use of immobilized microalgae for H_2 photoproduction was limited until recently by an extremely low yield of H_2 gas in algal cultures. As a result, the early research efforts were focused mainly on investigating the mechanisms of H_2 evolution in suspensions. A few dozen algal species were tested for their ability to photoproduce H_2 gas after the period of dark anaerobic adaptation (Boichenko and Hoffmann, 1994). It was discovered that some of them, but not all, are capable of direct water biophotolysis, a biochemical reaction resulting in simultaneous accumulation of molecular hydrogen and oxygen in the same volume:

$$H_2O + 2Fd_{ox} \leftrightarrow 2H^+ + 1/2O_2 + 2Fd_{red}, \quad (14.3)$$

$$2H^+ + 2Fd_{red} \leftrightarrow H_2 + 2Fd_{ox}. \quad (14.4)$$

The first reaction is typical to all oxygenic phototrophs, including plants and cyanobacteria. This process results in the release of oxygen in photosystem II (PSII) and simultaneous production of NADPH and ATP that are further utilized mainly in the CO_2 fixation pathway. The reaction (14.4) is possible only under anaerobic conditions. The role of this process in the physiology of green algae is still a matter of debate. Most probably, it serves as a regulatory valve preventing overreduction of photosynthetic apparatus in algae during their transition from dark anaerobic to light aerobic conditions (Appel and Schulz 1998; Boichenko et al. 2004). The reaction is driven by a special enzyme, [FeFe]-hydrogenase, that is extremely sensitive to O_2 (Ghirardi et al. 1997). The reaction proceeds at high initial rates and high light to hydrogen conversion efficiencies that raise a question about industrial applicability of the process in future (Boichenko et al. 2004). At the current point, however, water biophotolysis cannot be sustained due to a rapid (within seconds) inactivation of [FeFe]-hydrogenase by O_2 co-evolved in photosynthesis (Ghirardi et al. 1997).

The immobilization of green algae for long-term H_2 photoproduction has become possible after the discovery of partial inactivation of O_2–evolving activity in algal cells in the absence of some essential nutrients, mainly sulfur and phosphorus (Wykoff et al. 1998). Taking into account the experimental data obtained by Wykoff and co-authors, the collaborative group of researchers from UC Berkeley and National Renewable Energy Laboratory successfully applied a sulfur deprivation procedure to sustain H_2 production in *C. reinhardtii* cultures (Melis et al. 2000). In these experiments, the long-term H_2 photoproduction was possible due to a metabolic switch occurring in sulfur-deprived algal cells that separated temporarily the O_2-evolving (the reaction 14.3 above) and H_2-producting (the reaction 14.4) stages in the same culture (Ghirardi et al. 2000). The later studies showed that the same principle works for phosphorus- (Batyrova et al. 2012) and nitrogen-depleted (Philipps et al. 2012) microalgae. Although the overall efficiency of H_2 evolution in nutrient-deprived cultures was shown far below the capacity of direct biophotolysis (Melis 2007; Ghirardi et al. 2000), this approach allowed sustaining the process for several days (Melis et al. 2000; Kosourov et al. 2002). As a result, the nutrient-deprivation protocol has become a platform for testing the performance of a variety of algal mutants, growth conditions and other engineering factors, including different immobilization techniques.

The original sulfur deprivation procedure requires repetitive washing of cells in sulfur-free medium by centrifugation (Melis et al. 2000). Furthermore, H_2 photoproduction in the batch cultures can be repeated several times by rejuvenating the cells in sulfur-containing medium (Ghirardi et al. 2000). The fact that algae are suspended in a liquid phase makes it difficult to cycle the batch system between rounds of sulfur deprivation and sulfur re-addition without large energy input for the required centrifugation steps. In attempting to overcome this barrier and increase the overall duration of the process, two independent research groups immobilized sulfur-deprived *C. reinhardtii* cultures using different solid supports (Laurinavichene et al. 2006, 2008; Hahn et al. 2007).

Laurinavichene et al. (2006) immobilized a wild-type *C. reinhardtii* strain, 137C *mt*$^+$ and a non-motile mutant, CC-1036 *pf18 mt*$^+$ on glass fiber matrices having a linen-like structure. The alga with paralyzed flagella was supposed to have a better attachment to the glass surface. Both strains, however, exhibited similar immobilization properties. Different glass fiber matrices with different water absorption properties were tested and the best (TR-03) was selected for further work. All of these matrices are available in the industrial scale at very low cost. The authors used two methods for attachment. In the quick immobilization procedure, glass was activated by 3-(2-aminoethyl-aminopropyl)-trimethoxysilane (Tsygankov et al. 1994) and matrices were placed in cell suspensions for 2.5 h. This approach resulted in the glass matrices having below 60 mg total Chl per m^2. In the second procedure, algae were allowed to colonize glass surfaces in a natural way. The matrices were incubated with cells for about 2 weeks during the growth phase on a regular medium. This approach produced matrices with significantly higher cell densities (about 570 mg total Chl per m^2). Independently of the technique used, immobilization of algal cells on glass fiber matrices significantly increases the duration of H_2 photoproduction in sulfur-deprived algae (up to 4 weeks) with the specific rate similar to suspension cultures. Both approaches use the property of microalga to form biofilm. However, in the first one the biofilm formation was accelerated (see above). In the best case, the immobilized cells produced H_2 gas with the rate of about 6.5 μmol H_2 (mg Chl h)$^{-1}$. In comparison, sulfur-deprived suspension cultures produce H_2 with a maximum specific rate usually ranging from ~4 to ~6 μmol H_2 (mg Chl h)$^{-1}$ (Kosourov et al. 2002), although under the most favorable conditions, rates as high as 9.5 μmol H_2 (mg Chl h)$^{-1}$ have been observed (Kosourov et al. 2003). The average rate in algal cultures immobilized on glass fiber matrices on a per volume basis was around 4 ml H_2 (L_{PhBR} h)$^{-1}$ and the maximum was 9.2 ml H_2 (L_{PhBR} h)$^{-1}$ (Laurinavichene at al. 2006). The following studies with either a constant flow of the medium containing micromolar sulfate concentrations or cycling immobilized cells between minus and plus sulfate conditions improved the duration of H_2 production up to at least 3 months (Laurinavichene at al. 2008). Nevertheless, due to irregular colonization of glass fibers by the algal cells, the system showed significant physical and physiological heterogeneities in different parts of the matrix, resulting in irregular light and nutrient distributions. The authors found that algae had a very high photochemical activity in some parts of the matrix and evolved oxygen instead of hydrogen. Produced O_2 inhibited H_2 photoproduction activities in other algae. As a result, the efficient H_2 production in this system required continuous argon flow for mixing and gas removal.

Hahn et al. (2007) attached *C. reinhardtii* cells to the fumed silica particles. The study was based on the principle that the fixed cells can be cycled between sulfate reach and sulfate free environments via filtration, thus, eliminating the expensive centrifugation steps. Similar to Laurinavichene et al. approach, the immobilization was done through a natural colonization of algal cells on the silica particles during the growth phase. In the initial experiment, the authors also tried glass beads as a support without success. The study showed that the algae bound to the fumed silica particles produce H_2 gas at a very similar rate to free-floating algae. It is important to note here that these experiments were also done with a suspension of algae/silica particles that should not give any significant advantage in the light utilization as compared to free-floating algae. Unfortunately, the authors did not provide any information about the rates of H_2 production allowing the precise estimation of H_2 photoproduction efficiency. However, based on the total H_2 photoproduction yields anyone can conclude that the suspended silica particles carrying microalgae utilize light not better than the free-floating cells.

Although very cheap, natural immobilization techniques resulted only in a very slight improvement in the light to hydrogen conversion efficiency in the nutrient-deprived microalgae as compared to the suspension cultures (Ghirardi 2006), but led to a significant prolongation of H_2 photoproduction period (Laurinavichene at al. 2008). Trying to improve the light absorption properties of immobilized microalgae, Kosourov and Seibert (2009) entrapped *C. reinhardtii* cells within thin alginate films. The technique was based on the idea of thin layer cell immobilization into the thin nanoporous latex coatings (Flickinger et al. 2007; Lyngberg et al. 2001). The entrapment of phototrophic cells into the polymer allows a very precise control of the cell density inside the coatings and their thicknesses that, in the ideal case, must provide the immobilized cells with the best environment for light distribution. Indeed, Gosse and co-authors (2007) observed an improvement in the rate of H_2 photoproduction in phototrophic bacterium, *Rp. palustris*, entrapped within thin nanoporous latex coatings as compared to suspension cultures. Unfortunately, viability of green algae could not be maintained during the drying process inherent to regular latex film formation (JL Gosse, S Kosourov, M Seibert and MC Flickinger, unpublished, 2013). Therefore, Kosourov and Seibert (2009) used alginate for entrapment of H_2-producing *C. reinhardtii* cells. The switch to alginate significantly improved the cell viability and reactivity of the coatings, but decreased their mechanical stability, as expected. The mechanical stability of the system was improved by introducing a special template consisting of a polymer insect screen placed over the sticky side of a wide adhesive tape. In this configuration, the alginate films with entrapped algae survived for up to 1 month. They also demonstrated high cell densities (up to 2,000 μg Chl per mL of the matrix) resulting in the specific rate of H_2 evolution up to 12.5 μmol H_2 (mg Chl h)$^{-1}$, which is almost three-times higher than in suspensions (~4 to ~6 μmol H_2 (mg Chl h)$^{-1}$; Kosourov et al. 2002), but the rates in suspensions were obtained at significantly higher light intensities (200 μE m^{-2} s^{-1} PAR in suspensions *vs.* 62 μE m^{-2} s^{-1} PAR in alginate films). As a result, the conversion efficiency of incident light energy into energy of the H_2 gas in alginate films at 62 μE m^{-2} s^{-1} PAR was above 1.5 % for the period of the maximum H_2-production rate and was close to 1 % for the whole period of nutrient deprivation. These values are significantly higher than the values reported for immobilized on glass fibers (0.36 %) and suspension (0.24 %) *C. reinhardtii* cultures, but calculated at much higher light intensities: 120 and 200 μE m^{-2} s^{-1} PAR, respectively (Ghirardi 2006). The above efficiencies, however, are given only for comparison reasons and could not be used for the estimation of the real light to H_2 conversion efficiency in microalgal cultures since in all cases they were calculated in the presence of acetate in the medium.

As discussed above, entrapment of microbial cells into the alginate polymer protects them against adverse environmental conditions. Following this common property of gel entrapments, microalgae immobilized within alginate films showed a high resistance of their H_2-photoproducing system to inactivation by atmospheric oxygen. In well-mixed suspensions even trace amounts of O_2 in the PhBR headspace inhibit H_2 photoproduction (Ghirardi et al. 2000). In contrast, the algal cells entrapped in alginate films and placed in vials containing 21 % O_2 in the headspace evolved up to 67 % of the H_2 gas produced under anaerobic conditions (Kosourov and Seibert 2009). The lower susceptibility of the immobilized algal H_2-producing system to inactivation by O_2 was caused by the higher rate of respiration in the dense film and the capability of the alginate polymer itself to effectively separate the entrapped cells from O_2 in the liquid and headspace and restrict O_2 diffusion into the matrix. The alginate polymer, however, slows down diffusion of O_2 not only from the atmosphere to the film, but also from the film to the atmosphere. Under high light conditions, this affects immobilized algae oppositely preventing the efficient release of O_2 originated in PSII from the cells and decreasing the overall H_2-photoproduction performance of alginate films (Kosourov et al. 2011). In the paper mentioned above, the authors partly solved the problem by immobilizing the *C. reinhardtii* mutants with truncated light-harvesting antennae. Although the CC-4169 strain affected in the TLA1 gene also experienced photoinhibition under high light, it produced significantly more H_2 gas than the parental CC-425 strain. Unfortunately, CC-425 itself showed very low specific rates of H_2 photoproduction under all light intensities tested, thus, making difficult any comparison of the daughter CC-4169 strain with better H_2-producers. Nevertheless, this work was the first report on the increase performance of H_2 photoproduction under high light in the alga with truncated light-harvesting antenna complexes.

The low mechanical stability of the alginate polymer, especially in the presence of such chelating agents as phosphates that are very important for the cell metabolism, continued the efforts of the researchers to find a more stable and cheap material for microalgae immobilization. The most interesting advance in this direction was done recently with immobilization of cells into the latex polymer (Gosse et al. 2012). The authors described a latex wet coalescence method for gas-phase immobilization of microorganisms on a filter paper, which does not require drying for adhesion. In this approach, the mixture of algal cells or other microorganisms in latex is dropped on the top of the filter paper strip. Then, the paper strip is partly submerged into the medium allowing the coated cells staying in the vial headspace. Since the paper is wet throughout the experiment, the full latex polymerization does not occur, but latex does allow binding of the cells to the paper surface. This method is applicable for microorganisms that do not tolerate desiccation stress during latex drying. Interestingly, in contrast to a regular latex immobilization technique, wet *C. reinhardtii* coatings retain cell reactivity throughout the experiment (~250 h) as measured by oxygen gas evolution or CO_2 consumption (Gosse et al. 2012). More interestingly, *C. reinhardtii* cells showed even higher CO_2 consumption and O_2 evolution rates than the cyanobacterium, *Synechococcus* sp. PCC7002 placed under the same conditions. As an example, the *Chlamydomonas* and *Synechococcus* coatings consumed CO_2 at the rate of about 3.9 and 3.6 mmol CO_2 m^{-2} h^{-1} and produced O_2 at the rate of about 10.2 and 5.0 mmol O_2 m^{-2} h^{-1}, respectively. The coatings also tolerated 20 % CO_2 in the PhBR headspace. These observations indicate that the latex polymer itself is not toxic to the cells and that the viability of algal cells in the latex coatings depends drastically on the presence of water. The reactivity of wet coatings can further be improved by increasing the coating surface area and cell density in the latex emulsion. Another advantage of wet cell binding to the filter paper is that the technique significantly increases the transfer rates for the gases allowing faster consumption of CO_2 by the cells and faster release of O_2 and H_2 gases from the cells to the PhBR headspace. Since H_2 photoproduction in microalgae decreases significantly with increasing the H_2 partial pressure (Kosourov et al. 2012), the faster release of H_2 gas from the coatings may further improve the rates and yields of H_2 photoproduction. Taking into account the above findings, the latex wet coalescence method should be considered as a suitable approach for generation of H_2 gas by immobilized microalgae, especially under autotrophic conditions.

The vast majority of experiments on H_2 production by nutrient-deprived microalgae have been done so far with *C. reinhardtii* cultures. However, other species of green algae also produce H_2 gas under this condition (Winkler et al. 2002; Skjanes et al. 2008; Meuser et al. 2009). Some of these strains show very efficient H_2 photoproduction in an immobilized state. For example, Song et al. (2011) immobilized *Chlorella* sp. into the square ($0.5 \times 0.5 \times 0.5$ cm) pieces of agar.

Although the system was not perfect in terms of light and nutrients distribution as compared to the thin-layer immobilization approach, sulfur-deprived *Chlorella* cells produced H_2 gas with the maximum rate above 23 ml $H_2\,h^{-1}$ per L_{PhBR}. This rate, however, was observed for a short period of time (less than 10 h) in the presence of 30 mM glucose. Nevertheless, under these conditions immobilized *Chlorella* cultures yielded around 500 mL H_2 gas per L_{PhBR} for less than 40 h. Under photoautotrophic conditions, when only CO_2 was used as a carbon source, the alga yielded up to 160 ml $H_2\,h^{-1}$ per L_{PhBR} that corresponds to the average rate of ~3 ml $H_2\,h^{-1}$ per L_{PhBR}. The system also demonstrated the multiple cycles of H_2 photoproduction after periodic restorations of algal cultures in the full (sulfur-containing) medium both under photoheterotrophic (+glucose) and photoautotrophic (+CO_2) conditions. In the presence of glucose, authors observed up to ten cycles of H_2 production in immobilized *Chlorella* cultures with the average yield of 460–480 mL H_2 per L_{PhBR} for the each cycle. Information on other species of microalgae capable of H_2 photoproduction in immobilized state is very limited. There are data that the marine green alga, *Platymonas subcordiformis* produces H_2 gas when entrapped in alginate beads (Guan et al. 2003). Similar to *C. reinhardtii*, algae released H_2 gas in the two-stage process. However, the efficient H_2 photoproduction in *Platymonas* was only observed in the presence of carbonyl cyanide m-chlorophenylhydrazone (CCCP), an uncoupler of photophosphorylation (Guan et al. 2004).

From all the above-mentioned examples it becomes obvious that the immobilization approach can bring significant advances to H_2 gas photoproduction in microalgae. It not only improves the volumetric rate of hydrogen production, but also allows their easy cycling between nutrient-replete and nutrient deprived stages. In addition, immobilization simplifies a continuous flow of liquid media through the PhBR. As a result, microalgae produce H_2 gas more efficiently and for longer periods of time as compared to the suspensions. Nevertheless, at the current state the rates of H_2 photoproduction in immobilized algal cells are still not high enough for the industrial H_2 production systems to be viable. Here, the most important barrier is a high sensitivity of the process to molecular oxygen, either atmospheric or co-evolved in photosynthesis. Although entrapments of algal cells into the alginate polymer showed us how to defend the H_2-producing system from inactivation by atmospheric O_2, immobilization itself cannot protect the hydrogenase enzymes from O_2 originated in photosynthesis. The situation is more dramatic in autotrophic cells, where O_2 respiration depends solely on the level of stored carbohydrates (Tsygankov et al. 2006). In this case, the PhBR with immobilized algae should be constructed in the way allowing a fast release of O_2 from the cells to the atmosphere, but, if possible, preventing its back diffusion to the cells. The next issue, which is waiting for technological solution, is the direct dependence of H_2 photoproduction in microalgae on the H_2 partial pressure in the gas phase above the culture (Kosourov et al. 2012). Inhibition of H_2 photoproduction in algal cells by increasing levels of H_2 gas can be eliminated in the hybrid system (Dante 2005), where H_2 production is tightly coupled to its consumption in a fuel cell for electricity generation.

VI. Hydrogen Production by Immobilized Cyanobacteria

Cyanobacteria, phototrophic O_2-evolving prokaryotes, are other good candidates for generation of H_2 gas in the artificial systems. These organisms perform oxygenic photosynthesis using the same pathway as green algae and high plants, which includes four major multiprotein complexes: PSII, cytochrome b6f, PSI and ATP synthase. Similar to green algae, they split water in the light and store energy in the form of NADPH and ATP that are consumed in CO_2 fixation via the Calvin-Benson-Bassham cycle and used in other energy-dependent metabolic pathways. Cyanobacteria display a relatively wide range of morphological diversity, including unicellular, filamentous, and colonial forms (Tamagnini et al. 2007). Some filamentous strains form differentiated cells called heterocysts that are specialized in N_2 fixation.

The vast majority of cyanobacterial strains are capable to indirect water biophotolysis resulting in the production of H_2 gas either in the light or in the dark. The light-dependent process of H_2 production is usually linked to N_2 fixation and is catalyzed by the nitrogenase enzyme (see Eq. 14.1 above). The release of 1 mol of H_2 per 1 mol of fixed N_2 occurs only under optimal conditions (Tsygankov 2007). If the amount of nitrogen is insufficient, hydrogen is released at higher rates. Under the condition of complete absence of nitrogen, nitrogenases catalyze solely the reduction of protons to H_2, thus decreasing the ATP requirement of the process from 16 to 4 mol per mole of H_2 produced (see Eq. 14.2). Although H_2 photoproduction via nitrogenases is rather efficient process, the wild-type strains do not accumulate H_2 gas in the cultures under normal conditions due to the presence of uptake (*hup*-encoded) hydrogenases in the cells that recycle the produced H_2 gas (Tamagnini et al. 2002). In some non-N_2-fixing cyanobacteria, molecular hydrogen is produced in the light by the [NiFe]-bidirectional (*hox*-encoded) hydrogenase in a manner similar to H_2 production in green algae (see the reactions 14.3 and 14.4 above). However, in contrast to the green algal [Fe-Fe]-hydrogenases, the *hox*-encoded hydrogenases are NADH/NADPH-dependent enzymes (Vignais and Billoud 2007; Carrieri et al. 2011). Hydrogen photoproduction via [NiFe]-bidirectional hydrogenases in cyanobacteria is a transient phenomenon usually lasting less than 30 s in the light and is followed by H_2 uptake (Appel et al. 2000; Cournac et al. 2004). It serves as an electron valve for the disposal of low-potential electrons generated at the onset of illumination, thus preventing over-reduction of the photosynthetic electron-transport chains or for additional electron donation to the electron-transport chains in the presence of H_2 (Appel and Schulz 1998; Appel et al. 2000). In addition to the light-dependent H_2 evolution, some cyanobacteria are capable of releasing H_2 gas in the dark via the *hox*-encoded hydrogenases (Tsygankov 2007). This process regenerates $NADP^+$ from NADPH during fermentation allowing catabolism of endogenous carbohydrates to proceed (Carrieri et al. 2011). Some species, like *Gloeocapsa alpicola* and *Arthrospira maxima*, evolve significant amounts of H_2 gas via the *hox*-encoded hydrogenases, especially under the stress conditions (Troshina et al. 2002; Ananyev et al. 2008).

Due to a significant diversity of H_2 metabolism in cyanobacteria and extensive literature on the topic, in the current review we will consider only the most interesting and promising approaches used for immobilization of H_2-producing cyanobacteria. Starting from the natural colonization techniques, we should first mention a series of works on immobilization of the N_2-fixing heterocystous cyanobacterium *Anabaena variabilis* on hollow fibers (Markov et al. 1993, 1995). Entrapment of algal cells on hollow fibers has a numerous different advantages. The material is relatively cheap and non-toxic to the cells. It also has a very large surface-to-volume ratio, which allows the design of compact systems. Markov et al. (1995) tested the degree of cell immobilization on the different hollow fibers and found the better attachment of *Anabaena* cells to hydrophilic cellulosic over than hydrophobic polysulphone fibers. Without CO_2 re-additions, the immobilized cells produced H_2 gas at rates of 0.02–0.2 ml $H_2\,h^{-1}$ per mg dry weight for up to 5 months (Markov et al. 1993). However, the maximum rate was observed only in the first 3 days. The following decrease in the rate of H_2 photoproduction to a low steady-state level was caused by the lack of CO_2 in the system. Re-additions of CO_2 to the PhBR not only solved the problem, but also improved the rate up to 20 ml $H_2\,h^{-1}$ per mg dry weight (Markov et al. 1995). In this system, however, *Anabaena* cells did not produce H_2 gas during the photosynthetic, CO_2-consuming phase that usually lasts for few days. Such a two-phase system was maintained continuously for over 1 year. The importance of CO_2 for hydrogen photoproduction in N_2-fixing cyanobacteria emphasizes the role of photosynthetic apparatus of vegetative cells in satisfying the high-energy demand of nitrogenase-driven processes in heterocysts. The long-term experiments with *Anabaena* cells entrapped onto hollow fibers also demonstrated the importance of periodic re-additions of nitrogen to the system since it is required to support cell metabolism.

A variety of cyanobacteria show a high natural adhesion to glass. This property is used for the attachment of cyanobacterial cells to different glass substrates. When attached, cultures start growing and making a very thin biofilm on the surface of the glass. Using the high adhesion property of the cells, Serebryakova and Tsygankov (2007) successfully entrapped unicellular *Gloeocapsa alpicola* CALU 743 on the glass fiber matrices. When grown on a regular BG11 medium, cells uniformly covered the surface of glass within 7–8 days. The total density of the culture achieved ~765 mg Chl *a* per m^2 matrix that is significantly higher than in immobilized *C. reinhardtii* cultures (570 mg total Chl per m^2 of the same matrix). *Gloeocapsa alpicola* cells produce H_2 gas via the *hox*-encoded hydrogenase in the dark during fermentation of endogenous glycogen (Serebryakova et al. 1998). Nitrate limitation significantly improves H_2 production rates since under this condition cells accumulate glycogen up to 50 % of dry weight (Troshina et al. 2002). For efficient H_2 production, Serebryakova and Tsygankov (2007) applied a two-stage principle allowing effectively accumulate glycogen under nitrate starvation during the photosynthetic stage and utilize it for H_2 during the dark period. Up to ten cycles of H_2 production were demonstrated by the authors. The experiments were done under continuous flow of the medium (15 mL h^{-1}) with periodic (during the dark period) purging of the bioreactor by argon (400–800 mL h^{-1}). Increase in the argon flow rate improved the rate of H_2 production since the *hox-encoded* hyrogenase catalyzes the reversible process, which depends significantly on the H_2 partial pressure. The argon and medium flows also removed CO_2 and acetate from the system, thus preventing their negative influence on the process. Under the best conditions, *Gloeocapsa alpicola* cells immobilized on glass fiber matrices produced H_2 gas at the average rate of about 20 mL $H_2 h^{-1} L^{-1}{}_{matrix}$, which is significantly higher than the rates observed in suspensions (6 mL $H_2 h^{-1} L^{-1}$ with argon flow rate of 2 L h^{-1}).

The entrapment of cyanobacterial cells into the polymeric materials in the form of beads or films also significantly improves the volumetric rates of H_2 production. Here, the N_2-fixing cyanobacteria are good candidates for immobilization since the entrapment of the cells into the polymer matrix provides additional protection of their nitrogenase enzyme from inactivation by atmospheric O_2. A study of H_2 photoproduction by the non-heterocystous filamentous marine cyanobacterium, *Oscillatoria* sp. immobilized in rectangular (1 × 1 × 0.2 cm) pieces of 1.5 % agar showed that the rate and longevity of H_2 evolution increased significantly compared to the free cell suspensions (Phlips and Mitsui 1986). The authors observed the rates above 13 μL $H_2 g^{-1}$ dry weight h^{-1}. Immobilization sustained the process for at least 3 weeks and allowed to drive the process under outdoor light conditions. Interestingly, in contrast to other N_2-fixing cyanobacteria, *Oscillatoria* sp. Miami BG7 has a negligible H_2 uptake activity (Kumazawa and Mitsui 1985) that prevents recycling of H_2 gas accumulated in cells and in the agar matrix. As a result, cells photoproduce H_2 more efficiently. The enhanced H_2 photoproduction, as compared to the suspension cultures, was also observed in another non-heterocystous filamentous cyanobacterium, *Plectonema boryanum* immobilized in alginate beads (Sarkar et al. 1992). H_2 production in this strain was accompanied by the release of ammonia in the medium. Another interesting observation was that the additions of 24 % N_2 into the gas phase containing 72 % Ar and 4 % CO_2 stimulated the H_2 photoproduction rate twofold as compared to the PhBRs containing only argon and CO_2 in the headspace. Usually, the presence of N_2 inhibits H_2 evolution driven by the nitrogenase system (see the reactions 14.1 and 14.2 above). Although the authors did not explain this effect, one could suggest that stimulation was caused by the decreased diffusion of N_2 through the alginate matrix, limiting the steady-state level of N_2 inside the cells. Nitrogen at low concentrations, however, supports the cell metabolism and may prolong H_2 photoproduction period. Indeed, the authors observed such prolongation. In the presence of N_2, immobilized cultures produced H_2 gas for up to 12 days (7 days in Ar/CO_2 atm).

The cyanobacteria capable of H_2 release in the dark via a two-stage process were entrapped into the gels as well. Unicellular non-N_2-fixing cyanobacterium, *Microcystis aeruginosa* is good example (Rashid et al. 2009, 2012). Similar to *Gloeocapsa alpicola*, this organism accumulates considerable amounts of glycogen during the photosynthetic stage. Subsequent fermentation in the dark produces H_2 gas. In contrast to experiments with *G. alpicola*, where nitrate starvation was used, Rashid et al. (2009) applied sulfur deprivation to the cultures entrapped into 1.5 % agar pieces during the H_2 production stage. Since the rates of H_2 evolution were not very significant (1.5–2 mL H_2 h^{-1} L^{-1}), the authors introduced glucose to the medium. The externally added glucose improved the maximum H_2 production rate in the immobilized cultures up to 17 mL H_2 h^{-1} L^{-1} (34 mL H_2 h^{-1} $L^{-1}{}_{matrix}$). Under these conditions, the average rate was about 13 mL H_2 h^{-1} L^{-1}. The system also demonstrated the multiple cycles of H_2 production (Rashid et al. 2009, 2012).

The entrapment of cells in thin films instead of beads or rectangular pieces leads to a more efficient light utilization. Following this idea, Leino et al. (2012) immobilized cells of two N_2-fixing heterocystous cyanobacteria, *Calothrix* 336/3 and *Anabaena* PCC 7120 (wild-type and *ΔhupL* strains) within thin alginate films. In all cases, efficient H_2 photoproduction was observed in vials containing 6 % CO_2 in argon. Re-additions of CO_2 gas into vials were important for restoration of the H_2 photoproduction activity in cells and allowed several cycles of H_2 photoproduction to occur. Without CO_2 supplementations, the immobilized cultures of *Calothrix* and *Anabaena* stopped H_2 photoproduction after the first cycle. Cyanobacteria entrapped in alginate films release H_2 gas simultaneously with O_2 evolution. Simultaneous production of molecular H_2 and O_2 is typical to all heterocystous cyanobacteria and observed in suspensions as well (Tsygankov et al. 2002). As expected, the wild-type *Anabaena* strain having the uptake hydrogenase produced significantly less H_2 gas than the *ΔhupL* since alginate slows down the release of H_2 from the film to atmosphere, thus allowing re-cyclization of produced H_2 gas via the uptake hydrogenase, especially at high O_2 concentrations. On the contrary, *Calothrix* 336/3, which also possesses *hupSL* genes, in some cases demonstrated even higher the H_2 photoproduction activity than the *ΔhupL* mutant without uptake hydrogenase. Most probably, *Calothrix* cells do not have the real H_2-uptaking activity despite the presence of *hupSL* genes. Immobilization in alginate films had a positive effect on cell viability. All three strains, the *Calothrix* 336/3, wild-type *Anabaena* and *ΔhupL* cells, were viable for over 10 months in the initial nutrient media without additions of CO_2. Interestingly, the alginate films with entrapped cyanobacteria were more mechanically stable than the films with entrapped green algae. The maximum specific rates of H_2 photoproduction were 35, 9 and 30 μmol H_2 (mg Chl *a* h)$^{-1}$ for *Calothrix* 336/3, wild-type *Anabaena* and *ΔhupL* cultures, respectively. These rates are significantly higher than the rates observed in *C. reinhardtii* cultures immobilized approximately under the same conditions.

Another interesting technique developed for immobilization of H_2-producing cyanobacteria is the entrapment of cells within sol-gel silica matrices. This approach has never been reported successful for immobilization of H_2-producing green algae. However, some non-H_2-producing microalgae, like *Haematococcus pluvialis* survived through the immobilization procedure (Fiedler et al. 2007). Recently, Dickson et al. (2009) immobilized the cyanobacterium, *Synechocystis* sp. PCC6803 and its mutant, M55 deficient in the NDH-1 complex into silica matrices using tetraethoxysilane (TEOS), tetramethoxysilane (TMOS) and methyltriethoxysilane (MTES) as precursors in the sol-gel reaction. Glycerol and polyethylene glycol (PEG 400) were used as additives helping to increase porosity of the matrix and reduce osmotic stress on the encapsulated cells. H_2 photoproduction driven by the *hox*-encoded hydrogenase was measured under the light/dark exposure (2 min on/18 min off) for up to 5 days. The results showed that H_2 photoproduction rates from encapsulated cells in most cases were comparable to the rates in suspensions, confirming the activity of encapsulated cells. Although the process is currently far from optimized, it provides a proof of the concept demonstrating the ability to achieve measurable amounts of H_2 gas from sol-gel encapsulated cells.

Despite extensive investigations over the last few decades, a real potential of cyanobacterial species to produce H_2 gas has not been fully explored. If compared to other phototrophs, these organisms have several significant advantages. The first and the most important is that they are capable of producing H_2 gas under truly autotrophic conditions. Although green algae can do the same, at the current state efficient H_2 evolution in their cultures requires the presence of acetate or glucose in the medium. As a rule, H_2 photoproduction in N_2-fixing cyanobacteria is very stable to O_2 inactivation due to a range of different protection mechanisms. Some cyanobacteria can produce H_2 gas in the presence of atmospheric levels of O_2 and above. In addition, cyanobacteria can easily adapt to periodic light environment and drive H_2 production under solar light intensities. This makes the process possible under outdoor conditions. These advantages emphasize the importance of future research in this direction, including application of different immobilization techniques.

VII. Concluding Remarks

Although many photosynthetic microorganisms are capable of producing H_2 gas utilizing either water or organic substrates, so far, there are no any commercial applications of the process. One of the major bottlenecks for a large-scale H_2 generation using phototrophs is the low volumetric rate of H_2 photoproduction for all known pathways. From a practical point of view, however, the application of photosynthetic microorganisms is a great advantage since their cultures require only solar energy and relatively small amounts of other inputs for growth and operation. Presumably everyone will agree that a significant improvement in the rate of H_2 gas generation will be possible only after understanding all barriers limiting H_2 photoproduction in the particular group of photosynthetic microorganisms and generating mutants capable of overcoming these barriers. Nevertheless, there is still a space for technological improvements.

For example, the problem of inhomogeneous light distribution in the suspension culture with constantly increasing biomass can be partly solved by immobilizing the cells. As discussed above, thin-layer cell immobilization has already improved the light to H_2 conversion efficiency above 1 % in green algal cultures and slightly increased the specific rates of H_2 photoproduction both in green algae and purple bacteria. A combination of thin-layer immobilization with appropriate photobioreactor geometry significantly increased volumetric rate of hydrogen production, as was shown for purple bacteria (Tsygankov et al. 1994). The light to H_2 conversion efficiency can further be improved by co-immobilizing different mutants (for example, the wild-type strain and mutants with truncated light-harvesting antennae) or different organisms (for example, purple bacteria and green algae). In addition, cell immobilization significantly simplifies the culture maintenance in the PhBR and, thus, decreases the overall cost of the system operation. Nevertheless, the material for entrapping phototrophic cultures should be non-toxic to the cells, transparent, stable, available in industrial quantities, made from a renewable source and cheap. At this point, there are no such materials that satisfy all requirements. Therefore, the future efforts should also be concentrated on the screening of new materials and substrates for immobilization of photosynthetic microorganisms.

Acknowledgements

This work was supported by Russian Ministry of Science and Education (Agreement 8077).

References

Akkerman I, Janssen M, Rocha J, Wijffels RH (2002) Photobiological hydrogen production: photochemical efficiency and bioreactor design. Int J Hydrog Energy 27:1195–1208

Ananyev G, Carrieri D, Dismukes GC (2008) Optimization of metabolic capacity and flux through environmental cues to maximize hydrogen production by the cyanobacterium *Arthrospira* (*Spirulina*) *maxima*. Appl Environ Microbiol 74:6102–6113

Appel J, Schulz R (1998) Hydrogen metabolism in organisms with oxygenic photosynthesis: hydrogenases as important regulatory devices for a proper redox poising? J Photochem Photobiol B 47:1–11

Appel J, Phunpruch S, Steinmuller K, Schulz R (2000) The bidirectional hydrogenase of *Synechocystis* sp. PCC 6803 works as an electron valve during photosynthesis. Arch Microbiol 173:333–338

Azbar N, Kapdan IK (2011) Use of immobilized cell systems in biohydrogen production. In: Levin D, Azbar N (eds) State of the art and progress in production of biohydrogen. Bentham Science Publishers, Bussum, pp 228–250

Baca HK, Carnes E, Singh S, Ashley C, Lopez D, Brinker CJ (2007) Cell-directed assembly of bio/nano interfaces – a new scheme for cell immobilization. Acc Chem Res 40:836–845

Bai X, Ye ZF, Li YF, Ma YX (2010) Macroporous poly(vinyl alcohol) foam crosslinked with epichlorohydrin for microorganism immobilization. J Appl Polym Sci 117:2732–2739

Batyrova K, Tsygankov A, Kosourov SN (2012) Sustained hydrogen photoproduction by phosphorus-deprived *Chlamydomonas reinhardtii* cultures. Int J Hydrog Energy 37:8834–8839

Boichenko VA, Hoffmann P (1994) Photosynthetic hydrogen production in prokaryotes and eukaryotes: occurrence, mechanism, and functions. Photosynthetica 30:527–552

Boichenko VA, Greenbaum E, Seibert M (2004) Hydrogen production by photosynthetic microorganisms. In: Archer MD, Barber J (eds) Molecular to global photosynthesis: photoconversion of solar energy. Imperial College Press, London, pp 397–452

Brena BM, Batista-Viera F (2006) Immobilization of enzymes. In: Guisan JM (ed) Methods in biotechnology. Immobilization of enzymes and cells. Humana Press, Totowa, pp 15–30

Brodelius P, Vandamme EJ (1987) Immobilized cell systems. In: Kennedy JF (ed) Biotechnology, vol 7a: Enzyme technology. VCH Publication, New York, pp 405–464

Carnes EC, Lopez DM, Donegan NP, Cheung A, Gresham H, Timmins GS, Brinker CJ (2010) Confinement-induced quorum sensing of individual *Staphylococcus aureus* bacteria. Nat Chem Biol 6:41–45

Carrieri D, Wawrousek K, Eckert C, Yu J, Maness P-C (2011) The role of the bidirectional hydrogenase in cyanobacteria. Bioresour Technol 102:8368–8377

Carturan G, Campostrini R, Dire S, Scardi V, DeAlteriis E (1989) Inorganic gels for immobilization of biocatalysts. Inclusion of invertase-active whole cells of yeast (*Saccharomyces cerevisiae*) into thin layers of SiO_2 gel deposited on glass sheets. J Mol Catal 57:L13–L16

Chen CY, Chang JS (2006) Enhancing phototropic hydrogen production by solid-carrier assisted fermentation and internal optical-fiber illumination. Process Biochem 41:2041–2049

Cheng K-C, Demirci A, Catchmark JM (2010) Advances in biofilm reactors for production of value-added products. Appl Microbiol Biotechnol 87:445–456

Cournac L, Guedeney G, Peltier G, Vignais PM (2004) Sustained photoevolution of molecular hydrogen in a mutant of *Synechocystis* sp. strain PCC 6803 deficient in the type I NADPH-dehydrogenase complex. J Bacteriol 186:1737–1746

Dagher SF, Ragout AL, Sineriz F, Bruno-Barcena JM (2010) Cell immobilization for production of lactic acid: biofilms do it naturally. In: Laskin AI, Sariaslani S, Gadd GM (eds) Advances in applied

microbiology, vol 71. Elsevier Academic Press, San Diego, pp 113–148
Dante R (2005) Hypotheses for direct PEM fuel cells applications of photobioproduced hydrogen by *Chlamydomonas reinhardtii*. Int J Hydrog Energy 30:421–424
de-Bashan LE, Bashan Y (2010) Immobilized microalgae for removing pollutants: review of practical aspects. Bioresour Technol 101:1611–1627
Dickson D, Page C, Ely R (2009) Photobiological hydrogen production from *Synechocystis* sp. PCC 6803 encapsulated in silica sol–gel. Int J Hydrog Energy 34:204–215
Fedorov A, Tsygankov A, Rao KK, Hall DO (1998) Hydrogen photoproduction by *Rhodobacter capsulatus* immobilized on polyurethane foam. Biotechnol Lett 20:1007–1009
Fiedler D, Hager U, Franke H, Soltmann U, Bottcher H (2007) Algae biocers: astaxanthin formation in sol-gel immobilised living microalgae. J Mater Chem 17:261–266
Fissler J, Kohring GW, Giffhorn F (1995) Enhanced hydrogen production from aromatic acids by immobilized cells of *Rhodopseudomonas palustris*. Appl Biochem Biotechnol 44:43–46
Flickinger MC, Schottel JL, Bond DR, Aksan A, Scriven LE (2007) Painting and printing living bacteria: engineering nanoporous biocatalytic coatings to preserve microbial viability and intensify reactivity. Biotechnol Progr 23:2–17
Fukushima Y, Okamura K, Imai K, Motai H (1988) A new immobilization technique of whole cells and enzymes with colloidal silica and alginate. Biotechnol Bioeng 32:584–594
Garnham GW, Codd GA, Gadd GM (1992) Accumulation of cobalt, zinc and manganese by the estuarine green microalga *Chlorella salina* immobilized in alginate microbeads. Environ Sci Technol 26:1764–1770
Ghirardi ML (2006) Hydrogen production by photosynthetic green algae. Indian J Biochem Biophys 43:201–210
Ghirardi ML, Togasaki RK, Seibert M (1997) Oxygen sensitivity of algal H_2-production. Appl Biochem Biotechnol 63:141–151
Ghirardi ML, Zhang L, Lee JW, Flynn T, Seibert M, Greenbaum E (2000) Microalgae: a green source of renewable H_2. Trends Biotechnol 18:506–511
Gosse JL, Engel BJ, Rey FE, Harwood CS, Scriven LE, Flickinger MC (2007) Hydrogen production by photoreactive nanoporous latex coatings of nongrowing *Rhodopseudomonas palustris* CGA009. Biotechnol Progr 23:124–130
Gosse JL, Engel BJ, Hui JC-H, Harwood CS, Flickinger MC (2010) Progress toward a biomimetic leaf: 4,000 h of hydrogen production by coating-stabilized nongrowing photosynthetic *Rhodopseudomonas palustris*. Biotechnol Progr 26:907–918
Gosse JL, Chinn MS, Grunden AM, Bernal OI, Jenkins JS, Yeager C, Kosourov S, Seibert M, Flickinger MC (2012) A versatile method for preparation of hydrated microbial-latex biocatalytic coatings for gas absorption and gas evolution. J Ind Microbiol Biotechnol 39:1269–1278
Guan YF, Zhang W, Yu XJ, Deng MC (2003) Two-stage photo hydrogen production using immobilized marine green alga *Platymonas subcordiformis*. Abstracts of marine biotechnology: basics and applications. Biomol Eng 20:37–82
Guan YF, Zhang W, Deng MC, Jin MF, Yu XJ (2004) Significant enhancement of photobiological H_2 evolution by carbonylcyanide m-chlorophenylhydrazone in the marine green alga *Platymonas subcordiformis*. Biotechnol Lett 26:1031–1035
Gugerli R, Breguet V, von Stockar U, Marison IW (2004) Immobilization as a tool to control fermentation in yeast-leavened refrigerated dough. Food Hydrocoll 18:703–715
Guo CL, Zhu X, Liao Q, Wang Y-Z, Chen R, Lee D-J (2011) Enhancement of photo-hydrogen production in a biofilm photobioreactor using optical fiber with additional rough surface. Bioresour Technol 102:8507–8513
Hahn JJ, Ghirardi ML, Jacoby WA (2007) Immobilized algal cells used for hydrogen production. Biochem Eng J 37:75–79
Hall DO, Rao KK (1988) Immobilized photosynthetic membranes and cells for the production of fuels and chemicals. In: Gaber BP, Schnur JM, Chapman D (eds) Biotechnological applications of lipid structures. Plenum Press, New York/London, pp 225–245
Hallenbeck PC (1983) Immobilized microorganisms for hydrogen and ammonia production. Enzyme Microb Technol 5:171–180
Hatanaka Y, Kudo T, Miyataka M, Kobayashi O, Higashihara M, Hiyama K (1999) Asymmetric reduction of hydroxyacetone to propanediol in immobilized halotolerant microalga *Dunaliella parva*. J Biosci Bioeng 88:281–286
Jimenez-Perez MV, Sanchez-Castillo P, Romera O, Fernadnez-Moreno D, Perez-Martinez C (2004) Growth and nutrient removal in free and immobilized planktonic green algae isolated from pig manure. Enzyme Microb Technol 34:392–398
Kandimalla VB, Tripathi VS, Ju HX (2006) Immobilization of biomolecules in sol-gels: biological and analytical applications. Crit Rev Anal Chem 36:73–106
Kannaiyan S, Rao KK, Hall DO (1994) Immobilization of *Anabaena azollae* from *Azolla filiculoides* in polyvinyl foam for ammonia production in a

photobioreactor system. World J Microbiol Biotechnol 10:55–58

Kaya VM, Picard G (1995) The viability of *Scenedesmus bicellularis* cells immobilized on alginate screens following nutrient starvation in air at 100 percent relative humidity. Biotechnol Bioeng 46:459–464

Kosourov SN, Seibert M (2009) Hydrogen photoproduction by nutrient-deprived *Chlamydomonas reinhardtii* cells immobilized within thin alginate films under aerobic and anaerobic conditions. Biotechnol Bioeng 102:50–58

Kosourov S, Tsygankov A, Seibert M, Ghirardi ML (2002) Sustained hydrogen photoproduction by *Chlamydomonas reinhardtii*: effects of culture parameters. Biotechnol Bioeng 78:731–740

Kosourov S, Seibert M, Ghirardi ML (2003) Effects of extracellular pH on the metabolic pathways of sulfur-deprived, H_2-producing *Chlamydomonas reinhardtii* cultures. Plant Cell Physiol 44:146–155

Kosourov SN, Ghirardi ML, Seibert M (2011) A truncated antenna mutant of *Chlamydomonas reinhardtii* can produce more hydrogen than the parental strain. Int J Hydrog Energy 36:2044–2048

Kosourov SN, Batyrova K, Petushkova EP, Tsygankov A, Ghirardi ML, Seibert M (2012) Maximizing the hydrogen photoproduction yields in *Chlamydomonas reinhardtii* cultures: the effect of the H_2 partial pressure. Int J Hydrog Energy 37:8850–8858

Kourkoutas Y, Bekatorou A, Banat IM, Marchant R, Koutinas AA (2004) Immobilization technologies and support materials suitable in alcohol beverages production: a review. Food Microbiol 21:377–397

Krastanov A (1997) Continuous sucrose hydrolysis by yeast cells immobilized to wool. Appl Microbiol Biotechnol 47:476–481

Kumazawa S, Mitsui A (1985) Comparative amperometric study of uptake hydrogenase and hydrogen photoproduction activities between heterocystous cyanobacterium *Anabaena cylindrica* B629 and nonheterocystous cyanobacterium *Oscillatoria* sp. strain Miami BG7. Appl Environ Microbiol 50:287–291

Lau PS, Tam NFY, Wong YS (1998) Effect of carrageenan immobilization on the physiological activities of *Chlorella vulgaris*. Bioresour Technol 63:115–121

Laurinavichene TV, Fedorov AS, Ghirardi ML, Seibert M, Tsygankov AA (2006) Demonstration of sustained hydrogen photoproduction by immobilized, sulfur-deprived *Chlamydomonas reinhardtii* cells. Int J Hydrog Energy 31:659–667

Laurinavichene TV, Kosourov SN, Ghirardi ML, Seibert M, Tsygankov AA (2008) Prolongation of H_2 photoproduction by immobilized, sulfur-limited *Chlamydomonas reinhardtii* cultures. J Biotechnol 134:275–277

Leino H, Kosourov SN, Saari L, Sivonen K, Tsygankov A, Aro E-M, Allahverdiyeva Y (2012) Extended H_2 photoproduction by N_2-fixing cyanobacteria immobilized in thin alginate films. Int J Hydrog Energy 37:151–161

Leon R, Galvan F (1995) Glycerol photoproduction by free and Ca-alginate entrapped cells of *Chlamydomonas reinhardtii*. J Biotechnol 42:61–67

Liu L, Shang L, Guo S, Li D, Liu C, Qi L, Dong S (2009) Organic-inorganic hybrid material for the cells immobilization: long-term viability mechanism and application in BOD sensors. Biosens Bioelectron 25:523–526

Lyngberg OK, Ng CP, Thiagarajan V, Scriven LE, Flickinger MC (2001) Engineering the microstructure and permeability of thin multilayer latex biocatalytic coatings containing *E. coli*. Biotechnol Progr 17:1169–1179

Mallick N (2002) Biotechnological potential of immobilized algae for wastewater N, P and metal removal: a review. Biometals 15:377–390

Mallick N (2006) Immobilization of microalgae. In: Guisan JM (ed) Immobilization of enzymes and cells. Humana Press, Totowa, pp 373–391

Markov SA, Lichtl R, Rao KK, Hall DO (1993) A hollow-fiber photobioreactor for continuous production of hydrogen by immobilized cyanobacteria under partial vacuum. Int J Hydrog Energy 18:901–906

Markov SA, Bazin MJ, Hall DO (1995) Hydrogen photoproduction and carbon dioxide uptake by immobilized *Anabaena variabilis* in a hollow-fiber photobioreactor. Enzyme Microb Technol 17:306–310

Martens N, Hall EAH (1994) Immobilization of photosynthetic cells based on film-forming emulsion polymers. Anal Chim Acta 292:49–63

Matsunaga T, Takeyama H, Sudo H, Oyama N, Ariura S, Takano H, Hirano M, Burgess JG, Sode K, Nakamura N (1991) Glutamate production from CO_2 by marine cyanobacterium *Synechococcus* sp. using a novel biosolar reactor employing light-diffusing optical fibers. Appl Biochem Biotechnol 28:157–167

Melis A (2007) Photosynthetic H_2 metabolism in *Chlamydomonas reinhardtii* (unicellular green algae). Planta 226:1075–1086

Melis A, Zhang L, Forestier M, Ghirardi ML, Seibert M (2000) Sustained photobiological hydrogen gas production upon reversible inactivation of oxygen evolution in the green alga *Chlamydomonas reinhardtii*. Plant Physiol 122:127–136

Meunier CF, Dandoy P, Su BL (2010) Encapsulation of cells within silica matrixes: towards a new advance

in the conception of living hybrid materials. J Colloid Interface Sci 342:211–224
Meuser JE, Ananyev G, Wittig LE, Kosourov S, Ghirardi ML, Seibert M, Dismukes GC, Posewitz MC (2009) Phenotypic diversity of hydrogen production in chlorophycean algae reflects distinct anaerobic metabolisms. J Biotechnol 142:21–30
Moreno-Garrido I (2008) Microalgae immobilization: current techniques and uses. Bioresour Technol 99:3949–3964
Moreno-Garrido I, Campana O, Lubian LM, Blasco J (2005) Calcium alginate immobilized marine microalgae: experiments on growth and short-term heavy metal accumulation. Mar Pollut Bull 51:823–929
Pannier A, Oehm C, Fischer AR, Werner P, Soltmann U, Bottcher H (2011) Biodegradation of fuel oxygenates by sol-gel immobilized bacteria *Aquincola tertiaricarbonis* L108. Enzyme Microb Technol 47:291–296
Park IH, Rao KK, Hall DO (1991) Photoproduction of hydrogen, hydrogen-peroxide and ammonia using immobilized cyanobacteria. Int J Hydrog Energy 16:313–318
Philipps G, Happe T, Hemschemeier A (2012) Nitrogen deprivation results in photosynthetic hydrogen production in *Chlamydomonas reinhardtii*. Planta 235:729–745
Phlips J, Mitsui A (1986) Characterization and optimization of hydrogen production by a salt water blue-green alga *Oscillatoria* sp. Miami BG 7. II. Use of immobilization for enhancement of hydrogen production. Int J Hydrog Energy 11:83–89
Planchard A, Mignot L, Jouenne T, Junter G-A (1984) Photoproduction of molecular hydrogen by *Rhodospirillum rubrum* immobilized in composite agar layer/microporous membrane structures. Appl Microbiol Biotechnol 31:49–54
Rashid N, Song W, Park J, Jin H-F, Lee K (2009) Characteristics of hydrogen production by immobilized cyanobacterium *Microcystis aeruginosa* through cycles of photosynthesis and anaerobic incubation. J Ind Eng Chem 15:498–503
Rashid N, Choi W, Lee K (2012) Optimization of two-staged bio-hydrogen production by immobilized *Microcystis aeruginosa*. Biomass Bioenergy 36:241–249
Robinson PK (1998) Immobilized algal technology for wastewater treatment purposes. In: Wong Y-S, Tam NFY (eds) Wastewater treatment with algae. Springer and Landes Bioscience, New York, pp 1–16
Romanova YM, Didenko LV, Tolordava ER, Ginzburg AL (2011) Biofilms of pathogenic bacteria: a role in chronic of infectious process and search of agents of struggle. Vestnik RAMN (Russ) 10:31–39
Rooke JC, Meunier C, Leonard A, Su BL (2008) Energy from photobioreactors: bioencapsulation of photosynthetically active molecules, organelles, and whole cells within biologically inert matrices. Pure Appl Chem 80:2345–2376
Saetang J, Babel S (2009) Effect of leachate loading rate and incubation period on the treatment efficiency by *T. versicolor* immobilized on foam cubes. Int J Environ Sci Technol 6:457–466
Sarkar S, Pandey KD, Kashyap AK (1992) Hydrogen photoproduction by filamentous nonheterocystous cyanobacterium *Plectonema boryanum* and simultaneous release of ammonia. Int J Hydrog Energy 17:689–694
Sasikala K, Ramana CV, Rao PR (1992) Photoproduction of hydrogen from the waste-water of a distillery by *Rhodobacter sphaeroides* OU 001. Int J Hydrog Energy 17:23–27
Scholz W, Galvan F, de la Rosa FF (1995) The microalga *Chlamydomonas reinhardtii* CW-15 as a solar cell for hydrogen peroxide photoproduction: comparison between free and immobilized cells and thylakoids for energy conversion efficiency. Sol Energy Mater Sol Cell 39:61–69
Semple KT, Cain RB, Schmidt S (1999) Biodegradation of aromatic compounds by microalgae. FEMS Microbiol Lett 170:291–300
Serebryakova LT, Tsygankov AA (2007) Two-stage system for hydrogen production by immobilized cyanobacterium *Gloeocapsa alpicola* CALU 743. Biotechnol Progr 23:1106–1110
Serebryakova L, Sheremetieva M, Tsygankov A (1998) Reversible hydrogenase activity of *Gloeocapsa alpicola* in continuous culture. FEMS Microbiol Lett 166:89–94
Shi J, Podola B, Melkonian M (2007) Removal of nitrogen and phosphorus from wastewater using microalgae immobilized on twin layers: an experimental study. J Appl Phycol 19:417–423
Skjanes K, Knutsen G, Kallqvist T, Lindblad P (2008) H_2 production from marine and freshwater species of green algae during sulfur deprivation and considerations for bioreactor design. Int J Hydrog Energy 33:511–521
Smirnova TA, Didenko LV, Andreev AL, Alekseeva NV, Stepanova TV, Romanova YM (2008) Electron microscopic study of *Burkholderia cepacia* biofilms. Microbiology 77:55–61
Smirnova TA, Didenko LV, Azizbekyan RR, Romanova YM (2010) Structural and functional characteristics of bacterial biofilms. Microbiology 79:413–423
Song W, Rashid N, Choi W, Lee K (2011) Biohydrogen production by immobilized *Chlorella* sp. using cycles of oxygenic photosynthesis and anaerobiosis. Bioresour Technol 102:8676–8681

Stolarzewicz I, Bialecka-Florjanczyk E, Majewska E, Krzyczkowska J (2011) Immobilization of yeast on polymeric supports. Chem Biochem Eng Q 25:135–144

Tamagnini P, Axelsson R, Lindberg P, Oxelfelt F, Wunschiers R, Lindblad P (2002) Hydrogenases and hydrogen metabolism of cyanobacteria. Microbiol Mol Biol Rev 66:1–20

Tamagnini P, Leitao E, Oliveira P, Ferreira D, Pinto F, Harris D, Heidorn T, Lindblad P (2007) Cyanobacterial hydrogenases: diversity, regulation and applications. FEMS Microbiol Rev 31:692–720

Tekucheva DN, Laurinavichene T, Seibert M, Tsygankov A (2011) Immobilized purple bacteria for light-driven H_2 production from starch and potato fermentation effluents. Biotechnol Progr 27:1248–1256

Tian X, Liao Q, Liu W, Wang YZ, Zhu X, Li J, Wang H (2009) Photo-hydrogen production rate of a PVA-boric acid gel granule containing immobilized photosynthetic bacteria cells. Int J Hydrog Energy 34:4708–4717

Tian X, Liao Q, Zhu X, Wang Y, Zhang P, Li J, Wang H (2010) Characteristics of a biofilm photobioreactor as applied to photo-hydrogen production. Bioresour Technol 101:977–983

Torzillo G, Pushparaj B, Masojidek J, Vonshak A (2003) Biological constraints in algal biotechnology. Biotechnol Bioprocess Eng 8:338–348

Travieso L, Benitez F, Dupeiron R (1992) Sewage treatment using immobilized microalgae. Bioresour Technol 40:183–187

Travieso L, Benitez F, Weiland P, Sanchez E, Dupeyron R, Dominguez AR (1996) Experiments on immobilization of microalgae for nutrient removal in wastewater treatments. Bioresour Technol 55:181–186

Troshina O, Serebryakova L, Sheremetieva M, Lindblad P (2002) Production of H_2 by the unicellular cyanobacterium *Gloeocapsa alpicola* CALU 743 during fermentation. J Gen Microbiol 27:1283–1289

Tsygankov AA (2001a) Laboratory scale photobioreactors. Appl Biochem Microbiol 37:333–341

Tsygankov AA (2001b) Hydrogen production by purple bacteria: immobilized vs. suspension cultures. In: Miyake J, Matsunaga T, San Pietro A (eds) Biohydrogen 2. An approach to environmentally acceptable technology. Pergamon Press, Amsterdam, pp 229–244

Tsygankov A (2004) Hydrogen production by suspension and immobilized cultures of phototrophic microorganisms. Technological aspects. In: Miyake J, Igarashi H, Rogner M (eds) Biohydrogen III. Renewable energy system by biological solar energy conversion. Elsevier, Amsterdam, pp 57–74

Tsygankov AA (2007) Nitrogen-fixing cyanobacteria: a review. Appl Biochem Microbiol 43:279–288

Tsygankov AA, Hirata Y, Miyake M, Asada Y, Miyake J (1993) Immobilization of the purple nonsulfur bacterium *Rhodobacter sphaeroides* on glass surfaces. Biotechnol Tech 77:575–578

Tsygankov AA, Hirata Y, Miyake M, Asada Y, Miyake J (1994) Photobioreactor with photosynthetic bacteria immobilized on porous glass for hydrogen photoproduction. J Ferment Bioeng 77:575–578

Tsygankov AA, Fedorov AS, Laurinavichene TV, Gogotov IN, Rao KK, Hall DO (1998a) Actual and potential rates of hydrogen photoproduction by continuous culture of the purple non-sulphur bacterium *Rhodobacter capsulatus*. Appl Microbiol Biotechnol 49:102–107

Tsygankov AA, Fedorov AS, Talipova IV, Laurinavichene TV, Miyake J, Gogotov IN (1998b) Use of immobilized phototrophic microorganisms for wastewater treatment and simultaneous production of hydrogen. Appl Biochem Microbiol 34:362–366

Tsygankov AA, Fedorov AS, Kosourov SN, Rao KK (2002) Hydrogen production by cyanobacteria in an automated outdoor photobioreactor under aerobic conditions. Biotechnol Bioeng 80:777–783

Tsygankov AA, Kosourov SN, Tolstygina IV, Ghirardi ML, Seibert M (2006) Hydrogen production by sulfur-deprived *Chlamydomonas reinhardtii* under photoautotrophic conditions. Int J Hydrog Energy 31:1574–1584

Vignais PM, Billoud B (2007) Occurrence, classification and biological function of hydrogenases: an overview. Chem Rev 107:4206–4272

Vignais PM, Colbeau A, Willison JC, Jouanneau Y (1985) Hydrogenase, nitrogenase, and hydrogen metabolism in the photosynthetic bacteria. Adv Microb Physiol 26:155–234

Vincenzini M, Balloni W, Mannelli D, Florenzano G (1981) A bioreactor for continuous treatment of waste waters with immobilized cells of photosynthetic bacteria. Experientia 37:710–711

Vincenzini M, Materassi R, Tredici MR, Florenzano G (1982a) Hydrogen production by immobilized cells. I. Light-dependent dissimilation of organic substances by *Rhodopseudomonas palustris*. Int J Hydrog Energy 7:231–236

Vincenzini M, Materassi R, Tredici MR, Florenzano G (1982b) Hydrogen production by immobilized cells. II. H_2-photoevolution and wastewater treatment by agar-entrapped cells of *Rhodopseudomonas palustris* and *Rhodospirillum molischianum*. Int J Hydrog Energy 7:725–728

Vincenzini M, Materassi R, Sili C, Florenzano G (1986) Hydrogen production by immobilized cells. III. Prolonged and stable H_2 photoevolution by *Rhodopseudomonas palustris* in light-dark cycles. Int J Hydrog Energy 11:623–626

Wang YZ, Liao Q, Zhu X, Tian X, Zhang C (2010) Characteristics of hydrogen production and substrate consumption of *Rhodopseudomonas palustris* CQK 01 in an immobilized-cell photobioreactor. Bioresour Technol 101:4034–4041

Winkler M, Hemschemeier A, Gotor C, Melis A, Happe T (2002) [Fe]-hydrogenases in green algae: photofermentation and hydrogen evolution under sulfur deprivation. Int J Hydrog Energy 27:1431–1439

Wykoff DD, Davies JP, Melis A, Grossman AR (1998) The regulation of photosynthetic electrontransport during nutrient deprivation in *Chlamydomonas reinhardtii*. Plant Physiol 117:129–139

Zhang L, Huang G, Yu Y (1998) Immobilization of microalgae for biosorption and degradation of butyltin chlorides. Artif Cell Blood Substit 26:399–410

Zhu H, Suzuki T, Asada Y, Miyake J (1999a) Entrapment of *Rhodobacter sphaeroides* RV in cationic polymer/agar gels for hydrogen production in the presence of NH_4^+. J Biosci Bioeng 88:507–512

Zhu H, Suzuki T, Tsygankov A, Asada Y, Miyake J (1999b) Hydrogen production from tofu wastewater by *Rhodobacter sphaeroides* immobilized on agar gel. Int J Hydrog Energy 24:305–310

Zurrer H, Bachofen R (1985) Production of molecular hydrogen with immobilized cells of *Rhodospirillum rubrum*. Appl Microbiol Biotechnol 23:15–20

Chapter 15

Hydrogen Production and Possible Impact on Global Energy Demand: Open Problems and Perspectives

Davide Zannoni*
Department of Pharmacy and BioTechnology (FaBiT), University of Bologna, 40126 Bologna, Italy

Giacomo Antonioni, Dario Frascari
Department of Civil, Chemical, Environmental and Materials Engineering (DICAM), University of Bologna, Bologna, Italy

and

Roberto De Philippis
Department of Agrifood Production and Environmental Sciences, University of Florence, 50144 Florence, Italy

Institute of Chemistry of Organometallic Compounds (ICCOM), CNR, 50019 Sesto Fiorentino, Florence, Italy

Summary

The main goal of this Chapter is to take the reader to the unconventional concept that if hydrogen is used as an energy carrier, there are consistent benefits to be expected, depending on how hydrogen is generated. As it will be illustrated, the technical problems lying ahead of the creation of an apparent "Hydrogen Based Society" are of technical nature although we are all confident that they can be solved within a reasonable period of time.

*Author for correspondence, e-mail: davide.zannoni@unibo.it

D. Zannoni and R. De Philippis (eds.), Microbial BioEnergy: Hydrogen Production, Advances in Photosynthesis and Respiration 38, pp. 349–356, DOI 10.1007/978-94-017-8554-9_15,

I. Introduction

Let's start this final chapter recalling a few facts concerning the global energy requirement, namely: (i) although one third of the world's consumption of fossil fuels (coal, oil, natural gas) is consumed in producing electricity, this latter is not available to one quarter of our present world's population; (ii) although the tremendous effects on our climate and the environments by emissions of the current energy systems and the huge losses within the present production/distribution cycle, these are neither discussed nor questioned; (iii) it is fairly evident that we cannot longer afford our rising appetite for energy, unless we find a way to protect our fragile world.

The above summarized points (i) to (iii) cast some light on the way the public opinion is also kept in the dark as far as concern hydrogen production; hydrogen is in fact sold as "100 percent clean" although in most cases is made from fossil fuels in a process called "steam reforming". In spite of this, media representatives and administrators do not question the manufacturing process of hydrogen that is, conversely, a crucial aspect in the environment protection.

II. Hydrogen as Energy Carrier

Hydrogen is an energy carrier like electricity, and not a primary source. It must therefore be produced from some other form of energy. This conversion process is not 100 % efficient because some of the energy is always converted into thermal energy which is dissipated into the atmosphere. Hydrogen is currently manufactured from hydrocarbon sources or is extracted from water through electrolysis. These processes, however, cannot be considered the smartest ones. Indeed, the hydrogen derived from water is as unsustainable as the fossil fuels themselves. If made by renewable energy sources, the situation would be quite different because the resulting hydrogen would be a truly clean and green energy carrier. When thinking of renewable energy sources, most people imagine wind, photovoltaic and possibly some amount of hydro- or geothermal power. There are however alternative ways to generate hydrogen from renewable energies without the need for electrical power, such as for example: (i) Biological Photolytic Hydrogen (BPH) that takes advantage of microalgae or photosynthetic bacteria (Chaps. 4, 5, 11, 12, 13 and 14 of this book) that use photosynthesis to make hydrogen instead of sugars and/or oxygen, and (ii) Conversion of Biomass to Waste (CBW), in which hydrogen can be produced by pyrolysis (thermochemical conversion) but also by microbial anaerobic digestion (fermentation) of biomass resources such as agricultural or industrial residues (Chaps. 8 and 9 of this book). Indeed, by far the biggest source of renewable energy is biomass which includes agricultural and forestry products or cereal crops and waste from landscape conservation (Evers 2010). In this latter case, efficient engineering pretreatment technologies to convert lignocellulosic biomass into sugar-rich feedstock including hemicelluloses and cellulose that can be fermented directly to produce hydrogen, ethanol, and other high-valuable chemicals, are needed to make this process economically suitable.

However, an evaluation on the technical and economical feasibility should not only take into account the costs and ways of producing hydrogen or other bio-gas, e.g. methane. Indeed, the profitability of bio-hydrogen production also depends on the ways by which this product is used for generation of electricity and heat, as a transport fuel (see Sect. III). Owing to this, the full "production to end-use" chain for H_2 as energy carrier must be included in the evaluation.

Based on the present technologies, bio-hydrogen production is a process requiring large volumes such as wet biomass or bio-waste to be processed. These feed-stocks are

Abbreviations: BPH – Biological photolytic hydrogen; CBW Conversion of biomass to waste; CGH_2 – Compressed gaseous hydrogen; DOE – US Department of Energy; IEA.HIA – International Energy Agency. Hydrogen Implementing Agreement; LH_2 – Liquid hydrogen; LHV – Lower heating value not easily transported over long-distances which implies that small-scale production systems, located where bio-wastes or biomass sources are available, are most likely. This leads to two possible configurations for the "product end-use", namely: 1. hydrogen is used at the point of production, the so called 'stand-alone system', or 2. the gas is supplied to a grid for use at another location, the so called 'grid-connected production system'. Configuration 1. although limited by the fact that both energy demand and biomass streams may not be available at the same location, is the type of system that is presently applied for conversion of bio-methane (biogas or landfill gas) into electricity or heat by gas engine/generator units. A 'stand-alone system', however, also requires flexibility in the production capacity, implying extra costs for storage (see Sect. III), and this might suggest that in a long-term scale economical feasibility, it would be convenient to link bio-gas production to a grid infrastructure. For bio-hydrogen it is evidently relevant to determine whether a future 'hydrogen infrastructure' is likely.

III. Hydrogen Storage: An Open Problem

The development of safe and cost-effective techniques for H_2 storage represents a significant challenge for materials engineers. In particular, in the case of H_2-fed vehicles, the limited space available for fuel storage requires the development of solutions characterized by the attainment of high volumetric energy densities (intended as the amount of chemical energy stored in 1 L of tank). Indeed, while H_2 is an attractive energy vector thanks (among other interesting features) to its particularly lower high heating value (120 MJ/kg; Perry et al. 1997), its gaseous state at standard pressure and temperature, together with its particularly low molecular weight, make the attainment of a volumetric energetic density comparable to those of traditional fuels (e.g., 32 MJ/L for gasoline) a complex challenge. Indeed, the volumetric energetic density of hydrogen is 8.7 MJ/L for liquid H_2, but only 0.01 MJ/L for gaseous H_2 at 0.1 MPa (Sørensen 2005).

As of today, the most mature technologies for H_2 storage are compressed gaseous H_2 (CGH_2) and liquid H_2 (LH_2). In particular, CGH_2 at 700 bar is considered the state-of-the-art technology (Eberle et al. 2009). Starting the analysis from CGH_2, if one considers that an average vehicle requires about 5 kg of H_2 to travel 500 km, 350 bar is the minimum pressure for compressing that amount of H_2 within an acceptable volume (217 L), whereas at 700 bar the same amount of H_2 occupies 135 L. Compression above 700 bar is not cost-effective, given the significant increase in tank cost (Eberle et al. 2009). The energy consumed for compressing H_2 varies between 12 % of the H_2 lower high heating value (LHV) for compression at 350 bar and 15 % of the LHV at 700 bar. CGH_2 at 700 bar allows the attainment of a volumetric energy density of 4.7 MJ/L, a value seven times smaller than that typical of gasoline. The storage of gaseous H_2 at hundreds of bars requires special vessel designs. H_2 tanks are typically cylindrical, the second best geometry after the spherical one. As each increase in tank diameter requires an at least proportional increase in wall thickness, several small-diameter cylinders are typically employed instead of a single large-diameter one (Eberle et al. 2009). The construction of CGH_2 tanks requires the use of materials characterized by a very high tensile strength, a low density, the lack of reactivity with H_2 and a low H_2 diffusivity (Hibbler 2000). Carbon composites represent an interesting solution, which satisfies the first three requirements (Cumalioglu et al. 2008). Thanks to their low density (1,900 kg/m^3, versus 8,160 kg/m^3 for steel alloys), they allow the storage of 4.2 kg of H_2 at 700 bar in a 135 kg vessel, an acceptable extra weight for a medium-size car (von Helmolt and Eberle 2007). However, due to their high permeability for H_2, carbon composites need to be integrated with coatings or liners, that can be realized with metal alloys or with polymers, such as polyethylene covered with a graphite fiber epoxy layer

(Lessing 2003; Zuttel 2003). An interesting but more sophisticated solution is represented by electrolytic H_2 diffusion barriers, consisting in three polymeric layers acting respectively as the cathode, the electrolyte and the anode: leaking H_2 is decomposed at the anode into hydrogen ions and electrons; the former are fed back to the tank, whereas the latter are transferred to the inner cathodic layer where H_2 is produced again (Lessing 2003). The attainment of a low-permeability tank wall is crucial not only to minimize H_2 losses, but also to avoid H_2 embrittlement, which consists in a significant reduction in ductility of the tank caused by H_2 permeation and which can lead to tank failure significantly below the normal yield stress (Suresh 1998; United States Department of Energy 2004).

An alternative to CGH_2 is represented by liquid H_2. Considering that the H_2 critical point is at −240 °C and about 13 bar (Perry et al. 1997), H_2 is typically liquefied at −242 °C and 10 bar or −253 °C and 1 bar. The advantages of LH_2 in comparison to CGH_2 are the significantly thinner tank walls required – thanks to the lower storage pressure – and the higher energy densities attainable (about 9 MJ/L, versus 4.7 MJ/L for CGH_2 at 700 bar) (Eberle et al. 2009). On the other hand, the drawbacks of LH_2 are the significant energy requirement for H_2 liquefaction (30 % of the H_2 LHV) and the high thermal insulation requirements for minimizing H_2 evaporation. Indeed, heat transfer from the environment (occurring through both conduction and radiation) leads to H_2 evaporation, which in turn determines an increase in tank pressure if the vehicle is parked. If, instead, the vehicle fuel cell is working, H_2 consumption is typically higher than H_2 evaporation, and no pressure increase occurs (Cumalioglu et al. 2008). After a certain time of vehicle inactivity, the pressure increase must be controlled by a suitable H_2 venting system, which determines a H_2 loss to the environment called boil-off. H_2 venting typically starts after 3–4 days of vehicle inactivity (Reijerkerk 2004), but this time can be significantly increased by improving the insulation performances of the tanks. High insulation performances can be attained by the use of multiple layers (up to 40) of insulating materials such as fiber glass, alternated to foils acting as radiation shields (Eberle et al. 2009). To increase the thermal resistance, the space between the layers can be maintained under vacuum. A further improvement can be obtained by adding a layer of liquid air, which absorbs the heat coming from the environment by evaporating (Reijerkerk 2004). As liquid H_2 must evaporate before entering the fuel cell, the H_2 latent heat of vaporization can be taken from the evaporating air layer, thus maintaining the liquid state of air. Another solution to reduce H_2 losses during the periods of vehicle inactivity consists in storing LH_2 in tanks designed for resisting to high pressures (Aceves 2003). This would allow a rather long period of H_2 evaporation before the activation of the H_2 venting system. However, this solution requires specific adaptations of the H_2 tanks designed for CGH_2, as many carbon composites cannot operate at the low temperatures required for H_2 liquefaction (Aceves and Berry 1998). A further disadvantage of LH_2 is that it requires the availability of refueling stations specifically designed for maintaining H_2 in the liquid state from the main storage tank, along the filling pipe until the H_2 tank of the vehicle. As a result of the drawbacks of LH_2, car manufacturers have so far oriented their efforts towards the development of safe and cost-effective solutions for CGH_2 at 700 bar (Eberle et al. 2012).

An alternative to compression and liquefaction is represented by H_2 adsorption on high specific surface area materials. However, considering that at room temperature and atmospheric pressure no adsorbent has a significant H_2 storage capacity, H_2 adsorption has to be performed at low temperatures (−200 °C) and/or high pressures (20–30 bar) (Eberle et al. 2009). Cryoadsorption represents a very promising solution for H_2 storage, but a lot of research still needs to be done to turn it into a market technology. In the first place, even at −200 °C and 20–30 bar, the best adsorbents known today can attain a 5 % H_2 capacity on weight basis, whereas a capacity of at least 10 % would be necessary to make this a technology applicable to H_2-fed vehicles. Secondly, as H_2 has a heat of adsorption of 2–5 MJ/kg H_2, the storage of 5 kg of H_2 releases 10–25 MJ of heat, a very high amount that could be compensated only by the evaporation of 65–165 kg of liquid nitrogen. The availability of such high amounts of liquid nitrogen poses engineering challenges that are still far to be met in the field of automotive applications (Eberle et al. 2009).

Lastly, H_2 storage in the form of metal hydrides represents a potentially very interesting solution, although its practical implementation poses chemical and engineering challenges that are still far from being met. Unlike the H_2 storage solutions presented so far, hydrides represent a chemical storage technique: during the refueling step, hydrogen covalently bonds to a less electronegative compound through a hydrogenation reaction, whereas during the H_2 release step the metal hydride is dehydrogenated. Metal hydrides allow the attainment of volumetric energy densities ranging from 8.3 MJ/L for LH_2 to 18 MJ/L for AlH_3 (Graetz 2009). These values, significantly higher than those attainable through the compression or liquefaction of H_2, range between ¼ and ½ of the energy density of gasoline. On the other hand, as a result of the rather low H_2 weight content (2–5 %) of metal hydrides, the attainable gravimetric energy densities are rather low (2.5–30 MJ/kg, versus 120 MJ/kg for compressed or liquid H_2) (Graetz 2009). Research in this field has initially focused on the so-called reversible hydrides, a term that refers to the hydrides that can be regenerated through direct hydrogenation at "technically acceptable" pressures. They are typically compounds in which the hydrogen atoms are covalently bonded to a central atom in an anion complex, such as $[AlH_6]^{3-}$, $[BH_4]^-$, $[NH_2]^-$; the anion is stabilized by a cation such as Li, Mg or Zn. As the hydrogenation (refueling) process is typically exothermic, it causes the release of large amounts of heat (about 100 MJ for 5 kg of H_2). This implies that in the desired case of a rapid (3–5 min) refueling, about 0.5 MW need to be dissipated, which represents a very challenging engineering problem in a vehicle. As a significantly longer refueling time would not be considered acceptable, off-board hydrogenation should be taken into consideration; in this case, the refueling process would simply consist in the replacement of an exhaust "hydride battery" with a re-hydrogenated one. On the contrary, as the dehydrogenation process is typically endothermic, the hydride temperature can be maintained constant thanks to the waste heat released by the fuel cell. An alternative to reversible hydrides is represented by the so-called irreversible ones, whose regeneration through direct hydrogenation requires extremely high pressures (e.g., 7,000 bar at room temperature for AlH_3) (Graetz 2009). In these compounds, regeneration is practically feasible only through indirect routes (e.g. AlH_3 can be produced through a reaction of $LiAlH_4$ with $AlCl_3$). However, the disadvantages of a more complex regeneration are partly offset by the advantages related to their low decomposition enthalpy (Matus et al. 2007). Lastly, very recent studies explored the H_2 solid state storage in nanostructured hydrides, which might lead to significant advantages in terms of increased gravimetric energy density and decreased dehydrogenation temperature (Hanlon et al. 2012).

In summary, while H_2 compression is a relatively mature technology and H_2 liquefaction is not far from being a marketable solution, H_2 solid state storage in the form of hydrides presents the potential to outperform the physical storage approaches, but a great amount of research is still needed to improve the thermodynamic and kinetic performances of the storage materials and to meet the engineering challenges associated to the application of this solution for H_2 storage in automotive applications. On the basis of the remarkable progress made during the past 10 years and of the significant amount of research being conducted in this field, it is likely that H_2 solid state storage will become a marketable technology over the next 10 years.

IV. Safety Issues in the Use of Hydrogen as a Fuel

Since the famous fatal accident occurred in May 1937 at the German airship Hindenburg, which caught fire and was destroyed in New Jersey at the end of an Europe to USA trip causing 36 fatalities, the public perception of the risks of the use of hydrogen has been very high. Even if at present time the most probable cause of the accident has been identified in the use of a highly flammable compound for waterproofing the blimp cloth (Sørensen 2005), the public still has the perception of this risks. Thus, a large number of studies have been carried out in order to ensure the highest level of safety in the use of hydrogen as energy vector or as a reagent in industrial applications. The safety issues have been intensively investigated for all the activities connected with the use of hydrogen: its production, its utilization in industrial plants, its use in automotive applications, its storage and transport, its use in fuelling stations etc. A huge number of documents and papers on studies on hydrogen safety are now available and can be looked up at the web sites of DOE (US Department of Energy) (http://www.hydrogen.energy.gov/biblio_database.html); of the International Association for Hydrogen Safety (http://www.hysafe.info/), and of the International Energy Agency – Hydrogen Implementing Agreement (IEA–HIA), Annex 31 "Collaboration on Hydrogen Safety" (http://ieahia.org/).

Recapitulating the main conclusions of these studies, it is possible to state that: (i) the level of investigations on the safety of hydrogen is by far much deeper than the level of investigations made for most of the other fuels; (ii) hydrogen can be safely stored under high pressures, in metal hydrides (see Sect. III) and as cryogenic liquid; (iii) the problems related with diffusion, permeation and embrittlement effects of hydrogen in metallic materials used for its storage can be safely managed by proper design and materials choice; (iv) a number of stringent rules have been defined and are ready to be implemented for the safe use of hydrogen vehicles.

V. Economical and Political Issues

One of the most attractive feature of the production of hydrogen by means of biological processes is the possibility to obtain the gas at ambient conditions, avoiding both high temperature and pressure conditions along with the addition of polluting catalysts typical of the chemical processes currently under use. Another very interesting feature of biological hydrogen production is the possibility to decentralize the production of hydrogen in small-scale installations located close to the areas where biomass or wastes are available, thus avoiding the energy expenditure and the economical and environmental costs related with the transport of the fuels to the centralized energy production plants. Furthermore, it is worth stressing that the rates of hydrogen production obtained with the dark fermentation of real wastes reached practical levels, comparable with those of the production of bio-ethanol from lignin-cellulosic substrates (Hallenbeck and Ghosh 2009). Moreover, the possibility to use different kinds of wastes, deriving from various agricultural or industrial activities, would make possible to overcome the control that a limited number of Countries currently have on the energy sources of the Planet. Indeed, shifting from the use of geographically concentrated fossil fuels to the plentiful and more widespread biofuels would completely change the relations between energy-producing and energy-consuming nations, possibly turning today's importers into tomorrow's exporters (Yergin 1991; Dunn 2002). A localized energy production would have a positive impact on geopolitical stability, avoiding the concentration of the energy sources in the hands of few Countries.

Another important advantage of biofuels in general, and biological hydrogen among them, is their capability to guarantee a major energy security in terms of supply reliability, capillary and domestic distribution and readiness of availability (Demirbas 2009).

Biological hydrogen production has been evaluated as a sustainable, renewable and energetically efficient process (Manish and Banerjee 2008). A Net Energy Analysis carried out on four different bio-hydrogen production processes, namely (i) dark fermentation, (ii) photo-fermentation, (iii) combined dark and photo-fermentation and (iv) biocatalyzed electrolysis, pointed out that all of them are renewable processes. Indeed, in all these processes, hydrogen output was larger than the input of non-renewable energy (net energy ratio) and they reduced the green house gas emissions by 57–73 % as compared to a well established non biological process taken as a reference (steam methane reforming). Considering the whole balance, the most convenient process resulted to be the integration of dark and photo-fermentation, having the least green house gas emissions, the highest energy efficiency (energy output/energy input) and the highest net energy ratio among the four processes (Manish and Banerjee 2008).

VI. Conclusions

Biological hydrogen production, once the technological problems still on the table will be solved, will be an effective and very useful system for producing a non-polluting energy carrier having no impact on the environment. Indeed, in addition to its sustainability, a number of economic and environmental benefits of the biological hydrogen production can be envisaged: (i) the possibility of creating new job opportunities for rural labor in this field, (ii) the creation of a new route of investments for plants and equipments, (iii) a high competitiveness in the energy field, (iv) a noticeable reduction of dependence on fossil fuels, (v) a significant reduction in CO_2 emissions.

These advantages fit very well the economical and environmental strategies of many Countries all over the World, which favor the development of the "Green Economy" as a way to foster the economic growth and job creation, in response to the current severe economic crises, with an environmental-friendly outlook (Balat 2007; European Environment Agency 2012) with the ultimate goal of the creation of an apparent "Hydrogen Based Society".

Acknowledgements

Financing by the Italian Ministry of Agriculture, Food and Forestry (MIPAAF) under grant "Combined Production of Hydrogen and Methane from Agricultural and Zootechnical Wastes through Biological Processes (BIO-HYDRO)" is acknowledged.

References

Aceves SM (2003) Hydrogen storage in insulated pressure vessels. In: Hydrogen, fuel cells and infrastructure technologies, FY 2003 progress report

Aceves SM, Berry GD (1998) Thermodynamics of insulated pressure vessels for vehicular hydrogen storage. ASME J Energy Resour Technol 120: 137–142

Balat M (2007) An overview of biofuels and policies in the European Union. Energy Source Part B 2:167–181

Cumalioglu I, Ertas A, Ma Y, Maxwell T (2008) State of the art: hydrogen storage. J Fuel Cell Sci Technol 5:034001

Demirbas A (2009) Politics, economic and environmental impacts of biofuels: a review. Appl Energy 86:S108–S117

Dunn S (2002) Hydrogen futures: toward a sustainable energy system. Int J Hydrog Energy 27:235–264

Eberle U, Felderhoff M, Schuth F (2009) Chemical and physical solutions for hydrogen storage. Angew Chem Int Ed 48:6608–6630

Eberle U, Mueller B, von Helmolt R (2012) Fuel cell electric vehicles and hydrogen infrastructure: status 2012. Energy Environ Sci 5:8780–8798

European Environment Agency (2012) Environmental indicator report 2012; ecosystem resilience and resource efficiency in a green economy in Europe. Publications Office of the European Union, Luxembourg

Evers AA (2010) The Hydrogen Society: More than just a vision. Hydrogeit Verlag, Oberkrämer

Graetz J (2009) New approaches to hydrogen storage. Chem Soc Rev 38:73–82

Hallenbeck PC, Ghosh D (2009) Advances in fermentative biohydrogen production: the way forward? Trends Biotechnol 27:287–297

Hanlon JM, Reardon H, Tapia-Ruiz N, Gregory DH (2012) The challenge of storage in the hydrogen energy cycle: nanostructured hydrides as a potential solution. Aust J Chem 65:656–671

Hibbler RC (2000) Mechanics of materials. Prentice Hall, Englewood Cliffs

Lessing PA (2003) Low permeation liner for hydrogen gas storage tanks. In: Hydrogen, fuel cells and infrastructure technologies, FY 2003 progress report

Manish S, Banerjee R (2008) Comparison of biohydrogen production processes. Int J Hydrog Energy 33:279–286

Matus MH, Anderson KD, Camaioni DM, Autrey ST, Dixon DA (2007) Reliable predictions of the thermochemistry of boron–nitrogen hydrogen storage compounds. J Phys Chem A 111: 4411–4421

Perry RH, Green DW, Maloney JO (1997) Perry's chemical engineers' handbook, 7th edn. McGraw-Hill, New York

Reijerkerk CJJ (2004) Potential of cryogenic hydrogen storage in vehicles. In: Alternative fuels. Linde AG, Hoellriegelskreuth

Sørensen B (2005) Hydrogen and fuel cells: emerging technologies and applications. Elsevier Academic Press, Boston

Suresh S (1998) Fatigue of materials, 2nd edn. Cambridge University Press, Cambridge

United States Department of Energy (2004) Hydrogen posture plan: an integrated research, development, and demonstration plan. United States Department of Energy, Washington, DC

von Helmolt R, Eberle U (2007) Fuel cell vehicles: status 2007. J Power Sources 165:833–843

Yergin D (1991) The prize: the epic quest for oil, money, and power. Simon and Schuster, New York

Zuttel A (2003) Materials for hydrogen storage. Mater Today 6:24–33

Subject Index

D. Zannoni and R. De Philippis (eds.), Microbial BioEnergy: Hydrogen Production, Advances in Photosynthesis and Respiration 38, pp. 357–366, DOI 10.1007/978-94-017-8554-9,

MIX
Papier aus verantwortungsvollen Quellen
Paper from responsible sources
FSC® C105338

If you have any concerns about our products,
you can contact us on
ProductSafety@springernature.com

In case Publisher is established outside the EU,
the EU authorized representative is:
Springer Nature Customer Service Center GmbH
Europaplatz 3, 69115 Heidelberg, Germany

Printed by Libri Plureos GmbH
in Hamburg, Germany